AF342264

THE NITROGEN CYCLE AT REGIONAL TO GLOBAL SCALES

Edited by

ELIZABETH W. BOYER

State University of New York, Syracuse, New York, U.S.A.

&

ROBERT W. HOWARTH

The Ecosystems Center, Marine Biological Lab, Woods Hole, Massachusetts, U.S.A.

Reprinted from *Biogeochemistry*
Volumes 57/58 (2002)

KLUWER ACADEMIC PUBLISHERS
DORDRECHT/BOSTON/LONDON

A C.I.P. catalogue record for this book is available from the Library of Congress.

ISBN 1-4020-0779-5

Published by Kluwer Academic Publishers,
P.O. Box 17, 3300 AA Dordrecht, The Netherlands.

Sold and distributed in the U.S.A. and Canada
by Kluwer Academic Publishers,
101 Philip Drive, Norwell, MA 02061, U.S.A.

In all other countries, sold and distributed
by Kluwer Academic Publishers,
P.O. Box 322, 3300 AH Dordrecht, The Netherlands.

Printed on acid-free paper.

Table of Contents

Biogeochemistry **57/58**: vii–ix, 2002.

Foreword

This issue is the final report from the International SCOPE Project on Nitrogen Transport and Transformations: A Regional and Global Analysis. SCOPE (the Scientific Committee on Problems of the Environment, ICSU) authorized the Nitrogen Project as an 8-year effort between 1994 and 2002 because of the need to better understand how humans have altered nitrogen cycling globally and at the scale of large regions. Human activity has more than doubled the rate of formation of reactive nitrogen on the land surface of the earth, and the nitrogen cycle continues to accelerate. The distribution of this reactive nitrogen is not uniform, though, and some regions such as Europe and Asia have seen massive increases in reactive nitrogen, while other regions have seen little change.

The SCOPE Nitrogen Project has synthesized detailed information on the nature of the human alteration of the nitrogen cycle through a series of workshops over the past 8 years. These cumulatively have involved over 250 scientists from over 20 different nations. The results of previous workshops have been published in a series of special journal issues and reports that synthesize information on nitrogen in the North Atlantic Ocean and its watersheds (Howarth 1996), nitrogen cycling in Asia (Hong-Chi Lin et al. 1996; Mosier et al. 2000), nitrogen cycling in the temperate and tropical Americas (Townsend 1999), and nitrogen dissipation in the environment and emissions of nitrogen gases to the atmosphere (Tsurtua & Mosier 1997; Mosier et al. 1998). The Project also contributed to two reports from the Ecological Society of America on global alteration of the nitrogen cycle (Vitousek et al. 1997) and on nitrogen pollution of coastal ecosystems (Howarth et al. 2000) and to a report from the US National Research Council on coastal nutrient pollution (NRC 2000).

This issue presents the results from a series of 8 working group meetings held since late 1998 at the National Center for Ecological Analysis and Synthesis (NCEAS) in Santa Barbara, CA. These workshops were assembled to revisit key uncertainties in quantifying the global nitrogen budget, and to add more insight to nitrogen processes and nitrogen budgets in various world regions. These meetings were sponsored by the Mellon Foundation and by NCEAS. Support for preparing this volume and continuing core support for the SCOPE Nitrogen Project since 1994 have come from the Mellon Foundation and from an endowment given by Mr. David Atkinson to Cornell University. We are grateful for this support.

The first of set of papers in the issue work to advance our understanding of processes controlling nitrogen transport and transformations and their importance in the global nitrogen budget. Vitousek et al. explore hypotheses and models to explain the patterns of nitrogen fixation in terrestrial ecosystems, while Karl et al. review the rate of biological nitrogen fixation in the world's oceans. Both of these papers assess the extent to which human activity has affected nitrogen fixation. Neff et al. review mechanisms controlling atmospheric organic nitrogen, reviewing the importance of this poorly-understood, yet highly significant, component of atmospheric nitrogen deposition.

The next series of papers focuses on the northeastern (NE) USA, quantifying inputs of nitrogen from natural and anthropogenic sources, and detailing the fate of nitrogen inputs to the region. Boyer et al. present detailed nitrogen budgets for 16 large NE watersheds that drain to the North Atlantic Ocean. Mayer et al. provide additional evidence for understanding nitrogen sources in the NE region through the use of isotopic techniques. Complimentary papers estimate the amount of nitrogen stored in forests (Goodale et al.) and lost to denitrification in stream sediments (Seitzinger et al.) in the region. Van Breemen et al. provide a synthesis of the sources and sinks of nitrogen across the northeastern watersheds, revealing the fate of the excess nitrogen inputs to landscapes. Alexander et al. review the performance and uncertainty associated with published models of nitrogen export, applying them to the 16 NE watersheds. This synthesis provides an understanding of the sources of nitrogen in landscapes, and highlights how human activities impact nitrogen cycling in the NE region.

The final papers present regional and landscape-scale nitrogen budgets, working to further refine regional-scale nitrogen budgets and estimates of nutrient export in rivers to the coast. This section includes papers focusing on nitrogen dynamics in east Asia, a region that heavily influences the global nitrogen cycle due to its high population density. Xing et al present nitrogen budgets for China, while Bashkin et al. present nitrogen budgets for South Korea and the Yellow Sea basin. Johnes et al. present an analysis of nitrogen budgets and associated uncertainty estimates in the UK. Other papers in this section focus on nitrogen budgets for small watersheds in the U.S.A., including process studies in an alpine ecosystem (Sickman and Melack) and budgets for small undisturbed basins throughout the US (Lewis). In summary, Mosier et al. describe the global implications of human-accelerated nitrogen cycling, exploring the costs and benefits of excess nitrogen cycling as a result of human activity.

Previous Reports Affiliated with the International SCOPE Nitrogen Project:

Howarth RW (Ed) (1996) Nitrogen Cycling in the North Atlantic Ocean and its watersheds. Biogeochemistry 35: 1–304

Howarth RW, Anderson D, Cloern J, Elfring C, Hopkinson C, Lapointe B, Malone T, Marcus N, McGlathery K, Sharpley A & Walker D (2000) Nutrient pollution of coastal rivers, bays, and seas. Issues in Ecology 7: 1–15

Hong-Chi Lin, Shang-Synyng Yang, Tsu-Chang Hung & Chang-Hung Chou (Eds) (1996) The effect of Human Disturbance on the Nitrogen Cycle in Asia. Proceedings of the SCOPE/ICSU Nitrogen Workshop: SCOPE/Adaemica Sinica, Taipei. 136 pages

Mosier A, Abrahamsen G, Bouwman L, Bockman O, Drange H, Frolking S, Howarth R, Kroeze C, Oenema O, Smith K & Bleken MA (Eds) (1997) International workshop on dissipation of N from the human N cycle, and its role in present and future N20 emissions to the atmosphere. Nutrient Cycling in Agroecosystems 52: 1–312

Mosier A, Freney J, Galloway J, Minami K, Powlson D & Zhu ZL (Eds) (2000) The effect of human disturbance on the nitrogen cycle in Asia. Nutrient Cycling in Agroecosystems 57: 1–117

NRC; National Research Council (2000) Clean Coastal Waters: Understanding and Reducing the Effects of Nutrient Pollution. National Academy Press, Washington, DC. 405 pp

Townsend AR (Ed) (1999) New Perspectives on Nitrogen Cycling in the Temperate and Tropical Americas. Biogeochemistry 46: 1–293

Tsuruta H. & Mosier A (Eds) (1997) NO_x Emission from Soils and its influence on Atmospheric Chemistry. Nutrient Cycling in Agroecosystems 48: 1–160

Vitousek PM, Aber JD, Howarth RW, Likens GE, Matson PA, Schindler DW, Schlesinger WH & Tilman DG (1997a) Human alteration of the global nitrogen cycle: sources and consequences. Issues in Ecology 1: 1–15

Elizabeth W. Boyer, Co-Editor
State University of New York
Syracuse, New York, U.S.A.

and

Robert W. Howarth, Co-Editor
The Ecosystems Center, Marine Biological Lab
Woods Hole, Massachusetts, U.S.A.

International Scope Project on Nitrogen Transport and Transformations: A Regional and Global Analysis

Objectives and Activities

- to foster the necessary synergism between scientist of many disciplines (marine ecologists, forest ecologists, agricultural scientists, microbiologists, atmospheric chemists, oceanographers, hydrologists) in order to help develop new approaches for the study of nitrogen cycling;
- to refine the global nitrogen budget and develop regional budgets for selected key and contrasting regions of the world;
- to more fully understand the problems stemming from accelerated nitrogen cycling, and the inter-relationships among these problems.

Scientific Advisory Committee

Co-chairs:
Robert Howarth (USA)
John Freney (Australia)

Members:
Frank Berendse (The Netherlands)
Pornpimol Chaiwanakupt (Thailand)
Valery Kudeyarov (Russia)
Scott Nixon (USA)
Peter Vitousek (USA)
Zhu Zhao-liang (People's Republic of China)

Consultants:
Ragner Elmgren (Sweden)
James Galloway (USA)
Alan Townsend (USA)

Webpage: http://www.icsu-scope.org/

Biogeochemistry **57/58**: 1–45, 2002.
© 2002 *Kluwer Academic Publishers. Printed in the Netherlands.*

Towards an ecological understanding of biological nitrogen fixation

PETER M. VITOUSEK[1], KEN CASSMAN[2], CORY CLEVELAND[3], TIM CREWS[4], CHRISTOPHER B. FIELD[5], NANCY B. GRIMM[6], ROBERT W. HOWARTH[7], ROXANNE MARINO[7], LUIZ MARTINELLI[8], EDWARD B. RASTETTER[9] & JANET I. SPRENT[10]

[1]*Department of Biological Sciences, Stanford University, Stanford, CA 94305, U.S.A.;* [2]*Department of Agronomy, University of Nebraska, Lincoln, NE 68583, U.S.A.;* [3]*Department of EPO Biology, University of Colorado, Boulder, CO 80523, U.S.A.;* [4]*Environmental Studies, Prescott College, Prescott, AZ 86301, U.S.A.;* [5]*Department of Plant Biology, Carnegie Institute of Washington, Stanford, CA 94305, U.S.A.;* [6]*Department of Biology, Arizona State University, Tempe, AZ 85287, U.S.A.;* [7]*Ecology and Systematics, Cornell University, Ithaca, NY 14853, U.S.A.;* [8]*CENA, University of Sao Paolo, Piracicaba, SP Brazil;* [9]*Ecosystems Center, Marine Biological Laboratory, Woods Hole, MA 02543, U.S.A.;* [10]*Department of Biological Sciences, University of Dundee, Dundee DD1 4HN, Scotland, UK*

Key words: cyanobacteria, decomposition, grazing, legumes, models, nitrogen fixation, nitrogen limitation, phosphorus, shade tolerance, trace elements, tropical forest

Abstract. N limitation to primary production and other ecosystem processes is widespread. To understand the causes and distribution of N limitation, we must understand the controls of biological N fixation. The physiology of this process is reasonably well characterized, but our understanding of ecological controls is sparse, except in a few cultivated ecosystems. We review information on the ecological controls of N fixation in free-living cyanobacteria, vascular plant symbioses, and heterotrophic bacteria, with a view toward developing improved conceptual and simulation models of ecological controls of biological N fixation.

A model (Howarth et al. 1999) of cyanobacterial fixation in lakes (where N fixation generally increases substantially when N:P ratios are low) versus estuaries (where planktonic N fixation is rare regardless of N:P ratios) concludes that an interaction of trace-element limitation and zooplankton grazing could constrain cyanobacteria in estuaries and so sustain N limitation. Similarly. a model of symbiotic N fixation on land (Vitousek & Field 1999) suggests that shade intolerance, P limitation, and grazing on N-rich plant tissues could suppress symbiotic N fixers in late-successional forest ecosystems. This congruence of results raises the question – why do late-successional tropical forests often contain many potentially N-fixing canopy legumes, while N fixers are absent from most late-successional temperate and boreal forests? We suggest that relatively high N availability in lowland tropical forests permits legumes to maintain an N-demanding lifestyle (McKey 1994) without always being required to pay the costs of fixing N.

Overall, both the few simulation models and the more-numerous conceptual models of ecological controls of biological N fixation suggest that there are substantial common features across N-fixing organisms and ecosystems. Despite the many groups of organisms capable of fixing N, and the very different ecosystems in which the process is important, we suggest that these common controls provide a foundation for the development of regional and global models that incorporate ecological controls of biological N fixation.

Introduction

As this volume illustrates, the nitrogen cycle poses many challenges. None of these is more fundamental than the question "Why is the supply of N so important to the functioning of many terrestrial and aquatic ecosystems?" This challenge can be framed as follows:
- The supply of fixed N demonstrably limits the productivity, composition, dynamics, and diversity of many ecosystems, in that all of these change when N is added.
- Organisms with the capacity to fix N_2 from the vast quantity in the atmosphere are widespread and diverse.
- How can limitation by fixed N, an abundant supply of N_2 in the atmosphere, and the biological capacity to make use of N_2 coexist for long in any ecosystem?

A number of processes tend to reduce the biological availability of N in ecosystems, notably the strong link between organic N and recalcitrant C compounds in soils and sediments, and the mobility of N out of ecosystems by hydrologic and atmospheric pathways (especially leaching and denitrification). However, the capacity of biological nitrogen fixers to convert N_2 to organic N is substantial, often exceeding 100 kg ha^{-1} y^{-1} where symbiotic N fixers are abundant – more than enough to maintain N pools in ecosystems and to replenish N losses. Why then are N fixers not a dominant species in all systems where N limits primary productivity? And as a byproduct of their activity, why is N limitation not alleviated? These questions are fundamental to understanding why N limitation is prevalent in terrestrial and aquatic ecosystems.

In many systems, N fixers do drive the accumulation of fixed N on long time scales, bringing N supply close to equilibrium with other potentially limiting resources; in some systems, this equilibration is relatively rapid. For example, in most freshwater lakes, N-fixing cyanobacteria respond to N deficiency (when other resources are available) with increased growth and activity, and the supply of fixed N equilibrates stoichiometrically with that of another resource (normally P) (Schindler 1977). This rapid response to N deficiency is easy to understand in lakes; why is the response much slower (or nonexistent) in many other ecosystems, even allowing for differences in generation times of the dominant organisms? Or, asked another way, what constrains the presence, growth and/or activity of N fixers in N-limited ecosystems?

The answer to these questions is necessary (though not sufficient) to explaining why N limits productivity (and other processes) in many ecosystems; thus it represents a fundamental question about the biogeochemistry of ecosystems. At the same time, it has substantial implications for understanding components of human-caused global change. The extent of human alteration of the global N cycle is well documented (Galloway et al. 1995; Vitousek et al. 1997). It is clear that systems in which N is limiting can be altered substantially by anthropogenic N (e.g., Aber et al. 1998). Where N is not limiting, N deposition may have smaller effects on within-system processes but greater effects on N losses than it does where N is limiting (Hall & Matson 1999; Matson et al. 1999). Accordingly, the mechanisms that control ecosystem-level N limitation strongly determine the consequences of increased N deposition.

For another example, the consequences of increasing levels of atmospheric carbon dioxide depend on what controls the supply of N and other nutrients. Short-term experiments show that plant growth increases substantially under elevated CO_2, averaging a 30–40 percent increase for a doubling of CO_2, more or less independently of nutrient status. However, longer-term simulations of the effect of elevated CO_2 suggest that ultimately the enhancement will be much less, perhaps 5–10 percent, due to a reduction in N availability caused by immobilization of N in plant biomass and decomposing litter (Rastetter et al. 1997; McKane et al. 1997; Schimel et al. 1997). These simulations capture much of our understanding of N supply and ecosystem response – but generally they do not include the possibility that N fixation could be enhanced under elevated CO_2. A number of field experiments do show a positive response of symbiotic N fixers to elevated CO_2 (Arnone & Gordon 1990; Lüscher et al. 1998; Hungate et al. 1999). However, to predict the responses of N fixers to elevated CO_2, we should know what constrains their activity in N-limited systems today, and how any constraints will be altered by elevated CO_2. In the same way, we should understand how N fixation will be affected by altered hydrologic regimes, land use change, and other global changes.

Despite its importance in understanding ecosystems and in predicting how they will respond to change, biological N fixation is either under-represented or not represented in most ecosystem models. For example, the Century ecosystem model (Schimel et al. 1997) now calculates N fixation as a linear function of actual evapo-transpiration (AET); most other models do less. Similarly, the one data-based effort to extrapolate global rates of N fixation in terrestrial ecosystems correlated point estimates with AET, and extrapol-

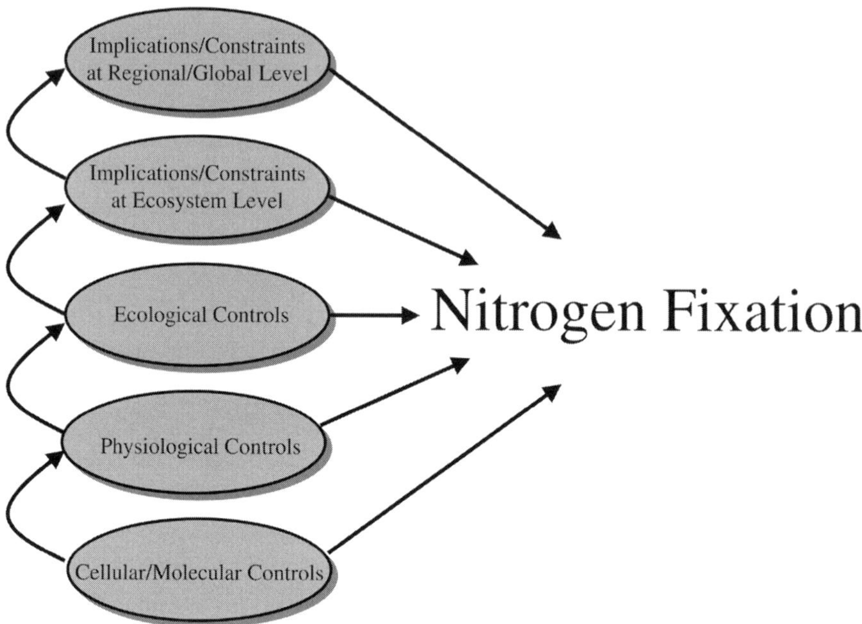

Figure 1. A hierarchy of explanation for patterns of nitrogen fixation. The focal level considered in this paper is the ecosystem level. Mechanistic explanation can be sought at cellular/molecular, physiological (whole organism), and sub-ecosystem levels (i.e., ecological controls). At the lowest level are controls at the sub-organismal level, including genetic control, enzyme synthesis, and other mechanisms. At the whole organism level, nitrogen fixers are subject to physiological controls that determine whether nitrogen fixation can occur; for example, oxygen concentrations or the ability to acquire molybdenum. In addition, the ability of nitrogen-fixing organisms to colonize or persist in a given environment is a function of competitive interactions, predation pressure, and availability of limiting nutrients. The third hierarchical level comprises this suite of ecological controls. At the ecosystem level, the patterns and balance of nitrogen inputs and outputs set constraints on the rates of nitrogen fixation, while at the final and highest level regional and global patterns of nitrogen fixation are controlled by patterns of land cover and use, biome distribution, global climatic patterns, and patterns of N deposition.

ated globally on that basis (Cleveland et al. 1999). While the correlation is useful for the latter purpose, it conveys little understanding of what controls N fixation, and how it is likely to change in the future.

We believe that a lack of information on the ecological regulation of N fixation is the foremost reason why understanding and modeling of N fixation at the ecosystem level lags behind our recognition of its importance. For any process that is important at a number of levels of organization, we need to identify a focal level, and seek mechanistic explanations at hierarchical levels below that focal level. Constraints to the process develop at higher levels in the hierarchy (Figure 1). For example, Hartwig (1998) considered the regulation of symbiotic N fixation at five levels – gene expression, biochemical, nodule, whole plant, and ecosystem. There is a wealth of information on the molecular biology of N fixation and its regulation and about N fixation on the whole-plant level, at least for a few well-studied (mostly crop) legumes. In contrast, our information on ecological controls on symbiotic N fixation in ecosystems other than crop production systems is substantially weaker. This lack of ecological information extends to free-living and heterotrophic N-fixing systems as well.

There are a number of good reasons why the understanding of ecological controls has lagged behind other levels of explanation. A lack of resources (to investigators) certainly contributes, but more importantly:

— Accurate measurements of N fixation rates are essential to testing hypotheses about controls of the process. However, it is extremely difficult to measure rates of N fixation accurately – especially in the field, especially working with long-lived perennial organisms such as N-fixing trees. It is possible to measure rates of N accumulation in ecosystems over relatively long periods of time, or to use acetylene reduction to detect the nitrogenase enzyme – but measuring fixation itself, in natural systems, is very difficult.

— There are many classes of biological N fixers that need to be considered in any thorough analysis. While understanding of the symbiotic system in a few legume crop plants is relatively advanced, much less is known about N fixation in non-agricultural legumes or in other N-fixing organisms, such as symbiotic cyanobacteria or free-living heterotrophic bacteria. In many ecosystems, the contribution of such organisms is significant and needs to be accounted for in any full understanding of N fixation.

— N fixation will need to be understood in a very wide variety of ecosystems, from estuaries to boreal forests. While there may be parallels in the ecological controls of N fixation, by different classes of organisms in different ecosystems, we cannot assume that the same controls are important everywhere.

— Ecologists have not studied or modeled the regulation of N fixation in any systematic way, with relatively few exceptions (e.g., Parrotta et al. 1996; Hartwig 1998; Howarth et al. 1999, and examples discussed below).

Our goal in this paper is to encourage the development of ecologically based understanding of N fixation in natural and managed ecosystems, and to work towards incorporating this understanding into ecosystem models. We review the nature of ecological controls of N fixation, summarize information on those controls for some of the better-understood classes of N-fixing organisms, and synthesize that information into relatively simple process-based models wherever possible. We also discuss other potentially important classes of fixers for which less is known of their capabilities and controls. Finally, we discuss a number of outstanding questions and uncertainties that we believe would reward concentrated effort to resolve them, and that would contribute substantially to understanding the regulation of N fixation.

Ecological controls

By ecological controls, we mean controls over the rate of N fixation that are or can be influenced by interactions between the N fixer and other organisms (excluding symbiotic partners, if any) and/or the N fixer and its environment. N fixers, like all other organisms, are subject to a very wide variety of biotic and abiotic controls; it can be too hot or too cold, too dry or too wet, too acid or too alkaline; there can be too many competitors for crucial resources, or too many grazers that restrict N fixers' distribution or abundance. We are particularly interested in such controls where they influence N fixers (or their activity) to a greater extent than they affect non-fixing organisms, because only in those circumstances will N fixers be constrained relative to other organisms.

Differential suppression of N fixers can occur where N fixers require a resource that other organisms do not need, or where they require more of a resource or less of another environmental factor than do non-fixers. It can also occur when N fixers experience systematically higher mortality than non-fixers, or when environmental conditions are outside the limits of adaptation for all N fixers. While requiring a resource that other organisms need much less of (e.g., molybdenum) could be considered a physiological rather than an ecological control, and conditions outside the bounds of all organisms might be considered an environmental constraint, we will treat these together. Also, the ecological distribution of N-fixing organisms may be wider than that of their ability to fix N (Hartwig 1998) – and we focus on the ability to fix N.

What are the general features of the N-fixation process that could lead to differential suppression of N fixers in some environments?

- N fixation is relatively energy-intensive.
- Nitrogenase enzymes are inactivated by O_2. Organisms maintain a delicate balance between the efficiency of using O_2 as an electron acceptor and the inactivation of nitrogenase, and free-living photosynthetic N fixers must segregate the O_2 they produce from their nitrogenase system.

- Most nitrogenases require molybdenum in order to function; many non-fixers require much less molybdenum. As discussed below, N fixers may also need more P, Fe, and/or other nutrients than other organisms.
- In most N-fixing organisms, the synthesis and/or activity of nitrogenase is inhibited by high levels of combined N.
- Many N-fixing organisms are rich in N compared to non-fixers and so may be grazed preferentially.

How can these overall differences between N fixers and other organisms translate into ecological controls of N fixation? In this analysis, we will consider three major groups of N fixers – free-living cyanobacteria, bacteria and cyanobacteria in symbiotic associations with plants, and heterotrophic bacteria. There are many other N fixers, including lichens with cyanobacterial phycobionts, bacteria in animal digestive systems, and many minerotrophic bacteria. These fixers are important to the metabolism of particular organisms and to the N budgets of particular ecosystems; lichens especially have been evaluated in a number of ecosystems (Fritz-Sheridan & Coxson 1988; Kurina & Vitousek 1999). However, symbiotic N fixers, free-living cyanobacteria, and heterotrophs are by far the most important contributors of fixed N in most ecosystems, and if we can understand and model what controls their rates of fixation, that will contribute substantially to explaining the interactions between N limitation and N fixation in most ecosystems globally.

Free-living cyanobacteria

Aquatic cyanobacteria

Lakes and estuaries. N-fixing cyanobacteria are among the most widespread and important N fixers on Earth. They are the major N fixers in fresh-water and marine systems, and they also grow and fix N in many terrestrial environments, from rainforests to deserts.

What regulates N fixation by cyanobacteria? What constrains it? These questions have been addressed most clearly in comparisons of the plank-tonic portions of lake and estuarine ecosystems. In temperate lakes of moderate to high productivity, the regulation of nitrogen fixation is relatively well understood (see review by Howarth et al. 1988b). When the ratio of nitrogen to phosphorus is high, little or no nitrogen fixation by planktonic cyanobacteria occurs as there is no competitive advantage to this energetically expensive process. However, when the ratio of nitrogen to phosphorus is low, certain species of cyanobacteria often dominate the planktonic community and fix nitrogen (e.g., Smith & Bennett 1999). While N may briefly limit phytoplankton production, cyanobacteria quickly add enough to bring N

8

availability into alignment with P availability – and hence P is the ultimate limiting nutrient in most freshwater lakes of moderate to high productivity (Schindler 1977). In addition to the ratio of N to P, other factors can regulate N fixation by plankton in lakes, including light and depth of the mixing zone, and perhaps grazing (Schaffner et al. 1994). Interestingly, N fixation by heterotrophic bacteria has never been observed as an important process in the water column of lakes (Howarth et al. 1988b), perhaps because heterotrophic bacteria cannot adequately protect the nitrogenase enzyme from the poisoning effect of O_2. Also, N fixation by planktonic cyanobacteria tends to be low in the water column of extremely oligotrophic lakes, many of which may therefore be N limited. The reasons for this require further study.

In contrast, most temperate estuaries are limited by N. Added N stimulates production, and can drive estuarine eutrophication. Added P does not stimulate N fixation in most estuaries; N-fixing cyanobacteria rarely are observed following loading by P. Why do N fixers respond to N limitation in productive lakes but (generally) not estuaries, with enormous consequences for the functioning of those systems? Many hypotheses have been proposed to account for this difference (Howarth et al. 1988b), but most attention has focused on two: (1) that the greater turbulence in estuaries breaks up cyanobacterial filaments and other aggregations, depriving nitrogenase of protection against O_2 (Paerl 1985); and (2) that molybdenum limitation caused by a stereochemical interference of molybdate uptake by sulfate suppresses N fixation (Howarth & Cole 1985; Marino et al. 1990).

Recently, Howarth et al. (1999) developed a simple simulation model for the growth of cyanobacteria in estuaries and lakes. The model (Figure 2) includes phytoplankton growth rates that can be decreased by a low supply of dissolved inorganic N and P (DIN, DIP), the potential for molybdenum limitation of cyanobacterial growth rate, and grazing by zooplankton. Rates of N fixation are modeled by simulating the particular characteristics of N-fixing filamentous cyanobacteria; N fixation takes place only in specialized cells called heterocysts that do not carry out photosynthesis, and so avoid the attendant inactivation of nitrogenase by oxygen. The energy needs for N fixation in each heterocyst are supported by many, in free-living cyanobacteria often 12–50, photosynthetic cells.

The model enables one to evaluate conditions that permit or inhibit the development of a bloom of N fixers; it does not calculate a mass balance of N, so it cannot yet evaluate the point at which enough N has been fixed to suppress further fixation. When the model is run for conditions representing the surface water of lakes, beginning with N concentrations that could limit the growth of phytoplankton, cyanobacteria rapidly respond to N deficiency with growth and substantial N fixation – just as is observed in most lakes

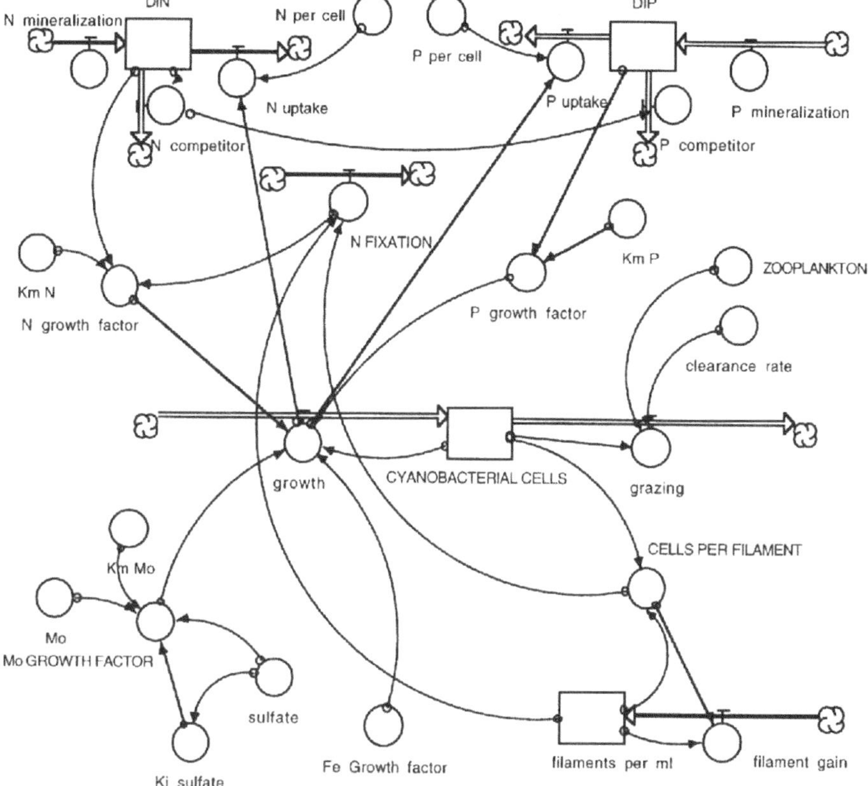

Figure 2. A graphical representation of the Howarth et al. (1999) model for controls on planktonic N fixation in lakes and estuaries.

(Figure 3A). When it is run for estuaries, however, no bloom occurs – again as is observed in most estuaries (Figure 3B). However, if grazing is turned off in the estuarine model, N-fixing cyanobacteria bloom. The growth of a bloom of N-fixing cyanobacteria is delayed relative to lakes, as a consequence of trace element limitation, but the delay is insufficient to prevent a bloom from developing within a season (Figure 3C). The effect of grazing is to cleave the growing cyanobacterial filament, and prevent the accumulation of enough photosynthetic cells to support the energetic requirements of N fixation in heterocysts.

The model describes a situation in which bottom-up (low availability of trace metals leading to slow growth rates) and top-down (grazing) controls interact to suppress the development of N-fixing blooms in estuaries. Neither alone suffices to explain the lack of N fixation (and so the pervasiveness of N limitation) in estuaries, but together they are sufficient.

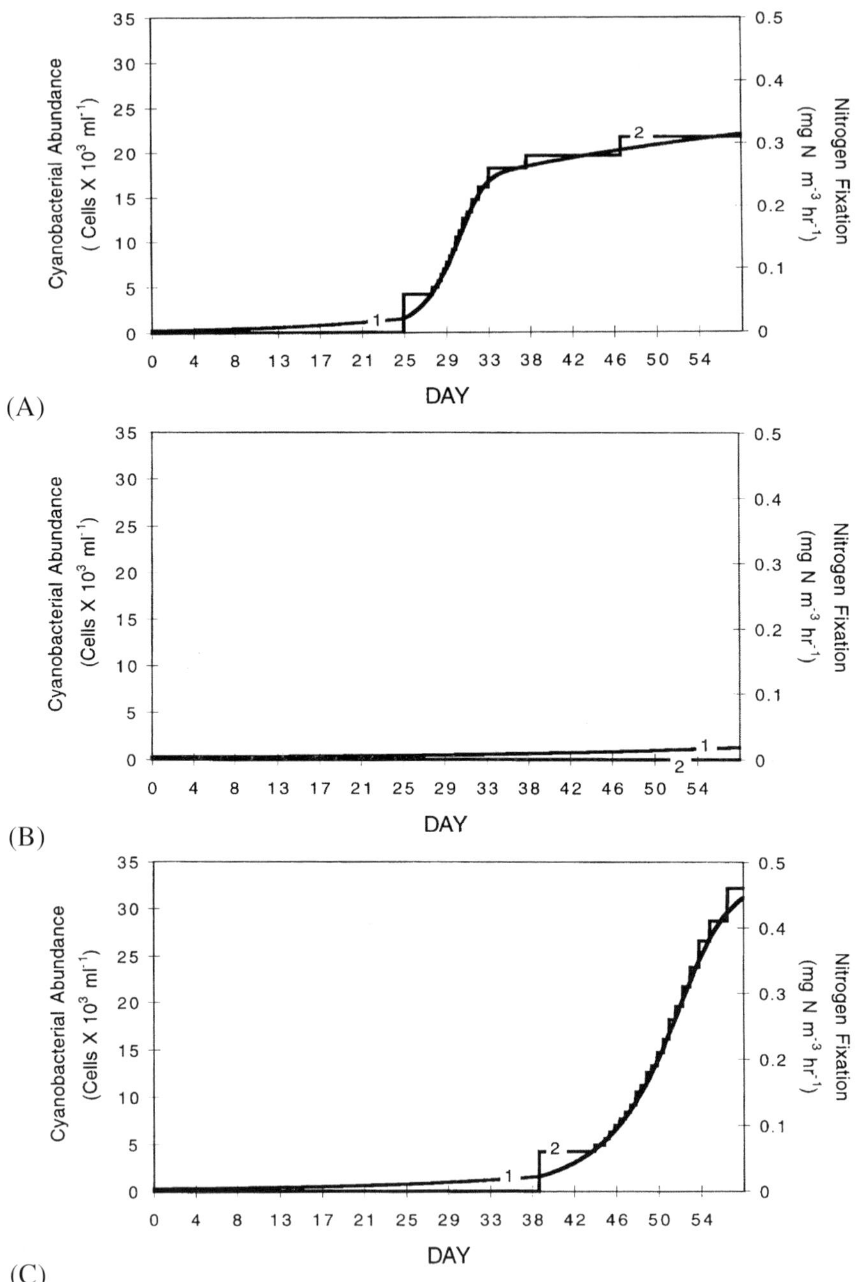

Figure 3. The abundance of N-fixing, heterocystous cyanobacteria and rates of nitrogen fixation under three different simulated conditions.
A. Average freshwater condition of Mo availability and zooplankton grazing.
B. Average seawater condition of Mo availability and zooplankton grazing as in A.
C. Average seawater condition of Mo availability, but zooplankton grazing set to zero.

Mesocosm experiments in estuarine water have validated portions of the model. When zooplankton populations were kept low by the activities of zooplanktivorous fish, planktonic cyanobacteria grew and fixed N, although their growth rate was slower than would typically be observed in freshwaters, probably because of the lower availabilities of trace metals. When zooplankton populations were more representative of estuaries, or when benthic filter feeding animals were added to the mesocosms, their grazing suppressed planktonic cyanobacteria (Marino et al, in preparation).

The model could be expanded to include feedbacks from N fixation to N availability, and from DIN concentrations to N fixation. There are two potential effects of increasing DIN availability on N fixation: (1) suppression of heterocyst formation and nitrogenase synthesis; and (2) loss of competitive advantage for the cyanobacteria. As DIN concentrations increase, other phytoplankton species can grow rapidly. To the extent these other phytoplankton are better competitors for DIP than are the N-fixing cyanobacteria, DIP is drawn down, potentially decreasing cyanobacterial growth.

To what extent can this or other models be extended to N fixation by other cyanobacteria in aquatic systems, or to cyanobacteria in other environments? Heterocystous cyanobacteria are widespread, but numerous non-filamentous cyanobacteria are known – and the model as it presently stands would not identify any constraints to them (over and above factors that influence all phytoplankton), other than trace metal limitation.

Benthic cyanobacteria. Both heterocystous and non-heterocystous free-living cyanobacteria colonize benthic sediments, forming extensive mats, turfs, or felts. These fixers are particularly successful in coastal marine ecosystems such as salt marshes, coral reefs, and intertidal zones. The constraints of high turbulence and control by grazing that operate in estuaries are less important in these types of marine habitats, and the availability of trace metals may be greater because the mats are underlain by reducing sediments (Howarth et al. 1988b). Consequently, rates of nitrogen fixation are among the highest reported for any assemblage of N fixers, from 100 to over 150 mg N_2 m^{-2} d^{-1} in some reports (Wiebe et al. 1975; Jones 1992). Indeed, water flow and grazing are believed to stimulate N fixation in coral reefs by reducing diffusion gradients and pruning turfs to actively growing cell layers (Wiebe et al. 1975; Carpenter et al. 1991; Williams & Carpenter 1997). Genera common in benthic marine assemblages include *Calothrix*, *Microcoleus*, *Lyngbya* and *Oscillatoria*. The last three genera are non-heterocystous and must rely on diel partitioning of N fixation and C fixation, and to some extent spatial partitioning of these processes, to avoid oxygen inhibition (Bebout et al. 1987; 1993; Villbrandt et al. 1991; Joye & Paerl 1993, 1994; Stal 1995).

12

Benthic cyanobacteria also exploit freshwater ecosystems, where they have received less attention than their planktonic counterparts. Mat-forming genera include *Anabaena* and *Oscillatoria*. Cyanobacteria also occur as epilithic periphyton; common genera include *Calothrix, Amphithrix, Dichothrix, Schizothrix, Rivularia, Nostoc*, and many others. Phosphorus is often the limiting nutrient in lakes, and fixers may become abundant when P loading increases, in shallow lake margins as well as for the plankton. In one study, N fixation by a periphyton community dominated by *Calothrix* was measured in three lakes receiving different whole-system enrichment treatments (P only, N+P, no addition); highest rates were found in the P-only lake (Bergmann & Welch 1990).

Control of cyanobacterial biomass by benthic grazers is evident in some lakes (Tuchman & Stevenson 1991; Schultze et al. 1996; McCollum et al. 1998), but these studies did not report effects of grazing on rates of N fixation. In one study, a snail (*Elimia* sp.) differentially cropped cyanobacteria, appearing to select food on the basis of growth form rather than by taxon (Tuchman & Stevenson 1991). If grazers select filamentous forms, this preference could control heterocystous-cyanobacterial N fixation in a way similar to that postulated by Howarth et al. (1999) for estuaries. Rates of N fixation by lake periphyton and benthic mats may therefore be subject to some of the same controls as for plankton (e.g., DIN and DIP availability and grazing, and perhaps trace metal availability), but may also be affected by physical factors such as wave action and light availability.

Flowing-water ecosystems. The few studies of nitrogen fixation in flowing-water systems suggest that cyanobacteria and N fixation are limited by some of the same factors limiting other lotic algae: low light in small, heavily shaded streams (Horne & Carmiggelt 1975) and high current velocity (small to mid-sized streams) or high turbidity (large rivers). Indeed, planktonic cyanobacteria do not occur at all in most streams. In desert streams, where shading is minimal and current velocity low, filamentous cyanobacteria often form extensive mats, and very high rates of N fixation have been observed therein (Grimm & Petrone 1997). These systems are characterized by very low DIN availability and abundant DIP, warm water temperatures, slow currents, and high light – all conditions conducive to N fixation. As with marine microbial mats, grazing and turbulence are not important controlling factors, and N fixation rates in *Anabaena* mats approach 150 mg N_2 m^{-2} d^{-1}. Mats can become very thick, and steep microgradients in the availability of dissolved gases and nutrients may form within them (Stal 1995). In contrast to the *Lyngbya* or *Microcoleus* mats typical of marine benthic habitats, N fixation in heterocystous mat-forming fixers like *Anabaena* peaks during daylight hours. Cyanobacterial mats may shift from being sources to sinks for fixed N when conditions conducive to denitrification occur deep within mats or during times of low oxygen availability (Joye & Paerl 1993, 1994). Thus one control on the importance of N fixation in aquatic, mat-forming cyanobacteria may be the competing process of denitrification, which can result in loss of up to 20 percent of the nitrogen fixed (Joye & Paerl 1994).

Another constraint to N fixation operates at a larger scale in streams: disturbance of assemblages by infrequent spates, to which cyanobacterial mats are particularly susceptible (Grimm & Fisher 1989). Littoral zones of lakes and marine intertidal zones also experience hydrodynamic disturbance. Export of cyanobacteria due to disturbance affects fixation through its impact on the standing biomass of fixers; it represents a larger-scale sporadic constraint on nitrogen fixation in frequently disturbed ecosystems.

Terrestrial cyanobacteria

The occurrence of cyanobacteria on land is related to their ability to colonize extreme environments that are not fully occupied by vascular plants. Free living cyanobacteria are most abundant relative to other vegetation in dry and/or cold regions such as deserts, grasslands and tundra. Cyanobacteria also regularly occur and fix nitrogen in wetter climates such as temperate or tropical forests, where they occur epiphitically on tree trunks and leaves, in bryophytes or on decomposing logs (Goosem & Lamb 1986, Crews et al. in press). However, their spatial extent and importance as nitrogen fixers appears to be greater in more extreme environments.

Habitat. In arid ecosystems, cyanobacteria typically occur in microphytic soil crusts that can include algae, fungi, bacteria as well as mosses, lichens and liverworts (Evans & Johansen 1999). In general, the extent to which surfaces of arid and semi-arid soils are occupied by these crusts is inversely related to the stone-free ground covered by vascular plants and their litter (Knapp & Seastedt 1986; West 1990; Eldridge & Greene 1994), but some soils are more conducive to colonization by microphytes than others. Soil crusts are generally more extensive in fine textured soils, presumably due to greater moisture holding capacity (Shield & Durell 1964; Graetz & Tongway 1986), and in soils with neutral to slightly alkaline pH (Stewart 1974; Hoffmann 1989).

Soil climate. The niche of terrestrial cyanobacteria is largely defined by their ability to endure (often in a dormant state) conditions of low moisture and/or extreme temperatures. Under these environmental conditions the growth of vascular plants is limited, which in turn allows light to reach the ground

14

and support the growth of cyanobacteria. However, cyanobacteria grow best under conditions of moderately high temperatures and adequate moisture; thus the success of cyanobacteria is not only attributable to their ability to endure climatic extremes, but also to their rapid physiological response to episodic and often brief favorable growing conditions. Moisture availability in particular has been shown to regulate rates of nitrogen fixation by cyanobacteria in tundra (Liengen & Olsen 1997), prairie (Kapustka & DuBois 1987), semi-arid grassland (Coxson & Kershaw 1983), and desert ecosystems (Rychert & Skujins 1974). Optimal moisture conditions for nitrogen fixation by cyanobacteria are near zero kPa (saturation), and fixation declines linearly with declines in soil water potential (Rychert & Skujins 1974; DuBois & Kapustka 1983; Kapustka & duBois 1987). Desiccated cyanobacteria have been shown to attain maximum rates of nitrogenase activity within 4–14 hours of rehydration (Coxson & Kershaw 1983; Kapustka & DuBois 1987). Temperature also influences rates of N fixation by terrestrial cyanobacteria, with maximum rates occurring between 19–35 °C depending on species and ecosystem (Rychert & Skujins 1974; Coxson & Kershaw 1983; Johansen & Rushforth 1985; Chapin et al. 1991; Liengen & Olsen 1997).

Available N, P and other elements. The extent to which combined N controls cyanobacterial N fixation in natural terrestrial ecosystems is not clear. Liengen and Olsen (1997) found N fixing activities by cyanobacteria to positively correlate with C:N ratios of the cyanobacterial layer in an Arctic tundra site. Work by Eisele et al. (1989), however, suggests that the inhibition of N fixation by combined N may be tied to the ratio of available N: available P, paralleling the behavior of cyanobacteria in many freshwater ecosystems and managed terrestrial ecosystems (Schindler 1977; Smith 1992). In a tallgrass prairie, Eisele et al. (1989) found that fire effectively lowered the ratio of available N:P, stimulating N fixation. The development of a stoichiometric approach to understanding elemental controls on fixation rates by terrestrial cyanobacteria (Eisele et al. 1989) provides a powerful framework for reconsidering other reports demonstrating the regulation of N fixation by combined N, available P or other nutrients (e.g., Wilson & Alexander 1979; Paul & Clark 1989; Chapin et al. 1991; Liengen & Olsen 1997; Crews et al., in press).

Disturbance by trampling and fire. The abundance and diversity of cyanobacteria in soils is strongly influenced by the frequency and intensity of two types of disturbance: fire and trampling. In semi-arid ecosystems that receive sufficient precipitation to accumulate a fuel load, fire has been shown to reduce populations of cyanobacteria (West 1990) (in contrast to the wetter tallgrass prairie discussed above). For example, Johansen et al. (1993) reported the disappearance of a common N-fixing species, *Nostoc commune*, from recently burned rangelands in the lower Columbia Basin. Trampling of soil microphytic communities by humans, vehicles and ungulates also has been shown to reduce the diversity and/or abundance of soil cyanobacteria in semi-arid ecosystems (Beymer & Klopatek 1992; Belnap et al. 1994). Within a year following a disturbance, free-living cyanobacteria re-colonize sites and dominate for years or decades until slower growing lichens, liverworts and other microphytes reestablish (Anderson et al. 1982; Evans & Belnap 1999).

The importance of N fixation by terrestrial cyanobacteria. Cyanobacteria have the potential to supply an appreciable amount of combined N to terrestrial ecosystems. Reviews report N fixation rates to range between 1–41 kg ha^{-1} yr^{-1} with the majority of studies ranging between 1–10 kg ha^{-1} yr^{-1} (Boring et al. 1988; West 1990; Warren 1995; Cleveland et al. 1999). Since the vascular plant communities in arid to semi-arid regions generally do not achieve canopy closure due to water limitation, cyanobacteria on soils have the potential to persist indefinitely. Even relatively low rates of N fixation could therefore have the potential to reduce or eliminate N limitation in arid ecosystems. Nitrogen limitation, however, has been well documented in most arid ecosystems studied (see Peterjohn & Schlesinger 1990; Hooper & Johnson 1999).

N limitation in arid ecosystems could be driven by N outputs that offset the relatively small but steady inputs from N fixation. With the exception of vertical leaching of nitrate through the soil profile, which is negligible, arid ecosystems experience the same avenues of N loss as more mesic systems, including erosion, ammonia volatilization and denitrification (Peterjohn & Schlesinger 1990; Crawford & Gosz 1982). The high pH of alkaline soils in arid regions can lead to high rates of N loss through ammonia volatilization. Evans and Boyd (as cited in Evans & Johansen 1999) measured high rates of NH_3 volatilization from microphytic crusts compared to soil where crusts had been removed. A number of studies suggest that episodic denitrification is the most important avenue of N loss in many arid ecosystems; indeed, relatively high rates of N loss associated with fluctuating soil moisture may contribute to the enriched $\delta^{15}N$ observed in dryland ecosystems (Austin & Vitousek 1998; Handley et al. 1999). Peterjohn and Schlesinger (1991) estimated annual N losses via denitrification to be greater than 7 kg ha^{-1} at the La Jornada LTER site in the Chihuahuan Desert. High rates of denitrification reported in arid ecosystems led Skujins (1981) to suggest that over 99 percent of the nitrogen fixed by cryptogamic crusts may be lost through this process,

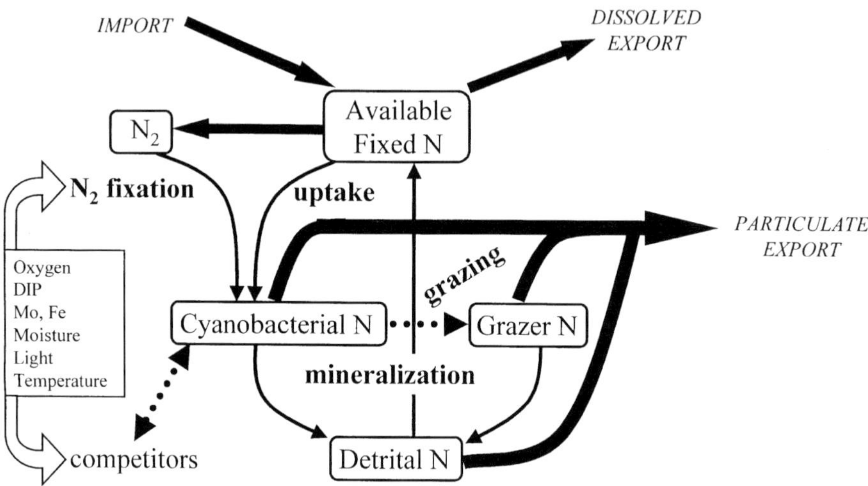

Figure 4. A generic model of the fluxes of nitrogen associated with free-living cyanobacterial nitrogen fixing systems. The relative importance of controls on these fluxes varies among the main system types considered: planktonic fixers, aquatic cyanobacterial mats, and terrestrial cyanobacterial crusts. Factors that control the rates of nitrogen transformations (thin black arrows), including oxygen, P, Fe, and Mo availability, light, temperature, and water availability, are shown as a box (ecophysiological controls) linked to N fixation (open arrows). At least some of these controls affect competitors as well as fixers. Grazing and competitive interactions (dotted arrows) are ecological controls that determine the presence/absence of nitrogen fixers and whether N fixation or uptake of fixed N is the predominant route of N acquisition. Export and denitrification are outputs (thick, black arrows) that affect the mass balance of N at the ecosystem level.

although there is evidence that vascular plants can acquire much of the N fixed by soil microphytic crusts (Evans & Ehleringer 1993).

Cyanobacteria in differing environments

Are there sufficient commonalities among the free-living cyanobacteria in oceans, estuaries, lakes, streams, and soils to develop a generic conceptual model of controls on their fixation of N? A start towards such a model is summarized in Figure 4. It includes both physiological and ecological controls, as well as higher level constraints, that apply to varying degrees across ecosystems. It suggests that we could could use N as a currency, allowing us to consider other N transformations (e.g., assimilation of DIN and denitrification) as factors potentially controlling N fixation. In general, availability of DIN, DIP, and micronutrients, and, for non-heterocystous cyanobacterial mats, the ability to spatially or temporally segregate oxygen production and N fixation, are the most important controls on rates of N fixation in free-living cyanobacteria. Competitors and grazers change the availability of nutrients; they may also exclude cyanobacteria altogether in some environments (e.g., shading of terrestrial free-living cyanobacteria by vascular plants).

Vascular plant-bacteria symbioses

The most important N fixers in terrestrial ecosystems are symbioses involving legumes and rhizobia, and the similar symbioses of a number of plants (*Alnus*, *Myrica*, *Rosaceae*, and others) with the actinomycete *Frankia*. These symbioses are well-studied on the biochemical and organismal levels. Areas dominated by symbiotic N fixers, whether agricultural or natural, can have very high rates of N fixation – often exceeding $100 \text{ kg N ha}^{-1} \text{ y}^{-1}$. Moreover, while woody legumes are most abundant in tropical regions, both herbaceous legumes and actinorhizal (*Frankia*-based) symbioses are widespread in temperate and even boreal regions. Many terrestrial ecosystems, especially at higher latitudes, are limited by N supply. Again, we ask why don't symbiotic N fixers respond to this N deficiency, and (ultimately) reverse it?

Vitousek and Field (1999) recently described a simple model for the ecosystem-level control of symbiotic N fixation. This model was designed for the same purpose as the estuarine model discussed above (Howarth et al. 1999) – to explore mechanisms that could keep N fixers from responding to N deficiency in ecosystems, and so could maintain limitation by N. Unlike the estuarine model, the terrestrial symbiotic fixer model is not individual-based; rather, it evaluates productivity and nitrogen acquisition by a generic non-fixer and a generic symbiotic N fixer. Also unlike the estuarine model, the terrestrial model is mass-balanced in N, including outputs as well as inputs, and it is possible for fixation to change the N status of an ecosystem with consequent effects on the relative abundance of fixers and non-fixers.

The model includes two pathways of N input: precipitation, which is controlled externally, and N fixation. It also includes two types of outputs: leaching and denitrification of the "excess available N" that remains within the soil after plants and microbes have taken up what they can, and losses of N by pathways that could continue even where and when N strongly limits biological processes in the ecosystem. Dissolved organic N loss may be an example of such a loss (Hedin et al. 1995; Vitousek et al. 1998; Campbell et al. 2000).

Initially, the model is based on a greater cost for N acquisition via fixation as opposed to uptake from inorganic N pools in soil. This cost has been evaluated biochemically; it conforms to the inverse relationship between the amount of available inorganic N supply and the proportion of plant N derived from N fixation by legumes (Allos & Bartholomew 1959), and to models of biological N fixation in terrestrial (Pastor & Binkley 1998) and marine (Tyrell 1999) ecosystems. The cost of N acquisition by the non-fixer is assumed to be dependent on N availability, being relatively low when N availability is high, but increasing once most of the available N in the soil has been taken up. In contrast, the cost of N fixation is assumed to be independent of N availability in the soil; it is higher than the cost of N acquisition from the soil as long as any appreciable available N remains in the soil. The model gives non-fixers priority for any available N in the soil; only if nearly all the inorganic N in the soil is exhausted and other resources remain available do N fixers grow and fix N. This last assumption represents too strong a constraint on N fixers, because they are known to be able to compete for available soil N. However, the point of the model is to ask why N fixers fail to respond to N deficiency – and if they do respond despite this rigid constraint, to identify other kinds of constraints that could keep N fixers out of N-limited ecosystems.

The model has been run for systems that start with no N or biomass, and accumulate N from either precipitation or the atmosphere. When the model is run without N fixation, N can only be accumulated slowly from very dilute concentrations in precipitation (prior to widespread alteration of the N cycle by humanity), and biomass accumulation is slow (Figure 5A). Adding the potential for N fixation causes a much more rapid accumulation of biomass and N, to the same equilibrium point (Figure 5C). These results are based on simulations of a system in which N can only be lost from an excess available N pool. If an additional pathway of loss is added that is independent of available N, then it is possible to get sustained N limitation in the absence of N fixation (Figure 5B). However, where N fixation can occur (under the conditions of this model), it has the capacity to overwhelm any losses of N, and maintain the system at the same equilibrium as occurs with losses of only excess available N (Figure 5D).

Constraining N fixation simply by giving non-fixers absolute priority for fixed N does not work to sustain N limitation, under the conditions in this model. What else could constrain symbiotic N fixation? A number of other pathways have been suggested (Vitousek & Howarth 1991; Vitousek & Field 1999), including:

(1) Reduced shade tolerance of symbiotic fixers, such that they are unable to enter closed canopy systems, even though if they would reach the canopy they could fix actively. (Disturbance will open the canopy, but disturbance also enhances N availability temporarily, thereby denying fixers an advantage at the one time in succession when sufficient light is available (Vitousek & Howarth 1991).)

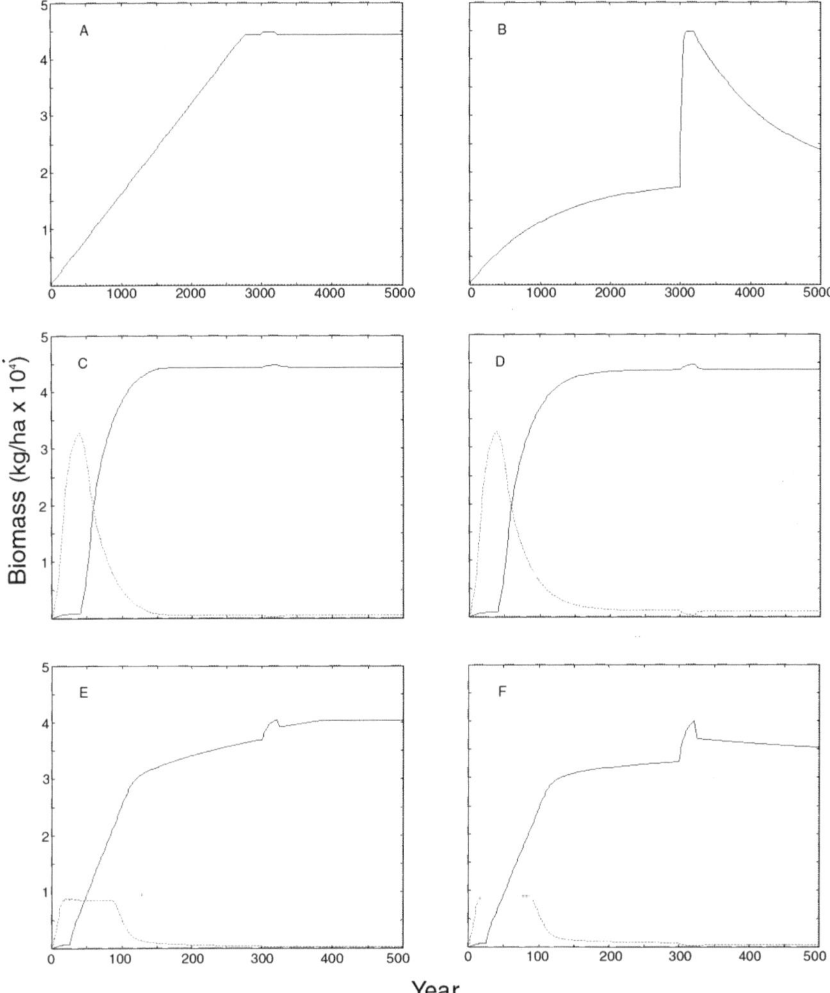

Figure 5. Simulated biomass of a symbiotic N-fixing plant (dashed line) and a non-fixer (solid line) over time, starting with no organic C or N in soil. In each, N fertilization was simulated for 20 years (after 3000 years in A and B, 300 years in C–F) to show the extent of N limitation. Revised from Vitousek and Field (1999).

A. No N fixation (inputs of N from atmospheric deposition only), and losses of N from an excess available N pool only (N remaining in available form after plants take up what they can).

B. No N fixation, an uncontrollable N loss in addition to excess available N.

C. Symbiotic N fixation can occur; it is controlled by a higher cost of N acquisition via fixation; losses of excess available N only.

D. Symbiotic N fixation, both pathways of loss.

E. Symbiotic N fixation can occur, but it is further constrained by shade intolerance, a greater P requirement, and preferential grazing; losses of excess available N only.

F. Symbiotic N fixation with the additional constraints; both pathways of N loss.

20

(2) A greater sensitivity of N-fixing organisms to limitations caused by a
deficient supply of P or another nutrient (Smith 1992; Cassman et al.
1993).

(3) Preferential grazing on the protein-rich tissues of N fixers, preventing
them from responding to N deficiency (e.g., Ritchie & Tilman 1995;
Ritchie et al. 1998). Many N fixers are chemically defended against
grazing, but that defense comes at the energetic cost of synthesizing
and maintaining defensive compounds.

The model was modified to include each of these possible controls; each is
capable of constraining N fixation substantially, and thereby delaying the
accumulation of biomass and N in simulations of developing ecosystems
(Vitousek & Field 1999). However, even when all three were combined, there
was relatively little effect on the equilibrium state of the system – as long
as N could only be lost from excess available N pools (Figure 5E). Adding
an additional leak of N yielded a simulated system that is substantially and
persistently limited by N at equilibrium, and in which N fixers are not able
to grow enough to offset that N deficiency (Figure 5F). This result is interesting because most conceptual models of ecosystems, in which N is lost
from excess available N pools and symbiotic N fixation is constrained by its
energetic cost, take us to Figure 5C – while much of the world (especially
in the temperate and boreal zones) appears to work more like Figure 5F. It
would be rewarding to understand the processes that take us from the scenario
illustrated in Figure 5C to that in Figure 5F.

Issues in symbiotic N fixation

Discussions among the authors of this chapter identified several ways that
our understanding of – and ability to model – symbiotic N fixation might be
improved. These include:

(1) A more realistic characterization of the energetics of N acquisition from
the soil versus by N fixation, and their consequences.

(2) A clear determination as to whether symbiotic N fixers in fact have a
systematically greater requirement for P, or less competitive ability to
acquire it, than do non-fixing plants.

(3) An explanation for the abundance of woody legumes in the canopy
of late-successional tropical forests, in contrast to the near absence of
symbiotic N fixers from closed-canopy forests elsewhere.

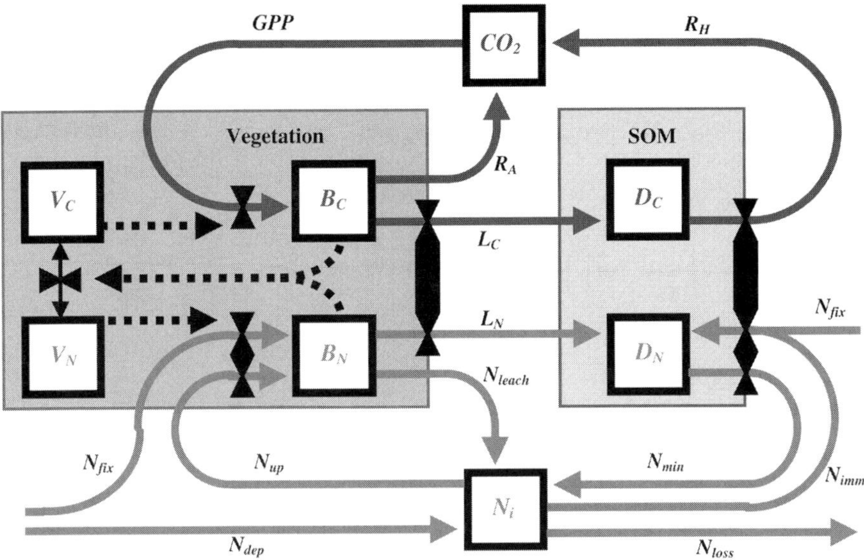

Figure 6. Structure for the Multiple Element Limitation (MEL) model (Rastetter et al. 1997) with inputs of N via biological fixation included. N fixation can occur when the energetic cost of acquiring N by investing in roots/mycorrhizae exceeds the cost of acquiring N via fixation.

Modeling costs of N-fixation

The costs and benefits of N fixation were assessed within the framework of the Multiple Element Limitation (MEL) model (Rastetter & Shaver 1992; Rastetter et al. 1997; Herbert et al. 1999), allowing the development of a theoretical model of symbiotic N fixation based on a resource optimization paradigm (Mooney & Gulmon 1979; Bloom et al. 1985; Chapin et al. 1987; Field et al. 1992; Rastetter & Shaver 1992; Rastetter et al. 1997; Herbert et al. 1999). Under this paradigm, resources within plants should be allocated so that growth is co-limited by all external resources; otherwise internal resources would be wasted by taking up some external resources in excess at the expense of not acquiring resources that are more limiting to growth. An implication of this paradigm is that N fixation should occur when the resource cost of N fixation is less than that of N uptake from other sources. We therefore develop our model by assessing the relative costs of N fixation versus N uptake. The structure of this model is outlined in Figure 6.

We assume that the only cost of N fixation to the plant is through the supply of C to the N-fixing symbiont and denote this cost as r_{Nfix} (g C g^{-1} N). This cost has been estimated at about 8 g C g^{-1} N for symbiotic N fixation in terrestrial ecosystems (Gutschick 1981). However, the value of r_{Nfix} might not be fixed. For example, if N fixation is limited by the availability of P or

22

of cofactors like Mo and Fe, then r_{Nfix} would have to be adjusted to reflect the cost of acquiring these cofactors.

We assess the cost of N uptake in terms of the C gain that would be realized if the resources expended on N uptake were reallocated toward photosynthesis. To make this assessment, we use a modified version of the coupled C and N uptake equations from the MEL model:

$$U_C = g_C \left(\frac{C_a}{k_C + C_a} \right) (1 - e^{-b_C R_I V_C}) \tag{1}$$

$$U_N = g_N \left(\frac{N_I}{k_N + C_I} \right) (1 - e^{-b_N R_I V_N}) \tag{2}$$

where U_C and U_N are the C and N uptake rates, C_a and N_I are the environmental concentrations of available inorganic C and N (i.e., CO2 and NH_4 + NO_3), R_I is a measure of the total amount of internal resources that can be allocated toward the uptake of external resources, and V_C and V_N are the fractions of R_I allocated toward C and N uptake, respectively. g_C and g_N are maximum uptake rates under saturating substrate concentrations and fully exploited canopy and soil volume; k_C and k_N are half-saturation constants for the Monod kinetics relating uptake to substrate concentration; and b_C and b_N are parameters controlling the diminishing return on uptake effort as the canopy and soil volumes become fully exploited by leaves and fine roots, respectively. For C, this diminishing return on uptake effort can be interpreted as a Beer's Law formulation of light extinction. For N, an analogous diminishing return on uptake effort is associated with the interference among fine roots as they become densely packed in the soil.

Under the optimization paradigm described above, V_C and V_N should adjust so that growth is equally limited by C and N. In the MEL model, this adjustment is achieved by comparing the C:N ratio of the simulated vegetation to an allometrically determined optimum C:N ratio. V_C and V_N are adjusted incrementally to drive the C:N ratio toward the optimum. For the present purposes, we must also consider uptake effort expended toward the acquisition of other resources (e.g., P). We represent the fraction of R_I allocated toward the acquisition of all these other resources as V_O and define a variable f such that $V_O = (f - 1)V_C$ (f will be a function of the availability of these other resources and C). Because the V_i represent fractions of the internal resources that can be allocated toward acquisition of external resources, we impose the restriction that $f V_C + V_N = V_O + V_C + V_N \equiv 1$.

With this relationship, a C cost of N uptake can be can be assessed in terms of the reallocation of uptake effort between N and C; under unchanged environmental conditions, the rate of C uptake must decrease as the rate of

N uptake increases when effort is reallocated. Thus, the cost of N uptake ($gCg^{-1}N$) can be assessed as the incremental decrease in C uptake per incremental increase in N uptake,

$$r_{Nup} = -\frac{dU_C}{dU_N} = \frac{dU_C}{dV_C}\frac{dV_C}{dV_N}\left(\frac{dU_N}{dV_N}\right)^{-1} \tag{3}$$

Because $fV_C + V_N \equiv 1$ and assuming that f does not change appreciably with changes in V_N,

$$\frac{dV_C}{dV_N} = -\frac{1}{f}.$$

From the uptake equations,

$$\frac{dU_C}{dV_C} = G_C b_C R_I e^{-b_C R_I V_C}$$

$$\frac{dU_N}{dV_N} = G_N b_N R_I e^{-b_N R_I V_N}$$

where $G_C = g_C\left(\frac{C_a}{k_C+C_a}\right)$ and $G_N = g_N\left(\frac{N_I}{k_N+N_I}\right)$. Thus, the cost of N uptake can be calculated as

$$r_{Nup} = \frac{G_C b_C}{fG_N b_N}e^{R_I(b_N V_N - b_C V_C)}. \tag{4}$$

If $r_{Nup} > r_{Nfix}$ then it is more economical to expend C fixing N than taking N up from inorganic sources in the soil. This equation implies that there should be a tendency towards fixing N it:

(1) the CO_2 concentration is high (G_C large) so that there is a high return on resource allocation towards the canopy,

(2) inorganic N concentrations in the soil are low (G_N small) so there is a low return on the allocation of resources towards N uptake from soil,

(3) the canopy is open (small $b_C R_I V_C$) so there is a high return on resource allocation to the canopy,

(4) the soil is well exploited by roots (large $b_N R_I V_N$) so there is a low return on allocation of resources towards N uptake from soil, and

(5) other resources like P are readily available so that the allocation of effort toward the uptake of these other resources is small relative to that allocated toward C acquisition (f is small).

Equation 4 automatically scales the relative costs and benefits of these factors.

The optimization paradigm implies that N fixation will occur if a further reallocation of uptake effort toward N acquisition will result in $r_{Nup} > r_{Nfix}$. The critical N uptake effort (V_{N*}) above which it becomes more economical to fix N rather than take it up can be calculated by setting Eq. 4 equal to r_{Nfix} and solving for V_N:

$$V_{N*} = \frac{f \ln\left(\frac{r_{Nfix} f G_N b_N}{G_C b_C}\right) + R_I b_C}{R_I(f b_N + b_C)} \tag{5}$$

Thus, in the adjustment scheme for uptake effort, V_N should never increase above V_{N*}. If the C:N ratio of the vegetation is still too high with $V_N = V_{N*}$, then an increase in N fixation (U_{Nfix}) will be more cost effective than an increase in V_N. The increase in U_{Nfix} can be formulated in much the same way as the incremental increases in V_N described above, except that the C cost of this increase is reflected in an increase in respiration ($R_{Nfix} = r_{Nfix} U_{Nfix}$) rather than in a decrease in V_C.

This approach to modeling N fixation is being implemented in MEL, for a range of ecosystems (Rastetter et al., in preparation). It offers a more mechanistic way of evaluating the energetic costs of N fixation than has been available to date. As such, it should be useful for analyzing the consequences of anthropogenic or natural changes in CO_2, N availability, and other resources.

Phosphorus requirement and acquisition

The importance of P supply and of N:P supply ratios in controlling rates of symbiotic N fixation is discussed widely (e.g., Smith 1992; Cassman et al. 1993; Crews 1993). What is the basis of the effect of P on N, and how general is it? There are at least two possible mechanisms, with rather different implications for the control of N fixation. One is that most systems are limited (ultimately) by P supply, with the supply of N adjusting to that of P in the long run (e.g., Walker & Syers 1976; Schindler 1977; Vitousek & Farrington 1997: Tyrrell 1999). If we add P to systems (or lose N without losing P), then N becomes limiting to organisms other than N fixers, and N fixers could have an advantage. The other alternative is that symbiotic N fixers systematically require more P or are less competitive in acquiring P than are non-fixers. In this case there will be a level of P availability at which N fixers are limited by P, but non-fixers are limited by N. The greater the disparity in P requirements between fixers and non-fixers, the greater the disequilibrium in supply of N versus P.

We see two reasons why symbiotic N fixers might have a greater requirement for P than non-fixers. The first is that building and/or maintaining the symbiosis, and/or the fixation process itself, requires more P than is needed by non-fixing organisms. Alternatively, fixers could have a P-demanding lifestyle, to go with, indeed to support, the N-demanding lifestyle (discussed below) that McKey (1994) proposes to be driving the legume-rhizobium symbiosis. To the extent that N is used metabolically (as opposed to its use in defensive compounds), a commitment to high N levels in organisms may entrain a commitment to high P levels as well.

Does N fixation per se involve a higher demand for P than other forms of N acquisition? This question can be considered in two parts, capital and running costs. The bacteroid component of infected cells has a high P content, partly for the adenosine nucleotides needed to provide energy used in the nitrogenase reaction (Sprent & Raven 1985). Bacteroids are enclosed in membranes that also have a high P content. In annual herbaceous species reliant on atmospheric N_2, the plant maintains about 5 percent of its dry weight as nodules, but in woody perennials the proportion decreases as more nutrients are recycled within the plant. The P concentration of nodules can range from 0.2–0.6% of DW depending on species and age (Allen et al. 1988), similar or slightly higher than that of leaves, which normally are a much higher fraction of total plant DW. Thus the P investment in nodules is a relatively small part of total plant P. Although fixation of N is energy intensive, P is derived from ATP recycling once the necessary ATP has been synthesized.

In order to show that legumes fixing N require more P than those reliant on mineral N, comparisons must be made on an appropriate basis. These can be done by examining the N × P interaction when plants are grown on N2 or mineral N with increasing levels of P supply (Robson 1983 and other publications from Robson's lab). If the interaction is negative, it suggests that plants fixing N_2 have a higher requirement for P than those growing on mineral N. A zero interaction suggests that the two forms of N are equally demanding of P and a positive interaction that high P levels may be inhibitory to N_2 fixation. Most of the interactions reported are negative. However, most studies have been of agricultural species, which are not adapted to low-nutrient soils. A zero interaction may be more common in natural environments and/or when woody plants are studied. For example Sanginga et al. (1995) found that nodulated *Gliricidia sepium*, a species widely used in agroforestry, only showed a response to P at the lowest level applied, which probably reflected the basic requirement for plant growth. *Acacia mangium* was found by Ribet and Drevon (1996) to have similar P requirements for growth on either N_2 or urea.

26

Another indicator of P requirement could be the P content of tissues other than nodules; if leaves of nitrogen fixing plants have higher P than those of non-fixing plants, this could reflect a greater requirement for P. Values in the literature fall into two groups, those from agricultural experiments and those from natural or unfertilised soils. Within these two groups legumes and non-legumes overlap. Although these data should be regarded with caution, since they were not gathered to address this particular question, they do suggest that not all legumes have an inherently higher requirement for P than do non-legumes.

The indisputable fact that many grasslands will only support nodulated legumes when they have sufficient available P may be a reflection, not only of the legumes' requirement for P, but also of inherent differences between grasses and legumes. These differences may influence P acquisition and concentrations. Simulation of P uptake from soil identifies root length density as the most sensitive parameter governing the rate of uptake (Barber 1984). While grasses have fibrous root systems that thoroughly exploit the top soil layer where P availability is greatest, many legumes have a tap root system with relatively less root length development in top soil. In addition, legume root nodules can represent a significant fraction of total root system biomass during early vegetative growth stages in low N supply environments. Hence, competition between nodule and root growth for assimilate supplied by the shoot can result in a reduction in the number and length of roots (Nutman 1948; Dart & Pate 1959; Cassman et al. 1980). Taken together, root system architecture and the effects of nodule growth on early root development may cause legumes to be less competitive for P against grasses that have dense, fibrous root systems, which may in turn help explain the need for P fertilizer applications to sustain legume components in grazed pastures (Smith 1992).

Sprent (1999) has summarized some of the ways in which legumes endemic to infertile soils can release P from normally unavailable sources. These include a greater overall allocation to roots (e.g., Binkley & Ryan 1998), possession of either or both of ecto- and arbuscular mycorrhizas, cluster (proteoid) roots and secretion of chelating agents which can separate P form inorganic complexes. All of these impose a carbon cost, and may be most suited to plants growing in a high light environment. These strategies are no different from those of other plants living in the same environments.

Overall, our understanding of the nature of the relationship between P supply and symbiotic N fixers remains incomplete. There is no question that some legume crops require large quantities of P, more when they acquire N by fixation than when they acquire it from soil. However, whether this greater P requirement extends to symbiotic N fixers that are adapted to infertile soils is unclear, as is the question of whether a greater requirement (where it exists) represents a P cost of N fixation per se or competitive disadvantages in P acquisition compared to non-fixers. Alternatively, legumes may have evolved a commitment to a more P- demanding lifestyle. Information on P concentrations in symbiotic N fixers and non-fixers in high diversity ecosystems, such as the patterns for N concentrations in potentially N-fixing legumes, non-fixing legumes, and non-legumes discussed in the next section, would be a useful first step towards addressing this last possibility.

Legumes in the canopy of tropical forests

Trees with the capacity to fix atmospheric N are very sparse or absent in the canopy of most late-successional temperate forests. However, legumes represent one of the most diverse and abundant families of higher plants in the canopy of many lowland tropical forests (e.g., Prance et al. 1976; Allen & Allen 1981; Moreira et al. 1992). A number of lines of evidence, including N concentrations in leaves and litterfall, rates of N mineralization, N trace gas emissions, and patterns of ^{15}N enrichment, suggest that soil N supply is relatively high in many lowland tropical forests – indeed that it may function as an excess nutrient there (Vitousek & Sanford 1986; Matson & Vitousek 1987; Keller & Reiners 1994; Martinelli et al. 1999; Matson et al. 1999). It is not hard to see why abundant canopy legumes might lead to high N availability – but it is more difficult to understand why potential N fixers persist (in abundance) in the biome in which N availability appears to be the greatest.

A possible explanation for the abundance of legumes in tropical forests can be developed based on McKey's (1994) suggestion that legumes in general have an N-demanding lifestyle – that they require higher concentrations of N than do plants in other families. He suggests that this greater requirement for N should be observed whether or not an individual plant is acquiring its N by fixation, and whether or not an individual species of legume even has the capacity to fix N.

McKey (1994) reviewed results of a number of studies; all reported higher concentrations of N in legumes than non-legumes. Moreover, while legumes in the subfamilies *Mimosoideae* and *Papilionoideae* (which most often support N-fixing symbioses) had greater N concentrations than those in the *Caesalpinioideae* (which generally do not), the *Caesalpinioideae* had higher N concentrations than did non-legumes. McKey noted that relatively few species were included in these comparisons, but suggested that the results are consistent with a greater demand for N by legumes – independent of N fixation.

Additional data on N concentrations in tropical forests can be applied to this question. In the lowland primary forest of the Samuel Reserve in Brazil, the average foliar N concentration of *Mimosoideae* species (2.73 ± 0.90%, n = 9) was significantly higher than that of *Caesalpinioideae* species (2.20 ± 0.60%, n = 44) which, in turn, was significantly higher than that of non-legume species (1.79 ± 0.50%, n = 254). Analyses of foliar N in Brazilian cerrado show that mimosoid legumes averaged 1.76 ± 0.14% (n = 39), papilionoid legumes averaged 1.93 ± 0.08% (n = 86), nodulated caesalpinioid legumes of the genus *Chamaecrista* 1.61 ± 0.07% (n = 287), non-nodulated caesalpinioids 1.75 ± 0.12% (n = 71), and non-legumes 1.28 ± 0.07% (n = 57) (Sprent et al. 1996). In an Amazonian inundation forest (*várzea*), the average foliar concentration of *Papilionoideae* species was 3.28 ± 0.19% (n = 11), *Mimosoideae* averaged 2.62 ± 0.45% (n = 7), Caesalpinioideae species averaged 2.45 ± 0.85% (n = 5), and non-legume species 1.96 ± 0.61% (n = 59). Although in the latter case N concentrations in *Caesalpinioideae* did not differ significantly from non-legumes, the pattern is consistent with McKey's explanation. Yoneyama et al. (1993) also found higher foliar concentrations of N in legume species in comparison to non-legumes in a study site near Manaus (central Amazon), regardless of nodulation capacity.

To what extent are the potentially N-fixing legumes present in the canopy of late-successional tropical forest deriving their abundant N from fixation? As noted earlier, it is difficult to measure rates of fixation in the field, especially in perennial vegetation. However, ^{15}N natural abundance can provide a qualitative means for identifying trees that derive a substantial fraction of their N via fixation. The rationale for this is that N fixers acquire some of their nitrogen from the air, which has a nitrogen isotopic composition (δ^{15}N) of 0‰. Where the N derived from soil has a very different isotopic composition, the isotopic composition of a fixing plant differs from that of a non-fixing plant. Relatively large differences between atmospheric and soil δ^{15}N are necessary for the success of this approach (Hogberg 1997; Handley & Scrimegour 1997).

Fortunately, the δ^{15}N of soil, and of trees that derive their N from soil, is highly enriched in many lowland tropical forests (Martinelli et al. 1999), so it should be possible to identify individual legumes that derive much of their N from fixation with some confidence. The comparison of average foliar δ^{15}N content in leguminous trees and non-fixing species from two areas of primary forest in the Amazon region indicated that only a few trees in "terra-firme" forests were fixing N. In the Samuel forest, only four individuals out of 34 had δ^{15}N significantly lower than the average foliar δ^{15}N value of non-fixers, while Yoneyama et al. (1993) found only one individual out of 18 with δ^{15}N significantly below the foliar δ^{15}N of non-fixers in a site near Manaus.

The facts that legume trees are abundant in tropical forests, support high N concentrations, and are not regularly fixing N suggests that symbiotic N fixation might occur only during temporary N shortage (McKey 1994). Bonnier and Brakel (1969), cited by Sylvester-Bradley (1980), suggested a similar explanation for the lack of nodulation in African primary forests. Moreover, measurements of nitrogen fixation in several areas of tropical forests in Brazil, carried out by Sylvester-Bradley et al. (1980), support this working hypothesis. The occurrence of nodules was rare in "terra- firme" primary forest in the central Amazon where N is highly available, sporadic in disturbed primary forests (roadside and small clearings) and secondary forests, and frequent in cultivated soils where N losses generally are substantial. Acetylene-reducing activity also showed a similar trend, being greater in perturbed than in pristine areas. Clearing, fire, and cultivation cause a temporary shortage of N that, in turn, could trigger N fixation by legumes.

With few exceptions, legumes are absent from the canopy of late- successional temperate forests, although herbaceous legumes fix N actively all the way to high arctic and alpine ecosystems. The absence of woody legumes could represent a phylogenetic constraint (Crews 1999), in that legumes originated and radiated in the tropics. However, woody actinorhizal species are widespread in temperate and boreal forests, if generally confined to early successional ecosystems. The extent to which actinorhizal species may have committed to an N-demanding lifestyle is worth investigating; recent phylogenetic work demonstrates that actinorhizal and rhizobial species share a clade (Soltis et al. 1995).

The Vitousek and Field (1999) model discussed above addressed a number of reasons why N fixers are absent from the canopy of late-successional temperate forests, where N is in short supply and an N-fixer might be expected to have a substantial advantage. Another reason that could contribute to this pattern is the lower overall availability of N in temperate forests – the very factor that it seems should give fixers an advantage. The striking difference in N availability between most lowland tropical forests and most temperate forests could be due, at least in part, to processes in addition to the abundance of legumes. One possibility is the more rapid rates of decomposition in the tropics; these cause rapid cycling of N through litter, while colder systems immobilize N in litter for much longer periods (Vitousek & Howarth 1991). Another possibility is that periods of high N availability in spring/early summer in temperate regions are congruent with periods of adequate soil moisture, while later-season N deficiency is congruent with low water availability. In contrast, water availability is more continuous in the lowland tropics. On a larger time scale, another factor could be the more advanced stage of soil development in many tropical ecosystems, whereas temperate and boreal soils are reset frequently by glaciation (Chadwick et al. 1999). Whatever the reason, the relatively low N availability in temperate forests could mean that N-demanding legumes can *only* persist there if they fix N, and therefore they can be excluded by the mechanisms discussed above. In contrast, the relatively high availability of soil N in lowland tropical forests could mean that legumes can grow there without fixing N, most of the time.

Other symbiotic systems

In addition to the nodulated rhizobial and actinorhizal symbioses considered above, there are a number of other N-fixing symbioses that include higher plants. In general, these are not known as well as rhizobial systems; as far as we know, there are no models describing ecosystem- level controls of fixation in these symbioses. Some of the major symbioses and what we know of them are summarized below. Unless otherwise referenced, this listing draws heavily upon Sprent and Sprent (1990).

- All species of the genus *Gunnera* are symbiotic with *Nostoc*. These may be important as a source of N in some environments, mainly in the montane tropics and the southern hemisphere, e.g., New Zealand, southern part of South America (Chile down to Tierra del Fuego), Falklands (Malvinas). These plants may be more-or-less obligate for N fixation; they are always observed with endophytes in nature. Estimates of N fixation in the field are almost lacking, but some approximations could be made on a biomass basis. We suspect that these systems are constrained similarly to other higher plant-based symbioses, although moisture stress may be relatively more important.
- Cycads have coralloid roots with endophytic cyanobacteria (*Nostoc* and some other genera) which can apparently fix enough N for the plant's needs. We know of no studies on environmental constraints, except that some cycads (e.g., *Macozamia reidlii*) are fire resistant. They are likely to be relatively drought tolerant, and, in view of the time when they evolved, they are likely to thrive as atmospheric carbon dioxide levels rise. Cycads effectively replace legumes (such as species of *Acacia* and some endemic Australian genera) in some open forest ecosystems in Australia and southern Africa.
- The aquatic fern genus *Azolla*, with six species native to tropical and warm temperate freshwaters, is the only pteridophyte known to have a nitrogen fixing symbiosis. Its cyanobacterial endophyte is usually called *Anabaena azollae*, although it is almost certainly a species of *Nostoc*, in common with most cyanobacterial symbioses. It has not yet been cultured axenically, making detailed analysis of its genes difficult. In natural conditions *Azolla* usually grows symbiotically, although when grown in the laboratory with high levels of added nitrate or ammonium

it excludes the cyanobacteria. The endosymbiont inhabits pouches on the underside of the dorsal lobe of the leaf. *Azolla* is capable of rapid growth, doubling its weight in seven days; it has been exploited for many years as a green manure in wetland rice production, but it can also be a weed. Agriculturally it is particularly satisfactory in areas where it is killed by high summer temperatures; otherwise its use is labor intensive.

There are also looser associations between plants and N fixers. For all of these, there is little good evidence of direct transfer of combined N from the fixing component to the plant. The alternative, that the plants receive N after bacterial cell death, may reduce the total quantity of N fixed, because there is no continuous export of ammonium that can be assimilated by plant cells (as in rhizobia and *Frankia*).

- Associative symbioses. This is taken here to include organisms on the outer cells of root cortices and on root surfaces, whose carbon sources vary from dead cells to root exudates. The latter could support significant fixation, but will be competed for by other soil microbes. Measuring N fixation in this environment is methodologically difficult, but long-term mass balance studies in a number of cases reveal a gap which might well be filled by N-fixing associative bacteria.

- Endophytic bacteria. These are species that are found inside the plant. There is much evidence, especially from land races of sugar cane, that significant nitrogen fixation may occur, and nitrogen-fixing organisms have been isolated (James & Olivares 1998; James 2000). In global terms, the significance of this source of fixation is unknown, but one cannot ignore the fact that some grasslands in Brazil (*Brachiaria* spp, *Paspalum notatum*) can grow well continuously without added N or associated legumes. Similarly, sugar cane in Brazil is generally grown with substantially less fertilizer N than the crop's N requirement. After much skepticism, the Brazilian work is now being extended to other countries. It is impossible to put meaningful values upon rates of fixation at present, much less on controls, but this source may well turn out to be significant in ecological terms.

Heterotrophic N fixation

The third major pathway of N fixation is by heterotrophic bacteria in soils and sediments. Heterotrophic fixation during the decomposition of plant litter might be expected to be important in terrestrial ecosystems, because most plants produce litter with ratios of C:N substantially greater than those required by most microorganisms. A C:N ratio of 20:1 would be low for a terrestrial plant, while 8:1 is more typical of heterotrophic bacteria. Often the ratio in plants is very much wider; leaf litter in low-N systems not infre-

32

quently has a C:N ratio of 150:1, and the ratio in wood is wider yet. These wide ratios suggest that N might ordinarily be in short supply for microorganisms, while energy (reduced C) should be abundant. Consequently, microbial growth (and decomposition) could be limited by N, giving heterotrophs that can fix N an advantage (as long as other resources are abundant).

How important is heterotrophic N fixation? The capacity to fix N is widespread in anaerobic and microaerophilic bacteria (Sprent & Sprent 1990), and most freshwater and marine sediments support measurable rates of fixation. However, the ecosystem-level contribution of N by this pathway is constrained by (1) the inefficiency of anaerobic metabolism, which reduces the quantity of N that can be fixed; and (2) the importance of denitrification in anaerobic environments. Rates of denitrification exceed rates of N fixation in those sediments where both processes have been measured (Howarth et al. 1988a). Nevertheless, the relatively high rates of fixation that often occur in anaerobic sediments even where ammonium concentrations are high (Howarth et al. 1988b) suggest that the energetic cost of fixation is relatively low where there is no need to protect nitrogenase against O_2.

The net contribution of heterotrophic N fixation to ecosystem N budgets may be greater in wetland soils. N budgets of flooded rice suggest that 50–100 kg N ha^{-1} y^{-1} may be added by N fixation (Cassman et al. 1995), and heterotrophic fixers contribute a substantial proportion of this total (Eskew et al. 1981). Similarly, high rates of heterotrophic fixation may support plant production in some natural wetlands, for example the annually-burned papyrus marshes along the margins of the Amazon River in Para, Brazil.

The capacity to fix N is more restricted taxonomically among heterotrophic bacteria in aerobic environments. Such organisms must be able to protect their nitrogenase against O_2, often by barriers to diffusion or by high rates of respiration or some combination thereof. This protection against O_2 can be energetically costly. In addition, the activity of heterotrophic N fixers may be restricted to a narrower range of environmental conditions than is decomposition; low soil pH in particular is associated with lower rates of N fixation in aerobic soils (Sprent & Sprent 1990). Nevertheless, rates of N fixation in the range of 1–5 kg N ha^{-1} y^{-1} have been reported for decomposing litter in a range of environments. These rates are far below what symbiotic N fixers or aquatic cyanobacteria can achieve, but in the long term they could be significant to the N budget of unpolluted sites.

Overall, where N supply limits rates of decomposition (and the growth of microbial populations), then N-fixing heterotrophs should have an advantage, and N fixation should be relatively high. A fundamental question is then – when and where does N supply limit rates of decomposition? There is a great deal of indirect evidence for N limitation to decomposition, including positive correlations between N concentration in litter and rates of decomposition, and the fact that a substantial quantity of N is immobilized by decomposers in many sites. However, experimental studies which add N to decomposing litter yield mixed results. N limitation is generally observed in agricultural systems that produce relatively decomposable litter. However, in natural systems N additions sometimes stimulate decomposition, more often have no effect, and occasionally even slow decomposition (Prescott 1995; Downs et al. 1996; Hobbie & Vitousek 2000).

Alternatively, the activity of decomposers could be controlled by carbon quality, more than by the supply of N or other nutrients. The litter of plants growing in low nutrient sites often contains much of its C is in lignin, soluble polyphenols, and other recalcitrant compounds. To the extent that the growth of microbes and the decomposition of litter are controlled by the abundance of such compounds, we would expect that N supply would not limit microbial growth or decomposition, and so heterotrophic N fixation would be very slow or absent. Where microbial activity is not proximally constrained by N supply, then N fixation represents an energetically costly activity without a substantial benefit.

A recent study evaluated the control of decomposition rate, and of associated heterotrophic N fixation, by C quality, N supply, and the supply of other nutrients in Hawaiian montane forests (Hobbie & Vitousek 2000; Vitousek & Hobbie, 2000). It made use of the litter of a single tree species that occupies a wide range of sites that differ in nutrient supply, and that has a wide range of litter chemistry (N concentrations from 0.19 to 0.90 percent, P from 0.013 to 0.27 percent, lignin from 11 to 28 percent). Moreover, long-term fertilization experiments with N, P, and all other essential elements, alone and in factorial combinations, were underway on several of the sites, allowing the effects of tissue chemistry and rates of external nutrient supply to be considered separately. Results of the study included:

- Decomposition of low-lignin litter was stimulated by additions of N, while the stimulation was small or absent for high-lignin litter.
- Rates of heterotrophic N fixation were several-fold higher in low-lignin than in high-lignin litter, with a strong positive correlation between the integrated quantity of N fixed during decomposition of unamended litter and the extent to which decomposition of that litter was stimulated by additions of N (Figure 7).
- P supply did not affect N fixation directly. While litter produced in some P-fertilized plots supported high tissue P concentrations and enhanced rates of N fixation, the increase in N fixation could be explained by decreased lignin concentrations in that litter.

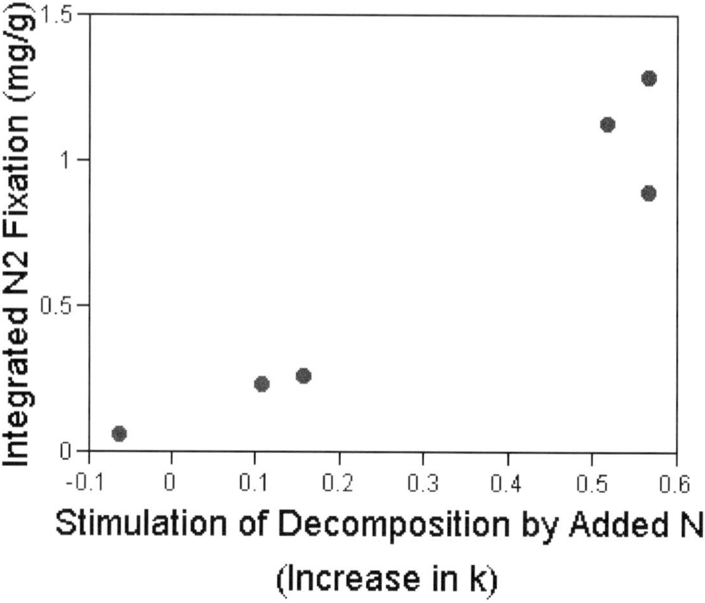

Figure 7. The relationship between N limitation to litter decomposition and heterotrophic N fixation in leaf litter from Hawaiian montane forests. The *x*-axis is the extent to which rates of decomposition (k, per year) are increased by additions of N fertilizer in a range of litter types, while the y-axis is the integrated N fixation during the course of decomposition for each litter type, in the absence of fertilization. From Vitousek and Hobbie (2000).

— Sites in which N supply demonstrably limited rates of forest growth had low C quality (high lignin) litter, with low rates of heterotrophic N fixation and rates of litter decomposition that responded little to added N. This disconnection between proximate N limitation to forest growth versus proximate C-quality limitation to decomposition and N fixation can keep heterotrophic N fixers from responding to N limitation, in effect sustaining N limitation to forest growth.

We have begun to summarize possible controls of heterotrophic N fixation in a simple model analogous to those for estuarine cyanobacteria and symbiotic higher plants; however, this model is at a more preliminary stage than the others. We consider three populations of decomposers. One population (#1) makes use of available C and N from litter (or the turnover of other microorganisms) at a C:N ratio of 16 (respiring half of the C for a cellular C:N ratio of 8). The second population is an N fixer. When C from litter plus microbial turnover is available at a ratio of C:N exceeding 16:1, this population (#2) can use C to fix N_2. The third population is a lignin-degrader; it can invest C available above the 16:1 ratio in the breakdown of lignin or polyphenol-

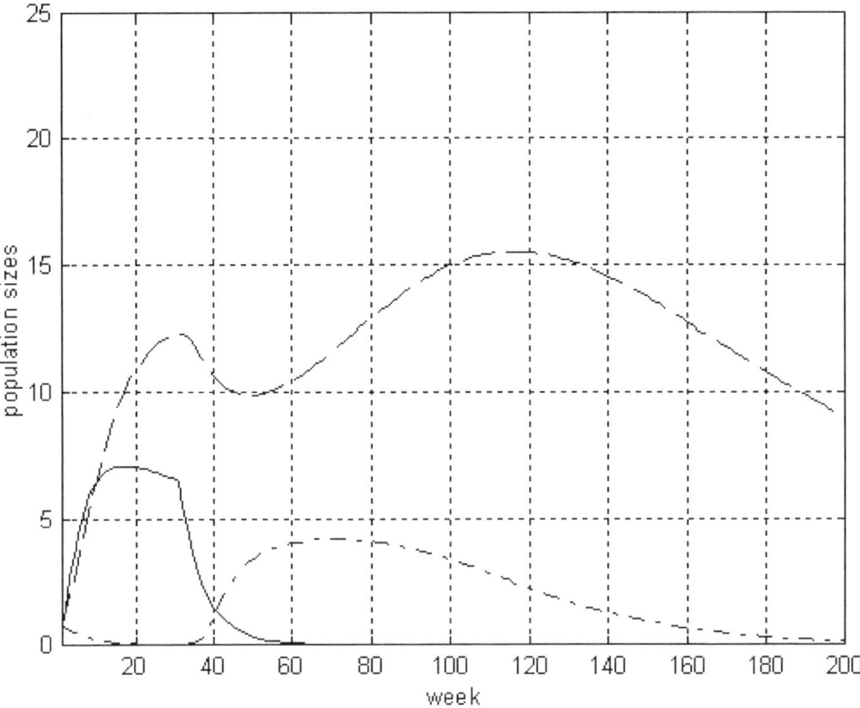

Figure 8. Simulated microbial populations during litter decomposition. The dashed line represents a microbial population that acquires N from labile pools in the substrate; the solid line represents a N-fixing population, and the dash-dot combination represents a lignin-degrading population.

protein complexes, thereby obtaining N. We assume that the relative success of populations 2 and 3 depend on the N yield from investing C in N fixation versus investing it in the breakdown of recalcitrant polyphenol-protein complexes.

For the model, litter consists of available C and N that can be used by microbes in a first-order decay process, and lignin C and lignin-associated N. A fraction of the available N associates with lignin as it becomes accessible to decomposers. All decomposers have the same stoichiometry and are grazed at the same rate.

This simple model yields an intuitively reasonable pattern of N fixation and lignin degradation over time. Population 1 dominates the decomposer community initially, followed by a pulse of the N fixer (#2), with the lignin-degrader growing once the concentration of N-associated lignin makes that N source rewarding (Figure 8). In practice, a peak in heterotrophic N_2 fixation some weeks to months following the initiation of decomposition has been observed in longitudinal studies of heterotrophic N_2 fixation in the field (Thompson & Vitousek 1997; Vitousek & Hobbie, in press). However, varying the initial lignin concentration in the model yields a pattern different from field observations – the more lignin that is incorporated in the simulated litter, the lower the quantity of lignin-associated N per unit of lignin C, and so the less rewarding (in terms of N yield) is lignin degradation. Consequently, more N is fixed in the model when more lignin is present – the opposite of field observations.

If our other assumptions are reasonable, then the effect of C quality (lignin concentration) on N fixation cannot be captured simply by evaluating the cost of N acquisition by fixation versus lignin degradation. Alternatively, if we assume that lignin (or compounds correlated with it) act directly to suppress N fixers in particular, we can generate the pattern observed in the field. However, we do not have a mechanistic basis for such an assumption. More information on the physiology and biochemistry of these functional groups of soil microbes will be required in order to carry this modeling approach further.

Conclusions

Overall, the ecological regulation of N fixation has a number of features in common, across the diverse N-fixing systems considered here. There is good evidence that non-N nutrients (P, in some cases Mo or Fe) can control the growth of cyanobacteria in lakes, estuaries, and terrestrial ecosystems, and also the growth of rhizobial and actinorhizal symbiotic systems. There is evidence that grazing disproportionately reduces the growth and activity of N fixers, in estuaries and rhizobial symbioses. Moreover, there is a suggestion that energetic constraints to N fixation may be more usefully considered qualitatively rather than quantitatively. For rhizobial and actinorhizal symbioses, the possibly greater shade intolerance of N fixers could have a larger effect on the N status of ecosystems in the long term than can their greater cost for N acquisition. For free-living heterotrophic fixers, the carbon (= energy) quality of their substrate is more important than its quantity in controlling rates of fixation. However, the treatment of N fixation in the multiple element limitation (MEL) model (Figure 6) is quantitative rather than qualitative, and Tyrrell's (1999) marine model similarly evaluates proximate N limitation to marine primary production in terms of relative costs of N acquisition by fixers and non-fixers.

More broadly, as Hartwig (1998) pointed out, N fixers may grow or actively fix N across a narrower range of environmental conditions than do non-fixers. To the extent that fixers are constrained by pH, drought, temperature, salinity, or other conditions, other organisms might be more likely to be limited by N under those conditions. However, our knowledge of the biology of N fixers is limited to a few relatively well-studied systems, and we are uncertain about the environmental tolerances of many fixers, particularly in natural systems.

Towards regional and global models

The similarities in controls of N fixation across very different groups of organisms suggest that ultimately it may be possible to incorporate physiological and ecosystem controls of N fixation into regional and global ecosystem models – an improvement that would allow such models to deal more realistically with the long-term consequences of global environmental change. However, a number of steps will be required before mechanistic controls of N fixation can be incorporated into regional and global models.

We believe that modeling the energetic controls of N fixation is closest to application. A framework for evaluating physiological components of these controls exists in marine, freshwater, and terrestrial ecosystems (Howarth et al. 1999;Tyrrell 1999, discussion of the multiple element limitation model [MEL] in this chapter). Moreover, light absorption by algal communities and plant canopies is calculated in existing ecosystem models, and also is accessible to direct remote sensing (Field et al. 1998). This information could be used to identify where N fixation by terrestrial cyanobacteria could be supported, and where shade-intolerant rhizobial or actinorhizal symbioses could colonize.

Other components of ecological controls over N fixation will be more difficult to incorporate in models. First, models will need to identify areas where N supply proximately limits NPP and other ecosystem processes – something that is now simply assumed to be true everywhere in many terrestrial ecosystem models. Second, the availability and dynamics of non-N nutrients, especially P, will need to be dealt with more realistically. P dynamics are incorporated in some but not all regional and global models (they are in Century but not TEM, for example), but even where they are included their treatment is sketchy in natural ecosystems, particularly forests. Other elements that could control the distribution and activity of N fixation, from major cations to trace elements like Mo, are not well represented in any terrestrial model. Finally, to the extent that herbivory represents an important control on the distribution and abundance of N fixers (Ritchie et al. 1998; Howarth et al. 1999), we are far from realistic models that can be applied on regional to global scales.

Modifying our models to include these controls will be difficult, but it should not be impossible. Moreover, these improvements to regional/global models are needed for reasons beyond understanding controls of N fixation,

38

particularly to deal effectively with the large portions of Earth where N appears *not* to be a proximate limiting resource.

While the effort to develop more mechanistic and realistic regional/global models continues, are there useful steps we can take to scale up our analysis of rates/patterns of N fixation in the meantime? We have such measures for NPP – correlations between climatic parameters and NPP are well worked out regionally and globally, and more directly we can use satellite remote sensing to measure light absorption by plants, and to drive models of photosynthesis and NPP (Sellers et al. 1997; Field et al. 1998). Are there comparable correlates/controls for N fixation?

Cleveland et al.'s (1999) synthesis of empirical studies of N fixation demonstrated that rates of fixation are correlated with calculated actual evapotranspiration (AET), across a range of biomes. Schimel et al. (1997) earlier estimated N fixation globally within the Century model using an assumed relationship between fixation and AET, although the slope of the relationship assumed by Schimel et al. is shallower than that suggested by the data in Cleveland et al.

This correlation of N fixation with climate may be the best that can be done globally, for now, but the empirical correlation is crude, not particularly strong, and there is no clear mechanism underlying the pattern. It is intriguing that the empirical analysis of Cleveland et al. (1999) identifies several biomes in which N fixation appears to be greater than expected based on AET, including deserts, arid shrublands, tropical savannas, and xenomorphic forest and woodland. All of these are open-canopied systems in which light availability at the soil surface could support cyanobacterial fixation and allow colonization by shade-intolerant symbiotic systems; all have fluctuating precipitation that could drive high levels of N losses. We think it likely that a hybrid between the empirical approach of Cleveland et al. (1999) and Schimel et al. (1997) on the one hand and the conceptual model approach outlined in this paper may provide useful interim predictions of N fixation, regionally and globally, while we work to improve the mechanistic basis of regional and global models.

Acknowledgments

This work was initiated as part of the International SCOPE N Project, which received support from both the Mellon Foundation and from the National Center for Ecological Analysis and Synthesis, and supported in part by grants to Stanford University from the USDA-NRI and the A. W. Mellon Foundation. Dan Binkley and an anonymous reviewer made useful comments on the manuscript. We thank Larry Bond for patiently preparing the manuscript for publication, and Douglas Turner for help with the figures.

References

Aber JD, McDowell W, Nadelhoffer K, Magill A, Berntson G, Kamakea M, McNulty S, Currie W, Rustad L & Fernandez I (1998) Nitrogen saturation in temperate forest ecosystems: hypothesis revisited. Bioscience 48: 921–934

Allen O & Allen E (1981) The Leguminosae: A Source Book of Characteristics, Uses, and Nodulation. University of Wisconsin Press, Madison, WI

Allen S, Raven JA & Sprent JI (1988) The role of long-distance transport in intracellular pH regulation in *Phaseolus vulgaris* grown with ammonium or nitrate as nitrogen source, or nodulated. J. Exp. Bot. 39: 513–528

Allos HF & Bartholomew WV (1959) Replacement of symbiotic fixation by available nitrogen. Soil Sci. 87: 61–66

Anderson DC, Harper KT & Holmgren RC (1982) Factors influencing development of cryptogamic soil crusts in Utah deserts. J. Range Manage. 35: 180–185

Arnone JA III & Gordon JC (1990) Effect of nodulation, nitrogen fixation and CO_2 enrichment on the physiology, growth and dry mass allocation of seedlings of *Alnus rubra* Bong. New Phytol. 116: 55–66

Austin AT & Vitousek PM (1998) Nutrient dynamics on a precipitation gradient in Hawai'i. Oecologia 113: 519–529

Barber SA (1984) Soil Nutrient Bioavailability: A Mechanistic Approach. John Wiley & Sons, New York

Bebout BM, Fitzpatrick MW & Paerl HW (1993) Identification of the sources of energy for nitrogen fixation and physiological characterization of nitrogen-fixing members of a marine microbial mat community. Appl. Environ. Microb. 59: 1495–1503

Bebout BM, Paerl HW, Crocker KM & Prufert LE (1987) Diel interactions of oxygenic photosynthesis and N_2 fixation (acetylene reduction) in a marine microbial mat community. Appl. Environ. Microb. 53: 2353–2362

Belnap J, Harper KT & Warren SD (1994) Surface disturbance of cryptobiotic soil crusts: nitrogenase activity, chlorophyll content, and chlorophyll degradation. Arid Soil Res. Rehab. 3: 1–8

Bergmann MA & Welch HE (1990) Nitrogen fixation by epilithic periphyton in small arctic lakes in response to experimental nitrogen and phosphorus fertilization. Can. J. Fish. Aquat. Sci. 47: 1545–1550

Beymer RJ & Klopatek JM (1992) Effects of grazing on cryptogamic crusts in pinyon-juniper woodlands in Grand Canyon National Park. Am. Midl. Nat. 127: 139–148

Binkley D & Ryan M (1998) Net primary production and nutrient cycling in replicated stands of *Eucalyptus saligna* and *Albizzia facultaria*. Forest Ecol. Manag. 112: 79–85

Bloom AJ, Chapin FS III & Mooney HA (1985) Resource limitation in plants, an economic analogy. Annu. Rev. Ecol. Syst. 16: 363–392

Boring LR, Swank WT, Waide JB & Henderson GS (1988) Sources, fates, and impacts of nitrogen inputs to terrestrial ecosystem: review and synthesis. Biogeochemistry 6: 119–159

Campbell JL, Hornbeck JW, McDowell WH, Buso DC, Shaley JB & Likens GE (2000) Dissolved organic nitrogen budgets for upland, forested ecosystems in New England. Biogeochemistry 49: 123–142

Carpenter RC, Hackney JM & Adey WH (1991) Measurements of primary productivity and nitrogenase activity of coral reef algae in a chamber incorporating oscillatory flow. Limnol. Oceanogr. 36: 40–49

Cassman KG, Whitney AS & Stockinger (1980) Root growth and dry matter distribution of soybean as affected by phosphorus stress, nodulation and nitrogen source. Crop Sci. 20: 239–244

Cassman KG, Singleton PW & Lindquist BA (1993) Input/output analysis of the cumulative soybean response to phosphorus on an Ultisol. Field Crop Res. 34: 23–36

Cassman KG, De Datta SK, Olk DC, Alcantara JM, Samson MI, Descalsota JP & Dizon MA (1995) Yield decline and the nitrogen economy of long-term experiments on continuous, irrigated rice systems in the tropics. In: Lal R & Stewart BA (Eds) Soil Management: Experimental Basis for Sustainability and Environmental Quality (pp 181–222). Lewis/CRC Publishers, Boca Raton

Chadwick OA, Derry LA, Vitousek PM, Huebert BJ & Hedin LO (1999) Changing sources of nutrients during four million years of ecosystem development. Nature 397: 491–497

Chapin DM, Bliss LC & Bledsoe LJ (1991) Environmental regulation of nitrogen fixation in a high arctic lowland ecosystem. Can. J. Bot. 69: 2744–2755

Chapin FS III, Bloom AJ, Field CB & Waring RH (1987) Plant responses to multiple environmental factors. Bioscience 37: 49–57

Cleveland CC, Townsend AR, Schimel DS, Fisher H, Howarth RW, Hedin LO, Perakis SS, Latty EF, VonFischer JC, Elseroad A & Wasson MF. (1999) Global patterns of terrestrial biological nitrogen (N_2) fixation in natural ecosystems. Global Biogeochem. Cycles 13: 623–645

Coxson DS & Kershaw KA (1983) Rehydration response of nitrogenase activity and carbon fixation in terrestrial *Nostoc* commune from *Stipa-Bouteloa* grassland. Can. J. Bot. 61: 2658–2668

Crawford CS & Gosz JR (1982) Desert ecosystems: Their resources in space and time. Environ. Conserv. 9: 181–195

Crews TE (1993) Phosphorus regulation of nitrogen fixation in a traditional Mexican agroecosystem. Biogeochemistry 21: 141–166

Crews TE (1999) The presence of nitrogen fixing legumes in terrestrial communities: evolutionary versus ecological considerations. Biogeochemistry 46: 233–246

Crews TE, Kurina LM & Vitousek PM. Organic matter and nitrogen accumulation and nitrogen fixation during early ecosystem development in Hawaii. Biogeochemistry, in press

Dart PJ & Pate JS (1959) Nodulation studies in legumes. III. The effects of delaying inoculation on the seedling symbiosis of barrel medic. Aust. J. Biol. Sci. 12: 427–456

Downs MR, Nadelhoffer KJ, Melillo JM & Aber JD (1996) Immobilization of a ^{15}N-labelled nitrate addition by decomposing forest litter. Oecologia 105: 141–150

DuBois JD & Kapustka LA (1983) Biological nitrogen influx in an Ohio relict prairie. Am. J. Bot. 70: 8–16

Eisele KA, Schimel DS, Kapustka LA & Parton WJ (1989) Effects of available P and N:P ratios of non-symbiotic dinitrogen fixation in tallgrass prairie soils. Oecologia 79: 471–474

Eldridge DJ & Greene RSB (1994) Microbiotic soil crusts: A review of their roles in soil and ecological processes in the rangelands of Australia. Aust. J. Soil Res. 32: 389–415

Eskew D, Eaglesham AR & App AA (1981) Heterotrophic [15]N_2 fixation and distribution of newly fixed nitrogen in a rice-flooded soil system. Plant Physiol. 68: 48–52

Evans RD & Belnap J (1999) Long-term consequences of disturbance on nitrogen dynamics in an arid ecosystem. Ecology 80: 150–160

Evans RD & Johansen JR (1999) Microbiotic crusts and ecosystem processes. Crit. Rev. Plant Sci. 18: 183–225

Evans RD & Ehleringer JR (1993) A break in the nitrogen cycle of arid lands: evidence from $\delta^{15}N$ of soils. Oecologia 94: 314–317

Field CB, Chapin FS III, Matson PA & Mooney HA (1992) Responses of terrestrial ecosystems to the changing atmosphere: a resource-based approach. Annu. Rev. Ecol. Syst. 23: 201–235

Field CB, Behrenfeld MJ, Randerson JT & Falkowski P (1998) Primary production of the biosphere: Integrating terrestrial and oceanic components. Science 281: 237–240

Fritz-Sheridan RP & Coxson DS (1988) Nitrogen fixation on a tropical volcano, La Soufriere (Guadeloupe). Lichenologist 20: 63–81

Galloway, JN, Schlesinger WH, Levy H II, Michaels A & Schnoor JL (1995) Nitrogen fixation: atmospheric enhancement – environmental response. Global Biogeochem. Cycles 9: 235–252

Goosem S & Lamb D (1986) Measurements of phyllosphere nitrogen fixation in a tropical and two subtropical rain forests. J. Trop. Ecol. 2: 373–376

Graetz RD & Tongway DJ (1986) Influence of grazing management on vegetation, soil structure, nutrient distribution and the infiltration of applied rainfall in a semi-arid chenopod shrubland. Aust. J. Ecol. 11: 347–360

Grimm NB & Fisher SG (1989) Stability of periphyton and macroinvertebrates to disturbance by flash floods in a desert stream. J. N. Am. Benthol. Soc. 8: 293–307

Grimm NB & Petrone KC (1997) Nitrogen fixation in a desert stream ecosystem. Biogeochemistry 37: 33–61

Gutschick VP (1981) Evolved strategies in nitrogen acquisition by plants. Am. Nat. 118: 607–637

Hall SJ & Matson PA (1999) Nitrogen oxide emissions after nitrogen additions in tropical forests. Nature 401: 152–155

Handley LL & Scrimgeour CM (1997) Terrestrial plant ecology and ^{15}N natural abundance: the present limits to interpretation for uncultivated systems with original data from a Scottish old field. Adv. Ecol. Res. 27: 133–212

Handley LL, Austin AT, Robinson D, Scrimgeour CM, Raven JA, Heaton THE, Schmidt S & Stewart GR (1999) The ^{15}N natural abundance ($\delta^{15}N$) of ecosystem samples reflects measures of water availability. Aust. J. Plant Physiol. 26: 185–199

Hartwig UA (1998) The regulation of symbiotic N2 fixation. A conceptual model of N feedback from the ecosystem to the gene expression level. Persepct. Plant Ecol. Evol. Syst. 1: 92–120

Hedin LO, Armesto JJ & Johnson AH (1995) Patterns of nutrient loss from unpolluted, old-growth temperate forests: evaluation of biogeochemical theory. Ecology 76: 493–509

Herbert DA, Rastetter EB, Shaver GR & Agren GI (1999) Effects of plant growth characteristics on biogeochemistry and community composition in a changing climate. Ecosystems 2: 367–382

Hobbie SE & Vitousek PM. (2000) Nutrient regulation of decomposition in Hawaiian forests: Do the same nutrients limit production and decomposition? Ecology 81: 1867–1877

Hoffmann L (1989) Algae of terrestrial habitats. Bot. Rev. 55: 77–105

Högberg P (1997) ^{15}N natural abundance in soil-plant systems. New Phytol. 137: 179–203

Hooper DU & Johnson L (1999) Nitrogen limitation in dryland ecosystems: Responses to geographical and temporal variation in precipitation. Biogeochemistry 46: 247–293

Horne AJ & Carmiggelt CJW (1975) Algal nitrogen fixation in California streams: seasonal cycles. Freshwater Biol. 5: 461–470

Howarth RW & Cole JJ (1985) Molybdenum availability, nitrogen limitation, and phytoplankton growth in natural waters. Science 229: 653–655

Howarth RW, Chan F & Marino R (1999) Do top-down and bottom-up controls interact to exclude nitrogen-fixing cyanobacteria from the plankton of estuaries? An exploration with a simulation model. Biogeochemisty 46: 203–231

Howarth RW, Marino R, Land J & Cole JJ (1988a) Nitrogen fixation in freshwater, estuarine, and marine ecosystems. 1. Rates and importance. Limnol. Oceanogr. 33: 669–687

Howarth RW, Marino R & Cole JJ (1988b) Nitrogen fixation in freshwater, estuarine, and marine ecosystems. 2. Biogeochemical controls. Limnol. Oceanogr. 33: 688–701

Hungate BA, Dijkstra P, Johnson DW, Hinkle CR & Drake BG (1999) Elevated CO_2 increases nitrogen fixation and decreases soil nitrogen mineralization in Florida scrub oak. Glob. Change Biol. 5: 797–806

James EK (2000) Nitrogen fixation in epiphytic and associative symbiosis. Field Crop Res. 65: 197–209

James EK & Olivares FL (1998) Infection and colonization of sugar cane and other graminaceous plants by endophytic diazotrophs. CRC Cr. Rev. Plant Sci. 17: 77–119

Johansen JR & Rushforth SR (1985) Cryptogamic soil crusts; Seasonal variation in algal populations in the Tintic mountains, Juab County, Utah. Great Basin Nat. 45: 14–21

Johansen JR, Ashley J & Rayburn WR (1993) Effects of rangefire on soil algal crusts in semiarid shrub-steppe of the lower Columbia Basin and their subsequent recovery. Great Basin Nat. 53: 73–88

Jones K (1992) Diurnal nitrogen fixation in tropical marine cyanobacteria: a comparison between adjacent communities of non-heterocystous *Lyngbya* sp. and heterocystous *Calothrix* sp. Brit. Phycol. J. 27: 107–118

Joye SB & Paerl HW (1993) Contemporaneous nitrogen-fixation and denitrification in intertidal microbial mats – rapid response to runoff events. Mar. Ecol.-Prog. Ser. 94: 267–274

Joye SB & Paerl HW (1994) Nitrogen cycling in microbial mats: rates and patterns of denitrification and nitrogen fixation. Mar. Biol. 119: 285–295

Kapustka LA & DuBois JD (1987) Dinitrogen fixation by cyanobacteria and associative rhizosphere bacteria in the Arapaho Prairie in the Sand Hills of Nebraska. Am. J. Bot. 74: 107–113

Keller M & Reiners WA (1994) Soil-atmosphere exchange of nitrous oxide, nitric oxide, and methane under secondary succession from pasture to forest in the Atlantic lowlands of Costa Rica. Global Biogeochem. Cycles 8: 399–409

Knapp AK & Seastedt TR (1986) Detritus accumulation limits productivity of tallgrass prairie. Bioscience 36: 662—68

Kurina LM & Vitousek PM (1999) Controls over the accumulation and decline of a nitrogen-fixing lichen, *Stereocaulon vulcani*, on young Hawaiian lava flows. J. Ecol. 87: 784–799

Liengen T (1999) Environmental factors influencing the nitrogen fixation activity of free-living terrestrial cyanobacteria from a high arctic area, Spitsbergen. Can. J. Microbiol. 45: 573–581

Liengen T & Olsen RA (1997) Nitrogen fixation by free-living cyanobacteria from different coastal sites in a high arctic tundra, Spitsbergen. Arctic Alpine Res. 29: 470–477

Lüscher A, Hendry GR & Nösberger J (1998) Long-term responsiveness to free air CO_2 enrichment of functional types, species and genotypes of plants from fertile permanent grassland. Oecologia 113: 37–45

Marino R, Howarth RW, Shamess J & Prepas E (1990) Molybdenum and sulfate as controls on the abundance of nitrogen-fixing cyanobacteria in saline lakes in Alberta. Limnol. Oceanogr. 35: 245–259

Marino R, Chan F, Howarth RW & Pace M (Manuscript in preparation)

Martinelli LA, Piccolo MC, Townsend AR, Vitousek PM, Cuevas E, McDowell W, Robertson GP, Santos OC & Treseder K (1999) Nitrogen stable isotope composition of leaves and soil: tropical versus temperate forests. Biogeochemistry 46: 45–65

Matson PA & Vitousek PM (1987) Cross-system comparison of soil nitrogen transformations and nitrous oxide fluxes in tropical forests. Global Biogeochem. Cy. 1: 163–170

Matson PA, McDowell WH, Townsend AR & Vitousek PM (1999) The globalization of N deposition: ecosystem consequences in tropical environments. Biogeochemistry 46: 67–83

McKane RB, Rastetter EB, Shaver GR, Nadelhoffer KJ, Giblin AE, Laundre JA & Chapin FS III (1997) Reconstruction and analysis of historical changes in carbon storage in arctic tundra. Ecology 78: 1188–1198

McCollum EW, Crowder LB & McCollum SA (1998) Complex interactions of fish, snails, and littoral zone periphyton. Ecology 79: 1980–1994

McKey D (1994) Legumes and nitrogen: the evolutionary ecology of a nitrogen-demanding lifestyle. In: Sprent JL & McKey D (Eds) Advances in Legume Systematics: Part 5 – The Nitrogen Factor (pp 211–228). Royal Botanic Gardens, Kew, England

Mooney HA & Gulmon SL (1979) Environmental and evolutionary constraints on the photosynthetic characteristics of higher plants. In: Solbrig OT, Jain S, Johnson GB & Raven PH (Eds) Plant Population Biology (pp 316–337). Columbia University Press, New York

Moreira FMS, Silva MF & Faria SM (1992) Occurence of nodulation in legume species in the Amazon region of Brazil. New Phytol. 121: 563–570

Nutman PS (1948) Physiological studies on nodule formation. I. The relation between nodulation and lateral root formation in red clover. Ann. Bot. N.S. 12: 81–96

Paerl HW (1985) Microzone formation: its role in the enhancement of aquatic N_2 fixation. Limnol. Oceanogr. 30: 1246–1252

Parrotta JA, Baker DD & Fried M (1996) Changes in dinitrogen fixation in maturing stands of *Casuarina equisetifolia* and *Leucaena leucocephala*. Can. J. Forest Res. 26: 1684–1691

Pastor I & Binkley D (1998) Nitrogen fixation and the mass balances of carbon and nitrogen in ecosystems. Biogeochemistry 43: 63–78

Paul EA & Clark FE (1989) Soil Microbiology and Biochemistry. Academic Press, London

Peterjohn WT & Schlesinger WH (1990) Nitrogen loss from deserts in the southwestern United States. Biogeochemistry 10: 67–79

Peterjohn WT & Schlesinger WH (1991) Factors controlling denitrification in a Chihuahuan desert ecosystem. Soil Sci. Soc. Am. J. 55: 1694–1701

Prance GT, Rodrigues WA & Silva MF (1976) Inventário florestal de um hectare de mata de terra-firme km 30 da estrada Manaus-Itacoatiara. Acta Amazon. 6: 9–35

Prescott CE (1995) Does nitrogen availability control rates of litter decomposition in forests? Plant Soil 143: 1–10

Rastetter EB & Shaver GR (1992) A model of multiple-element limitation for acclimating vegatation. Ecology 73: 1157–1174

Rastetter EB, Ågren GI & Shaver GR (1997) Responses of N-limited ecosystems to increased CO_2: A balanced-nutrition, coupled-element-cycles model. Ecol. App. 7: 444–460

Ribet J & Drevon J-J (1996) The phosphorus requirement of N_2-fixing and urea-fed *Acacia mangium*. New Phytol. 132: 383–390

Rychert RD & Skujins J (1974) Nitrogen fixation by blue-green algae-lichen crusts in the Great Basin Desert. Soil Sci. Soc. Amer. Proc. 38: 768–771

Ritchie ME & Tilman D (1995) Responses of legumes to herbivores and nutrients during succession on a nitrogen-poor soil. Ecology 76: 2648–2655

Ritchie ME, Tilman D & Knops JMH (1998) Herbivore effects on plant and nitrogen dynamics in oak savanna. Ecology 79: 165–177

Robson AD (1983) Mineral nutrition. In: Broughton WJ (Ed) Nitrogen Fixation Vol 3 Legumes (pp 36–55). Clarendon Press, Oxford

Sanginga N, Danso SKA, Zapata F & Bowen GD (1995) Phosphorus requirements and nitrogen accumulation by N_2-fixing and non-N2-fixing leguminous trees growing in low P soils. Biol. Fert. Soils 20: 205–211

Schaffner WR, Hairston NG & Howarth RW (1994) Feeding rates and filament clipping by crustacean zooplankton consuming cyanobacteria. Verh. Internat. Verein. Limnol. 25: 2375–2381

Schimel DS, Brassell BH & Parton WJ (1997) Equilibration of the terrestrial water, nitrogen, and carbon cycles. PNAS 94: 8280–8283

Schindler DW (1977) Evolution of phosphorus limitation in lakes. Science 195: 260–267

Schultze LS, Ferris FG, Sherwood-Lollar B & Gerits JP (1996) Ultrastructure and seasonal growth patterns of microbial mats in a temperate climate saline alkaline lake: Goodenough Lake, British Columbia, Canada. Canadian Journal of Microbiology 42: 147–161

Sellers PJ, Dickenson RE, Randall DA, Betts AK, Hall FG, Berry JA, Collatz GJ, Denning AS, Mooney HA, Nobre CA, Sato N, Field CB & Henderson-Sellers A (1997) Modeling the exchanges of energy, water, and carbon between continents and the atmosphere. Science 275: 502–509

Shields LM & Durrell LW (1964) Algae in relation to soil fertility. Bot. Rev. 30: 92–128

Skujins J (1981) Nitrogen cycling in arid ecosystems. In: Clark FE & Rosswall T (Eds) Terrestrial Nitrogen Cycles: Processes, Ecosystem Strategies and Management Impacts (pp 477–491). Ecological Bulletins (Stockholm) 33, Swedish Natural Science Research Council, Stockholm, Sweden

Smith VH (1992) Effects of nitrogen:phosphorus supply ratios on nitrogen fixation in agricultural and pastoral systems. Biogeochemistry 18: 19–35

Smith VH & Bennett SJ (1999) Nitrogen:Phosphorus supply ratios and phytoplankton community structure in lakes. Arch. Hydrobiol. 146: 37–53

Soltis DE, Soltis PS, Morgan DR, Swenson SM, Mullin BC, Dowd JM & Martin PG (1995) Chloroplast gene sequence data suggest a single origin of the predisposition for symbiotic nitrogen fixation in angiosperms. PNAS 92: 2647–2651

Sprent JI (1999) Nitrogen fixation and growth of non-crop legume species in diverse environments. Perspect. Plant Ecol. Evol. Syst. 2: 149–162

Sprent JI & Raven JA (1985) Evolution of nitrogen fixing symbioses. Proceedings of the Royal Society of Edinburgh B85: 215–237

Sprent JI & Sprent P (1990) Nitrogen Fixing Organisms. Chapman and Hall, London

Sprent JI, Geoghegan IE, Whitty PW & James EK (1996) Natural abundance of ^{15}N and ^{13}C in nodulated legumes and other plants in the Cerrado and neighboring regions of Brazil. Oecologia 105: 440–446

Stal LJ (1995) Physiological ecology of cyanobacteria in microbial mats and other communities. New Phytol. 131: 1–32

Stewart WDP (1974) Blue-green algae. In: Quispel A (Ed) North Holland Research Monographs. Front. Biol. 33: 202–237

Sylvester-Bradley R, Oliveira LA, Podestá Filho JA & St.John TV (1980) Nodulation of legumes, nitrogenase activity of roots and occurrence of nitrogen-fixing *Azospirillum* spp. in representative soils of Central Amazonia. Agro-Ecoystems 6: 249–266

Thompson MV & Vitousek PM (1997) Asymbiotic nitrogen fixation and decomposition during long-term soil development in Hawaiian montane rain forest. Biotropica 29: 134–144

Tuchman NC & Stevenson RJ (1991) Effects of selective grazing by snails on benthic algal succession. J. N. Am. Benthol. Soc. 10: 430–443

Tyrrell T (1999) The relative influences of nitrogen and phosphorus on oceanic primary production. Nature 400: 525–531

Villbrandt M, Krumbein WE & Stal LJ (1991) Diurnal and seasonal variations of nitrogen fixation and photosynthesis in cyanobacterial mats. Plant Soil 137: 13–16

Vitousek PM & Field CB (1999) Ecosystem constraints to symbiotic nitrogen fixers: a simple model and its implications. Biogeochemistry 46: 179–202

Vitousek PM & Farrington H (1997) Nutrient limitation and soil development: experimental test of a biogeochemical theory. Biogeochemistry 37: 63–75

Vitousek PM & Hobbie SE. The control of heterotrophic nitrogen fixation in decomposing litter. Ecology 81: 2366–2376

Vitousek PM & Howarth RW (1991) Nitrogen limitation on land and in the sea: How can it occur? Biogeochemistry 13: 87–115

Vitousek PM & Sanford RL Jr. (1986) Nutrient cycling in moist tropical forest. Annu. Rev. Ecol. Syst. 17: 137–167

Vitousek PM, Aber JD, Howarth RW, Likens GE, Matson PA, Schindler DW, Schlesinger WH & Tilman D (1997) Human alteration of the global nitrogen cycle: sources and consequences. Ecol. Applic. 7: 737–750

Vitousek PM, Hedin LO, Matson PA, Fownes JH & Neff J (1998) Within-system element cycles, input-output budgets, and nutrient limitation. In: Pace M & Groffman P (Eds) Successes, Limitations, and Frontiers in Ecosystem Science (pp 432–452). Springer-Verlag, Berlin

Walker TW & Syers JK (1976) The fate of phosphorus during pedogenesis. Geoderma 15: 1–19

Warren SD (1995) Ecological role of microphytic soil crusts in arid ecosystems. In: Allsopp D, Colwell RR & Harksworth DL (Eds) Microbial Diversity and Ecosystem Function (pp 199–209). CAB International

West NE (1990) Structure and function of microphytic soil crusts in wildland ecosystems of arid to semi-arid regions. Adv. Ecol. Res. 20: 179–223

Wiebe WJ, Johannes RE & Webb KL (1975) Nitrogen fixation in a coral reef community. Science 188: 257–259

Williams SL & Carpenter RC (1997) Grazing effects on nitrogen fixation in coral reef algal turfs. Mar. Biol. 130: 223–231

Williams SL & Carpenter RC (1998) Effects of unidirectional and oscillatory water flow on nitrogen fixation (acetylene reduction) in coral reef algal turfs, Keneohe Bay, Hawaii. J Exp. Mar. Ecol. Biol. 226

Wilson JT & Alexander M (1979) Effect of soil nutrient status and pH on nitrogen-fixing algae in flooded soils. Soil Sci. Soc. Am. J. 43: 936–939

Yoneyama T, Moraoka T, Murakami T & Boonkerd N (1993) Natural abundance of ^{15}N in tropical plants with emphasis on tree legumes. Plant Soil 153: 295–304

Biogeochemistry **57/58**: 47–98, 2002.
© 2002 *Kluwer Academic Publishers. Printed in the Netherlands.*

Dinitrogen fixation in the world's oceans

D. KARL[1]*, A. MICHAELS[2], B. BERGMAN[3], D. CAPONE[2],
E. CARPENTER[4], R. LETELIER[5], F. LIPSCHULTZ[6], H. PAERL[7],
D. SIGMAN[8] & L. STAL[9]
[1]*School of Ocean and Earth Science and Technology, Department of Oceanography,
University of Hawaii, Honolulu, Hawaii 96822, U.S.A.;* [2]*Wrigley Institute for Environmental
Studies, University of Southern California, Los Angeles, CA 90089-0371, U.S.A.;*
[3]*Department of Botany, Stockholm University, SE-106 91 Stockholm, Sweden;*
[4]*Romberg Tiburon Center, San Francisco State University, Tiburon, CA 94920, U.S.A.;*
[5]*College of Oceanic and Atmospheric Sciences, Oregon State University, Corvallis, OR
97331-5503, U.S.A.;* [6]*Bermuda Biological Station for Research, 17 Biological Station Lane,
Ferry Reach GE 01, Bermuda;* [7]*Institute of Marine Sciences, University of North Carolina –
Chapel Hill, Morehead City, NC 28557, U.S.A.;* [8]*Department of Geosciences, Princeton
University, Princeton, NJ 08544, U.S.A.;* [9]*Netherlands Institute of Ecology, Centre for
Estuarine & Coastal Ecology, AC Yerseke NL-4400, The Netherlands*
(*author for correspondence, e-mail: dkarl@soest.hawaii.edu; Fax: 808-956-5059)*

Key words: bacteria, biogeochemistry, climate, cyanobacteria, iron, nitrogen, oceanic N_2 fixation, phosphorus, *Trichodesmium*

Abstract. The surface water of the marine environment has traditionally been viewed as a nitrogen (N) limited habitat, and this has guided the development of conceptual biogeochemical models focusing largely on the reservoir of nitrate as the critical source of N to sustain primary productivity. However, selected groups of *Bacteria*, including cyanobacteria, and *Archaea* can utilize dinitrogen (N_2) as an alternative N source. In the marine environment, these microorganisms can have profound effects on net community production processes and can impact the coupling of C-N-P cycles as well as the net oceanic sequestration of atmospheric carbon dioxide. As one component of an integrated 'Nitrogen Transport and Transformations' project, we have begun to re-assess our understanding of (1) the biotic sources and rates of N_2 fixation in the world's oceans, (2) the major controls on rates of oceanic N_2 fixation, (3) the significance of this N_2 fixation for the global carbon cycle and (4) the role of human activities in the alteration of oceanic N_2 fixation. Preliminary results indicate that rates of N_2 fixation, especially in subtropical and tropical open ocean habitats, have a major role in the global marine N budget. Iron (Fe) bioavailability appears to be an important control and is, therefore, critical in extrapolation to global rates of N_2 fixation. Anthropogenic perturbations may alter N_2 fixation in coastal environments through habitat destruction and eutrophication, and open ocean N_2 fixation may be enhanced by warming and increased stratification of the upper water column. Global anthropogenic and climatic changes may also affect N_2 fixation rates, for example by altering dust inputs (i.e. Fe) or by expansion of subtropical boundaries. Some recent estimates of global ocean N_2 fixation are in the range of 100–200 Tg N ($1–2 \times 10^{14}$ g N) yr^{-1}, but have large uncertainties. These estimates are nearly an order of magnitude greater than historical, pre-1980 estimates, but approach modern estimates of oceanic denitrification.

48

Introduction

Nitrogen (N) is an essential major element for life, accounting for nearly 10% of the dry weight of most microbial cells in the sea. In organic tissue, N is distributed primarily in proteins and nucleic acids, but is also an important constituent of bacterial cell walls (as muramic acid), energy transfer compounds such as nucleotides, photosynthetic pigments including chlorophylls and phycobilins, nucleic acids, vitamins and selected storage products (e.g. cyanophycin granules).

The N cycle in the sea involves a complex series of primarily microbiological transformations including (Figure 1): (1) nitrate (NO_3^-) and nitrite (NO_2^-) reductions to nitrous oxide (N_2O), dinitrogen (N_2), ammonium (NH_4^+) and organic-N by any one of several independent assimilatory or dissimilatory processes, (2) NH_4^+ production from the decomposition of organic-N (ammonification), (3) NH_4^+ oxidation to NO_2^-, N_2O and NO_3^- (nitrification) and (4) N_2 reduction to NH_4^+ and organic-N (N_2 fixation). Most of these N pool interconversions affect the oxidation state of N, and therefore the free energies of the various molecules and compounds. Consequently, most of these biological interconversions are either energy-yielding (e.g. nitrification) or energy-demanding (e.g. nitrogen fixation) and are fundamental processes in microbial biosynthesis and bioenergetics.

Although the thermodynamically-favorable form of N at the pH and redox state of seawater is NO_3^-, the ocean is far from chemical equilibrium in this regard; N_2 is by far the dominant form (e.g. the N_2:NO_3^- ratio is ≥ 25 in deep ocean waters and is ≥ 100 in most surface ocean waters). The relative stability of the triple bond of N_2 ($N \equiv N$) and the continual production of N_2 by the process of bacterial denitrification both contribute to this chemical disequilibrium. Thus although N, as an element, is present in a nearly inexhaustible supply in the marine environment, N that is combined in chemically 'fixed' or 'reactive' compounds, either as oxidized [NO_2^-/NO_3^-] or reduced [NH_4^+/organic N] forms may limit organic productivity. The low nutrient, fixed N-starved habitats of the near surface waters of the open ocean provide a seemingly ideal niche for N_2-fixing microorganisms. It would appear that this potential niche is largely unoccupied; or is it?

The planktonic, prokaryotic microorganisms that are responsible for N_2 fixation are taxonomically, physiologically and ecologically diverse, including: (1) *Bacteria* (phototrophs, heterotrophs, chemolithotrophs), (2) heterocystous and non-heterocystous cyanobacteria and (3) *Archaea*. Some species such as *Trichodesmium* are conspicuous by their sometimes massive open ocean blooms, other species are more cryptic. Biochemical considerations and accumulating field evidence suggest that Fe bioavailability

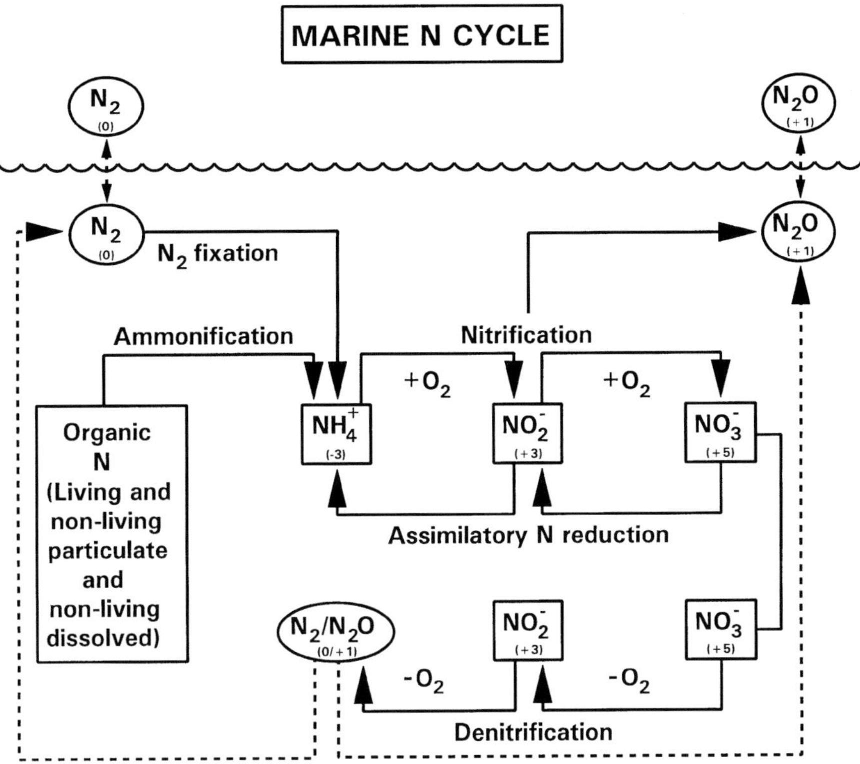

Figure 1. Schematic representation of the marine N cycle showing the major N pools and fluxes. Solid lines indicate transformations that are typically accompanied by direct or coupled energy release to the cell or organism or transformations that require an investment of energy and dotted lines indicate mass redistribution by physical-chemical processes such as gas exchange or water mass movements. Common terms that are assigned to selected pools or fluxes are also included. The small numbers in parentheses refer to the valence of N in each molecule or ion.

may control the distribution and abundance of N_2-fixing microorganisms in the sea. The primary pathway of Fe delivery to the upper oceans is via atmospheric deposition (e.g. dust), with upwelling and cross-shelf transport increasingly important in coastal and high productivity environments.

The global balance between denitrification, or the loss of bioavailable N ($NO_3^-/NO_2^- \rightarrow N_2$) and nitrogen fixation, or gain of bioavailable N ($N_2 \rightarrow NH_4^+/$organic-N), is a key for sustaining life in the sea on time scales of millennia. Furthermore, oceanic N_2 fixation may directly influence the sequestration of atmospheric carbon dioxide (CO_2) by providing a source of 'new' N to sustain the net production and export of organic matter from the euphotic zone. N_2-fixing microorganisms are involved in global feedbacks with the climate system and these feedbacks will exhibit complex dynamics on varying time-scales. The hypothesized feedback mechanisms will have the following component parts: the rate of N_2 fixation can impact the concentration of the greenhouse gas, carbon dioxide (CO_2), in the atmosphere on time-scales of decades (variability in surface biogeochemistry) to millennia (changes in the total NO_3^- stock from the balance of N_2 fixation and denitrification); CO_2 concentrations in the atmosphere can influence the climate; the climate system, in turn, can influence the rate of N_2 fixation in the oceans by controlling the supply of Fe associated with dust, and by influencing the stratification of the upper ocean. Humans also have a direct role in the current manifestation of this feedback cycle by their influence on dust production, through agriculture at the margins of deserts, and by our collective discharge of CO_2 into the atmosphere. These influences can lead to a cyclic feedback system, particularly on longer time-scales. Consequently, a large challenge in contemporary biogeochemical oceanography is to understand the molecular-to-global scale controls on N_2 fixation in the sea.

This chapter will review data on the physiological ecology of N_2-fixing marine microorganisms in an attempt to (1) re-assess the global ocean rates of biological N_2 fixation, (2) determine the major controls on rates of N_2 fixation, (3) analyze the impact of N_2 fixation on the oceanic carbon cycle, including carbon sequestration and (4) consider how human activity, including eutrophication, habitat alternation and climate change, might affect N_2 fixation in the world's oceans. In preparing this report, we have made use of several excellent and up to date reviews on N_2 fixation (Fay 1992; Gallon 1992; Gallon & Stal 1992; Stal 1995; Zehr 1995; Bergman et al. 1997; Zehr & Paerl 1998; Capone & Carpenter 1999; Paerl 2000; Paerl & Zehr 2000). A recent NATO Advanced Science Institutes Series volume devoted to 'Marine Pelagic Cyanobacteria: *Trichodesmium* and other Diazotrophs' provides a comprehensive summary of the process of N_2 fixation in the sea (Carpenter et al. 1992). More recently a workshop was convened at Catalina Island, USA to examine the conceptual and practical issues concerning the integration of N_2 fixation into global ocean carbon models (Hood et al. 2000). Our report will focus on the open ocean habitat and the eco-physiological controls on environmental N_2 fixation. Various methods for estimating local, regional and global scale rates of N_2 fixation will be presented along with the inherent assumptions and caveats. Finally, a research prospectus for the future will be presented and discussed.

Diversity of N_2-fixing microorganisms

N_2-fixing microorganisms are exclusively prokaryotic (including both *Bacteria* and *Archaea*); however, beyond that single distinguishing characteristic they show a remarkable diversity in form and function. Much of the research in the marine environment, especially that in the open ocean, has focused on the relatively conspicuous, non-heterocystous, filamentous cyanobacterium *Trichodesmium* (Carpenter & Romans 1991; Capone et al. 1997). These planktonic microorganisms are cosmopolitan in the low nutrient tropical and subtropical seas that dominate our planet and often form massive near-surface blooms (Carpenter & Capone 1992). Despite years of research focusing on this organism as the principal oceanic N_2 fixer, rigorous proof that *Trichodesmium*, and not the associated heterotrophic bacteria, actually fixed N_2 did not come until Zehr and McReynolds (1989) examined the associated *nifH* gene sequence and subsequent direct microscopic, immunochemical localization of nitrogenase in *Trichodesmium* cells (Paerl et al. 1989b; Bergman & Carpenter 1991).

Five *Trichodesmium* species have been identified based on cytomorphological (Janson et al. 1995) and 16S rDNA and *hetR* gene sequence analysis (Janson et al. 1999a): *T. thiebautii, T. erythraeum, T. tenue, T. hildebrandtii* and *T. contortum*. Among these groups, three main clades (*T. thiebautii* and *T. hildebrandtii, T. contortum* and *T. tenue,* and *T. erythraeum*) are present. These major groups often coexist in nature (Carpenter et al. 1993). At least three laboratory cultures of *Trichodesmium* are now available: strain NIBB1067 isolated from Kuroshio waters by Ohki and Fujita (1982), strain IMS101 isolated from North Atlantic coastal waters by Prufert-Bebout et al. (1993) and strain MACC0993 isolated from coastal waters near Qingdao, China by Haxo et al. (1987). NIBB1067 and IMS101 are most closely aligned to *T. erythraeum*.

In the field, the filamentous cyanobacterium *Trichodesmium* is polymorphic and can exist as free trichomes (single filaments of cells, about 100–200 cells long), or in one of two characteristic colony morphologies (Figure 2): (1) fusiform colonies composed of trichomes arranged in a generally parallel but often twisted orientation (also called tufts or rafts) or (2) spherical colonies composed of trichomes arranged in a generally radially symmetric pattern (also called puffs). For *Trichodesmium* sampled in the Kuroshio current off Japan, N_2 fixation (more specifically, acetylene reduction – hereafter, AR; see *'Direct field measurements of N_2 fixation'*) in free trichomes was only about 10% of the trichome-normalized rate measured in colonies (Saino & Hattori 1979). However, in a recent study conducted in the North Pacific subtropical gyre, the chl *a*-normalized free trichome vs. colony rates differed by only a factor of three (Letelier & Karl 1998). There is no

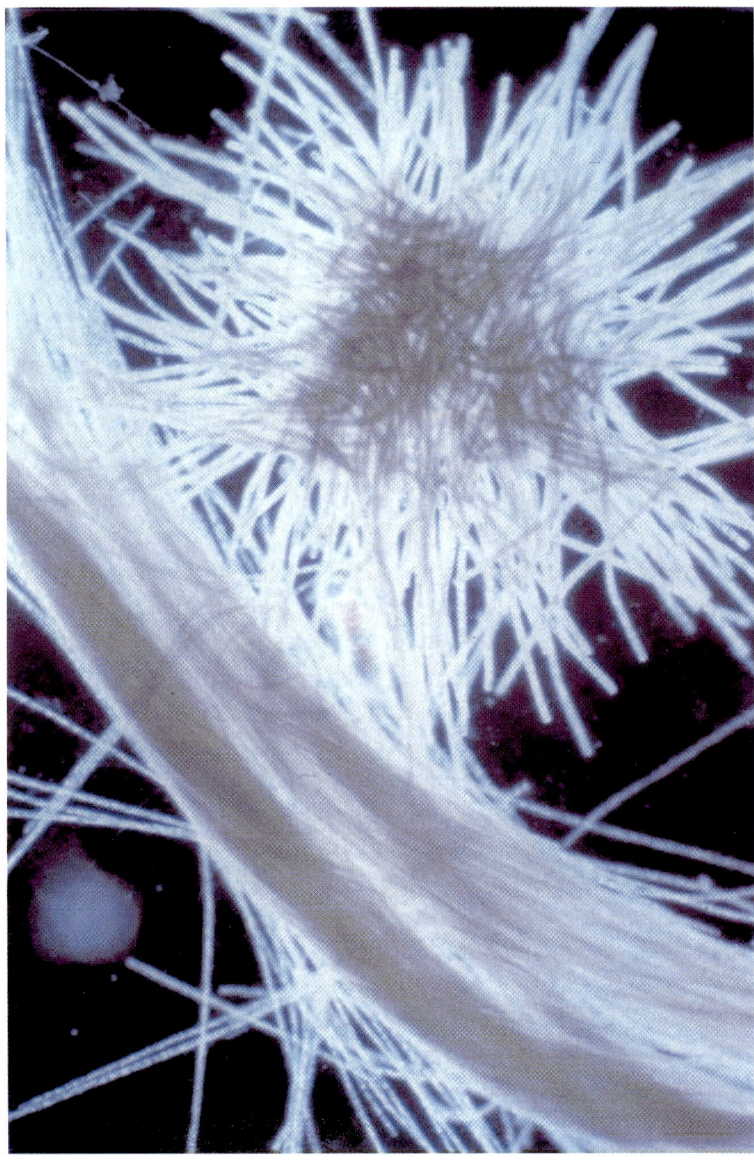

Figure 2. Morphological variability in field-collected samples of *Trichodesmium*. This light micrograph shows both the spherical (puff) colony morphology and fusiform (tuft) colony morphology. *Trichodesmium* can also be present as free trichomes, chains of approximately 100–150 cells (not shown). All three forms of *Trichodesmium* can co-exist in nature. Photo courtesy of Pernilla Lundgren and Birgitta Bergman, Stockholm University.

doubt that colony formation appears to enhance, but it is not a prerequisite for N_2 fixation.

First conducted by Christian Ehrenberg more than a century ago, field research on *Trichodesmium* has been promoted by the occurrence of prominent and extensive near surface ocean accumulations of colonies ('blooms'), especially during conditions of calm wind and sea (Figure 3). Its positive buoyancy (presence of gas vacuoles), high light-adapted photosynthetic apparatus and phosphorus-sparing effect (ability to grow with anomalously high N:P and C:P ratios) coupled with its high capacity for N_2 fixation and buoyancy control are ecologically-relevant adaptations for survival in marine environments that are chronically depleted in fixed N.

While *Trichodesmium* is undoubtedly the most well-studied marine N_2-fixing organism and perhaps one of the most important (Capone et al. 1997), it is not alone in the sea of diazotrophic microbes. Other known or suspected marine N_2 fixers include (see Postgate 1982; Benson 1985; Capone 1988; Sprent & Sprent 1990; Bergman et al. 1997; Paerl & Zehr 2000): (1) free-living (unicellular and filamentous), heterocystous and non-heterocystous, photoauto- and photoheterotrophic cyanobacteria including selected species of the genera *Synechococcus*, *Synechocystis*, *Oscillatoria*, *Aphanizomenen* and *Nodularia*, and two recently described non-heterocystous species, one unicellular (*Erythrosphaera marina*; Waterbury et al. 1988) and one filamentous (*Katagnymene* spp.; Lundgren et al. 2000), (2) anoxygenic photoautotrophic and photoheterotrohic *Bacteria*, (3) free-living, facultatively anaerobic pelagic chemoheterotrophic *Bacteria*, including *Vibrio diazotrophicus* (Guerinot et al. 1982; Guerinot & Colwell 1985; Urdaci et al. 1988), (4) chemoautotrophic *Bacteria*, including selected species of the genera including *Thiobacillus* and *Beggiatoa*, (5) obligately anaerobic *Bacteria* (e.g. *Desulfovibrio desulfuricans*) and *Archaea* (*Methanosarcina* spp.), (6) epiphytic cyanobacteria growing on pelagic Sargassum and other macroalgae (Carpenter 1972; Hanson 1977), (7) chemoheterotrophic bacteria growing as endosymbionts within mat-forming diatoms *Rhizosolenia castracanei* and *R. imbricata* var. *shrubsolei* (Alldredge & Silver 1982; Martinez et al. 1983), (8) the heterocystous photoautotrophic cyanobacterium, *Richelia intracellularis*, growing either as a free-living population or in a more common endosymbiotic association within several diatom genera including *Rhizosolenia* and the more ubiquitous *Hemiaulus*, or epiphytically with *Chaetoceros* (Mague et al. 1974; Venrick 1974; Villareal 1991; Janson et al. 1999b), (9) *Bacteria* living as ecto- and endosymbionts with marine invertebrates (Carpenter & Culliney 1975; Guerinot & Patriquin 1981; Proctor 1997), and (10) oxygenic, photoautotrophic cyanobacteria of the genera *Synechococcus* and *Synechocystis* growing as endosymbionts within several heterotrophic dinoflagellate

54

Figure 3. *Trichodesmium* bloom as viewed from space to the sea surface. [TOP] A massive *Trichodesmium* bloom in the Capricorn Channel of the southern Great Barrier Reef, Australia (near 22°50'S, 152°50'E) as viewed from the U.S. Space Shuttle flight STS-9, November 1983. This photo was taken using a hand-held 70 mm Hasselblad camera from a perspective 300 km above the Earth; the scale is approximately 1:850,000 (Reproduced from Kuchler & Jupp 1988); [BOTTOM] A massive *Trichodesmium* bloom in the North Pacific Subtropical Gyre (near 22°48'N, 158°11'W) as viewed from a U.S. Coast Guard C-130 aircraft (altitude 1.2 km), August 1989. The inset is a shipboard view of this same bloom from the deck of the *SSP Kaimalino* (photo credits: K. Louder and D. Hebel, inset).

genera and presumably supplying both C through photosynthesis and N through N_2 fixation to the host cells (Gordon et al. 1994). The continued use of direct optical and electron microscopic techniques, immunochemical and molecular procedures and novel isolation and culture methods are likely to reveal additional species of N_2-fixing microorganisms. At the present time it is not possible to determine the relative contributions of these various groups to global ocean N_2 fixation.

Field studies of laminated marine microbial mats and other benthic habitats have documented the presence of complex N_2-fixing microbial consortia with interrelated metabolic associations (Paerl & Pinckney 1996). A similar complexity is revealed upon close microscopic and physiological examination of field-collected *Trichodesmium* colonies (Paerl et al. 1989a; Siddiqui et al. 1992) suggesting that photoautotrophic-chemoheterotrophic syntrophy may be the rule rather than the exception for many natural microbial assemblages.

More recently, molecular methods have been used to ascertain the diversity of nitrogenase genes (*nifH*) in natural samples (see '***Nitrogenase Form and Function: Ecological Considerations***' section). Application of these methods to seawater samples collected from the Atlantic and Pacific Oceans has revealed an unexpected variation of *nifH* gene sequences suggesting the presence of previously undescribed N_2-fixing microorganisms (Zehr et al. 1998, 2000; Paerl & Zehr 2000). Picoplankton-sized ($< 2 \mu$m) organisms with *nifH* gene sequences included those with likely phylogenetic affiliations with α- and γ-proteobaceria, β-proteobacteria and unicellular cyanobacteria clades. Major phylotype differences were observed between ocean basins in waters of similar physical and chemical characteristics (Zehr et al. 1998); a majority of the Pacific Ocean *nifH* sequences aligned with group II genera *Myxosarcina* and *Xenococcus* whereas the Atlantic ocean samples revealed a greater preponderance of group I cyanobacteria. The most remarkable aspect of this study was the extremely high phylogenic diversity of *nifH* genes (Zehr et al. 1998, 2000); it is possible that some marine N_2-fixing microorganisms have, to date, evaded detection. The metabolic activities of these previously undescribed 'virtual' microbes may require a revision of current dogma and, more importantly, may help to balance the marine N cycle in open ocean, low nutrient habitats once systematic ecological studies have been conducted.

Nitrogen fixation by *Trichodesmium* now appears to be much more important that we had previously suspected, and most likely many other presently unknown diazotrophic microorganisms also contribute to the global ocean N budget. However, one thing seems to be clear: heterocystous cyanobacteria are quite rare in the marine environment and in most estuaries, while such organisms are very common in freshwater environments. They are occasionally common in a few brackish environments such as the Baltic Sea or very shallow estuaries during periods of low salinity (see Howarth et al. 1999). Even when present in marine environments, heterocystous cyanobacteria are usually the symbionts of other algae and, thus, not living in a typical marine environment. There is also no doubt that, considering the non-compatibility of N_2 fixation and oxygenic photosynthesis (see '*Oxygen*' section), heterocystous cyanobacteria are by far superior to non-heterocystous species with respect to diazotrophic growth.

Howarth et al. (1999) suggest that the balance between slow growth rates from trace element limitation (e.g. iron or molybdenum) and grazing losses limits the biomass of heterocystous cyanobacteria in most saline waters of estuaries waters and thus, by extension, the saline open ocean (see also Vitousek et al. this volume). The salinity-dependent process in that model is the hypothesized sulfate inhibition of molybdenum uptake (Howarth et al. 1999), however any balance of nutrient inhibition of growth and high grazing could yield this result. Under this scenario, *Richelia* and the other symbiotic heterocystous cyanobacteria, may persist and occasionally bloom in the open ocean if they experience a lower grazing loss with their host or if the host provides a trace nutrient environment that enhances symbiont growth. The balance of processes that control the abundance of these cyanobacteria is logically some complex mix (Howarth et al. 1999; Paerl & Zehr 2000), and the question of why heterocystous cyanobacteria are not more common in the oceanic environment remains an enigma. Understanding what factor makes these organisms unsuitable to proliferate in the sea, and how non-heterocystous diazotrophs evolved in the marine pelagic environment remain as important scientific challenges.

Nitrogenase form and function: ecological considerations

N_2 fixation is dependent upon the expression of an enzyme system, nitrogenase, which is a complex of highly conserved proteins among the various terrestrial and aquatic N_2-fixing prokaryotes. The most well studied nitrogenase enzyme system requires the activity of two related proteins (Postgate 1982): N_2 reductase, an Fe-protein (*nifH*) and dinitrogenase, an Fe-Mo protein (*nif*DK). Alternative vanadium and tungsten-requiring N_2 fixation systems have also been identified (Bishop et al. 1980; Fallik et al. 1991). Importantly, neither the detection of nitrogenase genes nor the presence of the coded proteins can be used to unambiguously determine rates of catalysis under *in situ* conditions. The synthesis and eventual expression of nitrogenase are ultimately determined by a broad range of physiological and ecological variables, including presence of fixed N compounds, oxygen concentration, availability of P and enzyme cofactors (notably Fe and Mo), a sufficient supply of energy and, perhaps, temperature. Several comprehensive reviews of nitrogenase enzyme structure and function have appeared (Broughton & Puhler 1986; Smith & Eady 1992; Dean et al. 1993; Kim & Rees 1994), so only a few key control mechanisms that relate to nitrogenase activity in *open ocean* ecosystems will be discussed here.

Oxygen

Molecular oxygen (O_2) is a potent inhibitor of nitrogenase synthesis and activity. A comprehensive discussion of the probable mechanisms of O_2 inactivation of nitrogenase has been presented by Gallon (1981, 1992), which should be read to fully appreciate the complexity of this otherwise straightforward enzyme-catalyzed reaction.

In most surface seawaters of the world's oceans, the O_2 concentration is either at or slightly above equilibrium with O_2 in the atmosphere (250–350 μm, depending upon sea surface temperature and salinity). These relatively high O_2 concentrations would, in theory, preclude nitrogenase activity in these habitats. The fact that *in situ* N_2 fixation does occur suggests at least one of the following: (1) the presence of an efficient O_2 protection or removal mechanism, (2) the presence of an altered, O_2 insensitive, form of nitrogenase or (3) a high rate of nitrogenase enzyme turnover (replacement). While the presence of an O_2 insensitive form of nitrogenase might be expected following more than 3.5 billion years of selection and evolution of marine N_2-fixing prokaryotes, there is presently no evidence for its existence.

In large heterocystous, filamentous cyanobacteria, there is evidence for a spatial separation of O_2-producing (photosynthetically-active) cells from the differentiated heterocysts which do not perform oxygenic photosynthesis. Furthermore, high respiration within the heterocysts and decreased permeability to dissolved O_2 of the cell surface all combine to reduce or eliminate O_2 inhibition of nitrogenase. However in non-heterocystous cyanobacteria, especially small unicells, structural modification is not possible so other behavioral or metabolic strategies of adaptation become important. Ironically, in the non-heterocystous cyanobacterium *Trichodesmium*, N_2 fixation is coupled to O_2 production via photosynthesis (Ohki & Fujita 1988).

It was initially thought that N_2 fixation in *Trichodesmium* was confined to the central portions of colonies, where net O_2 consumption would be favored (Fogg 1974; Carpenter & Price 1976; Bryceson & Fay 1981). It was suggested that these segregated, weakly pigmented and photosynthetically inactive internal cells might be analogous to the differentiated heterocysts (Carpenter & Price 1976). Independent studies using localized tetrazolium

58

salt reduction confirmed the presence of reduced microzones in *Trichodesmium* colonies (Bryceson & Fay 1981; Paerl & Bland 1982); microelectrode observations of O_2 gradients in *Trichodesmium* aggregates confirmed the presence of O_2-depleted microzones (Paerl & Bebout 1988). Both studies support a model of aggregation control of N_2 fixation.

A re-evaluation of the aggregation hypothesis using *Trichodesmium thiebautii* samples collected from the Caribbean Sea, however, failed to support key ecological predictions of the aggregation control model (Carpenter et al. 1990). Specifically, there were no differences in the distribution of photosystem I and II between central (protected) cells and O_2-unprotected cells near the periphery of the colonies, nor were there any pigmentation gradients. Diffusion model calculations revealed that respiration alone would be unlikely to maintain reduced intracolony O_2 concentrations.

Bergman & Carpenter (1991) first suggested that nitrogenase expression in *Trichodesmium* may be filament specific. More recently, it has been demonstrated that each trichome of *Trichodesmium* may contain one or more consecutively arranged cells containing nitrogenase in what appear to be differentiated cells. This spatial compartmentation of nitrogenase would provide a separation of photosynthetic and N_2-fixing activities that is necessary for optimum growth of the colony (Janson et al. 1994). Ultrastructural characterization of *Trichodesmium* cells, with and without nitrogenase, revealed significant differences that were consistent with the proposed functional interpretation (Fredriksson & Bergman 1997). Finally, whole cell immunochemical localization of nitrogenase (Lin et al. 1999) confirmed the spatial segregation model and it is likely that natural populations of filamentous N_2-fixing microorganisms have adopted a similar strategy for optimum growth and survival.

Additional evidence suggests that N_2 fixation in *Trichodesmium* must be sustained by intracellular adaptations including cellular differentiation, metabolic O_2 consumption, hydrogenase activity or rapid rates of nitrogenase synthesis and, therefore, turnover. Bergman et al. (1993) reported high levels of cytochrome oxidase in *Trichodesmium* and Kana (1993) reported rapid rates of oxygen cycling in field collected samples of *Trichodesmium*. Both of these studies are consistent with respiratory protection of nitrogenase as a possible adaptation. Stal and Krumbein (1985) suggested that a high rate of nitrogenase synthesis may provide a critical pool of enzyme necessary for N_2 fixation in *Oscillatoria*, and Capone et al. (1990) demonstrated a diel cycle in nitrogenase synthesis that was consistent with this hypothesis. The overall process of simultaneous N_2 fixation and photosynthetic oxygen production in *Trichodesmium*, however, remains enigmatic.

N_2 fixation has also been detected in free trichomes of *Trichodesmium* (Paerl 1994; Letelier & Karl 1998) which would not exhibit the hypothesized 'colony O_2 protection mechanism,' and in other small, unicellular prokaryotes exposed to O_2-saturated seawater (Mitsui et al. 1986). At low light levels ($<$ 100 μEinstein m^{-2} sec^{-1}), photosynthesis is subsaturated and oxygen evolution rates are low. This would enable respiration to remove photosynthetically-produced oxygen, minimizing the impact of oxygen on N_2 fixation but still providing reducing agents for N_2 fixation. However, in field populations N_2 fixation occurs even at high light levels. Saino and Hattori (1978) have reported a 200-fold difference between N_2 fixation in the day versus night for field-collected samples of *Trichodesmium* with maximum activity at highest light levels. With regard to O_2 inhibition of nitrogenase, it would seem that N_2-fixing chemoheterotrophic bacteria may have a distinct metabolic advantage over O_2-producing photoautotrophic N_2 fixers, provided these bacteria can find refuge in a low oxygen microzonal habitat (e.g. organic aggregate, biofilm, etc.). Heterotrophic bacteria fix N_2 at reasonably high rates in organic-rich, anoxic sediments even in the presence of large amounts of ammonium (Howarth et al. 1988), implying that much of the ecological cost and relative disadvantage of diazotrophy may be related to the difficulties associated with the oxygenic environment (Vitousek et al. 2002 and ***Energy limitation*** section).

Energy limitation

The microbiological fixation of N_2 demands a significant amount of cellular energy in the form of ATP, and an appropriate electron donor, usually as reduced ferredoxin. This energy is a mixture of the direct costs of N fixation and the energy requirements to maintain an appropriate cellular environment (e.g. low oxygen, trace-element incorporation, resynthesis of enzymes, etc). Sixteen moles of ATP per mole of N_2 are directly required. Most of this energy is required to split the dinitrogen molecule (4 ATP molecules for each pair of electrons), not to reduce it. The thermodynamic investment to reduce NO_3^- to NH_4^+ is actually larger than that needed to reduce N_2 to NH_4^+ The overall reaction, $3H_2 + N_2 \rightarrow 2NH_3$, is actually exothermic ($\Delta G° =$ -33.4 kJmol^{-1}; Sprent & Sprent 1990). The higher energetic costs associated with nitrate reduction may be one reason that some N_2 fixation can occur even when NO_3^- is present. Organisms that are designed for diazotrophy have few ecological reasons to invest in a switch to NO_3^- uptake when it is added to oligotrophic waters, however, they are highly unlikely to switch to diazotrophy when growing on nitrate. Furthermore, fixed N, in the form of NH_3, may diffuse out of the cell, where it is protonated to NH_4^+ and subsequently taken back into the cell by an energy-dependent uptake mechanism. Depending upon the environmental conditions, this futile cycle of NH_3 can account for a significant portion of the cell's energy budget for N_2 fixation.

Other important bioenergetic considerations include the costs of enzyme synthesis and regulation; the former may be especially acute for non-heterocystous species who tackle the oxygen inhibition problem by high nitrogenase turnover. The energetic costs of maintaining a low oxygen environment to minimize nitrogenase turnover may also be considerable (Vitousek et al. this volume). With regard to free-living, microaerophilic and anaerobic chemoheterotrophic *Bacteria* with the potential for N_2 fixation, having access to oxygen-free microzones or participating in net oxygen consuming reactions is a key constraint on plankton N_2 fixation. In such associations, substrate and energy limitation play important regulatory roles.

The ATP required to drive N_2 fixation is supplied through photophosphorylation, substrate level phosphorylation and oxidative phosphorylation, depending upon the species. Both sources of potential energy, light and bioavailable dissolved organic matter, are present in limiting concentrations in most marine environments. The ability to survive in high light environments (Li et al. 1980; Carpenter & Roenneberg 1995; Kana 1993; Subramaniam et al. 1999a, b), and its buoyancy may be important adaptations for N_2 fixation in *Trichodesmium* during periods of low turbulence.

Temperature control

While temperature, *per se*, does not restrict the growth of N_2-fixing microorganisms (e.g. nitrogenase activity has been detected at subzero temperatures in Antarctic soils; Davey & Marchant 1983), the global distribution of *Trichodesmium* appears well constrained by seawater temperature. With rare exception, most reported *Trichodesmium* blooms occur in subtropical and tropical marine habitats with surface water $\geq 25\,°C$ (Carpenter & Capone 1992). Nevertheless, individual colonies and free trichomes of *Trichodesmium* actively fix N_2 under *in situ* conditions of light and temperature to depths of at least 75 m in the subtropical North Pacific (Letelier & Karl 1998) corresponding to temperature of 21–23 °C. Because temperature and nitrate are significantly negatively correlated in the marine environment, it is not certain whether the global patterns of N_2 fixation versus surface water temperature derive from an inhibition of nitrogenase by low temperature or selection against N_2-fixing microorganisms under conditions of high ambient nitrate, or both.

N and P nutrient control

Despite its role as the marine 'model' for N_2 fixation, *Trichodesmium*, and probably most other diazotrophs, can grow on fixed N compounds, including both reduced and oxidized inorganic N and some forms of dissolved organic N (Ohki et al. 1986, 1991). In general, nitrogenase synthesis is repressed by NH_4^+ and is induced by depletion of fixed N substrates. However, careful laboratory studies conducted with *Trichodesmium* sp. (NIBB1067) have demonstrated that nitrogenase activity in cells grown on N_2 was not suppressed after 7-hr incubations with 2 mM $NaNO_3$ or 20 μM NH_4Cl, but was repressed by 0.5 mM urea (Ohki et al. 1991). *Trichodesmium* grown on NO_3^-, NH_4^+ or urea as a source of fixed N completely lacked the ability to fix N_2. The authors reported a complex pattern of regulation of the proteins of nitrogenase system, one that involved both transcriptional and post-transcriptional controls (Ohki et al. 1991). An independent laboratory study of N assimilation in *Trichodesmium* sp. (NIBB1067) revealed both a high affinity and a high uptake capacity for NH_4^+, urea and glutamate, and a low capacity for NO_3^-, but the culture was able to grow on NO_3^- as the sole source of fixed N (Mulholland et al. 1999). This latter study emphasized the inextricable links between N_2 fixation and N assimilation, and the complex patterns of intracellular regulation and control. Despite these careful laboratory studies, it is difficult to reproduce the conditions found in nature where the rapid recycling of generally low concentrations of N substrates ($[NO_3^-]$ typically ≤ 20 nM, $[NH_4^+]$ typically ≤ 40 nM, bioavailable $[DON] \geq 1$ μM, all with turnover rates of approximately 1 day) may result in different strategies of biochemical adaptation.

It is generally agreed that a low N:P ratio of available nutrients selects for N_2-fixing microorganisms (Niemi 1979). On the other hand, once a bloom of N_2-fixing microorganisms is established, the N:P ratio of the ambient dissolved and particulate matter pools increases dramatically as a result of an overproduction of fixed N and an efficient scavenging of bioavailable P (Karl et al. 1992). Consequently the longer-term signature for regions that support net N_2 fixation is a high (> 16:1), rather than a low, molar N:P ratio.

In contrast to the effects of N substrates on nitrogenase activity, very few studies have been conducted on potential control by P bioavailability. In most marine environments, N and P bioavailability are tightly coupled, so selection for N_2-fixing microorganisms will generally imply P limitation as well. During periods of intense N_2 fixation there must be a specific mechanism for P delivery to sustain net organic matter production. Several possibilities including atmospheric deposition, the passive upward flux of low density, P-enriched organic matter and vertical migration of *Trichodesmium* colonies have been suggested as potential mechanisms (Karl et al. 1992; Karl &

62

Tien 1997). Two additional physiological adaptations in *Trichodesmium* (and perhaps other N_2-fixing microorganisms) are the use of dissolved organic P (DOP) pools, and the ability to grow with an altered P cell quota. In most oligotrophic environments, DOP concentrations exceed the preferred substrate, orthophosphate, sometimes by 1–2 orders of magnitude. Induction of specific transport and hydrolytic enzymes, such as alkaline phosphatase, may be crucial for survival. A reduction of the P per cell quota, the so-called 'P-sparing effect,' is common for most microorganisms and may involve reduced intracellular pools of nucleotides and a lower nucleic acid content, especially RNA. Furthermore, an intensification of bioavailable P recycling rates under P limitation and a retention of P by an interdependent, remineralization-intensive food web may promote efficient N_2 fixation and microbial growth under P-controlled conditions. These strategies are likely to impact the growth rate of the population and, in the case of *Trichodesmium*, the low growth rates that have been measured under *in situ* conditions may be a manifestation of such a P-sparing strategy.

Fe bioavailability

Iron is a critical metal co-factor for nitrogenase (Howard & Rees 1996), and Fe bioavailability may be the most important overall control on oceanic N_2 fixation. Raven (1988) estimated that photolithoautotrophic growth using N_2 as the sole source of N requires two orders of magnitude more Fe per cell than for growth on NH_4^+. Fe has been shown to limit *Trichodesmium* N_2 fixation under field conditions (Rueter et al. 1992; Paerl et al. 1994). The supply of Fe to support N_2 fixation differs significantly between pelagic and benthic habitats. In the open ocean, Fe is generally depleted in the surface waters (Johnson et al. 1997). There is accumulating evidence to suggest that the delivery of Fe to the oceans in airborne dust may ultimately control the rate of N_2 fixation on the global ocean scale (Michaels et al. 1996; Falkowski 1997).

The processes controlling Fe availability in the upper ocean add several layers of biogeochemical complexity. In open ocean systems, Fe is supplied both by upwelling/mixing from below and from the atmosphere above. When Fe-enriched waters from beneath the euphotic zone are mixed into the surface ocean, they deliver a suite of other required major (C, N, P, Si) and trace elements in approximately the proper stoichiometry to sustain plankton production and coupled export. Under these conditions, there would be little selection for N_2 fixation due to the relatively high NO_3^-:Fe ratio in the upwelled waters. Alternatively, the dust-associated Fe flux is N-depleted, so it would select for N_2-fixing microorganisms and would ultimately serve to decouple the otherwise linked C-N-P-Si cycles in the sea. Consequently, atmospheric dust inputs to oligotrophic, open ocean ecosystems could alter community structure, bioelemental stoichiometry and the net sequestration of atmospheric carbon dioxide. These and other potential biogeochemical consequences of N_2 fixation are discussed later.

Most of the Fe associated with atmospheric dust is locked into inaccessible aluminosilicate lattices and only a small amount is released as bioavailable Fe after dust contacts seawater. Aqueous dissolution studies on Atlantic (Zhu et al. 1997) and Indian Ocean (Siefert et al. 1999) aerosols have found that only about 1% of the total Fe is released as Fe(II). The mechanisms controlling the release of Fe from dust are not well understood and several otherwise unrelated processes are potentially important: (1) partial dissolution of Fe (III) oxides by acidic aerosols (Keene & Savoie 1998), (2) photochemical reduction to Fe (II), especially in the presence of organic matter (Zhuang et al. 1992; Siefert et al. 1996; Zhu et al. 1997) and (3) organic ligand complexation (Gledhill & Berg 1994; Rue & Bruland 1995; Wu & Luther 1995). Rueter et al. (1992) have suggested that *Trichodesmium* colonies may intercept dust particles, facilitate dissolution and, hence, enhance Fe(II) flux. Little is known about the bioavailability of the various forms of Fe in seawater, but recent reports indicate that even some forms of colloidal and particulate Fe might be taken up by plankton assemblages. Algal phagotrophy of Fe-rich chemoheterotrophic bacteria is another possible physiological adaptation to life in the 'Fe-free' zone.

Molybdenum-sulfate antagonism

Molybdenum (Mo) is a another required cofactor of nitrogenase and, like Fe, has been proposed to limit N_2 fixation (Howarth & Cole 1985; Cole et al. 1993). Even though the concentration of Mo in seawater exceeds that in freshwater systems replete with diazotrophy, Howarth and Cole (1985) have proposed that the relatively high concentrations of sulfate (SO_4^{2-}) in seawater ($\sim$28 mM), a structural analogue of molybdate (MoO_4^{2-}), could compete with Mo uptake and inhibit N_2 fixation. Competitive inhibition of MoO_4^{2-} uptake by high SO_4^{2-} concentrations was demonstrated by Cole et al. (1993). However, the inhibition is not complete and, in their experiments, only a partial inhibition of the uptake of Mo would be predicted at sulfate and Mo concentrations comparable to seawater. In the brackish Baltic, increased concentrations of Mo did not stimulate increases in N_2 fixation, but decreases in sulfate did (Stal et al. 1999). Thus, there appears to be a very complex interplay between Mo, sulfate, N_2 fixation and other cellular processes in the diazotrophs that characterize fresh and brackish waters.

The sulfate-Mo competition hypothesis has been the basis of an elegant model (Howarth et al. 1999) to explain the strong gradient in N_2 fixation from freshwater lakes (high N_2 fixation rates) into estuaries (low rates and rare occurrence of heterocystous cyanobacteria). This model implicates a complex mix of processes, the balance of which control the amount of diazotrophy. Since a salinity gradient characterizes the differences among these ecosystems, the model fits the observations quite well. Sulfate concentrations vary with salinity and the increasing competitive reduction in Mo uptake as the salinity increases slows the growth rate of the cyanobacteria. At the salinity where grazing losses exceed growth rates, the net losses of biomass prevent blooms by the diazotrophs. Iron limitation is not concurrently explored in the model, in part, due to a lack of data on which to parameterize this process in estuaries. The authors recognize that any nutrient-like process that slows diazotroph growth rates in saline compared to fresh waters will yield a similar pattern (Howarth et al. 1999). This model also uses the relationship between heterocysts and filament length (influenced by grazing) to further impact growth rates. The apparent dominance of a non-heterocystous cyanobacterium in the open ocean may make this species less sensitive to the grazing-growth imbalance (Howarth et al. 1999) or provide an alternative explanation for the relative lack of heterocystous forms in the sea.

There are few direct measurements of Mo uptake in the open ocean. N_2-fixing potentials of some marine diazotrophs in the Fe-rich coastal Atlantic waters appear unaffected by this competition (Paerl et al. 1987; Paulsen et al. 1991). Oceanic diazotrophs undergo large changes in biomass under relatively similar Mo and sulfate concentrations suggesting that many other processes must have important ecosystem level effects, even if the sulfate-Mo competition still plays a role in inhibiting growth. Most likely, the small cellular Mo requirements for N_2 fixation are met through reduced, but sufficient, uptake and storage at the low growth rates that are implied by many field observations. In addition, recent research has shown the presence of alternative non Mo-requiring nitrogenases in bacterial and cyanobacterial diazotrophs (Bishop & Premakumar 1992). If such microbes are broadly distributed in nature (but there are no data to evaluate this at present), it would represent a selective mechanism by which Mo limitation could be circumvented. However, the final resolution of this debate will probably require detailed experimentation on the oceanic organisms that are currently adapted to life in this chronically Fe-depleted environment.

Substrate specificity and reaction by-products

All known nitrogenase systems studied to date exhibit a very low substrate specificity (Burris 1991). Although the physiological substrate is N_2, nitrogenase can also catalyze the reduction of many compounds that are structurally-related to N_2, including acetylene (C_2H_2), cyanide (CN) and N_2O. A shared characteristic of these alternate substrates is the presence of a N-N, N-O, N-C or C-C double or triple bond. The kinetic properties and reaction mechanisms vary considerably among these different substrates. Some investigators believe that the original function of nitrogenase in pre-Cambrian microorganisms was for detoxification rather than N_2 fixation; presumably there was ample fixed N (especially NH_4^+) available under early Earth conditions.

In addition to the above-mentioned substrates, nitrogenase also reduces protons to form hydrogen (H_2) during N_2 fixation. This is an obligate reaction and at least 25% (Simpson & Burris 1984) and generally a larger share of the flow of electrons through nitrogenase is used for H_2 production. There are several potentially important ecophysiological consequences of H_2 forma-tion. First, H_2 is a specific, competitive inhibitor of nitrogenase (Burris 1991) so elevated intracellular concentrations are unfavorable for N_2 fixation. The H_2 produced is either recycled intracellularly (i.e. oxidized and coupled to ATP formation via hydrogenase) or excreted into the surrounding environ-ment. Scranton (1983) has reported both hydrogenase activity and elevated ambient concentrations of H_2 following short-term incubation with field-collected *Trichodesmium* colonies. This coupled N_2 reduction/H_2 formation may have important ecological consequences especially considering the fact that most bacteria can oxidize H_2. In this way, the population of N_2-fixing microbes directly influences the larger microbial community.

More recently, it has been shown that *Azotobacter vinelandii* nitrogenase (and presumably other nitrogenases as well) can reduce both C-S and C-O bonds, including the conversion of carbonyl sulfide (COS) to carbon monoxide (CO) and hydrogen sulfide (H_2S) and the conversion of carbon dioxide (CO_2) to CO and water (Seefeldt et al. 1995). Further investiga-tion demonstrated that several COS analogues, including thiocyanate (SCN^-) can also be reduced leading to the formation of methane (CH_4) and H_2S (Rasche & Seefeldt 1997). The role of N_2-fixing microorganisms has not yet been evaluated as a potential source for the trace levels (typically nM) of CH_4, H_2S or CO that are ubiquitous in most tropical and subtropical marine environments.

Estimation of global ocean N_2 fixation

Chronic undersampling is a fact of life in oceanography (Platt et al. 1989) and still constrains the interpretation of most field data. In the case of N_2 fixation, neither spatial nor temporal uniformity can be assumed; much of the total N_2 fixation in the sea probably occurs during stochastic, heterogeneous

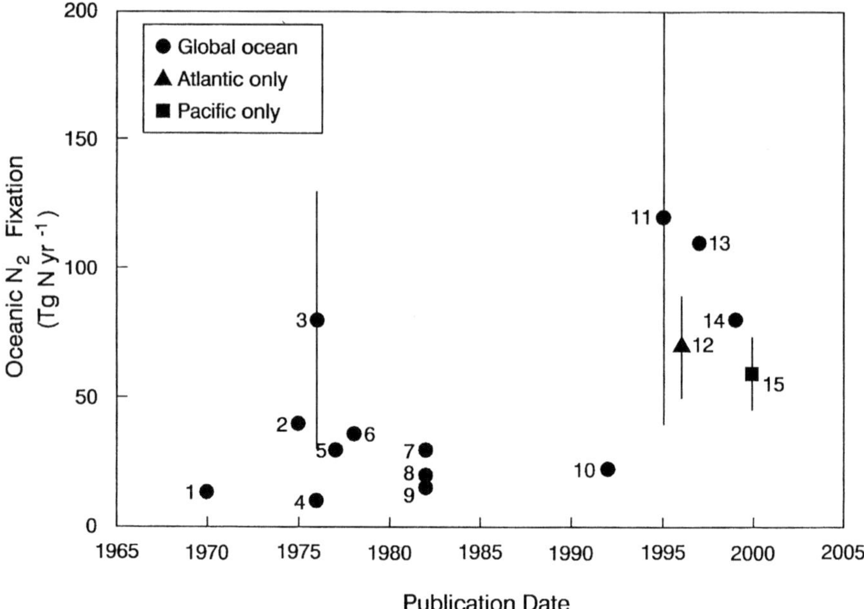

Figure 4. Basin scale and global ocean rates of N_2 fixation estimated by a variety of different methods plotted versus year of publication. These data were extracted from a recent review by Capone and Carpenter (1999) with more recent updates as indicated. The primary literature sources are: (1) Delwiche 1970, (2) Burns & Hardy 1975, (3) Soderlund & Svennson 1976, (4) McElroy 1976, (5) Delwiche & Likens 1977, (6) Paul 1978, (7) Soderlund & Rosswall 1982, (8) Capone & Carpenter 1982, (9) Fogg 1982, (10) Carpenter & Capone 1992, (11) Galloway et al. 1995, (12) Michaels et al. 1996, (13) Gruber & Sarmiento 1997, (14) Capone & Carpenter 1999 and (15) Deutsch et al. 2000. Although the individual estimates may have different integration time scales (see text), the full suite of estimates documents a potentially greater role for N_2 fixation than had been considered only a decade ago.

blooms that are not easily predicted or resolved by marine expeditionary field work. Furthermore, scientists who seek fundamental data on N_2 fixation in the sea usually target geographical regions and seasons when N_2 fixation is most likely to occur, and rarely produce data sets that can be statistically scaled to an unbiased average value for a particular biome. No existing N_2 fixation biogeochemical model captures the complexity of open ocean ecosystems and, as stated previously, we still have not even identified, much less isolated, all of the marine microorganisms with the genetic potential for N_2 fixation. Despite this sobering state of affairs, there exist several complementary approaches that can be used to assess rates of N_2 fixation in the global ocean (Figure 4). Much of the recent research, summarized below, has concluded that previous rates of global ocean N_2 fixation have been significantly underestimated.

Direct field measurements of N_2 fixation

Apart from the isolation of diazotrophs (e.g. Moore et al. 1921), N_2 fixation activity in the sea was initially discovered by Dugdale et al. (1961) in association with *Trichodesmium* colonies collected from the Sargasso Sea. Over the past 40 years numerous field studies have been conducted, many including direct estimation of N_2 fixation rates (see Capone et al. 1997 and Capone & Carpenter 1999). In our view, there is no substitute for direct field measurements if the stated objective is to estimate rates of N_2 fixation in the sea. This approach has provided many of the recent data sets that were necessary to question the existing dogma and to hypothesize an increased role for N_2 fixation in the world's oceans. They have also provided opportunities to sample plankton assemblages and to discover the previously unknown N_2-fixing species biodiversity (Zehr et al. 1998, 2000). More recently, the use of the polymerase chain reaction (PCR) and oligonucleotide probes designed to target the *nifH* gene have been used to detect N_2-fixing microorganisms and to assess the 'genetic potential' for N_2 fixation, whether or not N_2 fixation is actually occurring at the time of sampling (Kirshtein et al. 1993). Further development of these novel molecular methods may provide explicit links between the genetic potential and *in situ* rates of N_2 fixation in the sea (Zehr & Capone 1996; Zehr et al. 1996).

The first credible attempts to estimate rates of global ocean N_2 fixation resulted in a flux of 10–20 Tg N yr^{-1} (1 Tg = 10^{12} g) based primarily on the extrapolation of limited field measurements of N_2 fixation by *Trichodesmium* in the tropical Atlantic Ocean and Caribbean Sea (Capone & Carpenter 1982; Carpenter 1983) scaled globally using a historical data set for *Trichodesmium* abundance over larger portions of the world's oceans (Figure 4). A subsequent re-analysis of *Trichodesmium* abundance data for the tropical North Atlantic and Caribbean Sea increased this flux to 40–200 Tg N yr^{-1}, depending upon the regional boundaries that were selected (Carpenter & Romans 1991). Lipschultz and Owens (1996) provided a critical assessment of North Atlantic, basin-scale N_2 fixation rates. They concluded that the role of N_2 fixation, based on direct measurements of *Trichodesmium*, may have been overestimated and favor an estimate of approximately 15 Tg N yr^{-1}. Most recently, Capone and Carpenter (1999) have extrapolated average rates derived from field studies, which directly determined rates of N_2 fixation at diverse sites in the tropical oceans, across latitudinal bands adjusted for areas of upwelling and monsoonal periods. They derived a global estimate of about 80 Tg N per year, not accounting for input during bloom events. The primary limitation with the direct measurement approach is the spatial and temporal variability in N_2 fixation, relative to the measurement frequency. This makes the extrapolation of measured shipboard rates to regional and basin scale estimates uncertain. Then there are also the usual sampling and incubation problems that have plagued biological oceanographers for more than a century. By comparison, indirect geochemical estimates suggest that rates of N_2 fixation in the North Atlantic Ocean are near the high end of the range of direct estimates (see section on *'The N* parameter'*).

Most field studies of N_2 fixation have relied upon the acetylene reduction (AR) technique to estimate rates of N_2 reduction. Though indirect, the AR method is much more sensitive and much simpler than the alternative $^{15}N_2$ isotopic tracer method as traditionally applied (c.f. Montoya et al. 1996). Convenience and cost have generally dictated the choice of the AR method for most field studies. However, the advent of continuous flow isotope ratio mass spectrometry using multiple collectors has greatly improved the sensitivity of the $^{15}N_2$ uptake method, making it comparable to AR in some systems (Montoya et al. 1996). Zuckermann et al. (1997) have recently described a continuous on-line system for the measurement of AR using a laser-based photoacoustical detection of ethylene. This method is three orders of magnitude more sensitive than the standard gas chromatography-based AR method, but to our knowledge it has not yet been applied to field studies.

Two major assumptions must be made in the extrapolation of measured AR to rates of N_2 fixation: (1) selection of an appropriate C_2H_2 reduction to N_2 fixation stoichiometry and (2) extrapolation of nitrogenase enzyme activity (C_2H_2 reduction) to enzyme product (NH_4^+/amino acid) accumulation. Theoretically, three molecules of C_2H_2 are reduced for each N_2 molecule fixed (Stewart et al. 1967). However, empirical observations reveal significant deviations from theory, with C_2H_2:$^{15}N_2$ ratios ranging from 3.3:1 to 56:1 for samples collected from the North Pacific gyre (Mague et al. 1974). Furthermore, nitrogenase can produce H_2 from H_2O; this reaction accompanies N_2 reduction but not AR (Robson & Postgate 1980), so formation of H_2 can also affect the C_2H_2:N_2 reduction stoichiometry. Variation in the C_2H_2:N_2 stoichiometry may also be caused by failure of the second field assumption relating potential activity to actual *in vivo* product formation. If N_2-fixing microbial assemblages are nutrient (P, Fe, vitamin) limited then the potential rate for N_2 fixation, measured using the AR technique, may never be fully realized. Mague et al. (1974) demonstrated that the anomalously high C_2H_2:N_2 ratios (> 10–20:1) returned to the theoretical value of 3:1 following the addition of excess phosphate. Only under balanced growth conditions, they reasoned, would the AR and $^{15}N_2$ methods be expected to yield comparable estimates of N_2 fixation in field samples. It might, therefore, be possible to use simultaneous measurements of AR and $^{15}N_2$ reduction to assess the degree of nutrient limitation in natural populations, although to our knowledge this approach has not yet been attempted.

^{15}N isotopic abundance as an indicator of N_2 fixation

Nitrogen exists naturally as two stable isotopes, ^{14}N (99.634% by atoms) and ^{15}N (0.366% by atoms). Because the ^{15}N/^{14}N ratios of natural materials vary only slightly, they are expressed in δ-notation, where δ^{15}N (‰) = $((^{15}$N/^{14}N)$_{sample}$/$(^{15}$N/^{14}N)$_{standard}$ − 1) * 1000; the universal reference standard is atmospheric N_2. The nitrogen isotopes can be used to study the marine N cycle by examination of natural variations in the ^{15}N/^{14}N ratio, or by addition of tracers that are artificially enriched in ^{15}N. We focus in this section on natural isotopic variations. This subject was reviewed comprehensively by Owens (1987), although the field has evolved significantly since that time.

Two factors control the δ^{15}N of a given N pool: (1) the δ^{15}N of its source and (2) isotopic fractionation associated with its production and loss (with enzymatic reactions typically favoring the conversion of ^{14}N-bearing substrates). Dissolved N_2, the substrate for marine microbial N_2 fixation typically has a δ^{15}N of ~0.6‰, close to the value expected from equilibrium with atmospheric N_2. Microbial N_2 fixation has a small isotopic fractionation, so that the organic-N produced by N_2 fixation is only slightly depleted in ^{15}N compared to its substrate, with a δ^{15}N of 0 to −1‰ relative to atmospheric N_2 (Wada 1980; Wada & Hattori 1991; Carpenter et al. 1997). Mean oceanic nitrate, the largest reservoir of fixed N, has a δ^{15}N of ~5‰ (Sigman et al. 1997, 1999, 2000), so N_2 fixation should be discernible in marine systems as a source of ^{15}N-depleted N, providing a potential constraint on the relative importance of N_2 fixation in open ocean environments. Using the low δ^{15}N of newly fixed N_2, one could potentially trace its path through each of the important N pools, from its origin in the particulate (and dissolved) organic-N of the surface ocean, to its export from the euphotic zone as sinking particulate N, finally to its oxidation to nitrate in the underlying thermocline.

The δ^{15}N of particulate N in surface waters of oligotrophic basins tends to be low, consistent with a significant contribution of newly fixed N_2 to this pool (Saino & Hattori 1980, 1987). However, the isotopic effect of N recycling represents a competing alternative explanation for this low δ^{15}N (Checkley & Miller 1989; Altabet 1988). Zooplankton appear to release ammonium which has a lower δ^{15}N than their food source, making their tissues and solid wastes ~3‰ higher in δ^{15}N than their food source. The low-δ^{15}N ammonium is consumed by phytoplankton and thus retained in the surface ocean N pool, while the ^{15}N-enriched particulate N is preferentially exported, potentially leading to a lower δ^{15}N of surface particulate N in environments where recycled N is an important component of the gross N supply to phytoplankton. The low δ^{15}N values observed in suspended particulate N from Antarctica and other high latitude regions probably cannot be attributed to N_2 fixation, and thus are most likely due to N recycling. However, in the low-latitude, low-nutrient ocean surface, such as the Sargasso Sea and western tropical Pacific Ocean, the relative importance of N_2 fixation and N recycling in producing ^{15}N-depleted surface particles is uncertain.

One potential approach to discern the N isotopic effects of N_2 fixation and recycling on suspended particles is the coincident measurement of C isotopic composition. *Trichodesmium* has the highest δ^{13}C content of any phytoplankton species (–12.9‰, compared to –20 to –22‰ for most others) and the lowest δ^{15}N, as described above. Thus, Carpenter et al. (1997) suggest that dual isotopic tracer measurements of marine particulate organic matter may provide an unique tracer for *Trichodesmium* and of heterotrophs that incorporate their organic matter. Of course, the carbon isotopes will not track the newly fixed N of *Trichodesmium* through its subsequent reincorporation by other autotrophic plankton.

While N recycling can result in a decrease in δ^{15}N of the suspended particulate N pool due to the preferential export of ^{15}N-enriched material, only N_2 fixation represents a true source of ^{15}N-depleted fixed N. For this reason, the development of an annually integrated N isotope budget for a region of the ocean surface (that is, the study of fluxes rather than pools) should reveal whether N_2 fixation is an important component of the N budget, regardless of N recycling. This approach has been taken for the BATS region in the North Atlantic (Altabet 1988) and the HOT station in the North Pacific (Karl et al. 1997) using δ^{15}N data for the N sinking flux and assuming that: (1) fixed N input to the surface ocean is either as upwelled nitrate or via N_2 fixation and (2) the total export is well represented by the sinking particulate N collected in sediment traps. Altabet (1988) found no need for a N_2 fixation term to close the isotope budget at BATS, although subsequent changes in our understanding of the Sargasso Sea and in sediment trap processing methods may require a re-analysis of that budget.

By contrast, Karl et al. (1997) required a 25–50% contribution of newly fixed N_2 at Sta. ALOHA (22°45′N, 158°W), in agreement with other measures of N_2 fixation at that North Pacific subtropical gyre site. Karl et al. (1997) also observed a seasonal change in the δ^{15}N of sedimenting organic matter indicative of a systematic seasonal alternation between predominantly NO_3^- supported export production in winter, and predominantly N_2 supported export production during the more stratified summer period, although N_2 fixation was detected throughout the year. These sediment trap-based studies have, by necessity, ignored certain aspects of the N budget that might also be significant, including: (1) DON and its associated horizontal and vertical N fluxes, (2) atmospheric fixed N inputs and (3) N fluxes associated with zooplankton migration. As these terms are added, N isotope budgets will provide an increasingly rigorous constraint on the role of N_2 fixation in the N nutrition of the oligotrophic surface ocean.

Much of the sinking flux that exports fixed N from the surface ocean is oxidized to nitrate at thermocline depths. Thus, in regions where significant N_2 fixation occurs, newly fixed N is exported from the euphotic zone and should appear as ^{15}N-depleted nitrate in shallow subsurface waters. In the absence of N_2 fixation, since nitrate consumption is complete in oligotrophic surface waters, the δ^{15}N of the N export will converge on the δ^{15}N of the nitrate supply from the subsurface, so that the 'normal' pathway of N assimilation and dissimilation should not cause the δ^{15}N of nitrate to change from its 'preformed' value of $\sim$5‰. Thus, the δ^{15}N of nitrate in the thermocline should provide a measure of the addition of nitrate to the thermocline from the oxidation of newly fixed N_2. Combining such data with transient geochemical tracers should allow for the calculation of integrated (in time and in space) rates of N_2 fixation.

Liu et al. (1996) measured the N isotopic composition of nitrate in thermocline waters of the Kuroshio near Taiwan. Compared to the δ^{15}N measured for nitrate in deeper waters (+5.5 to 6.1‰ at 500–780 m), the δ^{15}N of nitrate in the overlying thermocline water was 1–3‰, indicating the input of ^{15}N-depleted N, probably from the oxidation of newly fixed N_2. The authors conclude that $40 \pm 15\%$ of the subeuphotic zone nitrate pool was derived from local N_2 fixation (Liu et al. 1996).

Similarly, Brandes et al. (1998) documented a pool of isotopically light nitrate in the thermocline waters of the central Arabian Sea. Based on a simple vertical mixing model, the authors concluded that 40% of the nitrate at 80 m was derived from local N_2 fixation. In the Arabian Sea, the N_2 fixation signal is complicated by local denitrification (Brandes et al. 1998). However, these two processes are theoretically separable if the isotope data are combined with $[NO_3^-]/[PO_4^{3-}]$ ratio data (see following section).

Finally, the δ^{15}N of nitrate also shows an upward decrease across the thermocline of Sargasso Sea in the oligotrophic North Atlantic, with δ^{15}N values decreasing from 5‰ at 800 m to 2.3‰ at 300 m (D. Sigman, unpublished data; Figure 5). As in the studies described above, this pattern can be interpreted as indicating the thermocline-depth nitrification of newly fixed N_2. Thermocline ventilation in the North Atlantic subtropical gyre is conducive to the calculation of a rate for processes that alter the chemistry of the thermocline. This approach has been used for estimation of N_2 fixation rates on the basis of $[NO_3^-]/[PO_4^{3-}]$ ratio data (Gruber & Sarmiento 1997; see following section), and the isotope data (Figure 5) imply that nitrate δ^{15}N data could be used for the same purpose.

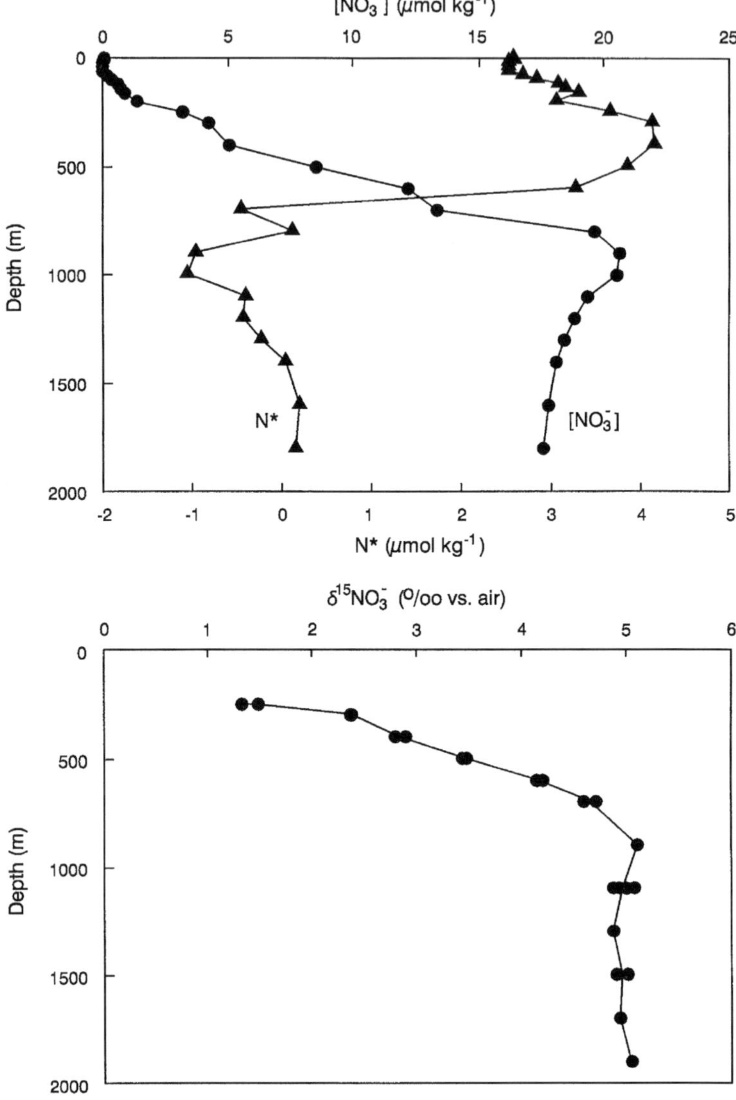

Figure 5. Depth profiles at the Bermuda Atlantic Time-series Study (BATS) station of (top) nitrate concentration and N* (see text), and (bottom) the $\delta^{15}N$ of nitrate. The data in (top) were collected as part of the BATS program and the data in (bottom) are from D. Sigman (unpublished results). Individual nitrogen isotope analyses are plotted as filled circles, and the trend line passes the mean value for replicate samples at each depth. The nitrate $\delta^{15}N$ below 800 m is close to the oceanic mean value of 5‰. The $^{15}N/^{14}N$ decrease toward the surface requires the addition of ^{14}N-rich nitrate to the upper water column, such as would result from the nitrification of newly fixed N. The nitrate $\delta^{15}N$ is lowest in the shallow thermocline, where the N* data suggest that the addition of newly fixed N is greatest in proportion to the nitrate concentration.

The N parameter*

In a recent report, Michaels et al. (1996) concluded that rates of N_2 fixation between 10–40°N latitude in the North Atlantic Ocean are on the order of 50–90 Tg N yr^{-1}. This value, for one of the Earth's smallest ocean basins, exceeds most previous estimates of *global* ocean N_2 fixation (Figure 4). Their estimate was based on an assessment of the patterns in the ratio of N to P in the upper thermocline, and on nutrient fluxes along isopycnal surfaces in the North Atlantic. Data were also presented for seasonal and interannual variability in ecosystem dynamics; the reported temporal variability makes a mass balance approach like this much more difficult to achieve. Despite these limitations, the authors clearly documented elevated ratios of nitrate-to-phosphate in the main thermocline of the central North Atlantic gyre, the Sargasso Sea, compared to surrounding waters. This had been noted previously by Fanning (1989, 1992), but to date, not fully explained. Michaels et al. (1996) went on to derive an anomaly parameter, which they called N*, as the concentration of N, in excess or in deficit of P, relative to the Redfield stoichiometry of 16N:1P (i.e. $N^* = [NO_3^-] - 16[PO_4^{3-}] + 2.72$). The constant, 2.72, was chosen to set the global ocean mean N* to zero, and implies that in the contemporary ocean, denitrification exceeds N_2 fixation. For samples collected in the upper 800 m of the Sargasso Sea, both near Bermuda and to the southeast in the core of the North Atlantic subtropical gyre, N* exceeds 2.72 implying net N_2 fixation. From independent information on the ventilation time-scale of each isopycnal surface, the N* inventory was extrapolated to an average rate of N_2 fixation, the integral of which was equivalent to 3.4–6.1 × 10^{12} moles excess nitrate yr^{-1}. They equated this nitrate excess to the net rate of N_2 fixation (i.e. gross N_2 fixation less any denitrification in the same water masses). Thus, gross N_2 fixation rates, as might be extrapolated from direct field measurements, could be even higher.

These larger than anticipated rates of N_2 fixation for the Sargasso Sea suggested that N_2 may be a major source of new N in this low nutrient habitat, accounting for > 50% of the annual particulate N export measured using free-drifting sediment traps. This mostly summertime input of fixed N in the absence of physical mixing processes is consistent, both in process and in amount, with the enigmatic summertime drawdown (net removal) of dissolved inorganic carbon that occurs in these waters in the apparent absence of new nutrients (Michaels et al. 1994). Consequently, this geochemical anomaly-based estimate of the rate of N_2 fixation appears robust and consistent with certain other complementary field observations.

Gruber and Sarmiento (1997) have taken this nutrient anomaly approach much further. In their more comprehensive treatment of the N* parameter, they redefined N* as a linear combination of NO_3^- and PO_4^{3-} ($N^* = 0.87$

74

[NO$_3$] $-$ 16 [PO$_4^{3-}$] $+$ 2.90 μmol kg^{-1}). As in the report by Michaels et al. (1996), positive N* values, they reasoned, would indicate regions where excess N, relative to P, had been regenerated. Their analyses concluded that the tropical and subtropical North Atlantic Ocean and the Mediterranean Sea were major sources of N, via N$_2$ fixation, whereas net N* in the subtropical gyre of the North Pacific Ocean was more often than not zero (Gruber & Sarmiento 1997). Gruber and Sarmiento (1997) calculated that the rate of N$_2$ fixation in the North Atlantic Ocean between 0°N and 50°N was 28 Tg N yr^{-1}, a value that is 2–3 times lower than that estimated by Michaels et al. (1996). However, 28 Tg N yr^{-1} is still larger than earlier estimates for the global ocean (Figure 4). They use a variety of assumptions to extrapolate N* to the global ocean, and derive an estimate of 110 Tg N yr^{-1}.

Deutsch et al. (2001) present a fixed N budget for the Pacific Ocean based on the recently completed World Ocean Circulation Experiment (WOCE) hydrographic data set. Using water mass age tracers, estimates of atmospheric deposition and riverine fluxes and N* calculations, they estimate rates of both denitrification and N$_2$ fixation assuming steady-state conditions. To achieve mass balance, they conclude that the N$_2$ fixation in the Pacific, north of 32°S is 59 $\pm$ 14 Tg N yr^{-1}. Based on the N* signals, N$_2$ fixation was enhanced in the western portion of the subtropical gyres (Figure 6), a result that is consistent with the hypothesis that iron supplied to the ocean via atmospheric dust deposition may be an important control.

Like all other indirect methods for estimating rates of N$_2$ fixation, the N* method has its unique limitations. First, it assesses net N$_2$ fixation not gross N$_2$ fixation. If a given habitat supports simultaneous N$_2$ fixation and denitrification of comparable rates, then N$_2$ fixation by this nutrient anomaly method would not be detected. The large areas of denitrification in the eastern tropical Pacific may be balanced by high rates of N$_2$ fixation elsewhere in the Pacific Ocean basin. There are strong gradients in N* across the Pacific, and the maintenance of these gradients implies a source of new nitrate to balance the denitrification in the east. This is an important limitation of the N* approach because net N$_2$ fixation 'neutral' ecosystems may function differently from those where N$_2$ fixation is truly absent.

The rates of N$_2$ fixation derived from N* are also dependent upon the assumption that diazotrophs produce biomass with an N:P ratio greater than the Redfield ratio (see '*N and P nutrient control*' section), thus accounting for the N excess regeneration anomaly. Furthermore, the rate of N$_2$ fixation derived from N* is highly sensitive to the reference organic matter N:P ratio that is selected. For instance, Gruber and Sarmiento (1997) used a molar value of 125N:1P derived from Karl et al. (1992). This is a relatively extreme bloom value, compared to more modest estimates for the N:P of *Trichodesmium*

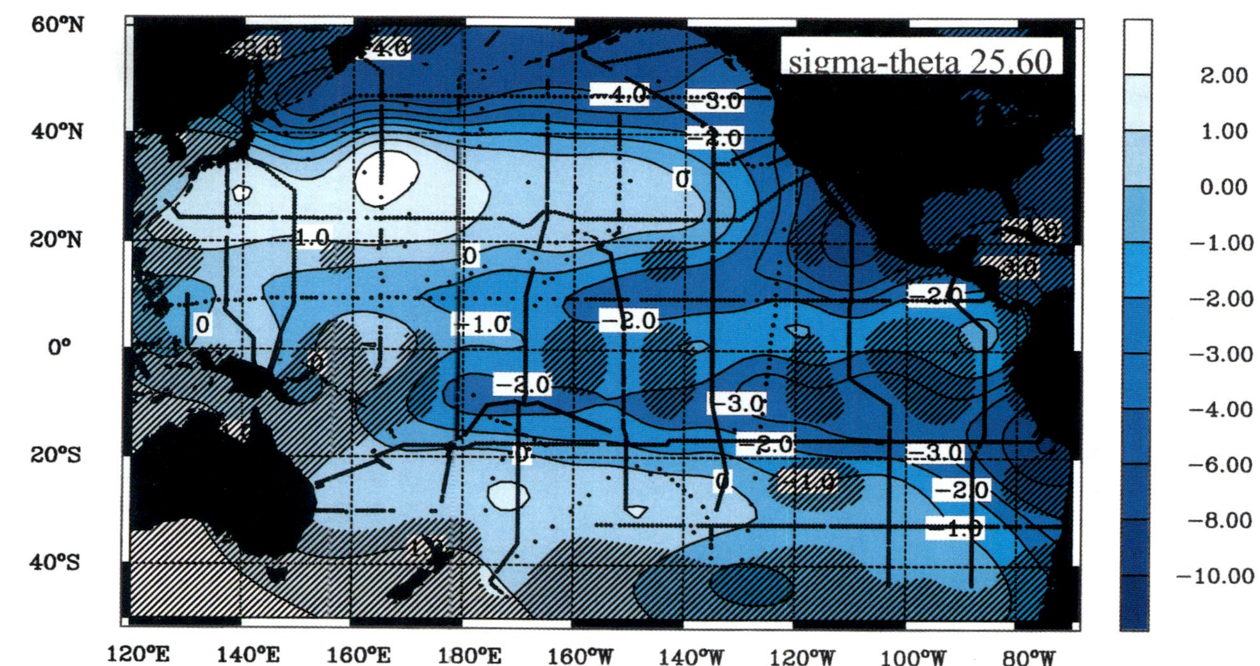

Figure 6. Map of N* (μmol kg^{-1}) in the thermocline on the sigma-theta = 25.6 isopycnal surface (located at approximately 200–300 m in central gyres shoaling to about 100 m in the east). This map was prepared using the objective mapping technique described by LeTraon (1990). Dashed areas indicate regions where the interpolation error in N* from this objective mapping procedure is > 20%. From: Deutsch et al. (2001).

which are in the range of 40 to 50 (Letelier & Karl 1996, 1998). As the assumed N:P of diazotroph biomass decreases to the canonical Redfield ratio of 16N:1P, the N_2 fixation rate derived from N* increases to infinity (see Figure 18 in Gruber & Sarmiento 1997). Therefore the N:P ratio assumption alone could more than account for much of the difference between the estimates of Michaels et al. (1996) and Gruber and Sarmiento (1997). Finally, by ignoring the dissolved organic matter (DON and DOP) pools one cannot accurately determine the true N:P stoichiometry of the dissolved nutrient pools. Despite these well founded criticisms, the N* parameter appears to be a robust qualitative, if not quantitative, indicator of the contribution of net N_2 fixation to the regional scale oceanic N cycle. It certainly has opened up a broader range of possibilities for the global scale of this process and its pattern across basins and with depth.

N:P stoichiometry of the suspended and exported particulate matter pools

There remains a major misconception about the stoichiometry of dissolved and particulate matter pools in the sea; more often that not, ambient pools have a N:P molar stoichiometry that deviates significantly from the 'expected' Redfield ratio of 16N:1P (Duarte 1992; Hecky et al. 1993). N_2 fixation is one of two major microbiological processes (the other being denitrification) that can influence oceanic N:P stoichiometry on global scales. In contrast to N_2 fixation, there is no comparable gas-phase to the phosphorus (P) cycle. Thus, N_2 fixation will either lead to variations in N:P stoichiometry or P supply will limit biological activity, or both.

At Sta. ALOHA in the oligotrophic North Pacific Ocean, the deviations from the nominal 16:1 N:P stoichiometry (Redfield ratio) are particularly intriguing (Figure 7). During the first two years of the HOT program, the mean N:P ratio for suspended particulate matter in the upper (0–100 m) water column was 15.3 (standard deviation [s.d.] = 3.1, n = 14), a value that was not significantly different from the Redfield prediction of 16.0 (Figure 7). Since 1991, however, there has been an increase in the molar N:P ratio of suspended particulate matter to a value greater than the expected Redfield stoichiometry (Figure 7). There is also much greater temporal variability and a greater overall range. Karl et al. (1997) suggested that the ecosystem N:P stoichiometry drifts out of a Redfield balance en route to phosphorus limitation as the supply of new N shifts from a limiting flux of nitrate from below the euphotic zone to the nearly inexhaustible pool of N_2 that is dissolved in the surface waters of the ocean. This shift to N_2 supported new and export production has significant consequences for biogeochemical cycling pathways and rates. The coherent temporal pattern observed for suspended N:P, with maxima in the summer periods, is consistent with relatively enhanced

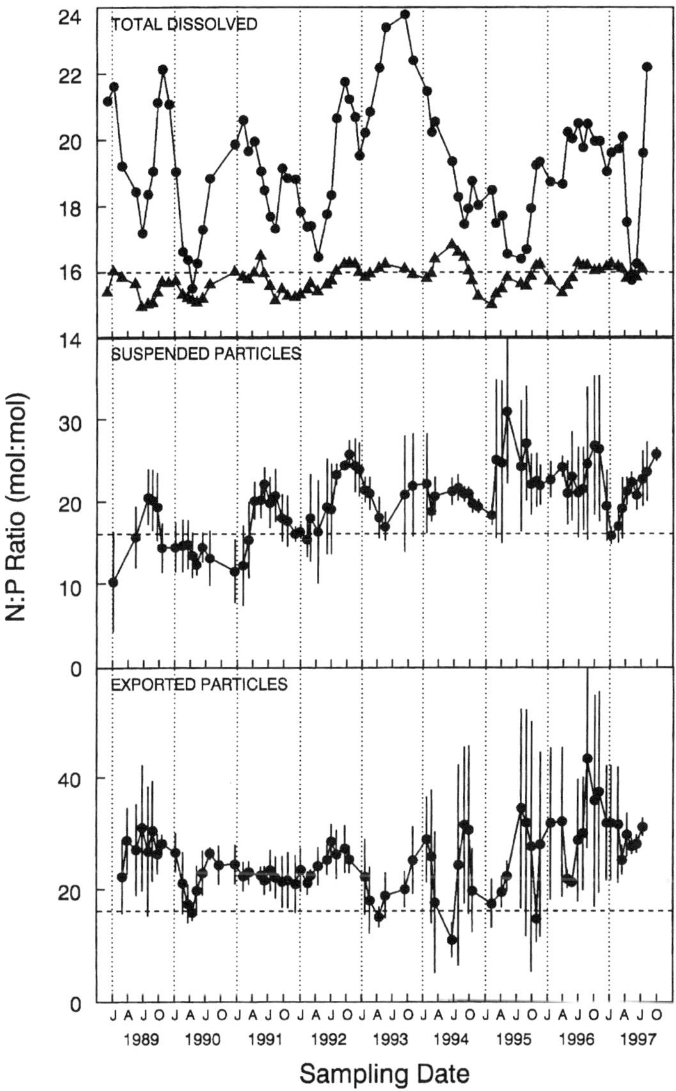

Figure 7. Time series of N and P analyses of dissolved and particulate matter, presented as N:P (mol:mol) ratios, for [TOP] dissolved matter, [CENTER] suspended particulate matter, and [BOTTOM] exported particulate matter. The top panel shows the 3-point running mean N:P ratios for 0–100 m (●) and 200–500 m (▲) portions of the water column. The center panel shows the 3-point running mean (±1 SD) for the average suspended particulate N:P ratio measured in the upper portion (0–100 m) of the water column on each cruise (depth-integrated particulate N ÷ depth-integrated particulate P). The bottom panel shows the 3-point running mean (± 1 SD) for the average N:P ratio of the sediment trap-collected particulate matter at the 150-m reference depth. The Redfield ratio (N:P = 16) is represented by a dashed line in all three panels. From Karl et al. (1997).

bioavailability of N. This most likely results from increased rates of N_2 fixation during periods of maximum water-column stratification when N_2 fixation is also likely to be greater.

The anomalously high N:P stoichiometry ($> 16:1$) of exported particulate matter at Sta. ALOHA (Figure 7) confirms an important prediction of the N_2 supported new production hypothesis model. The temporal variability of the exported matter N:P ratio, with lower, near Redfield ratio values in late winter and elevated ($> 20:1$) ratios throughout the remainder of the year is consistent with the previously mentioned seasonal model for an alternation between NO_3^- supported and N_2 supported new production (see '*^{15}N isotope abundance as an indicator of N_2 fixation*' section). The generally increasing trend in both the suspended particulate matter and the exported particulate matter N:P ratios with time, corresponding with a generally decreasing trend in bioavailable P (Karl et al. 1997), are strong independent lines of evidence for the role of N_2 fixation at this site. Based on a simple mass balance model, N_2 fixation at Sta. ALOHA supplied at least 32% of the new N for the period 1989–1995, with significant seasonal and interannual variability (Karl et al. 1997).

Karl et al. (1997, 2001) have also emphasized that the contemporary role of N_2 fixation in the marine N cycle at Sta. ALOHA must be greater than the recent past. The subeuphotic zone waters where the relatively high N:P exported matter is regenerated has not yet achieved the equilibrium N:P value that would be expected under steady-state export (Figure 7; Karl et al. 2001). A major implication of these data trends is that the N^* parameter is a time-variable quantity that may have increased significantly over the past several decades. These potentially dramatic changes in microbial community structure (selection for N_2-fixing prokaryotes) and rates and mechanisms of nutrient cycling may be related to large scale ocean-atmosphere interactions including, but not limited to, a change beginning in 1976 towards more frequent El Niño and fewer La Niña events (Trenberth & Hoar 1997; Karl 1999). In the Atlantic, significant variations in the total iron supply to the Sargasso Sea, caused by changes in the size of the Saharan desert, also imply a dramatic increase in N_2 fixation rates over the past four decades (Michaels et al. 1996). Clearly if the ocean is variable in time, or is exhibiting secular changes in response to climate forcing, it may be misleading to use historical data sets to map present-day conditions or to predict future trends. The dynamic, non-steady state behavior of the Pacific and, probably, Atlantic and Indian Oceans provides an ideal testing ground for the time-dependent behaviors of the hypothesized effects of climate variability on oceanic N_2 fixation (see '*The N_2 fixation-climate feedback hypothesis*' section).

Remote sensing of N₂ fixation

During the past two decades the development of novel ocean observation platforms, including instrumented ocean buoys and drifters and Earth-orbiting satellites, has improved our ability for continuous, large spatial scale surveillance of the world's oceans (Dickey 1991). These data, and in particular satellite remote sensing of ocean color, have revealed the presence of fairly coherent biogeochemical provinces characterized by relatively small horizontal gradients and well-defined boundaries (Longhurst 1998; Platt & Sathyendranath 1999). The presence of these biogeochemical biomes will undoubtedly facilitate regional and global-scale extrapolation of key ecological processes once the interprovince properties are reasonably well understood. At the very least, algorithms based on sea surface temperature, pigment content, wind and dynamic topography – all currently measured from space – could be used to help constrain rates of global ocean N_2 fixation.

It has previously been suggested that periods of calm seas (e.g. low wind, low turbulent mixing rates, low surface wave activity) favor *Trichodesmium* bloom formation (Carpenter & Price 1976). Karl et al. (1992) showed that when the North Pacific gyre sea state is low, there is a significant (> 1 °C) diurnal warming and cooling of the sea surface temperature. It is now possible to monitor sea surface temperature changes of this magnitude using satellite-based Advanced Very High Resolution Radiometer (AVHRR) sensors and these data could, in principle, reveal 'N_2 fixation-probable' regions of the ocean, especially the central gyres. Deployment of satellite-linked moored or free-drifting ocean buoys with thermistor chains, light and fluorescence probes and nutrient sensors could serve to ground-truth the basin-scale synoptic satellite view at key locations within each biome. Over time, an empirical predictive model of N_2 fixation rates might evolve.

When *Trichodesmium* colonies accumulate at the sea surface under calm water conditions, they are clearly visible from space. Dupouy et al. (1988) were the first to present data on this phenomenon based on a Nimbus-7 Coastal Zone Color Scanner (CZCS) image of a 90,000 km^2 *Trichodesmium* bloom near New Caledonia in the southwest Pacific Ocean. Although there was no ground-truth of this image, the CZCS spectral signature was presented as supporting evidence. They estimated that this single bloom could fix 7.2 × 10^9 g N in 10 days.

Borstad et al. (1992) and Dupouy (1992) later developed spectral-reflectance models applicable for surface ocean *Trichodesmium* blooms observed by the CZCS imager. At moderate colony densities, *Trichodesmium* and other cyanobacteria should be distinguishable from diatoms and dinoflagellates, provided high resolution spectral data are available (Borstad et al. 1992). Continued development of a *Trichodesmium*-specific remote sensing algorithm relied upon two unique physiological characteristics: (1) the presence of gas vacuoles and (2) the presence of the accessory pigment phycoerythrin (Subramaniam & Carpenter 1994). The former results in high reflectivity and the latter in a specific absorption of light at 550 nm. These two independent parameters can be used to distinguish *Trichodesmium* from most other marine phytoplankton. Further refinements in the reflectance model based on field measurements of the inherent optical properties of *Trichodesmium* colonies collected from the Caribbean Sea (Subramaniam et al. 1999a, b) currently provide a sophisticated empirical expression of surface ocean *Trichodesmium* blooms. Application of this optical model using visible and near-infrared sensors of the NOAA-12 AVHRR satellite mapped the near surface *Trichodesmium* distributions in the central Arabian Sea off the Somali coast in 1995; a major bloom that was observed was also intercepted by the R/V *Malcolm Baldridge* so ground truth data were available (Subramaniam et al. 1999a, b). In addition to the AVHRR imager, Tassan (1995) suggested that the Sea-viewing Wide Field-of-view Sensor (SeaWIFS) ocean color satellite might also be useful for detecting *Trichodesmium* at low, sub-bloom concentrations in open ocean habitats, but to our knowledge these model predictions have not yet been verified with field data. The utility of SeaWIFS, however, was verified during a recent series of *Trichodesmium* blooms in the Melanesian Archipelago. These blooms were both mapped by SeaWIFS imagery and sampled as part of the NSF-NASA Sensor Intercomparison and Merger for Biological and Interdisciplinary Ocean Studies (SIMBIOS) expedition in April 1998 (Dupouy et al. 2000).

Despite these successes, a major limitation with any remote ocean sensing application is the uncertainty in relating the surface ocean conditions of phytoplankton assemblage pigmentation or reflectance to surface ocean biomass and total euphotic zone-integrated population inventories. This would be a much more difficult task for relating the presence of a target N_2-fixing microorganism (i.e. *Trichodesmium*) to the *in situ* rate of N_2 fixation. Furthermore, ocean color imagery will not detect chemoautotrophic or chemoheterotrophic N_2-fixing *Bacteria* or *Archaea*; to the extent that they are important to the N budget, global N_2 fixation rates will be underestimated by the use of these remote sensing methods. Finally, N_2 fixation probably occurs in mid-ocean gyres throughout the year (Karl et al. 1997), so methods based simply on interrogation of sea surface blooms will have a built-in alias that is difficult to quantify. Regardless of their potential, it is simply impossible to conduct microbial ecology from space; however, remote sensing methods are likely to prove invaluable as a complementary approach to traditional ship-based investigations.

Human perturbations and climate variability: effects on oceanic N_2 fixation

Our current estimate of global ocean N_2 fixation (100–200 Tg N yr^{-1}) is similar to the rate of terrestrial N_2 fixation (estimated to be 90–130 Tg N yr^{-1}; Galloway et al. 1995), in the absence of human activities. However, the contemporary rate of terrestrial N_2 fixation is more than double this pre-industrial rate as a result of legume cultivation, energy demands and fertilizer production. As anthropogenic mobilization of N intensifies, fixed N fluxes to coastal and open oceans will likely increase, especially relative to P mobilization. This could impact contemporaneous rates of oceanic microbiological N_2 fixation, and could exacerbate N and P decoupling in open ocean habitats.

Oceanic areas of enhanced N_2 fixation are localized in the subtropical gyres and tropical seas, especially the tropical Atlantic, western Pacific and tropical Indian Oceans. Each of these regions is downwind of a major area of dust production, the Saharan Desert/Sahel, the Gobi Desert and the deserts bounding the Arabian Sea, respectively. For the Atlantic Ocean, the flux of atmospheric dust-derived Fe is comparable to that required to sustain the recent estimates of N_2 fixation in that ocean basin given our present understanding of the Fe requirements of *Trichodesmium* (Michaels et al. 1996). However, the current dust load is nearly four-fold higher than before the expansion of the Saharan desert in the early 1970s (Prospero & Nees 1986; Prospero et al. 1996). This fact alone suggests that contemporary rates of N_2 fixation in the North Atlantic Ocean may have been recently enhanced. On the other hand, human activity is presently causing a reduction in the dust plume from the Gobi desert as a result of an aggressive reforestation effort. However, we have no direct evidence that this has yet impacted N_2 fixation in the North Pacific Ocean.

In addition to anthropogenic influences on the fluxes of desert dust to the world's oceans, the natural climate system also causes large temporal variations. For example, marine sediment and ice core data both suggest that dust deposition was 2–20 times higher during the last glacial maximum than it is currently (Rea 1994; Cragin et al. 1977; compilation in Mahowald et al. 1999). These changes in dust deposition appear to be caused by changes in total global desert source area and atmospheric transport patterns (Joussaume 1993). During the current climate, the desert dust source areas lie mostly in subtropical regions (Husar et al. 1997). In the last glacial maximum, pollen and loess studies suggest that desert regions in mid- and high latitude Asia, North America and South America were significantly larger in extent (Liu et al. 1985; Beget 1996; Prentice & Webb 1998).

The historical imbalance between global oceanic N_2 fixation and denitrification is potentially sustained by anthropogenically-fixed N that is delivered to coastal and open ocean environments (currently estimated to be $\sim$59 Tg N yr^{-1}; Galloway et al. 1995). If N_2 fixation rates in the sea have been historically underestimated, as now appears to be the case (Figure 4), then there may well be a pool of 'missing N' or an additional sink for fixed N in the global ocean. Not unrelated to these considerations is the increasing burden of N_2O in the global atmosphere, and the role of the open ocean as a previously unrecognized source of N_2O. Dore et al. (1998) have recently suggested, based on dual ^{15}N and ^{18}O measurements of N_2O in the North Pacific Ocean, that bacterial nitrification rather than denitrification may be a major source for atmospheric N_2O. Nitrification is stimulated by N_2 fixation and the intensified flux of NH_4^+ to NO_3^- in the surface ocean. It now appears that both N_2 fixation and N_2O production may be linked to similar climate variables, such as dust deposition.

N_2 fixation and atmospheric CO_2

The oceans are both a source and a sink for atmospheric CO_2 and, on average, they are thought to absorb about 1–2 Gt C yr^{-1} (Tans et al. 1990; Siegenthaler & Sarmiento 1993; Takahashi et al. 1997). This uptake is a result of a combination of physical and biological processes. The physical processes (the solubility pump), involve the interaction of ocean circulation, the direct thermal effects on pCO_2 and the steady increase in atmospheric CO_2 over the past two centuries. Most of the global C models focus on the solubility pump because it is the one process where there is a clear mechanism leading to oceanic uptake of CO_2 in response to fossil fuel emissions. The biological pump (Longhurst & Harrison 1989) is less well understood and generally less well defined in global models. In its simplest form, surface organisms consume available nutrients and transport them to midwater depths via sinking particles or the mixing of dissolved organic matter. This surface drawdown of nutrients causes a depletion of total C in the surface waters and a concomitant decrease in pCO_2. Subsequent mixing re-introduces nutrients and C to the surface waters and, with simple stoichiometric assumptions, this balance results in little subsequent uptake of CO_2 as long as the mean surface nutrient concentrations remain the same.

N_2 fixation brings a new dimension to the ocean uptake of CO_2. On short time-scales, it adds a gaseous component to the N cycle. The creation of new reactive N in the euphotic zone and its potential to support a downward flux of C will be in excess of the upward fluxes of C by mixing. This should lower pCO_2 locally and sequester C on the time-scale of the ventilation of those waters. This mechanism is further accentuated by the relatively high ratios of

C:P and N:P in marine diazotrophs as evidenced by the anomalous dissolved nutrient ratios in areas of high N_2 fixation (see *'The N* parameter'* section). At an estimated rate of global ocean N_2 fixation of 100–200 Tg N yr^{-1} and a median C:N ratio of 11:1 for remineralization (as estimated from Sargasso Sea data sets), the annual amount of C transport could be about 1–2 Gt C yr^{-1}. Interannual fluctuations in N_2 fixation, or trends due to changing global climate, could be large enough to complicate interpretation of the record of changing atmospheric CO_2. As long as this nitrate remains in the ocean and the surface oceans remain depleted in nitrate, it will continue to sequester carbon in the deep sea. When denitrification removes the nitrate from the water, the subsequent ventilation of that water will result in an outgassing of the, now excess, CO_2 to the atmosphere.

The N_2 fixation-climate feedback hypothesis

On millennial time-scales, any imbalance between N_2 fixation and denitrification will change the total NO_3^- stock of the oceans (McElroy 1983; Codispoti 1989; Falkowski 1997). Increases in total oceanic NO_3^- should sequester C in the deep sea, provided bioavailable P is present and that N:P ratios of organisms can vary within narrow bounds. Decreases in oceanic NO_3^- should cause a gradual release of C to the atmosphere. If climate variations affect both N_2 fixation and denitrification on these time-scales, one might expect an increased dynamic amplitude in these coupled processes and the potential for both positive and negative feedback loops (Michaels et al. 2001).

The hypothesized feedback mechanism will have the following component parts (Michaels et al. 2001; Figure 8): the rate of N_2 fixation in the world's oceans, balanced against the denitrification rate, can have an impact on the concentration of the greenhouse gas, CO_2, in the atmosphere on time-scales of decades (variability in surface biogeochemistry) to millennia (changes in the total NO_3^- stock from the balance of N_2 fixation and denitrification); CO_2 concentrations in the atmosphere will influence the climate on the longer time-scales; and the climate system, in turn, can influence the rate of N_2 fixation in the oceans by controlling the supply of Fe on dust, and by influencing stratification of the upper ocean which also promotes N_2 fixation. Humans have a direct role in the feedback cycle by their influence on dust production, through agriculture at the margins of deserts, and by our own production of CO_2 into the atmosphere. Because of the interaction of the various parts of this system, keyed around the unique behavior and biogeochemistry of the prokaryotic microorganisms that can fix N_2, this feedback loop should exhibit complex behaviors on a variety of time-scales.

From a modeling perspective, the coupled N_2 fixation-climate hypothesis can be segregated by timescale. On interannual to decadal scales, the interac-

84

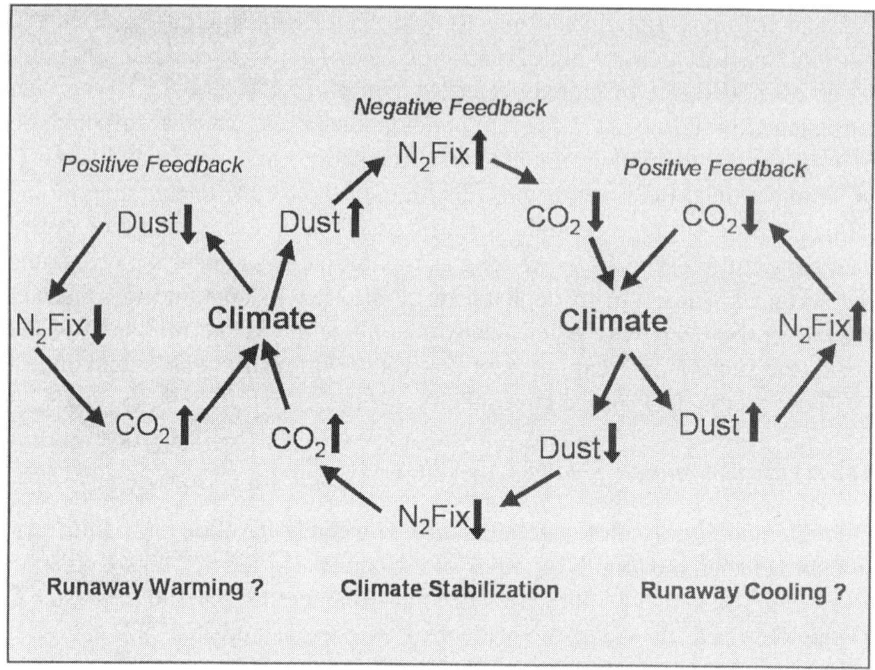

Figure 8. Schematic representation of potential global scale feedbacks between climate and N_2 fixation. Shown are key hypothesized roles of dust deposition as a positive effector for N_2 fixation and the crucial role of N_2 fixation in the potential for global ocean carbon sequestration (redrawn from Michaels et al. 2001).

tions among Fe deposition, climate (mostly ocean surface stratification) and N_2 fixation will be expressed as changes in the rates and community structure of marine ecosystems and will be reflected in the regional and global net air-sea exchange of CO_2. The relatively small resultant changes in oceanic N inventory and atmospheric CO_2 over these time periods do not have strong direct feedbacks on climate, simplifying the problem considerably. Nonlinearities in the dynamics of dust supply, bioavailable Fe release, diazotroph growth rates, bloom dynamics and export/remineralization processes will provide complex model outputs. However, these should still fall within some simpler bounds, namely: more Fe leads to more N_2 fixation leads to more C sequestration, and the inverse. Although denitrification rates may also vary on these time scales, the majority of that process occurs at depth. Thus, the resulting outgassing of CO_2 will be averaged over a longer time scale. The variability in the net impact will be dominated by the variability in N_2 fixation.

On millennial time scales, the changes in the total nitrate stock of the ocean are controlled by the balance of N_2 fixation and denitrification. Here the climate feedbacks will reach the full range of possible outcomes. If the relationship between high CO_2 and dust is positive, then a negative, stabilizing feedback will result (Michaels et al. 2001; Figure 8). If the converse relationship exists, then a positive feedback will drive the system towards either very low or very high CO_2 levels. In this case, some other process would have to temper the feedback, perhaps an interaction with the total availability of nitrate.

These processes can be studied in the existing framework of uncoupled and coupled ocean general circulation models (GCMs) and atmosphere-land surface models by incorporating the required dust and marine biogeochemical dynamics. The models, with full feedback dynamics, will undoubtedly reveal a variety of complex dynamics (in the mathematical sense of the term), but they may also be able to determine the role of this hypothesized feedback system in our global climate.

Summary and future prospectus

For nearly one hundred years oceanographers have studied the interactions between the photosynthetic production of organic matter and nutrient dynamics in the sea. Classical research efforts by H.W. Harvey, L.H.N. Cooper, A.C. Redfield and others established robust quantitative relationships between the nitrogen and phosphorus contents of phytoplankton cells in relationship to ambient nutrient levels. However, one unique feature of the coupled N-P cycles that has never been fully appreciated or quantified is the role of diazotrophy; the ability of certain microorganisms to use N_2 for cell metabolism and growth. N_2 fixation should 'force' marine ecosystems toward P-limitation.

Trichodesmium blooms are ubiquitous phenomena in tropical and subtropical oceanic waters and they are known to fix N_2 under *in situ* conditions. To date it has been difficult to quantify the importance of diazotrophy because of the stochastic nature of the blooms and, until recently, a lack of pure cultures for physiological studies. Recent budget estimates based upon seasonally- and interannually-averaged N imports to and exports from the epipelagic zone of the subtropical gyres of the North Atlantic and North Pacific Oceans suggest that diazotrophic production of fixed N may be an important source of new nitrogen for these open ocean biomes.

The revised estimates for the North Pacific subtropical gyre suggest that 30–50% of the N required to sustain particulate and dissolved matter export from the euphotic zone (the so-called 'new' N) is derived from N_2 fixation; the remainder is supplied by the vertical flux of NO_3^- from sub-euphotic zone waters. If these data extrapolations are verified by subsequent measurements, then our present conceptual models of ocean ecosystems will need to be revised. In this regard, we need to fully document both the phylogenetic diversity of N_2-fixing marine microorganisms and understand the breadth of their metabolic strategies for survival in the sea.

Regardless of the apparent importance of N_2 fixation to the global ocean N cycle, it is essential to emphasize that the field observations currently available were not designed to derive global estimates of N_2 fixation. For this reason, and also because there is a lack of physiological research on marine diazotrophs made under controlled environmental conditions, it is still difficult to constrain global ocean N_2 fixation at the present time. With additional field observations on N_2 fixation we may be able to characterize statistically the temporal and spatial distribution of N_2 fixation in the world's ocean. Combining this characterization with the study of the biogeochemical signature of N_2 fixation (elemental stoichiometry, N^*, $\delta^{15}N$) will improve our current estimates and refine our predictions regarding the coupling of climate variability and oceanic N_2 fixation. However, the mechanistic understanding to predict the effect of global change in N_2 fixation will probably require experimental manipulation at different biological levels.

Both conceptually and ecologically, N_2 supported new production is fundamentally different from NO_3^- supported new production even though the two processes were considered together in the original new versus regenerated N model of Dugdale and Goering (1967). For open ocean ecosystems it now appears that N_2 fuels both organic matter production and 'excess' NH_4^+ (or dissolved organic N; Karl et al. 1992; Capone et al. 1994; Glibert & Bronk 1994) production; the latter is mostly regenerated to NO_3^- in the euphotic zone. Consequently the previous paradigm of NO_3^- uptake being equivalent to new production, and NH_4^+ being equivalent to regenerated production, must be replaced by the realization that N_2 supports the production of new organic matter and 'new' NH_4^+; surface ocean NO_3^- pools, on the other hand, are mostly locally 'regenerated' in oligotrophic oceanic habitats.

This antithetical conceptualization has significant implication both for the design and interpretation of field experiments, and for the survival strategies of the resident microbial populations. When NO_3^- enters the euphotic zone from below by vertical advection and diffusion, it is delivered with a suite of other required major (e.g. C, P and Si) and trace (e.g. Fe) elements in the proper stoichiometry to sustain biological activity (Karl 1999; Cullen et al. 2001). However the process of N_2 fixation serves to decouple export from new nutrient import, which can lead to changes in the elemental stoichiometry of surface-ocean particulate and dissolved organic matter and selection for or against certain groups of microorganisms (see Karl 1999). Significant rates of N_2-based new production would eventually result in severe P and, perhaps, Si limitation because these vital nutrients are supplied from below. Furthermore, selective separation of the otherwise coupled N-P-Si cycles by vertically migrating microbial assemblages (Karl et al. 1992; Villareal et al. 1993, 1999) or positively buoyant particulate matter may further complicate these mass-balance considerations. These observations suggest that it may be inappropriate to assume that biogeochemical processes in open ocean ecosystems conform to the current new vs. regenerated dichotomy; a revised paradigm may be required (Karl 2000).

From research that has been conducted over the past several decades, N has emerged as the master variable for productivity and export modeling due to the perception that it was the production rate limiting nutrient. If current estimates of N_2 fixation are valid, then a re-assessment of this fundamental assertion must be made. In all likelihood, emphasis will shift to the role of P which has a much less complex cycle due to the absence of variable oxidation state chemistry and the lack of a significant biogenic gas phase, or to Fe. Although the debate on whether N or P ultimately limits marine productivity (see Codispoti 1989) will likely continue (e.g. Toggweiler 1999; Tyrrell 1999), it now appears almost certain that N_2 fixation must be considered as an ecologically relevant source of new N in the sea. Finally, the inextricable link between N_2 fixation in the world's oceans to climate variability and certain anthropogenic processes, suggests that predictable changes may occur in rates of N_2 fixation in regions such as those ranging from severely human-impacted to natural landscapes, seascapes and the pre-industrial bioelemental cycles.

Acknowledgements

The authors would like to thank Robert Howarth for his leadership on the international Scientific Committee on Problems of the Environment (SCOPE) Nitrogen Project, 'Nitrogen Transport and Transformations: A Regional and Global Analysis' which included this oceanic component as working group #3, and to the Mellon Foundation and the National Center for Ecological Analysis and Synthesis (NCEAS) for financial support. Curtis Deutsch kindly provided the Pacific N* map (Figure 6) from his unpublished paper. This is publication #5804 of the School of Ocean and Earth Science and Technology, #219 of the USC Wrigley Institute for Environmental Studies, #1604 of the Bermuda Biological Station for Research, Inc., #2707 of the Netherlands Institute of Ecology (NIOO-CEMO) and #672 of the U.S. Joint Global Ocean Flux Study Program. We also thank Roxanne Marino, Robert Howarth, Claudia Benitez-Nelson, Karin Björkman, and two anonymous reviewers for comments that substantially improved the manuscript. This paper is dedicated to the memory of Tim Mague, one of the pioneers of oceanic N_2 fixation.

References

Alldredge AL & Silver MW (1982) Abundance and production rates of floating diatom mats (*Rhizosolenia castracanei* and *R. imbricata* var. *shrubsolei*) in the Eastern Pacific Ocean. Mar. Biol. 66: 83–88

Altabet MA (1988) Variations in nitrogen isotopic composition between sinking and suspended particles: implications for nitrogen cycling and particle transformation in the open ocean. Deep-Sea Res. 35: 535–554

Beget JE (1996) Tephrochronology and paleoclimatology of the last interglacial-glacial cycle recorded in Alaskan loess deposits. Quat. Int. 34–36: 121–126

Benson DR (1985) Consumption of atmospheric nitrogen. In: Leadbetter ER & Poindexter JS (Eds) Bacteria in Nature, Volume 1: Bacterial Activities in Perspective (pp 155–198). Plenum Press, New York

Bergman B & Carpenter EJ (1991) Nitrogenase confined to randomly distributed trichomes in the marine cyanobacterium *Trichodesmium thiebautii*. J. Phycol. 27: 158–165

Bergman B, Siddiqui PJA, Carpenter EJ & Peschek GA (1993) Cytochrome oxidase: subcellular distrivution and relationship to nitrogenase expression in the nonheterocystous cyanobacterium *Trichodemium thiebautii*. Appl. Environ. Microbiol. 59: 3239–3244

Bergman B, Gallon JR, Rai AN & Stal LJ (1997) N_2 fixation by non-heterocystous cyanobacteria. FEMS Microbiol. Rev. 19: 139–185

Bishop PE, Jarlenski DML & Hetherington DR (1980) Evidence for an alternative nitrogen fixation system in *Azotobacter vinelandii*. Proc. Natl. Acad. Sci. USA 77: 7342–7346

Bishop PE & Premakumar R (1992) Alternative nitrogen fixation systems. In: Stacey G, Burris RH & Evans HJ (Eds) Biological Nitrogen Fixation (pp 736–762). Chapman and Hall, New York

Borstad GA, Gower JFR & Carpenter EJ (1992) Development of algorithms for remote sensing of *Trichodesmium* blooms. In: Carpenter EJ, Capone DG & Rueter JG (Eds) Marine Pelagic Cyanobacteria: *Trichodesmium* and other Diazotrophs (pp 193–210). Kluwer Academic Publishers, The Netherlands

Brandes JA, Devol AH, Yoshinari T, Jayakumar DA & Naqvi SWA (1998) Isotopic composition of nitrate in the central Arabian Sea and eastern tropical North Pacific: A tracer for mixing and nitrogen cycles. Limnol. Oceanogr. 43: 1680–1689

Broughton WJ & Pühler A (Eds) (1986) Nitrogen Fixation, Volume 4: Molecular Biology. Clarendon Press, Oxford

Bryceson I & Fay P (1981) Nitrogen fixation in *Oscillatoria* (*Trichodesmium*) *erythraea* in relation to bundle formation and trichome differentiation. Mar. Biol. 61: 159–166

Burns RC & Hardy RWF (1975) Nitrogen fixation in bacteria and higher plants. Springer-Verlag, New York

Burris RH (1991) Nitrogenases. J. Biol. Chem. 266: 9339–9342

Capone DG (1988) Benthic nitrogen fixation. In: Blackburn H & Sorensen J (Eds) Nitrogen Cycling in Coastal Marine Environments (pp 85–123). John Wiley, New York

Capone DG & Carpenter EJ (1982) Nitrogen fixation in the marine environment. Science 217: 1140–1142

Capone DG & Carpenter EJ (1999) Nitrogen fixation by marine cyanobacteria: Historical and global perspectives. Bull. Inst. Oceanogr. Monaco 19: 235–256

Capone DG, Ferrier MD & Carpenter EJ (1994) Cycling and release of glutamate and glutamine in colonies of the marine planktonic cyanobacterium, *Trichodesmium thiebautii*. Appl. Environ. Microbiol. 60: 3989–3995

Capone DG, O'Neil JM, Zehr J & Carpenter EJ (1990) Basis for diel variation in nitrogenase activity in the marine planktonic cyanobacterium *Trichodesmium thiebautii*. Appl. Environ. Microbiol. 56: 3532–3536

Capone DG, Zehr JP, Paerl HW, Bergman B & Carpenter EJ (1997) *Trichodesmium* a globally significant marine cyanobacterium. Science 276: 1221–1229

Carpenter EJ (1972) Nitrogen fixation by a blue-green epiphyte on pelagic *Sargassum*. Science 178: 1207–1209

Carpenter EJ (1983) Nitrogen fixation by marine Oscillatoria (*Trichodesmium*) in the world's oceans. In: Carpenter EJ & Capone DG (Eds) Nitrogen in the Marine Environment (pp 65–103). Academic Press, New York

Carpenter EJ, Bergman B, Dawson R, Siddiqui PJA, Soderback E & Capone DG (1992) Glutamine synthetase and nitrogen cycling in colonies of the marine diazotrophic cyanobacterium, *Trichodesmium* spp. Appl. Environ. Microbiol. 58: 3122–3129

Carpenter EJ & Capone DG (1992) Nitrogen fixation in *Trichodesmium* blooms. In: Carpenter EJ, Capone DG & Rueter J (Eds) Marine Pelagic Cyanobacteria: *Trichodesmium* and Other Diazotrophs (pp 211–217). Kluwer Academic Publishers, The Netherlands

Carpenter EJ, Chang J, Cottrell M, Schubauer J, Paerl HW, Bebout BM & Capone DG (1990) Re-evaluation of nitrogenase oxygen-protective mechanisms in the planktonic marine cyanobacterium *Trichodesmium*. Mar. Ecol. Prog. Ser. 65: 151–158

Carpenter EJ & Culliney JL (1975) Nitrogen fixation in marine shipworms. Science 187: 551–552

Carpenter EJ, Harvey HR, Fry B & Capone DG (1997) Biogeochemical tracers of the marine cyanobacterium *Trichodesmium*. Deep-Sea Res. 44: 27–38

Carpenter EJ, O'Neil JM, Dawson R, Capone DG, Siddiqui PJA, Roenneberg T & Bergman B (1993) The tropical diazotrophic phytoplankter *Trichodesmium*: biological characteristics of two common species. Mar. Ecol. Prog. Ser. 95: 295–304

Carpenter EJ & Price CC, IV (1976) Marine *Oscillatoria* (*Trichodesmium*): Explanation for aerobic nitrogen fixation without heterocysts. Science 191: 1278–1280

Carpenter EJ & Roenneberg T (1995) The marine planktonic cyanobacteria *Trichodesmium* spp.: photosynthetic rate measurements in the SW Atlantic Ocean. Mar. Ecol. Prog. Ser. 118: 267–273

Carpenter EJ & Romans, K (1991) Major role of the cyanobacterium *Trichodesmium* in nutrient cycling in the North Atlantic Ocean. Science 254: 1356–1358

Checkley DM Jr & Miller CA (1989) Nitrogen isotope fractionation by oceanic zooplankton. Deep-Sea Res. 36: 1449–1456

Codispoti L (1989) Phosphorus versus nitrogen limitation of new and export production. In: Berger WH, Smetacek VS & Wefer G (Eds) Productivity in the Ocean: Present and Past (pp 377–394). John Wiley & Sons, New York

Cole JJ, Lane JM, Marino R & Howarth RW (1993) Molybdenum assimilation by cyanobacteria and phytoplankton in freshwater and salt water. Limnol. Oceanogr. 38: 25–35

Cragin JH, Herron MM, Langway Jr. CC & Klouda G (1997) Interhemispheric comparison of changes in the composition of atmospheric precipitation during the late Cenozoic era. In: Dunbar MJ (Ed) Polar Oceans, Proceedings of the Polar Oceans Conference (pp 617–641). Arctic Institute of North America, Calgary, Alberta

Cullen JJ, Franks PJS, Karl DM & Longhurst A (2001) Physical influences on marine ecosystem dynamics. In: Robinson AR, McCarthy JJ & Rothschild BJ (Eds) The Sea, vol. 12, in press

Davey A & Marchant HJ (1983) Seasonal variation in nitrogen fixation by *Nostoc commune* Vaucher at the Vestfold Hills, Antarctica. Phycologia 22: 337–385

Dean DR, Bolin JT & Zheng L (1993) Nitrogenase metalloclusters: Structures, organization, and synthesis. J. Bact. 175: 6737–6744

Delwiche CC (1970) The nitrogen cycle. Sci. Amer. 223: 137–146

Delwiche C & Likens G (1977) Global chemical cycles and their alteration by man. Dahlem Konferenzen, Berlin

Deutsch CA, Gruber NP, Key RM, Sarmiento JL & Ganachaud A (2001) Denitrification and N_2 fixation in the Pacific Ocean. Global Biogeochem. Cycles 15: 483–506

Dickey TD (1991) The emergence of concurrent high-resolution physical and bio-optical measurements in the upper ocean and their applications. Rev. Geophys. 29: 383–413

Dore JE, Popp BN, Karl DM & Sansone FJ (1998) A large source of atmospheric nitrous oxide from subtropical North Pacific surface waters. Nature 396: 63–66

Duarte CM (1992) Nutrient concentration of aquatic plants: Patterns across species. Limnol. Oceanogr. 37: 882–889

Dugdale RC & Goering JJ (1967) Uptake of new and regenerated forms of nitrogen in primary productivity. Limnol. Oceanogr. 12: 196–206

Dugdale RC, Menzel DW & Ryther JH (1961) Nitrogen fixation in the Sargasso Sea. Deep-Sea Res. 7: 298–300

Dupouy C (1992) Discoloured waters in the Melanesian archipelago (New Caledonia and Vanuatu). The value of the NIMBUS-7 Coastal Zone Colour Scanner observations. In: Carpenter EJ, Capone DG & Rueter JG (Eds) Marine Pelagic Cyanobacteria: *Trichodesmium* and other Diazotrophs (pp 177–191) Kluwer Academic Publishers, The Netherlands

Dupouy C, Neveux J, Subramaniam A, Mulholland MR, Montoya JP, Campbell L, Carpenter EJ & Capone DG (2000) Satellite captures *Trichodesmium* blooms in the southwestern tropical Pacific. Eos 81: 13, 15, 16

Dupouy C, Petit M & Dandonneau Y (1988) Satellite detected cyanobacteria bloom in the southwestern tropical Pacific. Int. J. Remote Sens. 9: 389–396

Falkowski PG (1997) Evolution of the nitrogen cycle and its influence on the biological sequestration of CO_2 in the ocean. Nature 387: 272–275

Fallik E, Chan Y-K & Robson RL (1991) Detection of alternative nitrogenases in aerobic gram-negative nitrogen-fixing bacteria. J Bact. 173: 365–371

Fanning KA (1989) Influence of atmospheric pollution on nutrient limitation in the ocean. Nature 339: 460–463

Fanning KA (1992) Nutrient provinces in the sea: Concentration ratios, reaction rate ratios, and ideal covariation. J. Geophys. Res. 97: 5693–5712

Fay P (1992) Oxygen relations of nitrogen fixation in cyanobacteria. Microbiol. Rev. 546: 340–373

Fogg GE (1974) Nitrogen fixation. In: Stewart WDP (Ed) Algal Physiology and Biochemistry (pp 560–582). Blackwell, Oxford

Fogg GE (1982) Nitrogen cycling in sea waters. Phil. Trans. R. Soc. London 296: 299–576

Fredriksson C & Bergman B (1997) Ultrastructural characterization of cells specialized for nitrogen fixation in a non-heterocystous cyanobcterium, *Trichodesmium*. Protoplasma 197: 76–85

Gallon JR (1981) The oxygen sensitivity of nitrogenase: a problem for biochemists and micro-organisms. Trends Biochem. Sci. 6: 19–23

Gallon JR (1992) Reconciling the incompatible: N_2 fixation and O_2. Tansley review No. 144. New Phytol. 122: 571–609

Gallon JR & Stal LJ (1992) N2 fixation in non-heterocystous cyanobacteria: An overview. In: Carpenter EJ, Capone DG & Rueter JG (Eds) Marine Pelagic Cyanobacteria: *Trichodesmium* and other Diazotrophs (pp 115–139). Kluwer Academic Publishers, The Netherlands

Galloway JN, Schlesinger WH, Levy II H, Michaels A & Schnoor JL (1995) Nitrogen fixation: Anthropogenic enhancement-environmental response. Global Biogeochem. Cycles 9: 235–252

Gledhill M & Berg CMGVd (1994) Determination of complexation of iron(III) with natural organic complexing ligands in sewater using cathodic stripping voltammetry. Mar. Chem. 47: 41

Glibert PM & Bronk DA (1994) Release of dissolved organic nitrogen by marine diazotrophic cyanobacteria, *Trichodesmium* spp. Appl. Environ. Microbiol. 60: 3996–4000

Gordon N, Angel DL, Neori A, Kress N & Kimor B (1994) Heterotrophic dinoflagellates with symbiotic cyanobacteria and nitrogen limitation in the Gulf of Aqaba. Mar. Ecol. Prog. Ser. 107: 83–88

Gruber N & Sarmiento JL (1997) Global patterns of marine nitrogen fixation and denitrification. Global Biogeochem. Cycles 11: 235–266

Guerinot ML & Colwell RR (1985) Enumeration, isolation, and characterization of N_2-fixing bacteria from seawater. Appl Environ. Microbiol. 50: 350–355

Guerinot ML & Patriquin DG (1981) The association of N_2-fixing bacteria with sea urchins. Mar. Biol. 62: 197–207

Guerinot ML, West PA, Lee JV & Colwell RR (1982) *Vibrio diazotrophicus* sp. nov., a marine nitrogen-fixing bacterium. Int. J. Syst. Bact. 32: 350–357

Hanson RB (1977) Pelagic *Sargassum* community metabolism: Carbon and nitrogen. J. Exp. Mar. Biol. Ecol. 29: 107–118

Haxo FT, Lewin RA, Lee KW & Li M-R (1987) Fine structure and pigments of *Oscillatoria* (*Trichodesmium*) aff. *Thiebautii* (Cyanophyta) in culture. Phycologia 26: 443–456

Hecky RE, Campbell P & Hendzel LL (1993) The stoichiometry of carbon, nitrogen, and phosphorus in particulate matter of lakes and oceans. Limnol. Oceanogr. 38: 709–724

Hood RR, Michaels AF & Capone DG (2000) Answers sought to the enigma of marine nitrogen fixation. Eos, Trans. Amer. Geophys. Un. 81: 133, 138, 139

Howard JB & Rees DC (1996) Structural basis of biological nitrogen fixation. Chem. Rev. 96: 2965–2982

Howarth RW, Chan F & Marino R (1999) Do top-down and bottom-up controls interact to exclude nitrogen-fixing cyanobacteria from the plankton of estuaries: explorations with a simulation model. Biogeochem. 46: 203–231

Howarth RW & Cole JJ (1985) Molybdenum availability, nitrogen limitation and phytoplankton growth in natural waters. Science 229: 653–655

Howarth RW, Marino R, Lane J & Cole JJ (1988) Nitrogen fixation in freshwater, estuarine, and marine ecosystems. 2. Biogeochemical controls. Limnol. Oceanogr. 33: 688–701

Husar RB, Prospero JM & Stowe LL (1997) Characterization of tropospheric aerosols over the oceans with the NOAA advanced very high resolution radiometer optical thickness operational product. J. Geophys. Res. 102: 16889–16909

Janson S, Bergman B, Carpenter EJ, Giovannoni SJ & Vergin K (1999a) Genetic analysis of natural populations of the marine diazotrophic cyanobacterium *Trichodesmium*. FEMS Microbiol. Ecol. 30: 57–65

Janson S, Carpenter EJ & Bergman B (1994) Compartmentalization of nitrogenase in a non-heterocystous cyanobacterium *Trichodesmium contortum*. FEMS Microbiol. Lett. 118: 9–14

Janson S, Siddiqui PJA, Walsby AE, Romans K, Carpenter EJ & Bergman B (1995) Cytomorphological characterization of the planktonic diazotrophic cyanobacteria *Trichodesmium* spp. from the Indian Ocean and Caribbean and Sargasso Seas. J. Phycol. 31: 463–477

Janson S, Wouters J, Bergman B & Carpenter EJ (1999b) Host specificity in the *Richelia*-diatom symbiosis by *hetR* gene sequence analysis. Environ. Microbiol. 1: 431–438

Johnson KJ, Gordon RM & Coale KH (1997) What controls dissolved iron concentrations in the world ocean? Mar. Chem. 57: 181

Joussaume S (1993) Paleoclimatic tracers: An investigation using an atmospheric general circulation model under ice age conditions – 1. Desert dust. J. Geophys. Res. 98: 2767–2805

Kana TM (1993) Rapid oxygen cycling in *Trichodesmium thiebautii*. Limnol. Oceanogr. 38: 18–24

Karl DM (1999) A sea of change: Biogeochemical variability in the North Pacific subtropical gyre. Ecosystems 2: 181–214

Karl DM (2000) A new source of 'new' nitrogen in the sea. Trends in Microbiol. 8: 301 (Comment section)

Karl DM, Björkman KM, Dore JE, Fujieki L, Hebel DV, Houlihan T, Letelier RM & Tupas LM (2001) Ecological nitrogen-to-phosphorus stoichiometry at Station ALOHA. Deep-Sea Res. II 48: 1529–1566

Karl DM, Letelier R, Hebel DV, Bird DF & Winn CD (1992) *Trichodesmium* blooms and new nitrogen in the north Pacific gyre. In: Carpenter EJ, Capone DG & Rueter JG (Eds) Marine Pelagic Cyanobacteria: *Trichodesmium* and other Diazotrophs (pp 219–237). Kluwer Academic Publishers, The Netherlands

Karl D, Letelier R, Tupas L, Dore J, Christian J & Hebel D (1997) The role of nitrogen fixation in biogeochemical cycling in the subtropical North Pacific Ocean. Nature 388: 533–538

Karl DM & Tien G (1997) Temporal variability in dissolved phosphorus concentrations in the subtropical North Pacific Ocean. Mar. Chem. 56: 77–96

Keene WC & Savoie DL (1998) The pH of deliquesced sea-salt aerosol in polluted marine air. Geophys. Res. Lett. 25: 2181–2184

Kim J & Rees DC (1994) Nitrogenase and biological nitrogen fixation. Biochem. 33: 389–397

Kirshtein JD, Zehr JP & Paerl HW (1993) Determination of N_2 fixation potential in the marine environment: application of the polymerase chain reaction. Mar. Ecol. Prog. Ser. 95: 305–309

Kuchler DA & Jupp DLB (1988) Shuttle photograph captures massive phytoplankton bloom in the Great Barrier Reef. Int. J. Remote Sensing 9: 1299–1301

Letelier RM & Karl DM (1996) Role of *Trichodesmium* spp. in the productivity of the subtropical North Pacific Ocean. Mar. Ecol. Prog. Ser. 133: 263–273

Letelier RM & Karl DM (1998) *Trichodesmium* spp. physiology and nutrient fluxes in the North Pacific subtropical gyre. Aquat. Microb. Ecol. 15: 265–276

LeTraon PY (1990) A method for optimal analysis of fields with spatially variable mean. J. Geophys. Res. 95: 13543–13547

Li WKW, Glover HE & Morris I (1980) Physiology of carbon assimilation by *Oscillatoria thiebautii* in the Caribbean Sea. Limnol. Oceanogr. 25: 447–456

Lin S, Henze S, Lundgren P, Bergman B & Carpenter EJ (1999) Whole-cell immunolocalization of nitrogenase in marine diazotrophic cyanobacteria, *Trichodesmium* spp. Appl. Environ. Microbiol. 64: 3052–3064

Lipschultz F & Owens NJ (1996) An assessment of nitrogen fixation as a source of nitrogen to the North Atlantic Ocean. Biogeochem. 35: 261–274

Liu K-K, Su M-J, Hsueh C-R & Gong G-C (1996) The nitrogen isotopic composition of nitrate in the Kuroshio Water northwest of Taiwan: Evidence for nitrogen fixation as a source of isotopically light nitrate. Mar. Chem. 54: 273–292

Liu T, An Z, Yuan B & Han J (1985) The loess-paleosol sequence in China and climatic history. Episodes 8: 21–28

Longhurst A (1998) Ecological Geography of the Sea. Academic Press, San Diego, California

Longhurst AR & Harrison WG (1989) The biological pump: profiles of plankton production and consumption in the upper ocean. Prog. Oceanog. 22: 47–123

Lundgren P, Soederbaeck E, Carpenter EJ & Bergman B (2000) Nitrogen fixation and nitrogenase in *Katagnymene* spp., a non-heterocystous marine cyanobacterium. Submitted to J. Phycol.

Mague TH, Weare NM & Holm-Hansen O (1974) Nitrogen fixation in the North Pacific Ocean. Mar. Biol. 24: 109–119

Mahowald N, Kohfeld KE, Hansson M, Balkanski Y, Harrison SP, Prentice IC, Schulz M & Rodhe H (1999) Dust sources and deposition during the last glacial maximum and current climate: A comparison of model results with palaeodata from ice cores and marine sediments. J. Geophys. Res. in press

Martinez L, Silver MW, King JM & Alldredge AL (1983) Nitrogen fixation by floating diatom mats: A source of new nitrogen to oligotrophic ocean waters. Science 221: 152–154

McElroy MB (1976) Chemical processes in the solar system: a kinetic perspective. In: Herschbach D (Ed) MTP International Review of Science (pp 127–211). Buttersworth, London

McElroy MB (1983) Marine biological controls on atmospheric CO_2 climate. Nature 302: 328–329

Michaels AF, Bates NR, Buesseler KO, Carlson CA & Knap AH (1994) Carbon-cycle imbalances in the Sargasso Sea. Nature 372: 537–540

Michaels AF, Karl DM & Capone D (2001) Redfield stoichiometry, new production and nitrogen fixation. Oceanography (Special JGOFS edition), in press

Michaels AF, Olson D, Sarmiento JL, Ammerman JW, Fanning K, Jahnke R, Knap AH, Lipschultz F & Prospero JM (1996) Inputs, losses and transformations of nitrogen and phosphorus in the pelagic North Atlantic Ocean. Biogeochem. 35: 181–226

Mitsui A, Kumazawa S, Takahashi A, Ikemoto H, Cao S & Arai T (1986) Strategy by which nitrogen-fixing unicellular cyanobacteria grow photoautotrophically. Nature 323: 720–722

Montoya JP, Voss M, Kaehler P & Capone DG (1996) A simple, high precision tracer assay for dinitrogen fixation. Appl. Environ. Microbiol. 62: 986–993

Moore B, Whitley E & Webster TA (1921) Studies of photo-synthesis in marine algae – 1. Fixation of carbon and nitrogen from inorganic sources in sea water. 2. Increase of alkalinity of sea water as a measure of photo-synthesis. Proc. Roy. Soc. Lond. B 92: 51–58

Mulholland MR, Ohki K & Capone DG (1999) Nitrogen utilization and metabolism relative to patterns of N_2 fixation in cultures of *Trichodesmium* NIBB 1067. J. Phycol. 35: 977–988

Niemi A (1979) Blue-green algal blooms and N:P ratios in the Baltic Sea. Acta Bot. Fenn. 110: 57–61

Ohki K & Fujita Y (1982) Laboratory culture of the pelagic blue-green alga *Trichodesmium thiebautii*: Conditions for unialgal culture. Mar. Ecol. Prog. Ser. 7: 185–190

Ohki K & Fujita Y (1988) Aerobic nitrogenase activity measured as acetylene reduction in the marine non-heterocystous cyanobacterium *Trichodesmium* spp. grown under artificial conditions. Mar. Biol. 98: 111–114

Ohki K, Rueter JG & Fujita Y (1986) Cultures of the pelagic cyanophytes *Trichodesmium erythraeum* and *T. thiebautii* in synthetic medium. Mar. Biol. 91: 9–13

Ohki K, Zehr JP, Falkowski PG & Fujita Y (1991) Regulation of nitrogen fixation by different nitrogen sources in the marine non-heterocystous cyanobacterium *Trichodesmium* sp. NIBB1067. Arch. Microbiol. 156: 335–337

Owens NJP (1987) Natural variations in ^{15}N in the marine environment. Adv. Mar. Biol. 24: 389–451

Paerl HW (1994) Spatial segregation of CO_2 fixation in *Trichodesmium* sp.: Linkage to N_2 fixation potential. J. Phycol. 30: 790–799

Paerl HW (2000) Physical-chemical constraints on cyanobacterial growth in the oceans. International Symposium on Marine Cyanobacteria and Related Organisms, Institut Oceanographique, Paris

Paerl HW & Bebout BM (1988) Direct measurement of O_2-depleted microzones in marine *Oscillatoria*: relation to N_2 fixation. Science 241: 442–445

Paerl HW, Bebout BM & Prufert LE (1989a) Bacterial associations with marine *Oscillatoria* sp. (*Trichodesmium* sp.) populations: Ecophysiological implications. J. Phycol. 25: 773–784

Paerl HW & Bland PT (1982) Localized tetrazolium reduction in relation to N_2 fixation, CO_2 fixation, and H_2 uptake in aquatic filamentous cyanobacteria. Appl. Environ. Microbiol. 43: 218–226

Paerl HW, Crocker KM & Prufert LE (1987) Limitation of N_2 fixation in coastal marine waters: relative importance of molybdenum, iron, phosphorus and organic matter availability. Limnol. Oceanogr. 32: 525–536

Paerl HW & Pinckney JL (1996) A mini-review of microbial consortia: their roles in aquatic production and biogeochemical cycling. Microb. Ecol. 31: 225–247

Paerl HW, Priscu JC & Brawner DL (1989b) Immunochemical localization of nitrogenase in marine *Trichodesmium* aggregates: Relationship to N_2 fixation potential. Appl. Environ. Microbiol. 55: 2965–2975

Paerl HW, Prufert-Bebout L & Guo C (1994) Iron-stimulated N_2 fixation and growth in natural and cultured populations of the planktonic marine cyanobacterium *Trichodesmium* sp. Appl. Environ. Microbiol. 60: 1044–1047

Paerl HW & Zehr JP (2000) Marine nitrogen fixation. In: Kirchman DL (Ed) Microbial Ecology of the Oceans (pp 387–426). Wiley-Liss

Paul EA (1978) Contribution of nitrogen fixation to ecosystem functioning and nitrogen fluxes on a global basis. Ecol. Bull. 26: 282–293

Paulsen DM, Paerl HW & Bishop PE (1991) Evidence that molybdenum-dependent nitrogen fixation is not limited by high sulfate in marine environments. Limnol. Oceanogr. 36: 1325–1334

Platt T, Harrison WG, Lewis MR, Li WKW, Sathyendranath S, Smith RE & Vezina AF (1989) Biological production of the oceans: the case for a consensus. Mar. Ecol. Prog. Ser. 52: 77–88

Platt T & Sathyendranath S (1999) Spatial structure of pelagic ecosystem processes in the global ocean. Ecosystems 2: 384–394

Postgate JR (1982) The Fundamentals of Nitrogen Fixation. Cambridge University Press, Cambridge

Prentice IC & Webb III T (1998) BIOME 6000: Reconstructing global mid-Holocene vegetation patterns from paleoecological records. J. Biogeogr. 25: 995–1005

Proctor LM (1997) Nitrogen-fixing, photosynthetic, anaerobic bacteria associated with pelagic copepods. Aquat. Microbiol. Ecol. 12: 105–113

Prospero JM, Barrett K, Church T, Dentener F, Duce RA, Galloway JN, Levy II H, Moody J & Quinn P (1996) Atmospheric deposition of nutrients to the North Atlantic basin. Biogeochemistry 35: 27–73

Prospero JM & Nees RT (1986) Impact of the North African drought and El Niño on mineral dust in the Barbados trace winds. Nature 320: 735–738

Prufert-Bebout L, Paerl HW & Lassen C (1993) Growth, nitrogen fixation, and spectral attenuation in cultivated *Trichodesmium* species. Appl. Environ. Microbiol. 59: 1367–1375

Rasche ME & Seefeldt LC (1997) Reduction of thiocyanate, cyanate, and carbon disulfide by nitrogenase: Kinetic characterization and EPR spectroscopic analysis. Biochem. 36: 8574–8585

Raven JA (1988) The iron and molybdenum use efficiencies of plant growth with different energy, carbon and nitrogen sources. New Phytol. 109: 279–287

Rea DK (1994) The paleoclimatic record provided by eolian deposition in the deep sea: The geologic history of wind. Rev. Geophys. 32: 159–195

Robson RL & Postgate JR (1980) Oxygen and hydrogen in biological nitrogen fixation. Ann. Rev. Microbiol. 34: 183–207

Rue EL & Bruland KW (1995) Complexation of iron(III) by natural ligands in the central North Pacific as determined by a new competitive ligand equilibrium/absorptive cathodic stripping voltammetry method. Mar. Chem. 50: 117–138

Rueter JG, Hutchins DA, Smith RW & Unsworth NL (1992) Iron nutrition of Trichodesmium. In: Carpenter EJ, Capone DG & Rueter JG (Eds) Marine Pelagic Cyanobacteria: *Trichodesmium* and other Diazotrophs (pp 289–306). Kluwer Academic Publishers, The Netherlands

Saino T & Hattori A (1978) Diel variation in nitrogen fixation by a marine blue-green alga, *Trichodesmium thiebautii*. Deep-Sea Res. 25: 1259–1263

Saino T & Hattori A (1979) Nitrogen fixation by *Trichodesmium* and its significance in nitrogen cycling in the Kuroshio area and adjacent waters. Proc. 4th CSK Symp. Tokyo

Saino T & Hattori A (1980) ^{15}N natural abundance in oceanic suspended particulate matter. Nature 283: 752–754

Saino T & Hattori A (1987) Geographical variation of the water column distribution of suspended particulate organic nitrogen and its ^{15}N natural abundance in the Pacific and its marginal seas. Deep-Sea Res. 34: 807–827

Scranton MI (1983) The role of the cyanobacterium *Oscillatoria* (*Trichodesmium*) *thiebautii* in the marine hydrogen cycle. Mar. Ecol. Prog. Ser. 11: 79–87

Seefeldt LC, Rasche ME & Ensign SA (1995) Carbonyl sulfide and carbon dioxide as new substrates, and carbon disulfide as a new inhibitor, of nitrogenase. Biochem. 34: 5382–5389

Siddiqui PJA, Bergman B & Carpenter EJ (1992) Filamentous cyanobacterial associates of the marine planktonic cyanobacterium *Trichodesmium*. Phycologia 31: 326–337

Siefert RL, Johansen AM & Hoffman MR (1999) Chemical characterization of ambient aerosol collected during the southwest monsoon and intermonsoon seasons over the Arabian Sea: Labile-Fe(II) and other trace metals. J. Geophys. Res. 104: 3511–3526

Siefert RL, Webb SM & Hoffmann MR (1996) Determination of photochemically available iron in ambient aerosol. J. Geophys. Res. 101: 14441–14449

Siegenthaler U & Sarmiento J (1993) Atmospheric carbon dioxide and the oceans. Nature 365: 119–125

Sigman DM, Altabet MA, McCorkle DC, Francois R & Fischer G (1999) The $\delta^{15}N$ of nitrate in the Southern Ocean: Nitrate consumption in surface waters. Global Biogeochem. Cycles 13: 1149–1166

Sigman DM, Altabet MA, McCorkle DC, Francois R & Fischer G (2000) The $\delta^{15}N$ of nitrate in the Southern Ocean: Nitrogen cycling and circulation in the ocean interior. J. Geophys. Res. in press

Sigman DM, Altabet MA, Michener RH, McCorkle DC, Fry B & Holmes RM (1997) Natural abundance-level measurement of the nitrogen isotopic composition of oceanic nitrate: An adaptation of the ammonia diffusion method. Mar. Chem. 57: 227–242

Simpson FB & Burris RH (1984) A nitrogen pressure of 50 atmospheres does not prevent evolution of hydrogen by nitrogenase. Science 224: 1095–1097

Smith BE & Eady RR (1992) Metalloclusters of the nitrogenases. Eur. J. Biochem. 205: 1–15

Soderlund R & Rosswall T (1982) The nitrogen cycles. In: Hutzinger O (Ed) The Natural Environment and the Biogeochemical Cycles (pp 61–81). Springer-Verlag, New York

Soderlund R & Svensson BH (1976) The global nitrogen cycle. In: Svensson B & Soderlund R (Eds) Nitrogen, Phosphorus and Sulphur-Global Cycles (pp 23–73). SCOPE Report No. 7, Ecological Bulletin No. 21, NFR., Stockholm

Sprent JI & Sprent P (1990) Nitrogen Fixing Organisms: Pure and Applied Aspects. Chapman and Hall, New York

Stal LJ (1995) Physiological ecology of cyanobacteria in microbial mats and other communities. Tansley Review No. 84. New Phytol. 131: 1–32

Stal LJ & Krumbein WE (1985) Oxygen protection of nitrogenase in the aerobically nitrogen fixing, non-heterocystous cyanobacterium *Oscillatoria* sp. Arch. Microbiol. 143: 72–76

Stal LJ, Staal M & Villbrandt M (1999) Nutrient control of cyanobacterial blooms in the Baltic Sea. Aquat. Microb. Ecol. 18: 165–173

Stewart WDP, Fitzgerald GP & Burris RH (1967) *In situ* studies on N_2 fixation using the acetylene reduction technique. Proc. Natl. Acad. Aci. USA 58: 2071–2078

Subramaniam A & Carpenter EJ (1994) An empirically derived protocol for the detection of blooms of the marine cyanobacterium *Trichodesmium* using CZCS imagery. Int. J. Remote Sensing 15: 1559–1569

Subramaniam A, Carpenter EJ & Falkowski PG (1999a) Optical properties of the marine diazotrophic cyanobacteria *Trichodesmium* spp. II. Reflectance model for remote sensing. Limnol. Oceanogr. 44: 618–627

Subramaniam A, Carpenter EJ, Karentz PG & Falkowski D (1999b) Optical properties of the marine diazotrophic cyanobacteria *Trichodesmium* spp. I. Absorption and spectral photosynthetic characteristics. Limnol. Oceanogr. 44: 608–617

Takahashi T, Feely RA, Weiss RF, Wanninkhof RH, Chipman DW, Sutherland SC & Takahashi TT (1997) Global air-sea flux of CO_2: An estimate based on measurements of sea-air pCO_2 difference. Proc. Natl. Acad. Sci. USA 94: 8292–8299

Tans PP, Fung IY & Takahashi T (1990) Observational constraints on the global atmospheric CO_2 budget. Science 247: 1431–1438

Tassan S (1995) SeaWiFS potential for remote sensing of marine *Trichodesmium* at sub-bloom concentration. Int. J. Remote Sensing 16: 3619–3627

Toggweiler JR (1999) An ultimate limiting nutrient. Nature 400: 511–512

Trenberth KE & Hoar TJ (1997) El Niño and climate change. Geophys. Res. Lett. 24: 3057–3060

Tyrrell T (1999) The relative influences of nitrogen and phosphorus on oceanic primary production. Nature 400: 525–531

Urdaci MC, Stal LJ & Marchand M (1988) Occurrence of nitrogen fixation among *Vibrio* spp. Arch. Microbiol. 150: 224–229

Venrick EL (1974) The distribution and significance of *Richelia intracellularis* Schmidt in the North Pacific Central Gyre. Limnol. Oceanogr. 19: 437–445

Villareal TA (1991) Nitrogen-fixation by the cyanobacterial symbiont of the diatom genus *Hemiaulus*. Mar. Ecol. Prog. Ser. 76: 201–204

Villareal TA, Altabet MA & Culver-Rymsza K (1993) Nitrogen transport by vertically migrating diatom mats in the North Pacific Ocean. Nature 363: 709–712

Villareal TA, Pilskaln C, Brzezinski M, Lipschultz F, Dennett M & Gardner GB (1999) Upward transport of oceanic nitrate by migrating diatom mats. Nature 397: 423–425

Vitousek PM, Cassman K, Cleveland C, Crews T, Field CB, Grimm NB, Howarth RW, Marino R, Martinelli L, Rastetter EB & Sprent JI (2002) Towards an ecological understanding of biological nitrogen fixation. Biogeochemistry 57/58: 1–45

Wada E (1980) Nitrogen isotope fractionation and its significance in biogeochemical processes occurring in marine environments. In: Goldberg ED, Horibe Y & Saruhashi K (Eds) Isotope Marine Chemistry (pp 375–398). Uchida-Rokakuho, Tokyo

Wada E & Hattori A (1991) Nitrogen in the Sea: Forms, Abundances, and Rate Processes. CRC Press, Boca Raton, FL.

Waterbury JB, Watson SW & Valois FW (1988) Temporal separation of photosynthesis and dinitrogen fixation in the marine unicellular cyanobacterium: *Erythrospira marina*. Eos 69: 1089

Wu J & Luther GW (1995) Complexation of Fe(III) by natural organic ligands in the northwest Atlantic Ocean by competitive ligand equilibration method and kinetic approach. Mar. Chem. 50: 159–177

Zehr JP (1995) Nitrogen fixation in the marine environment: Why only *Trichodesmium*? In: Joint IR (Ed) Molecular Ecology of Aquatic Microbes (pp 335–364). Springer-Verlag, Berlin

Zehr JP, Braun S, Chen YB & Mellon MT (1996) Nitrogen fixation in the marine environment: Relating genetic potential to nitrogenase activity. J. Exp. Mar. Biol. Ecol. 203: 61–73

Zehr JP & Capone DG (1996) Problems and promise of assaying the genetic potential for nitrogen fixation in the marine environment. Microb. Ecol. 32: 263–281

Zehr JP & McReynolds LA (1989) Use of degenerate oligonucleotides for amplification of the *nifH* gene from the marine cyanobacterium *Trichodesmium thiebautii*. Appl. Environ. Microbiol. 55: 2522–2526

Zehr JP, Mellon MT & Zani S (1998) New nitrogen-fixing microorganisms detected in oligotrophic oceans by amplification of nitrogenase (*nifH*) genes. Appl. Environ. Microbiol. 64: 3444–3450

Zehr JP & Paerl H (1998) Nitrogen fixation in the marine environment: Genetic potential and nitrogenase expression. In: Cooksey KE (Ed) Molecular Approaches to the Study of the Ocean (pp 285–301). Chapman and Hall, London

Zehr JP, Carpenter EJ & Villareal TA (2000) New perspectives on nitrogen-fixing microorganisms in tropical and subtropical oceans. Trends in Microbiol. 8: 68–73

Zhu XR, Prospero JM & Millero FJ (1997) Diel variability of soluble Fe(II) and soluble total Fe in North African dust in the trade winds at Barbados. J. Geophys. Res. 102: 21297–21305

Zhuang G, Yi Z, Duce RA & Brown PR (1992) Link between iron and sulfur suggested by the detection of Fe(II) in remote marine aerosols. Nature 355: 537–539

Zuckermann H, Staal M, Stal LJ, Reuss J, Hekkert SL, Harren F & Parker D (1997) On-line monitoring of nitrogenase activity in cyanobacteria by sensitive laser photoacoustic detection of ethylene. Appl. Environ. Microbiol. 63: 4243–4251

Note added in proof

See pages 517–519.

Biogeochemistry **57/58**: 99–136, 2002.
© 2002 *Kluwer Academic Publishers. Printed in the Netherlands.*

The origin, composition and rates of organic nitrogen deposition: A missing piece of the nitrogen cycle?

JASON C. NEFF[1]*, ELISABETH A. HOLLAND[2], FRANK J. DENTENER[3], WILLIAM H. MCDOWELL[4] & KRISTINA M. RUSSELL[5]
[1]*The Natural Resource Ecology Lab, Colorado State University, Fort Collins, CO 80523, U.S.A.;* [2]*Max Planck Institute for Biogeochemistry, Jena, Germany, 07745;* [3]*Joint Research Centre, Environment Institute, TP280, I-21020 Ispra (Va), Italy;* [4]*Department of Natural Resources, University of New Hampshire, Durham, NH 03824 U.S.A.;* [5]*Department of Environmental Sciences, University of Virginia, Charlottesville, VA 22903 U.S.A. (*Author for correspondence)*

Abstract. Organic forms of nitrogen are widespread in the atmosphere and their deposition may constitute a substantive input of atmospheric N to terrestrial and aquatic ecosystems. Recent studies have expanded the pool of available measurements and our awareness of their potential significance. Here, we use these measurements to provide a coherent picture of the processes that produce both oxidized and reduced forms of organic nitrogen in the atmosphere, examine how those processes are linked to human activity and how they may contribute to the N load from the atmosphere to ecosystems. We summarize and synthesize data from 41 measurements of the concentrations and fluxes of atmospheric organic nitrogen (AON). In addition, we examine the contribution of reduced organic nitrogen compounds such as amino acids, bacterial/particulate N, and oxidized compounds such as organic nitrates to deposition fluxes of AON. The percentage contribution of organic N to total N loading varies from site to site and with measurement methodology but is consistently around a third of the total N load with a median value of 30% (Standard Deviation of 16%). There are no indications that AON is a proportionally greater contributor to N deposition in unpolluted environments and there are not strong correlations between fluxes of nitrate and AON or ammonium and AON. Possible sources for AON include byproducts of reactions between NO_x and hydrocarbons, marine and terrestrial sources of reduced (amino acid) N and the long-range transport of organic matter (dust, pollen etc.) and bacteria. Both dust and organic nitrates such as PAN appear to play an important role in the overall flux of AON to the surface of the earth. For estimates of organic nitrate deposition, we also use an atmospheric chemical transport model to evaluate the spatial distribution of fluxes and the globally integrated deposition values. Our preliminary estimate of the magnitude of global AON fluxes places the flux between 10 and 50 Tg of N per year with substantial unresolved uncertainties but clear indications that AON deposition is an important aspect of local and global atmospheric N budgets and deserves further consideration.

Introduction

The first measurements of organic nitrogen in precipitation were made in the 1800s and have continued through the 20th century (Smith 1872; Fonselius 1954; Wilson et al. 1959). These early measurements and subsequent studies have shown that organic forms of nitrogen appear in both wet and dry deposition in many locations (Hendry and Brezonik 1980; Lewis 1981; Rendell et al. 1993; Cornell et al. 1995; Eklund et al. 1997). Despite the growing number of studies documenting organic N in deposition, its sources and magnitudes remain unclear. As a result, the role of organic N in atmospheric N deposition is not widely discussed and the majority of deposition measurements focus on inorganic N species. National deposition networks such as the National Acid Deposition Program (NADP) in the U.S. and the European Monitoring and Evaluation Programme (EMEP) in Europe have monitored ammonium and nitrate deposition in precipitation for many years but have little information on organic N deposition. As a result, it has been difficult to make quantitative evaluations of the role of AON in deposition budgets for large-scale compilations of atmospheric N fluxes due to limited data (Prospero et al. 1996).

Some of the functions of organic nitrogen in atmospheric chemistry are clearly established. Organic nitrate formation can be an important process in the nighttime boundary layer when NO_2 accumulates in the absence of sunlight (Munger et al. 1988). Organic nitrates such as PAN are also important to long range atmospheric N transport (Roberts 1990; Lelieveld and Dentener 2000). The role of organic N compounds in the total atmospheric N budget and deposition, however, is substantially less clear. A group of atmospheric chemistry models which all use similar reaction schemes suggest that, in many cases, peroxyacetyl nitrates (PAN) and the organic nitrates constitute more than 50% of NO_y (total odd nitrogen including NO_x, HNO_3, HONO, HO_2NO_2, NO_3, radical NO_3^-, PAN, N_2O_2, organic nitrates) which is the primary source of N deposition of oxidized species (Hagulstaine et al. 1998; Klonecki, personal communication). But it is unclear how much these organic nitrates appear at the surface as wet or dry deposition or whether other processes/species are involved. The term atmospheric organic nitrogen includes a broad array of chemical compounds that differ widely in their function and reactivity. This complexity has hampered progress and complicated analysis of the role of AON in chemistry and deposition. There are also significant problems of contamination in AON measurements (Gorzelska et al. 1997; Church et al. 1999; Russell 1999). The chemical species that make up AON include compounds such as organic nitrates, amino acids and other organic acids with atmospheric lifetimes that range from seconds to weeks (Herlihy et al. 1987; Buhr et al. 1990; Gorzelska and Galloway 1990; Dentener and Crutzen 1994; Brassuer et al. 1998).

Conceptually, AON can be divided into three types of nitrogen: organic nitrates, reduced AON and biological/terrestrial AON. The first two categories are broad chemical descriptions but are useful because reduced and organic nitrate forms of AON can be thought of in conceptually distinct ways. Organic nitrates are the oxidized end products of reactions of hydrocarbons with NO_x ($NO + NO_2$) in polluted air masses. These forms of AON are generally formed *in situ* from the constituent molecules following reaction with light (Roberts 1990). The second category of AON primarily includes gas phase or aerosol amine N and urea that can be injected into the atmosphere in marine or agricultural environments but is not formed *in situ* due to the oxidizing conditions of the atmosphere (Milne and Zika 1993; Schade and Crutzen 1995). The last category of AON refers to biological and particulate forms of organic N including bacteria, dust particles and pollen all of which can be found in high concentrations in the lower levels of the atmosphere but have their sources in terrestrial environments (Bovallius et al. 1978; Lindemann et al. 1982; Jones and Cookson 1983; Littmann 1997).

Organic nitrate, reduced and biological/particulate contributions to AON have been reasonably well documented and, as a result, it is becoming increasingly apparent that organic N plays an important role in the chemistry of the atmosphere. Organic nitrates may also be an important input of N to ecosystems. This conclusion is further supported by an increasing number of detailed, analytically rigorous studies of nitrogen deposition that indicate significant organic contributions to deposited N (e.g. Gorzelska et al. 1992; Cornell et al. 1995). Despite the well-documented presence of AON in deposition, there remains substantial uncertainty regarding the quantitative role of AON in N deposition fluxes and in the global cycle of N. Are AON fluxes important contributors to atmospheric N loading in terrestrial and aquatic ecosystems? Are AON fluxes large enough to warrant routine analysis in deposition monitoring programs as these programs evolve to meet new challenges? Is AON a substantial global flux of N that should be considered in evaluations of the global N cycle? These questions remain difficult to answer but are important to developing a complete understanding of the N cycle at both small and large scales.

Methods

In this paper, we review published reports of all forms of organic N deposition including total organic N, oxidized forms of AON and reduced forms of AON. We discuss the implications of these results for estimates of N deposition and discuss the various pathways by which organic forms of N enter or are created in the atmosphere. We then evaluate the relationships between modeled and measured AON using the TM3 model of atmospheric chemistry and transport (Dentener et al. 1999) and use the model combined with our literature review to provide an estimated range of the organic nitrate and PAN contributions to global AON deposition.

Data selection and analysis

The measurement of AON involves many different sampling methods and multiple analytical techniques. The measurement of individual organic compounds such as amino acids or organic nitrates is too detailed to fully discuss here and has been addressed, at least in part, elsewhere (Roberts 1990; Russell 1999). In this review, we evaluate the potential contribution of reduced N, biological/particulate N and organic nitrates to the total organic N flux from the atmosphere. This differentiation is a conceptual simplification that provides insight into the possible controls over the total AON concentrations observed in deposition measurements. The biological and particulate N grouping may include both oxidized or reduced forms of N but is differentiated from the other two classes by the presence of a distinct biological form (e.g. bacteria) or particulate form that originates in terrestrial systems (e.g. dust, pollen). Sample molecules and definitions for these classifications are shown in Table 1. The biological AON definition is useful because of the established presence of organisms and dust in the atmosphere and in precipitation and the intriguing possibility that these materials could contribute to the net organic N flux from the atmosphere (Bovallius et al. 1978; Lindemann et al. 1982; Jones and Cookson 1983; Littman 1997). For each of these classes of organic N, we discuss the controls over the production and deposition of these compounds in order to better evaluate the regulation of the atmospheric organic N cycle.

We have compiled 41 measurements in which the total organic N concentrations in precipitation or wet/bulk deposition fluxes were presented. There are very few measurements of dry deposition of organic N containing gases, and so we cannot discuss these fluxes in detail here. However, we do discuss measurements of gas-phase organic nitrates such as PAN in order to examine the role that these gas phases organic nitrates may play in the deposition of organic N globally. It should be understood that the N wet deposition measurements presented in this paper represent the net flux of organic N from the atmosphere to ecosystems and contain a unspecified mixture of organic nitrates, amino acids, bacteria and other forms of organic N. One of the main goals of this paper is to link these N deposition measurements to the potential N deposition from the array of potential contributors to deposition.

Table 1. Examples of atmospheric organic nitrogen. Note that the bacterial and particulate N classes may also include reduced and oxidized N.

	General type of organic N	Example Molecules	Source
Bacterial nitrogen			
Reduced organic N larger than 0.2 microns	Multiple dissolved and particulate amine forms		Atmospheric Bacteria
Particulate nitrogen			
Particles larger than 0.45–1 micron	Multiple particulate organic nitrates or amine N forms		Atmospheric dust and organic debris
Reduced organic N			
Soluble N with sizes smaller than 0.45 microns	Amine N	Serine, Glycine, Alanine, Valine, Methylamine, Ethylamine and others	Oceanic aerosols, terrestrial emissions (agricultural sites or natural vegetation)
Oxidized organic N			
Soluble N with sizes smaller than 0.45 microns	Nitric acid esters	Methyl, ethyl, propyl and butyl nitrates	Atmospheric Reactions
	Nitric acid diesters	Ethylene & propylene glycol	Atmospheric Reactions
	Hydroxy nitric acid esters	Nitrooxy ethanol and propanol	Atmospheric Reactions
	Peroxynitric acid esters	Trifluoro methyl and dimethyl ethyl	Atmospheric Reactions
	Peroxycarboxylic nitric anhydrides	PAN	Atmospheric Reactions

For all the total organic N deposition measurements, the organic N (solution) concentration or flux data were presented alongside either total inorganic N or NH_4^+ and NO_3^- data. There are two types of organic N fluxes that we discuss in this review; total organic N (TON), which includes both particulate and dissolved forms of organic N and dissolved organic N (DON), which does not include the particulate component. The difference between TON and DON is operational and defined by the size of particle that will flow through a filter but both types of measurements are based on analyses performed on soluble (or at least filterable) material in deposition collectors. For the studies that reference DON, the samples have been filtered following collection. Because the pore size of the filter is not always provided in the referenced papers, we assumed DON represented the material that passed through a 0.45–1 micron filter (the most common size range). Unfiltered samples were assumed to contain both particulate and dissolved organic N and are represented as TON.

In addition to different definitions for the types of organic N in precipitation, there are multiple measurement techniques, sampling frequencies and field collection methodologies. DON (or TON) is measured as the difference between the total N in solution (determined via the oxidation or reduction of the organic compounds) and the sum of the initial NH_4^+, NO_2^- and NO_3^-. The conversion of organic N to inorganic N is generally carried out by chemical (persulfate), *uv* or high temperature oxidation of the solution. We also reference dissolved organic carbon concentrations and these measurements are generally made with high temperature oxidation techniques. These different approaches have varying oxidizing efficiencies but generally result in the conversion of at least 80% of the organic species to inorganic N in freshwater and seawater samples (Walsh 1989; Scudlark et al. 1998). The majority of the measurements described in this paper were made using persulfate oxidation-based conversion of organic N to nitrate or high temperature combustion.

As with analytical techniques, there are a variety of sampling approaches. Wet deposition measurements can be made for individual rain events on regular daily or weekly intervals. There are potential problems with storage of organic samples in field collectors in that organic N may be converted to inorganic N due to microbial decomposition or that organic N may be produced due to microbial activity. There is also substantial potential for contamination in samples that are left in the field for long periods of time. It is not readily apparent whether these post-deposition changes would lead to increased or decreased TON concentrations in collection devices. For event sampling of precipitation, where the potential for external contamination is minimized, organic N compounds may react with the materials in sampling equipment, storage containers or may volatilize from sampling vessels (Scudlark et al. 1998; Church 1999). For these reasons, there is potential for underestimation of organic N concentrations in event-based sampling. Most of the recent (after 1980) measurements discussed here were made either with event samples or with sub-weekly sampling intervals that should minimize the possibilities of N transformations or contamination post-collection. In some cases, samples were analyzed immediately following collection. The potential errors in DON deposition measurements, and options for mitigation have been discussed in more detail in Gorzelska et al. (1997), Russell (1999) and Church (1999). From the standpoint of this review, we note that measurement methodology does not appear to affect AON concentration measurements in a systematic manner. There were substantial, measurable quantities of DON in precipitation for every case study that we examined. We exclude some data points from analysis for reasons discussed below, but have maintained as large a data set as possible.

To examine the role of AON in total N fluxes, we analyzed published data for the fractional contribution of organic N to total N concentrations or fluxes. If both AON fluxes and solution concentration data were available, we use the flux data to estimate the proportional contribution of organic N to the total N concentration. However for Russell et al. 1998 and Malmquist 1978; we use concentration averages instead of fluxes because these particular data sets do not span an entire year. We also did not include the DON cloud deposition estimate for Chile (Weathers et al. 1998) and the DON dry deposition estimate for Canada (Simpson and Hemens 1978) because they are not directly comparable to wet or bulk deposition estimates compiled from other studies. We used the Statistica software package for Windows to carry out regression analyses and to estimate mean, median, standard deviation, and standard errors for the data (Statsoft Inc. Tulsa, OK, U.S.A.). For some analyses, we have classified a site as unpolluted or polluted. These classifications are based on the evaluation of the inorganic N flux for a site (Prospero et al. 1996) and/or the geographic location of a site (marine vs. continental).

Dry and wet deposition

AON can be deposited in both wet and dry forms. We discuss both types of deposition but there are some important distinctions in the two types of measurements. All but one of the flux measurements discussed here are made on soluble N from deposition collectors. These forms of deposition could include N deposited in either wet or dry form. In a bulk collector, for example, a certain amount of dry deposition will fall into the bucket and then be quantified as soluble N. Actual deposition rates of dry AON are difficult to obtain. Estimates of deposition can be made from measured atmospheric concentrations combined with deposition velocities but for most of the specific organic N species (e.g. amino acids or organic nitrates), it is difficult to find published deposition velocities, and therefore difficult to calculate a deposition rate (Dentener et al. 1999). Accordingly, we present data on the concentrations of specific dry AON species and discuss the ways that these dry AON species may contribute to the total N flux. We do not use any of these dry deposition calculations to estimate the fractional contribution of organic compounds to total N fluxes. For calculations of the average organic contribution to concentrations or fluxes, we present data summaries from the bulk or wet deposition collectors discussed above.

Modeled estimates of organic N fluxes

The deposition of organic nitrogen is strongly dependent on the distributions of sources, the lifetime and chemistry of N compounds and atmospheric transport. The lifetime of many reactive N compounds in the atmosphere is quite short. The global average lifetime of NO_x is on the order of one day while NH_x (NH_3 + NH_4^+) is between 1 and 5 days depending on NH_3 concentrations and other factors (Dentener et al. 1999). These global average lifetimes do not reflect the tremendous variability in lifetimes with season and height (Brasseur et al. 1999). For example, the lifetime of NO_x in the summer is on the order of a few hours to a day in the lower few kilometers of the troposphere, but extends to 4–7 days in the upper troposphere. The formation of organic nitrates provides a temporary reservoir of active nitrogen. Some of these organic nitrates have lifetimes of a week or more and so can be transported to the remote atmosphere where they can provide a source of NO_x. Peroxyacetyl nitrate (PAN) is the predominant organic nitrate produced in the atmosphere via NO_x, non-methane hydrocarbon reactions and has a much longer lifetime in the upper than in the lower troposphere (Bertman and Roberts 1992; Moxim et al. 1996; Nouaime et al. 1998; Roberts et al. 1998). Collectively, understanding the interaction between the lifetime of these compounds, their speciation and associated solubility are critical to understanding the spatial distribution of organic nitrate deposition. There are currently no atmospheric models that simulate the chemistry and transport of reduced or particulate AON. However, there are models that represent the chemistry of organic nitrates and these models may be useful tools for predicting the distribution and trends in this portion of the total AON flux.

Organic nitrate chemistry is represented in several models of global atmospheric chemistry (Stand and Hov 1994; Zimmerman 1987; Houweling et al. 1998). We use one of these models to provide insight into the potential transport and fluxes of the oxidized organic N species particularly as it relates to human activity and sources of precursor compounds. We use the TM3 model which is a 3D chemical transport model that includes representation of the atmospheric photo-oxidant, N and S cycles including NO_x, NO_y and reduced N (NH_3 and $(NH_4)_2SO_4$). The model incorporates globally-gridded source fields of NO_x and NH_3 and a series of detailed chemical reaction sequences (Dentener and Crutzen 1994; Houweling et al. 1998; Dentener et al. 1999; Lelieveld and Dentener 2000) including non-methane hydrocarbon and oxidized organic N reactions (Houweling et al. 1998) and NO_x source fields from Yienger and Levy (1995). The model has a resolution of 3.75 × 5 degrees and 19 vertical hybrid levels, 5 of which are located in the boundary layer. Dry deposition is parameterized following Ganzeveld and Lelieveld (1995), wet deposition using the parameterization of Guelle et al. (1999).

The deposition rates of organic nitrates, including PAN, are difficult to represent in models because uncertainty in NMHC reaction schemes, and the lack of published organic nitrate dry deposition rates and deposition velocities. One of the key uncertainties with regard to the wet deposition of organic nitrates is representation of their solubility and the efficiency of removal by precipitation. The solubility of individual organic nitrate compounds can span two orders of magnitude depending on the individual compound (Houewling et al. 1998). Uncertainties in wet and dry deposition parameters, such as surface characteristics, precipitation scavenging, and atmospheric stability limit the accuracy of simulation for even the better characterized compounds NO_x and HNO_3 . A detailed analysis of TM3 model errors in NO_x, NO_y and PAN representations is discussed in Houweling et al. (1998). In general, the model generally agrees within a factor of two with published PAN concentration data (Houweling et al. 1998). We present model based estimates of net global organic nitrate deposition as a comparison against our data compilation in order to evaluate the potential role of the organic nitrates and PAN in measured AON deposition. It is important to remember that these model-based estimates are preliminary and should be considered an upper limit on organic nitrate deposition.

Results and discussion

Reduced organic N compounds

The presence of reduced organic N in atmospheric aerosols and in deposition has been noted at multiple locations and like the distribution of oxidized organic N, appears to be a global phenomenon, albeit a poorly studied one. There are multiple types of reduced nitrogen that are found in the atmosphere or in deposition, including urea and amino acid N in aerosol and precipitation samples over oceans (Mopper and Zika 1987; Gorzelska and Galloway 1990; Spitzy 1990; Milne and Zika 1993; Cornell et al. 1998) and over continental environments (Fonselius 1954; Munczak 1960; Sidle 1964; Gorzelska et al. 1994). In contrast to the organic nitrates, considerably less is known about the mechanisms that lead to the production and emission of reduced N to the atmosphere or about the global transport patterns of this material. There is, however, enough information on the dynamics of these compounds to speculate on their role in the global AON cycle.

108

Unlike oxidized AON, reduced AON compounds are not produced *in situ* under the oxidizing conditions of the atmosphere. The most likely source of these compounds, which include urea, free amino acids and other methyated amines, is direct injection into the atmosphere from various sources. The types of reduced AON that may be expected in precipitation are shown in Table 1 and the potential sources of reduced AON include oceanic, agricultural and biomass burning emissions. The oceanic source of amino acid N to the atmosphere involves the direct injection of amino acid containing aerosols into the atmosphere during the bursting of bubbles at the air/sea interface (Mopper and Zika 1987; Milne and Zika 1993). The other potential source of amine compounds to the atmosphere is agricultural systems where a variety of aliphatic amines can be emitted from animal husbandry operations (Mosier et al. 1973; Hutchinson et al. 1982; Schade and Crutzen 1995) and from biomass burning (Lobert et al. 1991). The compounds that may be emitted from these sources include methylamine (MMA), dimethylamine (DMA) and trimethylamine (TMA) among others. In localized areas above animal feedlots, the concentrations of MMA, DMA and TMA can range from 20–280 pptv and may exceed NH_3 concentrations in the same air mass (Schade and Crutzen 1995).

Once in the atmosphere, many reduced N species react quickly. Compounds such as MMA and DMA are highly soluble and are very reactive with the OH radical (Schade and Crutzen 1995). Other radical species such as nitrate, perhydroxyl and superoxide molecules all may function as effective scavengers of reduced N compounds (Milne and Zika 1993). The general reactivity of these compounds suggests that they are not favorable candidates for long range atmospheric transport but are rather more likely to be locally emitted and deposited (consistent with the global lifetime of 1–5 days detailed by Dentner and Crutzen 1994). There is some indication of this type of behavior in comparisons of amino acid concentration in rains over the Gulf of Mexico where concentrations reached 13–15 μM compared to concentrations of 0.3 to 0.5 μM over Miami, Florida, further away from the presumed ocean source (Mopper and Zika 1987). The concentrations of amino acids in marine air range are generally in the low (1–10) pptv range but can range as high as 150 pptv (Mopper and Zika 1987; Gorzelska et al. 1994). In certain locations, such as the Gulf of Mexico, amino acid concentrations rival those of NH_4^+ and NO_3^- (Mopper and Zika 1987) but in other locations such as continental Canada, aerosol amine nitrogen concentrations were less than 5% of the sum of the inorganic N concentrations (Gorzelska et al. 1994). One pattern that is clear is that in most cases the concentrations of free amino acids in aerosol form are substantially less than PAN concentrations and more similar to the concentrations of the (non PAN) organic nitrates shown in

Table 2. The substantial variability in both aerosol amine content and amino N in rain samples in the above samples likely reflects the localized nature of these sources as well as short range transport of these compounds.

At the global scale there have been two estimates of global fluxes for reduced organic N compounds. Milne and Zika (1993) estimate an amine N flux to oceans at 0.6 Tg of N per year. Schade and Crutzen (1995) place the global MMA emission flux from animal husbandry operations at 0.15 Tg of N per year in comparison with a biomass burning source of 0.06 Tg N per year. If these two estimates are representative of the dominant reduced organic N fluxes, then the global deposition flux of reduced AON would be under 1 Tg of N per year. However, without additional information, it is premature to rule out additional sources or forms of reduced AON or to make more detailed estimates of the global reduced-AON flux. At local scales, however, the reduced AON contribution to N deposition may be substantial, particularly for areas influenced by marine air and areas downwind of agricultural operations.

Biological and particulate atmospheric N

The atmosphere contains not only aerosol and dissolved organic nitrogen but also airborne microorganisms, often associated with particulate material (Bovallius et al. 1978). This atmospheric flora contains N that could contribute to estimates of AON fluxes made from unfiltered samples of wet deposition or even to dry deposition measurements of total organic N. The role of bacteria in the AON budget is wholly unknown and there are no data on the N flux that might be associated with bacterial deposition. For dust deposition of N, there is not an extensive deposition data set, however there is enough information to evaluate the potential contribution of these materials to the AON deposition flux.

The concentration of bacteria in the atmosphere varies broadly between concentrations as low as 10 bacteria per m^3 over the open ocean to thousands of bacteria per m^3 in urban and agricultural environments (Bovallius et al. 1978; Lindemann et al. 1982; Jones and Cookson 1983). Concentrations of bacteria in the atmosphere increase with disturbances such as agricultural activity and dust storms and decline during rain events (Bovallius et al. 1978). Most airborne bacteria are associated with particles and may be transported long distances as a result (Bovallius et al. 1978). To make a preliminary estimate of how much N may be contained in airborne bacteria, we take a relatively high estimate of urban bacterial density of 3000 bacteria per m^3 of air. Assuming a radius of 0.5 microns, each bacterium would have a volume of approximately 5×10^{-19} m^3. With a solid to liquid ratio of 0.2, an average N content of 10%, and a density of 1000 kg m^{-3}, each bacterium would contain

Table 2. Concentrations of non-PAN organic nitrates for several sites and urban/non-urban PAN average concentrations for 48 studies contained in Roberts (1990). Standard Errors (SE) for the Roberts (1990) study means are shown. Concentration means are presented whenever possible with ranges presented when means were not available. Unpolluted (P) and Polluted (U) conditions were designated by the authors of this manuscript.

Site	Season (Sp, S, A, W)	Latitude, Longitude	Compound type	Concentration (pptv)	Unpolluted (U) Polluted(P)	Reference
North Pacific Ocean	Sp, S		C3–C7 alkyl nitrates	10	U	Atlas 1988
Alaska USA	Sp	64 N, 150 W	C2–C6 alkyl nitrates	34	U	Beine et al. 1996
Hawaii		20 N, 156 W	C3–C5 alkyl nitrates	7	U	Atlas and Schauffler 1991
South Africa	Sp	29 S, 24 E	C3–C5 alkyl nitrates	17	U	DeKock and Anderson 1994
Northwestern Canada	Sp	82 N, 62 W	C3–C7 alkyl nitrates	34–128	U	Muthuramu et al. 1994
Rural Ontario – Clean	S	45 N, 79 W	C3–C6 alkyl nitrates	8–50	U	O'Brien et al. 1995
			C2–C4 hydroxy nitrates & 1,2 dinitrooxybutane	5–20	U	

Location	Season	Coordinates	Species	Concentration		Reference
Polluted			C3–C6 alkyl nitrates	10–105	P	
			C2–C4 hydroxy nitrates & 1,2 dinitrooxybutane	1–35	P	
Alabama, USA	S	32 N, 86 W	C2-C5 alkyl nitrates	52	P	Bertman et al. 1995
Germany	Sp	51 N, 9 E	C1–C5 alkyl nitrates	1–230	P	Flocke et al. 1991
British Columbia	S	49 N, 123 W	C3–C6 alkyl nitrates	11–210	P	O'Brien et al. 1995
Rural Pennsylvania	S	40 N, 77 W	C2–C5 alkyl nitrates	2-200	P	Buhr et al. 1990
Urban Airmasses			PAN	780 140 SE		Roberts (1990)
Non-Urban Airmasses			PAN	380 50 SE		Roberts (1990)

1×10^{-14} g of N (Paul and Clark, 1996). With a boundary layer depth of 500 m, there would be 1.5×10^6 bacteria over a m^2 of land surface with a total N content of $1.5\ 10^{-8}$ g of N m^{-2}. Assuming a lifetime of 5 days and a removal coefficient of $2 \times 10^{-6}\ s^{-1}$, the annual flux of bacterial AON would be 9.5×10^{-6} kg of N ha^{-2}. Even with substantially higher bacterial densities, removal coefficients or bacterial N content, these hypothesized deposition rates appear to be inconsequential to the total N load to an ecosystem. While these calculations could certainly be improved upon with additional information, it does not appear that biological AON is a significant component of either local or global AON fluxes. It is also important to note that there are other forms of airborne biological N including insects. Given that precipitation-sampling equipment is open to the atmosphere, and the large amount of N that would be in an insect relative to precipitation, the possibility of organic contamination to sampling stations is significant. Techniques such as event sampling, filtration and visual inspection of precipitation samples will minimize this type of contamination, however both field studies and reviews such as this paper cannot rule out these forms of contamination at the present time.

The role of dust in the element cycling of ecosystems has received significant attention (c.f. Swap et al. 1992; Littmann 1997; Ramsperger et al. 1998a), however most of the discussion of element deposition in dust has focused on inorganic cations and anions. As with other forms of deposition, there appear to be significant organic matter inputs associated with dust deposition which may contribute to the organic N flux in certain types of measurements. The potential impact of dust to AON deposition measurements is limited to sampling designs that capture bulk (as opposed to wet only) deposition and to measurements made on unfiltered samples (although dust associated organic N could dissolve in contact with water). Dust particles can range in size from 0.1 to over 20 microns and as a result, they may fall into the operationally defined (filter pore size based) cutoff for dissolved organic nitrogen (Tegen et al. 1994; Ramsperger et al. 1998a). There is not sufficient published information to carry out a detailed evaluation of the role of different dust size classes in AON fluxes, however there are indications that there is substantial organic content on dust particles. In studies of the southwestern Argentinean Pampa, Ramsperger et al. (1998a) observed annual dust loading of 400–800 kg ha^{-1} yr^{-1}. Of this dust input, approximately 6–8% of the material was organic matter with a %N content of 0.5–0.8% for an annual deposition flux of $2 - 6.4$ kg N ha^{-1} (Ramsperger et al. 1998b). In an Israeli sand dune ecosystem, Littman (1997) found total N loading rates of 0.48 kg N ha^{-1} yr^{-1} with only 5% of the flux attributed to inorganic N compounds and in dust samples from West Africa, Ramsperger et al. (1998b) observed % N contents (by mass) of 0.3–1.4.

The role of dust in the AON budget is particularly important because global dust emissions are increasing as the result of human disturbance (Tegen et al. 1995). These increases almost certainly lead to an increased organic N load in the atmosphere and likely contribute substantially to bulk TON loading in deposition measurements. There are both local and remote sources of dust for atmospheric transport and a complete review of this topic is impossible here but it is clear that dust could be a significant contributor toward AON deposition. Dust deposition estimates range between 100–2000 kg ha^{-1} yr^{-1} (Rabenhorst et al. 1984; Swap et al. 1992; Ramsperger et al. 1998a). Assuming small (2%) organic fractions and relatively low N content (1%), a conservative approximation of dust AON fluxes leads to an estimate of 0.02 to 0.4 kg dust AON ha^{-1} yr^{-1}. The studies mentioned above indicate that these numbers could be substantially higher.

Organic nitrates

Organic nitrates occur in both polluted and remote portions of the atmosphere. These compounds are produced through photochemically mediated reactions of non-methane hydrocarbons (NMHC) and NO_x (NO and NO_2). There are a variety of compounds and reaction sequences that are important in the formation of oxidized AON and a detailed review and discussion of these reactions is in Roberts (1990). A greatly simplified version of the NMHC/NO_x reactions is shown in Figure 1. There have been numerous field studies of the reactions of NMHC and NO_x focusing on the role of these compounds in photochemical reactions and the production of compounds such as PAN (Bertman and Roberts 1992; Williams et al. 1993; Nouaime et al. 1998; Roberts et al. 1998).

There are both natural and anthropogenic sources of NMHCs and NO_x which may contribute to the production of organic nitrates. Natural sources of NMHC's are dominated by plant production of isoprene which accounts for 40% or more of the global hydrocarbon emissions, but also includes a variety of additional higher molecular weight volatile organic carbon (VOC) compounds (Guenther et al. 1995). Isoprene and other plant emitted hydrocarbon emissions increase sharply with temperature and the correlation between temperature and production rates results in a global flux that is tropically dominated and which should be strongly seasonal in temperate climates (Sharkey et al. 1991; Guenther et al. 1995; Keller and Lerdau 1999). In northern latitudes the atmospheric NMHC budget is dominated by fossil fuel combustion (Guenther et al. 1995). Estimates of the pre-industrial NO_x budget place the pre-industrial global flux of NO_x from soils, biomass burning, lightning, NH_3 oxidation and stratospheric injection at approximately 12 Tg of N which is roughly 30 percent of the contemporary

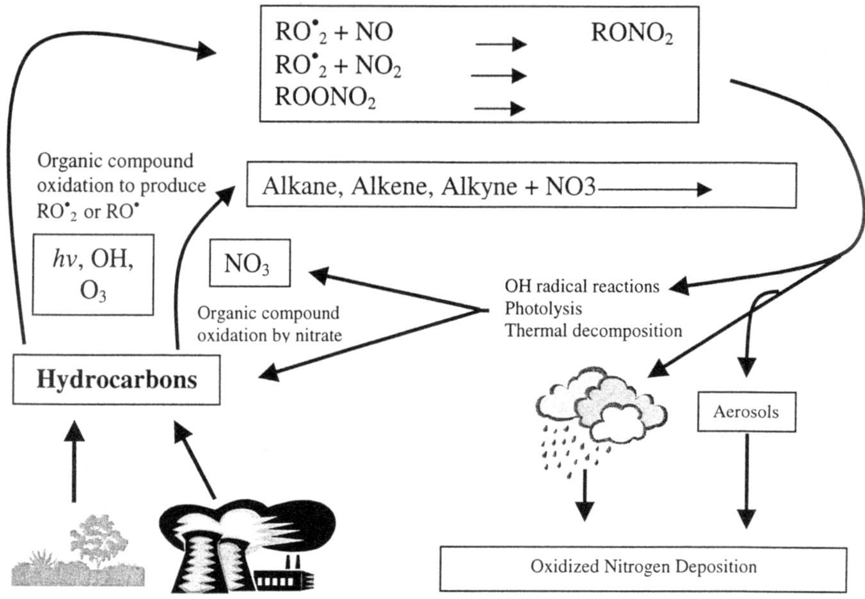

Figure 1. Simplified gas phase formation reactions and removal processes for organic, oxidized N species.

36 Tg NO_x budget (Holland et al. 1999). This three-fold increase in the global NO_x budget is due primarily to the combustion of fossil fuels. Once NMHC and NO_x compounds enter the atmosphere and begin the reaction sequences shown in Figure 1, there are additional factors that may influence organic nitrate deposition rates. The atmospheric lifetime of PAN is sensitive to temperature with longer PAN lifetimes at lower temperatures. This sensitivity should lead to longer transport distances of PAN in the northern latitudes relative to the tropics and during the winter relative to the summer (Houweling et al. 1998).

The large natural sources of NMHCs and NO_x suggest that there was a significant pool of oxidized AON in the pre-industrial world. In a review of the atmospheric cycle of organic nitrates, Roberts (1990) summarizes a series of PAN measurements from urban and non-urban air masses that we summarize with other non-PAN organic nitrates in Table 2. Concentrations of PAN in unpolluted sites are roughly 50 % of polluted sites (Table 2). This increase is generally consistent with the ratio of natural to anthropogenic fluxes of NMHCs and NO_x discussed above. In non-urban sites, there are a handful of measurements of the relative concentration of PAN to the total N concentration in the atmosphere with an average proportional PAN contribution of 15–30% for measurements in Colorado and Canada (Bottenheim et al.

1984; Anlauf et al. 1986; Fahey et al. 1986; Pierson et al. 1987; Daum et al. 1989). The contribution of PAN to total atmospheric N is related to the age of the air mass and its distance from anthropogenic inputs that increase the overall contribution of inorganic N vs. PAN (Roberts et al. 1990).

There is substantially less known about other (non-PAN) organic nitrates but there are a number of compounds that, in addition to PAN, may contribute a sizable fraction to the total oxidized AON load. Table 1 shows the concentrations for several urban and non-urban sites with organic nitrate concentrations that range from the low part per trillion by volume (pptv) range to highs of 200 pptv. Like PAN, the concentrations of non-PAN organic nitrates are higher in polluted or urban areas and lower concentration in more remote areas. Without additional studies, it will remain difficult to evaluate the controls over the concentrations of these compounds in the atmosphere.

The evidence for the importance of chemical speciation is indicated by the contrasting spatial patterns of PAN and non-PAN organic nitrates deposition predicted from the TM3 model. PAN concentrations and deposition are highest in the industrialized Northern Hemisphere and the long distance transport of PAN over the Northern Atlantic Ocean is clearly shown (Figure 2(a)). PAN fluxes range between 0.1 and 2 kg N ha^{-1} yr^{-1} with average deposition between 0.3 and 0.6 kg ha^{-1} yr^{-1} for the north eastern U.S. and Europe and a global flux of 2.5 Tg (Table 3). For dry organic nitrate deposition, there are also high rates of deposition in North America and Europe but tropical regions in Africa and South America also show significant deposition (Figure 2(b)). In comparison to PAN, the dry organic nitrates deposition rates are approximately one half to two thirds of the PAN fluxes (Table 3). Of the three compounds shown in Figure 2, the wet organic nitrates have the most tropically oriented deposition pattern with peak deposition rates in the northwestern Amazon and the Congo region of Africa (Figure 2(c)). Interestingly, the highest of these deposition rates match PAN deposition in the northern hemisphere but the range of transport over the tropical regions is much shorter. The combination of warm temperatures and high rainfall appears to limit the transport of organic nitrate out of the tropical regions while PAN concentration are elevated in the entire 30–45 degree N band.

Globally, the TM3 model estimates that the sum of organic nitrates and PAN contributes 9.1 Tg of N to the total annual deposition flux of 92.9 Tg N. Our feeling is that this estimate represents an upper limit of organic nitrate deposition given the uncertainties in the model. However, there is also debate over the magnitude of terrestrial NO$_x$ emissions and if higher terrestrial emissions fields were used in the model (e.g. Davidson and Kingerlee 1997) then higher organic nitrate deposition would follow. Modeled PAN concentration ranges roughly correspond with PAN data for urban and non-urban air masses

A. PAN Deposition (Dry)

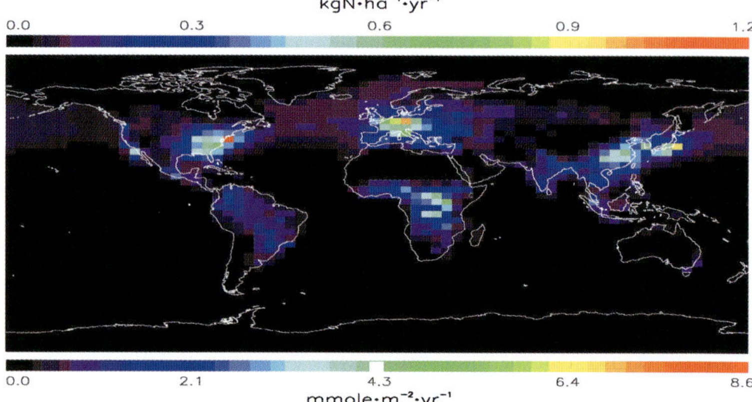

B. Organic Nitrate Deposition (Dry)

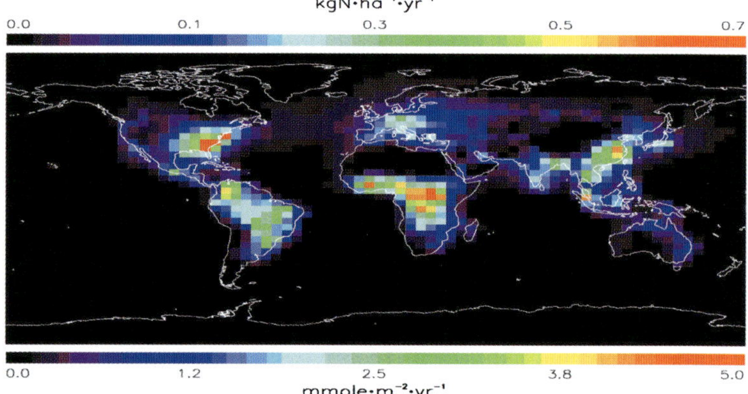

C. Organic Nitrate (Wet)

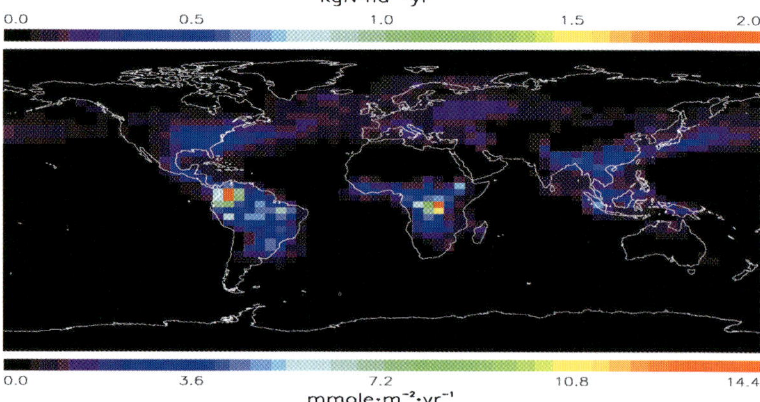

Figure 2. Annual deposition of organic nitrogen from the TM3 model. Panel A shows the PAN deposition (in dry form), panel B shows dry organic nitrates and panel C shows wet organic nitrate deposition.

Table 3. Annual global N fluxes from the TM3 Model and estimates of organic vs. inorganic loading in Teragrams of N (Tg).

	N Species	Global N deposition (Tg N)
Inorganic Species		
	NO_x (g)	6.4 Tg
	HNO_3^- (g)	10.6 Tg
	NH_4^+ (g)	3.6 Tg
	NH_3 (g)	20.8 Tg
	HNO_3^-	20.3 Tg
	NH_4^+	13.6 Tg
	NH_3	8.5 Tg
Organic Species		
	Organic Nitrates (g)	2.7 Tg
	PAN (g)	2.5 Tg
	Organic Nitrates	3.9 Tg
Total Modeled Deposition		
	Inorganic	83.8 Tg
	Fraction Organic	9.8 %
Modeled organic flux estimate	Oxidized Species only	9.1 Tg

shown in Table 2. It is difficult to evaluate the non PAN, organic nitrate contribution to oxidized AON, however, both the model and the data summarized in Table 2, suggest that other organic nitrates may be sizable contributors to the total oxidized AON flux. These calculations also highlight the potential for oxidized AON to contribute significantly to global N deposition calculations. In a subsequent section, we compare these model estimates of organic nitrate loading to measured estimates of organic nitrogen in deposition fluxes.

Total AON deposition rates

In contrast to most measurements of atmospheric AON concentrations, most deposition measurements do not differentiate between different AON compounds and instead present a total TON or DON deposition rate. There have been a number of TON/DON deposition measurements made over a wide range of ecosystems and these studies show consistently large contributions of organic compounds to the total N deposition flux or concentration. These measurements are summarized in Table 4 and indicate that the organic

Table 4. Summary of AON and DIN fluxes. Bulk deposition is specified as B and Wet only measurements as W. TON is total organic nitrogen based on unfiltered samples. DON represented dissolved organic samples that are stated as filtered or presented as dissolved. Data presented as $NH_4:NO_3:AON$ ratios (1 number = organic N, 2 numbers = inorganic N:organic N, 3 numbers = $NH_4:NO_3:DON/TON$). Percent composition is determined from fluxes wherever available and from concentration ratios if no flux data was presented.

Study	Site	Time	Latitude:Longitude	Meas Type	Concentration Ratios (mg/L)	Flux Ratios kg/ha/yr	% composition	Reference Number
1.	Minnesota	1971–1973	44 N, 93 W	B/TON		2.7:2.3:**2.3**	36:32:**32**	Verry and Timmons 1977
2.	Rhode River, Chesapeake Bay	1973–1991	38 N, 76 W	B/TON	0.28:0.47:**0.29**	3.12:5.18:**3.45**	27:44:**29**	Jordan et al. 1995
3.	Central Amazon Basin, Brazil	1983–1985	3 S, 60 W	W/DON		3.58:**2.45**	60:**40**	Williams et al. 1997
4.	Lewes, DE	1993–1994	38 N, 75 W	W/TON	0.19:0.30:**0.09**	14.8:23.5:**8.7***	32:50:**18**	Russell et al. 1998, Scudlark et al. 1998
5.	Gainesville, FL	1976–1977	29 N, 82 W	W/TON	0.1:0.19:**0.41**	1.2:2.3:**5.0**	14:27:**59**	Hendry and Brezonik 1980
		1976–1977	29 N, 82 W	B/TON	0.12:0.23:**0.47**	1.5:2.8:**5.7**	15:28:**57**	

Table 4. Continued

6.	Harp Lake, Ontario	1974	45 N, 79 W	Snow/TON	0.14:0.19:**0.16**		28:39:**33**	Nichols et al. 1978
				W/TON		10.1:**4.4**	70:**30**	
7.	Alberta, Canada	1975–1976	54 N, 113 W	W/TON	0.26:0.25:**0.15**	1.34:0.67:**1.11**	43:21:**36**	Caiaza et al. 1978
				D	0.19:0.03:**0.39**	5.56:0.48:**8.16**	40:3:**57**	
8.	New Zealand	1958	41 S, 175 E	Snow/TON	0.14:0.001:**0.11**	nd	56:0:**44**	Wilson 1959
9.	Taupo, New Zealand	1981	41 S, 174 E	W/DON	**0.13**	**2.2**		Timpereley et al. 1985
		1982				**1.0**		
	Japan		36 N, 138 E		**0.24**	**6.3**		
10.	Kampala, Uganda	≈1960	0 N, 32 E	W/TON	1.12:1.25:**2.73**	nd	22:25:**53**	Visser 1964
11.	Lake Valencia, Venezuala	1976–1978	10 N, 67 W	B/DON	nd	2.43:1.28:**1.33**	48:26:**26**	Lewis 1981
12.	Southern Chile	1987–1994	51 S, 71 W	Cloud/TON		1:1:**8****	10:10:**80**	Weathers et al. 1998
13.	Como Creek, CO	1975–1978	35 N, 105 W	B/DON	0.19:0.27:**0.18**	1.25:1.80:**1.22**	29:42:**29**	Grant and Lewis 1979

Table 4. Continued

Study	Site	Time	Latitude:Longitude	Meas Type	Concentration Ratios (mg/L)	Flux Ratios kg/ha/yr	% composition	Reference Number
14.	La Selva Biological Station, Costa Rica	1992– 1993– 1994–	10 N, 84 W	W/DON	0.12:0.07:**0.09**	3.6:2.4:**2.5** 2.6:1.8:**1.0** 5.0:6.6:**3.4**	42:28:**30** 48:33:**19** 33:44:**23**	Eklund et al. 1997
15.	UEA, United Kingdom	≈1994	54 N, 2 W	W/DON	0.97:**0.25**		80:**20**	Cornell et al. 1995
	Czech Republic		49 N, 15 E	Snow/DON	0.27:**0.1**		73:**27**	
	North Carolina		35 N, 79 W	W/DON	0.48:**0.13**		79:**21**	
	Maraba, Amazonia, Brazil		5 S, 49 W	W/DON	1.08:**0.31**		78:**22**	
	Recife, Brazil		8 S, 35 W	W/DON	0.13:**0.04**		75:**25**	
	Bermuda		32 N, 64 W	W/DON	0.15:**0.22**		41:**59**	
	Tahiti		17 S, 149 W	W/DON	0.04:**0.18**		16:**84**	
	NE Atlantic 50–80N		50–58 N, 20 W	W/DON	0.05:**0.11**		38:**62**	
	NE Atlantic 32–50N		32–50 N, 20 W	W/DON	0.04:**0.08**		33:**67**	
16.	Sweden, Goteborg	1977	57 N, 12 E	W/TON		**5.8:4.4:1.3*****	50:38:**11**	Malmquist 1978
17.	Harvard Forest, MA	1993– 1994	42 N, 72 W	W/DON		3.5:5.1:**0.6**	37:56:**7** 1996	Currie et al.
18.	SE Wyoming (pine forest)	1979– 1982	41 N, 105 W	B/TON	0.2:0.3:**0.27**		26:39:**35**	Fahey et al. 1985

Table 4. Continued

19.	Sierra Nevada, California	1987–1993	36 N, 118 W	W/DON			1.11:0.94:**1.33**	33:28:**39**	Chorover et al. 1994
20.	Manaus, Brazil	≈1980	3 S, 60 W	W/DON			1.8:0.4:**3.9**	30:6:**64**	Brinkman 1983
21.	Central Amazon Basis Lake Calado	1984–1985	3 S, 60 W	W/DON			2.4:1.3:**2.5**	39:20:**41**	Lesack and Melack 1996
22.	Southern North Sea	1988–1989	50–55 N, 0–5 E	W/DON	0.52:0.77:**0.09**			38:56:**6**	Rendell et al. 1993
23.	South Africa, Durban	1976–1977	30 S, 31 E	W/DON	0.37:0.34:**0.57**	4.22:3.95:**10.79**		22:21:**57**	Simpson and Hemens 1978
24.	Emerald Lake, Sequoia Nat. Park, CA, U.S.A.	1985–1998	36 N, 118 W	B/DON			1.0:1.6:**0.7**	30:48:**21**	Sickman et al. *this issue*
25.	HJ Andrews Forest, OR, U.S.A.	1977–1998		B W			0.33:0.46:**0.64** 0.41:0.69:**0.16**	23:32:**45** 32:55:**13**	Vanderbilt et al. *in review*

Table 4. Continued

Study	Site	Time	Latitude:Longitude	Meas Type	Concentration Ratios (mg/L)	Flux Ratios kg/ha/yr	% composition	Reference Number
26.	Hubbard Brook, NH, U.S.A.	1995–	42 N, 72 W	B/DON		2.6:5.5:**1.9**	26:55:**19**	Campbell et al.
	Cone Pone, NH, U.S.A.	1997	42 N, 72 W	B/DON		3.0:3.9:**1.4**	36:36:**16**	2000
	Sleepers River, VT, U.S.A.		44N, 72 W	B/DON		3.1:4.3:**1.8**	34:47:**19**	
27.	Morehead City, NC, U.S.A.	1995	29 N, 82 W	W/DON	**0.08**	1.96:**0.69**	74:**26**	Peierls and Paerl 1997
	Total Number Studies				**26**	**31**	**41**	
	Mean Value				**0.30 mg/L**	**3.1 kg N/ha/yr**	**34.0 %**	
	Median Value				**0.17 mg/L**	**2.2 kg N/ha/yr**	**30.0 %**	
	Standard Error				**0.10 mg/L**	**0.5 kg N/ha/yr**	**2.6 %**	
	Standard Deviation				**0.51 mg/L**	**2.8 kg N/ha/yr**	**16.7 %**	

*Extrapolated from the average flux of 60 daily measurements.

**Estimated from a cloudwater deposition estimate of 10 kg N/ha/yr of total N loading and an average 80% contribution of organic N to deposited N.

***Annual flux estimated from the arithmetic mean of 5 months of sampling.

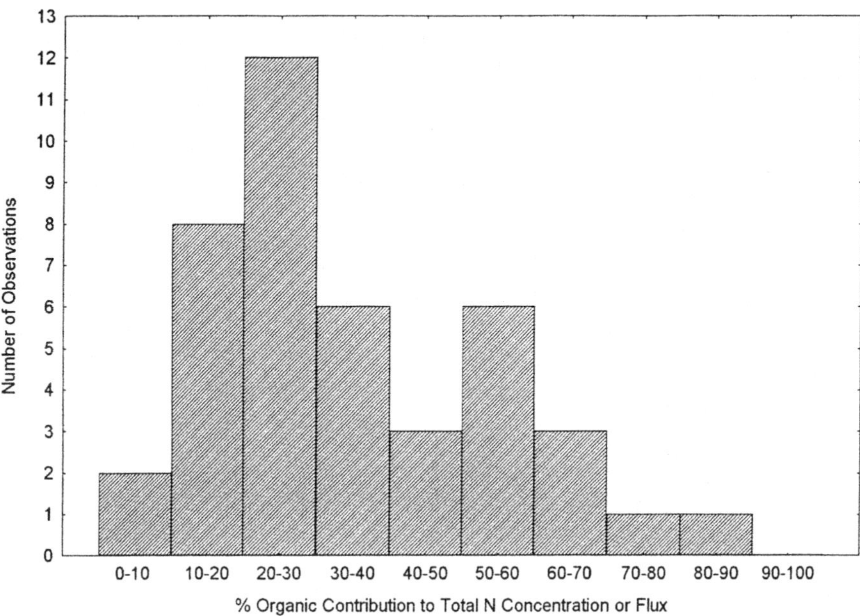

Figure 3. Histogram of the % organic contribution to deposition concentrations or fluxes from data listed in Table 4.

fraction of deposition can contribute between 7 and 80% of the total N in deposition. Although the range of the organic fraction in precipitation is quite large, 60% of the measurements in Table 4 have an organic fraction between 10% and 40%. Table 4 contains both TON and DON fractions and it would be reasonable to expect that unfiltered samples (TON) would contain a greater organic fraction than filtered samples (DON) due to the contribution of particulate organic N. On average the DON measurements in Table 4 have an organic fraction of 33% (plus/minus 19%) and TON measurements are 41% (plus/minus 18%) of total solution N. Although the mean contribution of TON to total N in deposition is higher than for DON, these two types of deposition samples are not different from one another statistically (t-test, t=1.24, p=0.21). The fractional distributions of percent organic N deposition are shown in Figure 3. Considering all the fractional organic contributions together, on average 34% of deposited N is in organic form globally with an average organic N concentration of 0.3 mg L^{-1} in deposition. Median, standard error, standard deviation values for the organic contributions to deposition are given in Table 4.

Fluxes of organic N in precipitation also show a large range of rates from 0.6 to 10.9 kg organic N ha^{-1} yr^{-1} with a median deposition rate of 2.2 kg organic N ha^{-1} yr^{-1} (Table 4). One significant and unanswered issue

124

in understanding AON deposition is whether AON deposition is impacted by human activity. The deposition of inorganic N constituents has increased substantially in most areas of the industrial world (Prospero et al. 1996; Vitousek et al. 1997). N deposition may lead to increased productivity and substantially increased rates of C sequestration under certain conditions and these calculations are relatively sensitive to the amount of N deposited to ecosystems (Townsend et al. 1996; Holland et al. 1997). In addition, high rates of inorganic N deposition fluxes can lead to the acidification of streams and ecosystems and over time, may lead to declines in terrestrial productivity (Agren and Bosatta 1988; Aber et al. 1991). Over oceans, the role of both organic and inorganic N deposition has been evaluated and both types of N deposition may play important roles in the regulation of primary productivity (Paerl et al. 1990; Paerl 1995). There are some indications that the deposited DON in precipitation is relatively labile and in this sense, it would behave similarly to inorganic N in its effects on the biotic activity of an ecosystem (Timperley et al. 1985; Seitzinger and Sanders 1999; Paerl et al. 1990; Paerl 1995; Herlihy et al. 1997). However given the diversity of potential AON compounds, some deposition of recalcitrant compounds unlikely to be rapidly broken down by microbial activity is possible. In this case, the effects of AON deposition on short-term biotic activity would be less significant. Further investigations are needed to evaluate these possibilities and whether there are differences in the decomposability of marine and continental AON.

Anthropogenic Influences

If the patterns and flux estimates shown in Tables 2 and 4 are correct, then organic compounds contribute substantially to both local and global N deposition rates. Assuming this is true, the question becomes whether these fluxes have a distinct geographic distribution or have exhibited changes over time. Both questions speak to the issue of human influences on AON fluxes. This issue is important because the role of AON in the global atmospheric N budget and ecosystem dynamics is closely related to whether AON fluxes are a stable background component of the N cycle or are increasing alongside the inorganic N species. If AON fluxes are impacted by human activity, then the changing deposition rates will exacerbate the impacts of N deposition on the dynamics of ecosystems and represent a human perturbation to the N cycle that has not received attention in recent evaluations of the global N cycle (c.f. Vitousek et al. 1997).

From a theoretical perspective, AON concentrations should increase, at least in the industrialized world, in response to increased NO_x concentrations due to combustion and agricultural activities, and increased amine emissions from agricultural systems. In addition, AON fluxes may reflect the elevated

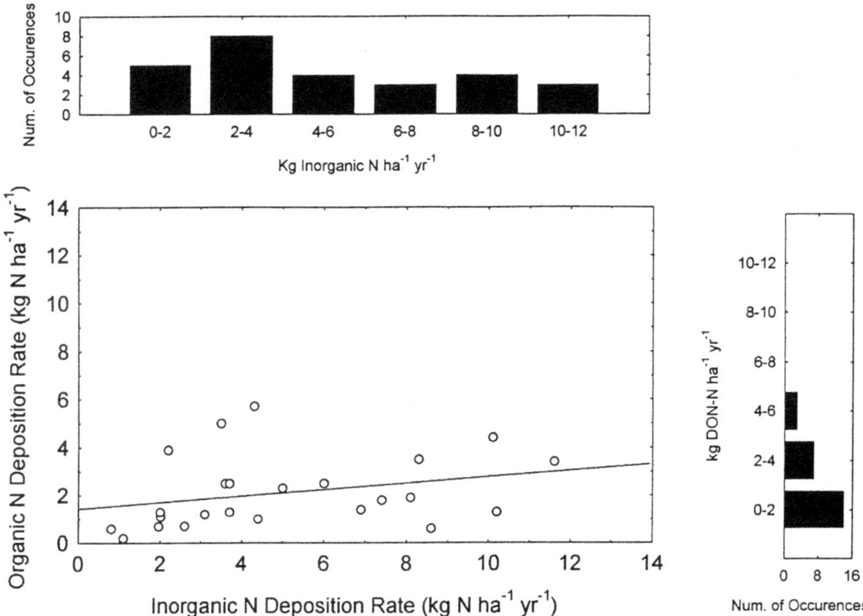

Figure 4. Inorganic vs. Organic N deposition regression and histograms based on the data in Table 4.

concentrations of atmospheric dust described above. For the organic nitrates, the work by Roberts (1990) and the data summarized in Table 2, indicate that organic nitrate concentrations are higher in urban vs. rural environments. Reduced forms of organic N should also be higher near agricultural areas. It is not as obvious, however, if these increased concentrations near areas of human activity will translate into higher concentrations of organic N in wet or bulk deposition.

To examine the relationship between inorganic and organic forms of N deposition, we evaluated the data listed in Table 4. Comparing AON and inorganic fluxes from Table 4, there is no correlation (at the $p < 0.05$ level) between DON flux and NH_4^+ flux (multiple regression analysis (MRA): $r^2 = 0.09$, $F = 2.1$, $P = 0.1$) and no correlation between DON flux and NO_3^- flux (MRA: $r^2 = 0.00$, $F = 0$, $P = 0.9$). In general, as the total deposition flux increases, the fractional contribution of DON to the total N flux decreases. This is shown in Figure 4 where DON fluxes generally remain below 4–6 kg N ha⁻¹ yr⁻¹ while inorganic N fluxes rise as high as 12 kg N ha⁻¹ yr⁻¹. If the sources of AON are related to sources of inorganic N and affected by human activities such as industrial activities, then inorganic N and AON should follow similar long-term trajectories. But with currently available

126

data, there is a discrepancy between patterns in AON fluxes in precipitation and the observed concentrations of gaseous and aerosol forms of organic N.

There is relatively little multi-year deposition data for AON. In a study at the Chesapeake Bay, inorganic N fluxes increased from 1979 to 1995 while TON showed a slight downward trend (Jordan et al. 1995). In a relatively unpolluted site in the Sierra Nevada of California, there were no suggestions of a temporal change in DON deposition rates between the mid 1980s and the present (Sickman et al. this issue). Similarly, long term (1983–1992) DON deposition rates at the H.J. Andrews Experimental Forest in Oregon have not increased over time, in contrast to NO_3^- deposition which shows increasing deposition rates in the same period (K. Vanderbilt, Pers. Com.). There also appear to be substantial interannual variations in DON rates in sites where multiple years of data are available and this year to year variation in organic N loads can be on the order of 200 to 300 percent for sites in Costa Rica and in the Eastern U.S. (Jordan et al. 1995; Eklund et al. 1997). These results suggest that AON fluxes have not increased during the past decade, at least not when considered as a bulk organic N in wet deposition. However, until there is more information on the flux rates and concentrations of both bulk AON and its underlying constituents, the long-term trajectory of organic N deposition and the impacts of humans on AON fluxes will remain unknown.

Spatial patterns

The modeled results shown in Figure 2, and the organic nitrate concentration data in Table 2 suggest that organic nitrates might be a likely candidates for the organic nitrogen commonly observed in precipitation. To evaluate whether modeled organic nitrates fluxes could account for the measured AON fluxes in Table 4, we carried out point comparisons of some of the more well documented studies in Table 4 against the results of organic N deposition from the TM3 model for the equivalent grid cell. While this is a very coarse comparison, it allows a rough assessment of the possibility that measured AON fluxes are made up primarily of oxidized AON compounds. For sites in the North-Eastern U.S., the TM3 model predicts organic nitrate loading rates of approximately 0.44 kg N ha−1 yr−1. In comparison, the fluxes for Cone Pond, NH, Sleepers River, VT, USA, Hubbard Brook and Harvard Forest DON deposition estimates are 0.6, 1.4, 1.8 and 1.9 kg N ha^{-1} yr^{-1} respectively (Campbell et al. 2000; Currie et al. 1996). For the La Selva research station in Costa Rica, the TM3 model prediction is 0.9 kg N ha^{-1}yr^{-1} and the measured deposition values range between 1 and 3.4 kg DON-N ha^{-1} yr^{-1} (Eklund et al. 1997). For the Sierra Nevada in California, TM3 predicted oxidized AON fluxes are 0.05 kg DON-N ha^{-1} yr^{-1} compared to measured deposition fluxes of 0.7 to 1.33 kg DON-N ha^{-1}

yr^{-1} (Chorover et al. 1994; Sickman et al. this issue). In the Brazilian and Venezuelan studies listed in Table 4, measured DON fluxes ranged between 1.3 for the Venezuelan site (Lewis, 1982) and 2.5 – 3.9 kg AON-N ha^{-1} yr^{-1} for the Brazilian Amazon (Brinkman 1983; Lesack and Melack 1993) compared to model AON deposition estimates of 0.08 kg AON ha^{-1} yr^{-1} for Venezuela and 1.1 kg AON-N ha^{-1} yr^{-1} for the Brazilian Amazon. The large differences in modeled deposition fluxes between Brazil and Venezuela and between the western and eastern U.S. result from the transport patterns and source field distributions in the model. Because the Venezuelan and western U.S. sites primarily receive marine air (at the grid scale of the model), they are far from the source fields for NO$_x$ and hydrocarbons that lead to larger deposition of organic nitrates in regions affected by either natural (Amazon) or industrial (eastern U.S.) emissions.

Based on the patterns in oxidized organic N as well as modeled emissions scenarios, we would expect to see the largest PAN deposition rates in continental environments near to large NO$_x$/hydrocarbon source areas. In contrast, other organic nitrates would be likely to dominate more remote continental sites, particularly in tropical regions. AON fluxes in marine air masses should be smaller than in continental environments due to distance from organic nitrate source fields. Over marine environments, it seems likely that the bulk of the organic species in precipitation would be in amine N form as the result of N fluxes associated with aerosol formation at the ocean/air interface. Particulate deposition could contribute substantially to certain types of measured fluxes but should not be a large factor in the wet only DON measurements in Table 4.

Regardless of the composition of AON, and assuming that the organic nitrate deposition estimates from the TM3 model are an upper bound on the possible flux, then there remains a missing organic N contribution to most measurements of organic N in precipitation or bulk deposition samples. There are several possibilities for what could lead to this missing contribution. First, the measurements may be in error. However, although the problems with AON measurements are significant, it is difficult to accept that out of 41 studies listed in Table 4, which span a range of both analytical and measurement approaches, each of the studies is systematically biased. On the contrary, the fractional contribution of AON to the total N load is surprisingly consistent across a range of sites and over many years. A second possibility is that there is a gap in our understanding of the potential sources of AON that leads to an underestimate of the role of reduced N species in the AON flux, particularly for more unpolluted or marine influenced areas. A third possibility is that there are other sources and types of organic compounds in the atmosphere that are simply not accounted for in existing studies. At

128

this point, we cannot rule out either the second or third possibilities described here.

Organic C:N ratios:

The ratio of C:N in AON may help in identifying sources of AON. Of the various potential contributors to the atmospheric organic flux, many of the compounds have somewhat different C:N ratios. For example, bacterial C:N ratios generally vary between 5 and 10 while soils vary between 15 and 30 (Holland and Coleman 1987; Paul and Clark 1996). Amino acid C:N ratios vary from 3 (Glycine, Alanine & Valine) to higher numbers for aliphatic amino acids, which can have larger C:N ratios of 5–16 (Beyer and Walter 1997). For the organic nitrates such as nitric acid esters and hydroxy nitric acid esters C:N ratios vary between 2 and 12 with most compounds within the 2–6 range. PAN has a C:N ratio of 2 and the other peroxycarboxylic nitric anhydrides vary between 7 and 11 (Roberts 1990). Organic C:N ratios for several studies are shown in Table 5 and average 12.3. There is a large range in AON C:N ratios from 8.4 to 21.9 but the most notable feature of these values is that they are clearly higher than C:N ratios of free amino acids or PAN. These measurements are also higher than the majority of the C:N ratios of organic nitrates and higher than many of the aliphatic amino acid compounds. While AON deposition certainly contains a mix of compounds and could indeed contain some contribution of organic nitrates, PAN or amino acids, these data also suggest that there is a contribution of higher C:N ratio material such as that observed in soils or plant material or some other, as yet unidentified, contributor to the AON flux.

Conclusions

This data compilation suggests that fluxes of organic nitrogen in deposition may be substantial contributors to the total atmospheric N flux. There are high concentrations and fluxes of AON for a variety of measurements including bulk, wet and event collections and for both filtered and unfiltered samples. There are no clear indications that AON fluxes are related to inorganic N deposition rates. Nor do we find any obvious relationship between AON fluxes and the relative unpolluted or polluted, marine or continental nature of the deposition sites. On average AON fluxes or concentrations contribute one third of the total N in precipitation but the available data are skewed toward relatively clean regions of the atmosphere.

There are several possible sources of AON including organic nitrates, reduced organic N, dust and biological AON. Of these sources, only the

Table 5. Organic Carbon Fluxes and DOM C:N ratios

Site	Time	DOC Flux kg ha^{-2} yr-1	Organic C:N ratio	Reference
Gainesville, Florida, U.S.A.	1976–77	63.2	12.6	Hendry and Brezonik 1980
	1976–77	115.3	20.2	
Lake Valencia, Venezuela	1976–1977	13.4	10.1	Lewis 1981
Como Creek, CO, U.S.A.	6/75 – 10/78	10.29	8.4	Grant and Lewis 1981
Southeast WY, U.S.A.	1979–1982		10	Fahey et al. 1985
Rhode River, Chesapeake Bay, U.S.A.	1973–1991	39.9	11.4	Jordan et al. 1995
Harvard Forest, MA, U.S.A.	1993–1994	13.8	21.6	Currie et al. 1996
La Selva Research Station,	92–93	27.9	11.2	Eklund et al. 1997
Costa Rica	93–94	21.9	21.9	
	94–95	35.6	5.4	
Hubbard Brook, NH	1995–	15.5	8.1	Campbell et al. 2000
Cone Pond, NH	1997	14.1	10	
Sleepers River, VT	17.2	9.5		
Mean DOM C:N Ratio			**12.3**	
Standard Deviation			**5.4**	

organic nitrates species, including PAN, and dust fluxes appear to have sufficient atmospheric concentrations or fluxes to account for a large portion of the observed AON flux at most sites. Amine forms of N are likely important to AON over marine environments but are unlikely to have sufficient concentrations to account for the deposition of AON, even in sites influenced by marine air. Our preliminary comparisons of modeled (organic nitrate and PAN) and measured AON deposition rates suggest that there are sources of AON that are not accounted for in current measurements of any of the AON classes described above. The C:N ratios of atmospheric organic deposition suggest that there is some contribution of material with C:N ratios greater than 10 to deposition fluxes which cannot be explained by either amine N or organic nitrate N. Possible sources of these higher C:N ratio material could include dust for bulk, unfiltered AON samples and other soil or plant based materials however, there is no data with which to evaluate this possibility.

The role of AON in the global N cycle and the atmospheric N budget appears to be significant. Preliminary model based estimates of the organic nitrate and PAN contributions to N deposition suggest that approximately 10 Tg of AON may be deposited annually. Model data comparisons and our data compilation suggest that this estimate may represent the low end of the total AON flux. Given contemporary reactive atmospheric N sources of 48–164 Tg of N ($NH_3 + NO_x$) and our data based median fractional AON contribution to deposition of 30%, we suggest that global AON fluxes could be as high as 14–50 Tg. While the results of this study indicate that the global AON fluxes are almost certainly higher than traditionally assumed, substantial work is needed before these flux estimates can be taken for anything other than a first order approximation. However, they highlight the need for increased attention to measurements and evaluations of organic N deposition.

Acknowledgements

This work was initiated as part of the International SCOPE N Project, which received support from both the Mellon Foundation and from the National Center for Ecological Analysis and Synthesis. The work was supported by the Mellon Foundation, the Max Planck Institute for Biogeochemistry and The National Center for Atmospheric Research (NCAR). NCAR is operated by the University Corporation for Atmospheric Research under the sponsorship of the National Science Foundation. T. Zbacnik was instrumental in the review of literature & J. Sultzman aided in model analysis. J. Baron and two anonymous reviewers provided helpful comments on an earlier version of this manuscript.

References

Aber JD, Melillo JM, Nadelhoffer KJ, Pastor J & Boone R (1991) Factors controlling nitrogen cycling and nitrogen saturation in northern temperate forest ecosystems. Ecological Applications 1: 303–315

Agren GI & Bosatta R (1988) Nitrogen Saturation of the terrestrial ecosystem. Environmental Pollution 54: 185–197

Anlauf KG, Bottenheim JW, Brice KA & Wiebe HA (1986) A comparison of summer and winter measurements of atmospheric nitrogen and sulfur compounds. Water Air Soil Pollution 30: 153–160

Atlas E (1988) Evidence for C3 alkyl nitrates in rural and remote atmospheres. Nature 331: 426–428

Atlas E & Schauffler S (1991) Analysis of Alkyl Nitrates and Selected Halocarbons in the Ambient Atmosphere using Charcoal Preconcentration Technique. Environmental Science and Technology 25: 61–67

Beine HJ, Jaffe DA, Blake DR, Atlas E & Harris J (1996) Measurements of PAN, alkyl nitrates, ozone, and hydrocarbons during spring in the interior Alaska. Journal of Geophysical Research 101: 12, 613–12, 619

Bertman SB, Roberts JM, Parrish DD, Buhr MP, Goldan PD, Kuster WC & Fehsenfeld FC (1995) Evolution of alkyl nitrates with air mass age. Journal of Geophysical Research 100: 22, 805–22, 813

Beyer H & Walter W (1997) Organic Chemistry. Albion Publishing, Chichester, UK (p 1037)

Bottenheim JW, Brice KA & Anlauf K (1984) Discussion of a Lagrangian trajectory model describing long-range transport of oxides of nitrogen, the incorporation of PAN in the chemical mechanism and supporting measurements of PAN and nitrate species at rural sites in Ontario, Canada. Atmospheric Environment 18: 2609–2619

Bovallius A, Bucht B, Roffey R & Anas P (1978) Three-year investigation of the natural airborne bacterial flora at four localities in Sweden. Applied and Environmental Microbiology 35: 847–852

Brasseur GP, Haugulstaine DA, Walters S, Rasch PJ, Mueller JF, Granier C & Tie XX (1998) MOZART, a global transport model for ozone and related chemical tracers. 1. Model description. Journal of Geophysical Research 103: 28265–28289

Brasseur GP, Orlando JJ & Tyndall GS (1999) Atmospheric Chemistry and Global Change (Topics in Environmental Chemistry). Oxford University Press, Oxford, UK (p 654)

Brinkmann WLF (1983) Nutrient balance of a central Amazonian rain forest: comparison of natural and man-managed systems. International Association of Hydrological Sciences Publication 140: 153–163

Buhr MP, Parrish DD, Norton RB, Fehsenfeld FC, Sievers RE (1990) Contribution of Organic Nitrates to the Total Reactive Nitrogen Budget at Rural Eastern U.S. Site. Journal of Geophysical Research 95: 9809–9816

Caiazza, R, Hage KD & Gallup D (1978) Wet and Dry Deposition of Nutrients in Central Alberta Water, Air and Soil Pollution 9: 309–314

Campbell JL, Hornbeck JW, McDowell JW, Buso DC, Shanley JB & Likens GE (2000) Dissolved organic nitrogen budgets for upland, forested ecosystems in New England. Biogeochemistry 42: 123–142

Chorover J, Vitousek PM, Everson DA, Esperanza AM & Turner D (1994) Solution chemistry profiles of mixed-conifer forests before and after fire. Biogeochemistry 26: 115–144

Church TM (1999) Atmospheric organic nitrogen deposition explored at workshop. EOS Transactions, American Geophysical Union 80: 355, 360

Cornell S, Rendell A & Jickells T (1995) Atmospheric inputs of dissolved organic nitrogen to the oceans. Nature 76: 243–246

Currie WS, Aber JD, McDowell WH, Boone RD & Magill AH (1996) Vertical transport of dissolved organic C and N under long-term N amendments in pine and hardwood forests. Biogeochemistry 35: 471–505

Davidson EA & Kingerlee W (1997) Global inventory of nitric oxide emissions from soils. Nutrient Cycling in Agroecosystems 48: 37–50

Daum PH, Kelly TJ, Tanner RL, Tang X, Anlauf K, Bottenheim J, Brice KA & Wiebe HA (1989) Winter measurements of trace gas and aerosol composition at a rural site in southern Ontario. Atmospheric Environment 23: 161–173

DeKock AC & Anderson CK (1994) The Measurement of C3–C5 Alkyl Nitrates at a Coastal Sampling Site in the Southern Hemisphere. Chemosphere 29: 299–310

Dentener FJ & Crutzen PJ (1994) A global 3D model of the ammonia cycle. Journal of Atmospheric Chemistry 19: 331–369

Dentener FJ, Feichter J & Jeuken A (1999) Simulation of the transport of Radon222 using on-line and off-line global models at different horizontal resolutions: a detailed comparison with measurements. Tellus 51B: 573–602

Eklund TJ, McDowell WH & Pringle CM (1997) Seasonal variation of tropical precipitation chemistry: La Selva, Costa Rica. Atmospheric Environment 31: 3903–3910

Fahey DW, Hubler G, Parrish DD, Williams EJ, Norton RB, Ridley BA, Singh HB, Liu SC & Fehsenfeld FC (1986) Reactive nitrogen species in the troposphere: measurements of NO, NO_2, HNO_3, particulate nitrate, peroxyacetyl nitrate (PAN), O_3 and total reactive odd nitrogen (NO_y) at Niwot Ridge Colorado. Journal of Geophysical Research 91: 9781–9793

Fahey TJ, Yavitt JB, Pearson JA & Knight DH (1985) The nitrogen cycle in lodgepole pine forests, Southeastern Wyoming. Biogeochemistry 1: 257–275

Flocke F, Volz-Thomas A & Kley D (1994) Measurements of Alkyl Nitrates in Rural and Polluted Air Masses. Atmospheric Environment 25A: 1951–1960

Fonselius S (1954) Amino Acids in Rainwater. Tellus 6: 90

Ganzeveld L & Lelieveld J (1995) Dry deposition parameterization in a chemistry general circulation model and its influence on the distribution of reactive trace gases. Journal of Geophysical Research 100: 20999–21012

Gorzelska K, Galloway JN, Watterson K & Keene WC (1992) Water-Soluble Primary Amine Compounds in Rural Continental Precipitation. Atmospheric Environment 26A: 1005–1018

Gorzelska D, Scudlark JR & Keene WC (1997) Dissolved organic nitrogen in the atmospheric environment. In Atmospheric Deposition of Contaminants to the Great Lakes and Coastal Waters (Baker (Ed)). SETAC Press, Pensacola, FL, U.S.A. (p 477)

Gorzelska K, Talbot RW, Klemm C, Lefer B, Klemm O, Gregory GL, Anderson B & Barrie LA (1994) Chemical composition of the atmospheric aerosol in the troposphere over the Hudson Bay lowlands and Quebec-Labrador regions of Canada. Journal of Geophysical Research 99: 1763–1779

Gorzelska K & Galloway JN (1990) Amine Nitrogen in the Atmospheric Environment Over the North Atlantic Ocean. Global Biogeochemical Cycles 4: 309–333

Grant MC & Lewis WM (1982) Chemical loading rates from precipitation in the Colorado Rockies. Tellus 34: 74–88

Guelle W, Balnkansi YJ, Dibb JE, Schulz M & Dulac F (1988) Wet deposition in a global size dependent aerosol transport model. 2. Influence of the scavenging scheme on PB-210 vertical profiles, surface concentrations and deposition. Journal of Geophysical Research 103: 28, 875–28891

Guenther A, Hewitt CN, Erickson D, Fall R, Geron C, Graedel T, Harley P, Klinger L, Lerdau M, McKay WA, Pierce T, Scholes B, Steinbrecher R, Tallamraju R, Taylor J & Zimmerman P (1995) A global model of natural volatile organic compound emissions. Journal of Geophysical Research 100: 8873–8892

Haugulstaine DA, Brasseur GP, Walters S, Rasch PJ, Mueller JF, Emmons LK & Carrol MA (1998) MOZART, a global chemical transport model for ozone and related chemical tracers. 1. Model results and evaluation. Journal of Geophysical Research 103: 28291–28335

Hendry CD & Brezonik PL (1980) Chemistry of precipitation at Gainesville, Florida. Environmental Science and Technology 14: 843–849

Herlihy LJ, Galloway JN & Mills AL (1987) Bacterial Utilization of Formic and Acetic Acid in Rainwater. Atmospheric Environment 21: 2397–2402

Holland EA & Coleman DC (1987) Litter placement effects on microbial and organic matter dynamics in an agroecosystem. Ecology 68: 425–433

Holland EA, Dentener FJ, Braswell BH & Sulzman JM (1999) Contemporary and pre-industrial global reactive nitrogen budgets. Biogeochemistry 46: 7–43

Holland EA, Braswell BH, Lamarque JF, Townsend AR, Sultzman JM, Muller JF, Dentener F, Brasseur G, Levy HI, Penner JE & Roelofs G (1997) Variations in the predicted spatial distribution of atmospheric nitrogen deposition and their impact on carbon uptake by terrestrial ecosystems. Journal of Geophysical Research 102: 15, 849–15, 866

Houweling S, Dentener F & Lelieveld J (1998) The impact of nonmethane hydrocarbon compounds on tropospheric photochemistry. Journal of Geophysical Research 103: 10, 673–10, 696

Hutchinson GL, Mosier AR & Andre CE (1983) Ammonia and amine emissions from a large cattle feedlot. Journal of Environmental Quality 11: 288–293

Jones BL & Cookson JT (1983) Natural atmospheric microbial conditions in a typical suburban area. Applied and Environmental Microbiology 45: 919–934

Jordan TE, Correll DL, Weller DE & Goff NM (1995) Temporal Variation in Precipitation Chemistry on the Shore of the Chesapeake Bay. Water, Air, and Soil Pollution 83: 263–284

Keller M & Lerdau M (1999) Isoprene emission from tropical forest canopy leaves. Global Biogeochemical Cycles 13: 19–31

Lelieveld J & Dentener F (in press). What controls tropospheric ozone? Journal of Geophysical Research

Lesack LF (1993) Export of nutrients and major ionic solutes from a rain forest catchment in the Central Amazon Basin. Water Resources Research 29: 743–758.

Lewis WM Jr. (1981) Precipitation chemistry and nutrient loading in precipitation in a tropical watershed. Water Resources Research 17: 169–181

Lindemann J, Constantinidou HA, Barchet WR & Upper CD (1982) Plants as Sources of Airborne Bacteria, Including Ice Nucleation-Active Bacteria. Applied and Environmental Microbiology 44: 1059–1063

Littmann T (1997) Atmospheric input of dust and nitrogen int the Nizzana sand dune ecosystem, north-western Negev, Isreal. Journal of Arid Environments 36: 433–457

Lobert JM, Scharffe DH, Hao WM, Kuhlbusch TA, Warneck P & Crutzen PJ (1991) Experimental evaluation of biomass burning emissions: Nitrogen and carbon containing compounds. In: J.S. Levine (Editor), Global Biomass Burning: Atmospheric, Climatic and Biospheric Implications. MIT Press, Cambridge, MA

Malmquist PA (1978) Atmospheric fallout and street cleaning – effects on urban storm water and snow. Progress in Water Technology 10: 495–505

Milne PJ & Zika RG (1993) Amino Acid Nitrogen in Atmospheric Aerosols: Occurrence, Sources and Photochemical Modification. Journal of Atmospheric Chemistry 16: 361–398

Mopper K & Zika RG (1987) Free amino acids in marine rains: evidence for oxidation and potential role in nitrogen cycling. Nature 325: 246–249

Mosier AR, Andre CE & Viets FG Jr. (1973) Identification of aliphatic amines volatilized from cattle feedyard. Environmental Science and Technology 7: 642–644

134

Moxim WJ, Levy II H & Kasibhatla PS (1996) Simulated global tropospheric PAN: Its transport and impact on NO_x. Journal of Geophysical Research 101: 12621–12638

Munger J, Fan SM, Bakwin PS, Goulden ML, Goldstein AH, Coleman AS & Wofsy SC (1988) Regional budgets for nitrogen oxides from continental sources: Variations of rates for oxidation and deposition with season and distance from source regions. Journal of Geophysical Research 103: 8355–8368

Muthuramu K, Shepson PB, Bottenheim JW, Jobson BT, Niki H & Anlauf KG (1994) Relationships between organic nitrates and surface ozone destruction during Polar Sunrise Experiment 1992. Journal of Geophysical Research 99: 25, 369–25, 378

Nicholls KH & Cox CM (1978) Atmospheric nitrogen and phosphorus loading to Harp Lake, Ontario, Canada. Water Resources Research 14: 589–592

Nouaime G, Bertman SB, Seaver C, Elyea E, Huang H, Shepson PB, Starn TK, Riemer DD, Zika RG & Olszyna K (1998) Sequential oxidation products from tropospheric isoprene chemistry: MACR and MPAN at a NO_2^- rich forest environment in the southeastern United States. Journal of Geophysical Research 103: 22, 463–22, 471

O'brien JM, Shepson PB, Wu Q, Biesenthal T, Bottenheim JW, Wiebe HA, Anlauf KG & Brickell P (1995) Production and Distribution of Organic Nitrates, and their Relationships to Carbonyl Compounds in an Urban Environment. Atmospheric Environment 31: 2059–2069

Paerl HW (1995) Coastal eutrophication in relation to atmospheric nitrogen deposition: current perspectives. Ophelia 41: 237–259

Paerl HW, Rudek J & Mallin MA (1990) Stimulation of phytoplankton production in coastal waters by natural rainfall inputs: nutritional and trophic implications. Marine Biology 107: 247–254

Paul EA & Clark FE (1996) Soil microbiology and biochemistry. Academic Press, San Diego (p 340)

Peierls BJ & Paerl HW (1997) Bioavailibility of atmospheric organic nitrogen deposition to coastal phytoplankton. Limnology and Oceanography 42: 1819–1823

Pierson WR, Brachaczek WW, Gorse RA, Japar SM, Norbeck JM & Keeler GJ (1987) Acid rain and atmospheric chemistry at Allegheny Mountain. Environmental Science and Technology 21: 679–691

Prospero JM, Barrett K, Church T, Dentener F, Duce RA, Galloway JN, Levy II H, Moody J & Quinn P (1996) Atmospheric deposition of nutrients to the North Atlantic Basin. Biogeochemistry 35: 27–75

Rabenhorst MC, Wilding LP & Gardner CL (1984) Airborn dusts in the Edwards Plateau. Soil Science Society of America Journal 48: 621–627

Ramsperger B, Peinemann N & Stahr K (1998a) Deposition rates and characteristics of aeolian dust in the semi-arid and sub-humid regions of the Argentinean Pampas. Journal of Arid Environments 39: 467–476

Ramsperger B, Herrmann L & Stahr K (1998b) Dust characteristics and source-sink relations in eastern West Africa (SW-Niger and Benin) and South America (Argentinean Pampas). Z. Pflanzenernahr. Bodenk. 161: 357–363

Rendell AR, Ottley CJ, Jickells TD & Harrison RM (1993) The atmospheric input of nitrogen species to the North Sea. Tellus 45B: 53–63

Roberts JM (1990) The atmospheric chemistry of organic nitrates. Atmospheric Environment 24A: 243–287

Roberts JM, Williams J, Baumann K, Buhr MP, Goldan PD, Holloway J, Hubler G, Kuster WC, McKeen SA, Ryerson TB, Trainer M, Williams EJ, Fehsenfeld FC, Bertman SB, Nouaime G, Seaver C, Grodzindsky G, Rodgers M & Young VL (1998) Measurements of PAN, PPN, and MPAN made during the 1994 and 1995 Nashville Intensives of the Southern Oxidant Study: Implications of regional ozone production from biogenic hydrocarbons. Journal of Geophysical Research 103: 22, 473–22, 490

Russell, KM, Galloway JN, Macko SA, Moody JL & Scudlark JR (1998) Sources of Nitrogen in Wet Deposition to the Chesapeake Bay Region. Atmospheric Environment 32: 2453–2465

Russell KM (1999) A review of the current state of knowledge of atmospheric organic nitrogen. The Chemist 76: 12–17

Schade GW & Crutzen PJ (1995) Emission of aliphatic amines from animal husbandry and their reactions: potential source of N_2O and HCN. Journal of Atmospheric Chemistry 22: 319–346

Scudlark JR, Russell KM, Galloway JN, Church TM & Keene WC (1998) Organic nitrogen in precipitation at the Mid-Atlantic U.S. Coast-methods evaluation and preliminary measurements. Atmospheric Environment 32: 1719–1728

Seitzinger SP & Sanders RW (1999) Atmospheric inputs of dissolved organic nitrogen stimulate estuarine bacteria and phytoplankton. Limnology and Oceanography 44: 721–370

Sharkey TD, Singsaas EL, Vanderveer PJ & Geron C (1996) Field measurements of isoprene emission from trees in response to temperature and light. Tree Physiology 16: 649–654

Sickman JO, Melack JM & Leydecker A (2002) Regional analysis of inorganic nitrogen yield and retention in high-elevation ecosystems of the Sierra Nevada and Rocky Mountains. Biogeochemistry 57/58: 341–374

Sidle AB (1967) Amino acid content of atmospheric precipitation. Tellus 19: 128–135

Simpson DE & Hemens J (1978) Nutrient budget for a residential stormwater catchment in Durban, South Africa. Progress in Water Technology 10: 631–643

Smith RA (1872) Air and rain. The beginnings of a chemical climatology. Longmans, Green and Co, 600 pp. London

Spitzy A (1990) Facts of Modern Biogeochemistry. Amino Acids in Marine Aerosol and Rain (pp 313–317). Springer-Verlag

Strand A & Hov O (1994) A two-dimensional global study of tropospheric ozone production. Journal of Geophysical Research 99: 22, 877–22, 895

Swap R, Garstang M, Greco S, Talbot R & Kallbert P (1992) Saharan dust in the Amazon Basin. Tellus 44B: 133–149

Tegen I & Fung I (1995) Contribution to the atmospheric mineral aerosol load from land surface modification. Journal of Geophysical Research 100: 18707–18726

Timperley MH, Vigor-Brown RJ, Kawashima M & Ishigami M (1985) Organic Nitrogen Compounds in Atmospheric Precipitation: Their Chemistry and Availability to Phytoplankton. Canadian Journal of Fishery and Aquatic Science 42: 1171–1177

Townsend AR, Braswell BH, Holland EA & Penner JE (1996) Spatial and temporal patterns in potential terrestrial carbon storage resulting from deposition of fossil fuel derived nitrogen. Ecological Applications 6: 806–814

Vanderbilt KL & Lajtha K (in review) Annual and seasonal patterns of nitrogen dynamics at the H.J. Andrews Experimental Forest, Oregon. Biogeochemistry

Verry ES & Timmons DR (1977) Precipitation nutrients in the open and under two forests in Minnesota. Canadian Journal of Forest Resources 7: 112–119

Visser SA (1964) Origin of Nitrates in Tropical Rainwater. Nature 201: 35–36

Vitousek PM, Aber JD & Tilman DG (1997) Human alteration of the global nitrogen cycle: sources and consequences. Ecological Applications 7: 737–752

Walsh TW (1989) Total dissolved nitrogen in seawater: a new high-temperature combustion method and comparison with photo-oxidation. Marine Chemistry 26: 295–311

Weathers KC, Lovett GM, Likens GE & Caraco NFM (1998) Cloud Water in Southern Chile: Whence Come the Nutrients? In Schemenauer RS and J Bridgman (Eds), First International Conference on Fog and Fog Collection (pp 313–315). IDRC, Ottawa, Canada
Williams MR & Melack JM (1997) Solute export from forested and partially deforested catchments in the central Amazon. Biogeochemistry 38: 67–102
Wilson AT (1959) Organic Nitrogen in New Zealand Snows. Nature 183: 318–319
Yienger JJ & Levy II H (1995) Empirical model of global soil-biogenic NO_x emissions. Journal of Geophysical Research 100: 11, 447–11, 464
Zimmerman PH (1987) A handy global tracer model, Sixteenth NATO/CCMS International Technical Meeting on Air Pollution Modeling and Its Application, New York

Biogeochemistry **57/58:** 137–169, 2002.

Anthropogenic nitrogen sources and relationships to riverine nitrogen export in the northeastern U.S.A.

ELIZABETH W. BOYER[1], CHRISTINE L. GOODALE[2], NORBERT A. JAWORSKI[3] & ROBERT W. HOWARTH[4]
[1] *State University of New York, College of Environmental Science and Forestry;* [2] *Carnegie Institution of Washington, Department of Plant Biology;* [3] *U.S. Environmental Protection Agency (retired);* [4] *Cornell University, Department of Ecology and Evolutionary Biology*

Key words: anthropogenic, atmospheric deposition, eutrophication, food production, nitrogen, nitrogen budget, nitrogen fixation, rivers

Abstract. Human activities have greatly altered the nitrogen (N) cycle, accelerating the rate of N fixation in landscapes and delivery of N to water bodies. To examine relationships between anthropogenic N inputs and riverine N export, we constructed budgets describing N inputs and losses for 16 catchments, which encompass a range of climatic variability and are major drainages to the coast of the North Atlantic Ocean along a latitudinal profile from Maine to Virginia. Using data from the early 1990's, we quantified inputs of N to each catchment from atmospheric deposition, application of nitrogenous fertilizers, biological nitrogen fixation, and import of N in agricultural products (food and feed). We compared these inputs with N losses from the system in riverine export.

The importance of the relative sources varies widely by catchment and is related to land use. Net atmospheric deposition was the largest N source (>60%) to the forested basins of northern New England (e.g. Penobscot and Kennebec); net import of N in food was the largest source of N to the more populated regions of southern New England (e.g. Charles & Blackstone); and agricultural inputs were the dominant N sources in the Mid-Atlantic region (e.g. Schuylkill & Potomac). Over the combined area of the catchments, net atmospheric deposition was the largest single source input (31%), followed by net imports of N in food and feed (25%), fixation in agricultural lands (24%), fertilizer use (15%), and fixation in forests (5%). The combined effect of fertilizer use, fixation in crop lands, and animal feed imports makes agriculture the largest overall source of N. Riverine export of N is well correlated with N inputs, but it accounts for only a fraction (25%) of the total N inputs. This work provides an understanding of the sources of N in landscapes, and highlights how human activities impact N cycling in the northeast region.

Introduction

Human activities have greatly altered the nitrogen (N) cycle, accelerating the rate of N fixation in landscapes and delivery of N to water bodies (Galloway et al. 1995; Howarth et al. 1996; Smil 1997; Vitousek et al. 1997; Caraco

138

& Cole 1999). In most estuaries, over-enrichment of N leads to eutrophication, presently the greatest pollution problem in coastal marine waters of the United States (NRC 2000). Over 40% of the estuaries in the U.S. are degraded from eutrophication, with particularly severe problems in the New England and mid-Atlantic regions (Bricker 1999). Nitrogen loadings in major U.S. rivers have increased during recent decades (e.g. Stoddard 1991; Turner & Rabalais 1991; Puckett et al. 1995; Jaworski et al. 1997). Most N delivered to coastal waters in the U.S. comes from non-point sources in the landscape, with agricultural sources and atmospheric deposition being major contributors (Howarth et al. 1996; Smith et al. 1997; Goolsby et al. 1999; Castro et al. 2000). Understanding the sources of N loadings is essential to developing nutrient management strategies.

To examine relationships between N inputs and riverine N export, we established N budgets for 16 catchments in the northeast (NE) U.S.A. These basins encompass a range of climatic variability and are major drainages to the coast of the North Atlantic Ocean along a latitudinal profile from Maine to Virginia. Nitrogen budgets were established by quantifying all new inputs of N to each catchment, where 'new' refers to N that is either newly fixed within or transported into each catchment. Budget terms included inputs of N from atmospheric deposition, fertilizer use, net imports in food and feed, and biological fixation in agricultural areas and in forests. The total net inputs were compared with N losses from the system in riverine export.

Our N budgets allow us to assess the importance of N sources, highlighting how human activities have impacted N cycling in the NE region. The relative importance of the input terms varied widely by catchment and is related to land use. Over the combined area of the catchments, net atmospheric deposition was the largest single source input (31%), followed by imports of N in food and feed (25%), fixation in agricultural lands (24%), fertilizer use (15%), and fixation in forests (5%). Riverine export of N is well correlated with N inputs, but represents only a fraction (25%) of the total N inputs, with inputs exceeding outputs. This implies that large percentages of the N inputs are stored (e.g. in vegetation, soil, or groundwater) or lost (e.g. denitrified) in the catchment.

Methods

Study area

We selected sixteen river basins (Figure 1) draining to the NE coast of the U.S.A. The catchments include the Penobscot, Kennebec, Androscoggin, and Saco Rivers flowing into the Gulf of Maine; the Merrimack and Charles

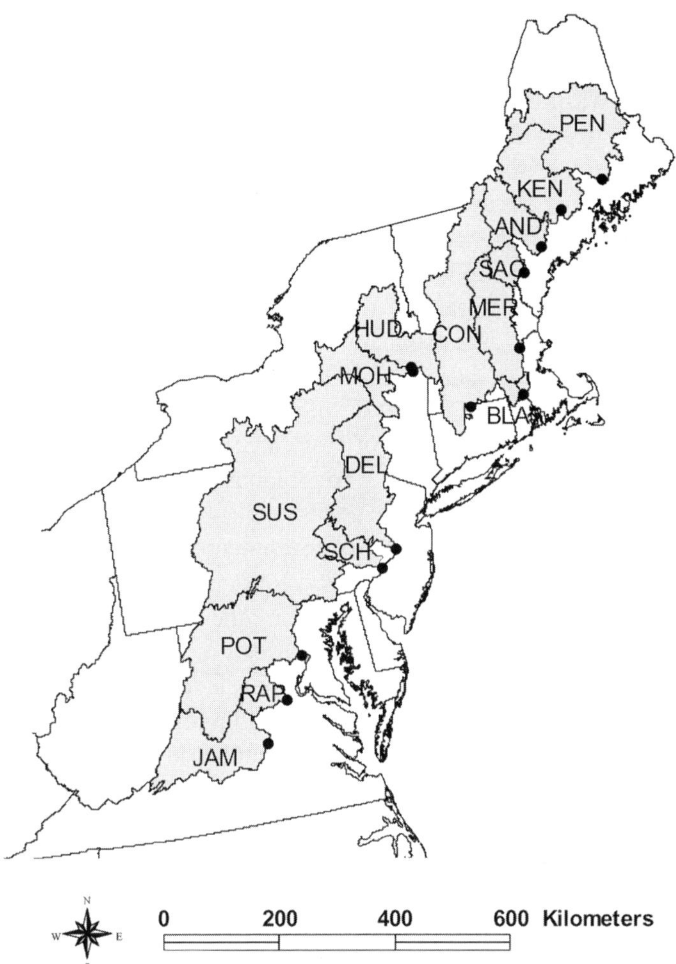

Figure 1. The basin boundaries are delineated upstream of USGS stations (denoted with black circles) where streamflow and water quality characteristics were measured. From north to south, the catchments include: Penobscot (PEN), Kennebec (KEN), Androscoggin (AND), Saco (SAC), Merrimack (MER), Charles (CHA), Blackstone (BLA), Connecticut (CON), Hudson (HUD), Mohawk (MOH), Delaware (DEL), Schuylkill (SCH), Susquehanna (SUS), Potomac (POT), Rappahannock (RAP), and James (JAM).

flowing into Massachusetts Bay; the Blackstone flowing into Narragansett Bay, the Connecticut flowing into Long Island Sound; the Mohawk and Upper Hudson flowing into the Hudson Estuary; the Delaware and Schuylkill flowing into Delaware Bay; and the Susquehanna, Potomac, Rappahannock, and James rivers flowing into Chesapeake Bay. We focused specifically on portions of the catchments upstream from individual USGS gaging stations, where long-term measurements of streamflow and water quality exist (Alexander et al. 1998). These stations are generally located upstream from the major coastal population centers of Portland, Boston, Providence, New York, Philadelphia, Washington D.C., and Richmond. We delineated catchment boundaries upstream of the gaging stations based on topography. The catchments cover a total area of approximately 250,000 km^2, and range in size from 475 km^2 (Charles) to 70,189 km^2 (Susquehanna).

Spatial data describing land use, population, and climate were aggregated to the scale of catchments using GIS software by weighting each county- or grid- estimate by the fraction of land area that is included within the catchment boundaries. Land use maps were obtained on a 30-m grid from the National Land Cover Database (NLCD), which represents land use during the early 1990's (MRLC 1995). County-level population data were obtained from the 1990 U.S. census (U.S. Dept. of Commerce 1990). Long-term monthly temperature and precipitation data were obtained on a half-degree grid from Kittel et al. (1997).

The combined landscapes of the 16 catchments were 72% forested, 19% agricultural, and 3% urban in the early 1990's, although land use varied greatly by catchment (Table 1). Portions of the Rappahannock, Potomac, Susquehanna, Schuylkill and Mohawk basins support intense crop and animal production, while the Penobscot and Kennebec largely support industrial timber production. Over 14 million people lived within the catchment boundaries in 1990, averaging 58 people km^{-2}. Population densities were highest in the Charles, Schuylkill, Blackstone, and Merrimack catchments with 556, 293, 276, and 143 people km^{-2}, respectively, and lowest in the Penobscot and Kennebec with only 8 and 9 people km^{-2}.

Climate data were obtained on a half degree grid for 1988–1993 from the VEMAP-II historical climate reconstruction (Kittel et al. 1997). Estimated average annual precipitation during 1988-1993 ranged from 930 to 1260 mm yr^{-1} with a mean of 1110 mm yr^{-1} (Table 1). Annual runoff during 1988–1993 ranged from 330 mm yr^{-1} in the Potomac to 670 mm yr^{-1} in the Saco. Evapotranspiration, as estimated from the annual water budget, varied with regional differences in mean temperature, from 44–50% of precipitation on the cool northern catchments (the Penobscot, Saco, Merrimack, and Connecticut), to more than 65% of precipitation on Potomac and Rappahannock. On average across the 16 catchments, evapotranspiration was 570 mm yr^1 or approximately 50% of precipitation.

Nitrogen budgets

We constructed N budgets that represent conditions during the early 1990's, following the approach put forth by Howarth et al. (1996). We quantify new inputs of N to each catchment, most of which are derived from human activ-

Table 1. Watershed characteristics

River Basin	Abbreviation	USGS[1] station	Area km²	Persons[4] #km²	Mean[2] Temp °C	Precip[2] mm/yr	Flow[1,3] mm/yr	Land Use[5] (% watershed area)					
								Forest	Agric.	Urban	Wetl.	Water	Other
Penobscot	PEN	1036390	20109	8	4.3	1075	588	83.8	1.5	0.4	5.2	6.2	3.0
Kennebec	KEN	1049265	13994	9	4.3	1085	566	79.6	5.9	0.9	3.6	6.4	3.6
Androscoggin	AND	1059000	8451	17	4.6	1151	640	84.6	4.8	1.1	3.4	4.6	1.5
Saco	SAC	1066000	3349	16	5.8	1218	672	87.4	3.6	0.8	3.9	3.1	1.1
Merrimack	MER	1100000	12005	143	7.4	1148	589	74.7	7.8	8.7	3.1	5.0	0.8
Charles	CHA	1103500	475	556	9.7	1207	583	59.3	8.4	22.2	7.2	2.5	0.5
Blackstone	BLA	1112500	1115	276	9.0	1260	651	63.3	8.1	17.6	6.8	3.4	0.8
Connecticut	CON	1184000	25019	65	6.3	1160	642	79.0	9.0	4.0	4.7	2.2	1.1
Hudson	HUD	1357540	11942	32	6.6	1126	622	80.8	10.4	2.7	2.5	3.4	0.2
Mohawk	MOH	1357500	8935	54	6.8	1142	548	63.1	28.0	4.7	2.6	1.5	0.1
Delaware	DEL	1463500	17560	85	8.7	1131	547	74.7	16.7	3.3	2.5	2.4	0.4
Schuylkill	SCH	1474500	4903	293	10.6	1134	488	48.1	38.4	10.2	0.7	1.2	1.5
Susquehanna	SUS	1578310	70189	54	8.9	1022	487	66.7	28.5	2.4	0.5	1.1	0.8
Potomac	POT	1646500	29940	63	11.3	985	328	60.8	34.6	2.6	0.5	0.7	0.8
Rappahannock	RAP	1668000	4134	24	12.6	1045	360	61.3	35.9	1.4	0.2	0.4	0.7
James	JAM	2035000	16206	24	10.1	934	407	80.6	15.6	1.4	0.6	0.7	1.1
Total Area	Total	—	248326	—	—	—	—	72.2	19.3	2.9	2.1	2.4	1.1

Watersheds listed in order from north to south. [1]USGS stream gaging station number; [2]Average temperature and precipitation for 1988–1993 from Kittel et al. 1997; [3]Average streamflow for water years 1988–1993 from USGS daily values; [4]U.S. Census 1990; [5]National land-cover database for the early 1990's from MRLC 1995.

ities: net atmospheric deposition, fertilizer application, agricultural and forest biological N fixation, and the net import of N in food and feed. Animal waste (manure) and human waste (sewage) are not considered as new inputs, as they represent recycling within a region; both of these terms are accounted for in our estimates of N transferred in food and feed. We compare the total N inputs to N exported in riverine streamflow. Throughout this paper, all graphs and tables show trends for the catchments arranged in geographical order from north to south, and all N fluxes are expressed in terms of kg N per km^2 of catchment area per year; for readers more accustomed to hectares, 100 kg km^{-2} yr^{-1} = 1 kg ha^{-1} yr^{-1}.

Input: Net atmospheric deposition

Associated with industrial, automotive, and biogenic N emissions, rates of N deposition in the eastern U.S. are the highest in the country, providing significant N inputs to our 16 catchments (NADP 2000). We considered wet and dry deposition of NO_y (NO_3^- and HNO_3), NH_x (NH_4^+ and NH_3), and AON (atmospheric organic nitrogen) in our budgets. To avoid double-accounting of N, we wanted to exclude all N that is both emitted and re-deposited within the catchment boundaries. Therefore, we quantify the new, net atmospheric deposition of NO_y, NH_x, and AON to each catchment via atmospheric deposition as described below.

Inorganic N deposition
Wet deposition of NO_3^- and NH_4^+ is measured regularly at a network of monitoring stations across the US called the National Atmospheric Deposition Program/National Trends Network (NADP/NTN). We chose data from 1991 to be representative of atmospheric deposition in the early 1990's since it was an average year for annual precipitation over the combined area of the catchments during the period 1988–1993. We obtained annual precipitation-weighted wet deposition values for 1991 for all stations in the New England and the Mid-Atlantic states from the NADP/NTN electronic database (NADP 2000). Using GIS software, we plotted the annual values observed at each sampling location, kriged the values to create isopleth maps, overlaid the catchment boundaries, and finally calculated the average value of wet deposition of NO_3^- and NH_4^+ for each catchment.

Inferential estimates of dry deposition can vary widely, due largely to different assumptions regarding deposition velocity values for various N species. For our budgets, we compared three methods for quantifying dry deposition (Figure 2(a)). First, we applied the commonly-used method of Lovett and Lindberg (1993), who observed that in eastern North America, total (wet + dry) N deposition (from NO_y and NH_x) is approximately twice measured wet deposition: *[total deposition (kg ha^{-1} yr^{-1}) = –0.72 + 2.07 * wet deposition (kg ha^{-1} yr^{-1}); R^2 = 0.91]*. However, the data on which this equation is based (from the Integrated Forest Study, Johnson & Lindberg 1991) included just 2 low-elevation sites from the NE, and one of those (Maine) had a very limited data set (G. Lovett, personal communication).

Ollinger et al. (1993) observed that in the eastern US, spatial patterns of dry deposition do not correlate directly with patterns of wet deposition, leading to substantial differences in the ratio of wet to dry deposition across the region. They used deposition data from NADP/NTN and other monitoring networks to derive linear regressions predicting the concentrations of both wet and dry N species as a function of latitude and longitude in the northeastern U.S. Wet deposition of NO_3^- and NH_4^+ was estimated by multiplying by precipitation, and dry deposition of HNO_3 vapor and NO_3 and NH_4 aerosols was estimated by multiplying by deposition velocity constants (Ollinger et al. 1993). As a second method to consider, we applied this model to 13 of our 16 catchments using the precipitation data reported in Table 1. The Potomac, James and Rappahannock catchments are outside of the latitudinal range in which the regression equations are suitable.

Estimates of dry deposition using the Ollinger et al. (1993) model are lower than those obtained by Lovett and Lindberg (1993) or Holland et al. (1999), largely because of differences in the deposition velocities used in these models. As a third method of calculating dry N deposition, we take advantage of additional information regarding deposition velocities in our study area. Lovett and Rueth (1999) compiled several years' worth of data from 7 sites in the NE and report updated deposition velocities for HNO_3 vapor (2.14 cm s^{-1}) and NO_3 and $(NH_4)_2SO_4$ aerosols (0.12 cm s^{-1}). The reported deposition velocity for HNO_3 vapor, a significant form of dry N deposition in the eastern U.S. is substantially higher than that used by Ollinger et al. (1993) (1.3 cm s^{-1}). As a third approach, we combined the spatial model of Ollinger et al. (1993) with the revised deposition velocities reported in Lovett and Rueth (1999).

Estimates of total (wet + dry) N deposition are generally similar using the three methods (Figure 2(a)). For our N budgets we chose the third method, which most reflects current understanding of dry deposition to the region (G. Lovett, personal communication). This method – the spatial model of Ollinger et al. (1993) updated with deposition velocities of Lovett and Rueth (1999) – provides estimates of the NO_y and NH_x components of total N deposition (Figure 2(b)). For the Potomac, James and Rappahannock catchments (which are out of the latitudinal range in which the Ollinger et al. regression equations apply), we used the first method of Lovett and Lindberg (1993) to quantify deposition.

144

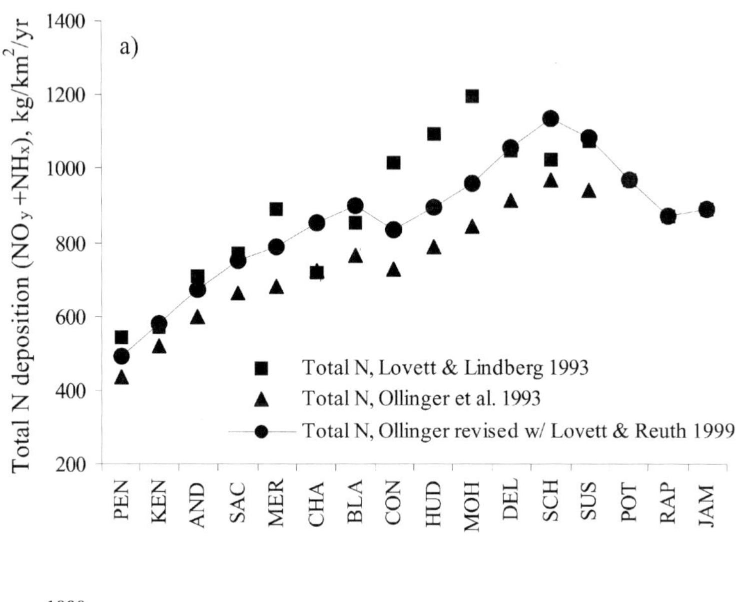

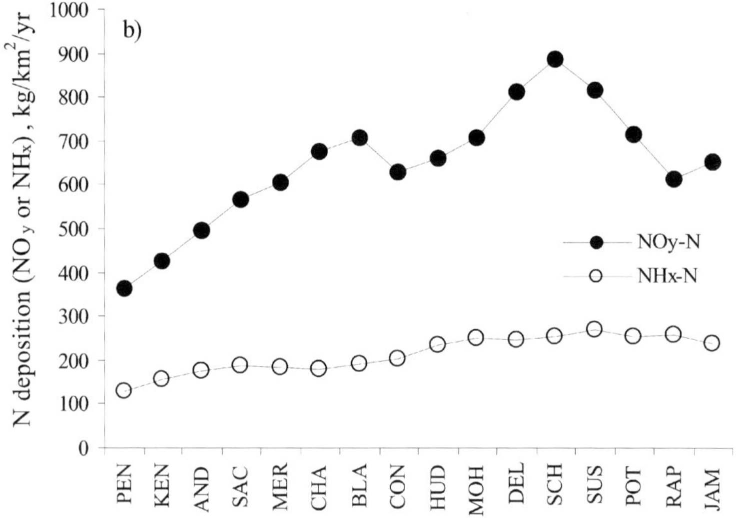

Figure 2. (a) Comparison of 3 models describing total (wet + dry) depositional inputs of inorganic N (NO_y + NH_x) to each catchment. Values indicated by dashed lines were used in budget calculations. (b) Portions of the total N input to each catchment from NO_y-N (in filled circles) and NH_x-N (in hollow circles), calculated using the revised Ollinger model.

Net NH$_x$ Input

Approximately 90% of NH$_x$ in the atmosphere comes from agricultural sources (Dentener & Crutzen 1994), with major emissions coming from animal wastes (manure) and lesser contributions from volatilization of fertilizers. Because NH$_x$ is short-lived in the atmosphere, with residence times ranging from hours to a few weeks (Fangmeier et al. 1994), NH$_x$ may redeposit within the same region from which it was emitted (Schlesinger & Hartley 1992; Prospero et al. 1996). Therefore, several studies over large spatial scales have simply assumed that NH$_x$ deposition reflects local recycling and have ignored it as a new input (Howarth et al. 1996; Jordan & Weller 1996; Castro et al. 2000). The volatilization and deposition cycle may be complete over the scale of a large region, but it is unlikely that this recycling is complete over shorter distances, and several recent studies have highlighted long-range transport of NH$_x$ (Dentener & Crutzen 1994; Galperin & Sofiev 1998). For example, with an area of 32,820 km^2, Belgium is almost 1/2 the size of the Susquehanna basin and is larger in size than all of our other study catchments. Source-receptor matrices produced by the Program for Monitoring and Evaluation of the Long-Range Transmission of Air Pollutants in Europe show that Belgium received transboundary imports of NH$_x$ deposition from more than 7 nearby countries (EMEP 2001). As illustrated by this example, our catchments may be too small in size for complete recycling of NH$_x$ to occur within the basin boundaries. Therefore, we explicitly consider both inputs of NH$_x$ in atmospheric deposition and outputs of NH$_x$ from volatilization to quantify the net depositional input of NH$_x$ to each catchment.

We estimated total (wet + dry) input of NH$_x$ using the revised Ollinger spatial model, as described above. Volatilization from animal waste (manure) is estimated based on NH$_x$ emission factors that have been developed for animal populations (Table 2). These factors are highly variable because they vary with agricultural management practices and with the size, dietary intake, and excretion of each animal, all of which can vary substantially over space and time. We estimated NH$_x$ volatilization for each catchment (Figure 3) based on 10 different published sets of emission factors. For our budgets, we used factors from Battye et al. (1994) because they are current estimates that are recommended to describe agricultural management practices in the U.S.A.

Fertilizers, especially those applied as urea, are potentially volatilized. Data on fertilizer types used in each catchment are described below (see 'input: nitrogenous fertilizer use'). We estimated volatilization losses (Figure 3) as a percentage of fertilizers applied: 15% of urea, 2% of ammonium nitrate, 2.5% of nitrogen solutions (mixed urea and ammonium

Table 2. Ammonia emission rates from animals*, kg N animal^{-1} yr^{-1}

Animal	Cass et al. 1982 *cited in Battye et al. 1994*	ApSimon et al. 1987	Buijsman et al. 1987	Kruse et al. 1989 *cited in Battye et al. 1994*	Moller & Schieferdecker 1989	Asman 1990 *cited in Battye et al. 1994*	Lee 1994 *cited in Battye et al. 1994*	Battye et al. 1994	Bouwman et al. 1997	van der Hoek 1998
Beef cattle	—	15.87	—	—	22.10	—	—	**18.83**	7.80	11.76
Dairy cattle	27.00	15.87	14.80	19.31	22.10	20.70	27.00	**18.83**	20.40	23.44
Young cattle	—	15.87	—	—	22.10	—	—	**10.72**	7.80	11.76
Pigs & hogs	4.50	2.35	2.30	2.86	5.20	3.96	1.60	**4.20**	4.00	3.98
Sheep	2.70	2.20	2.55	2.68	3.00	1.57	0.70	**2.77**	0.64	1.10
Goats	—	—	—	—	—	—	—	**5.26**	0.58	1.10
Horses	33.00	25.99	7.73	31.60	15.00	10.30	—	**10.03**	7.60	6.58
Chickens (layers)	0.24	0.19	0.21	0.23	0.22	0.26	0.12	**0.20**	0.20	0.30
Chickens (broilers)	0.24	0.19	0.21	0.23	0.22	0.26	0.12	**0.14**	0.20	0.30
Turkeys	0.66	0.19	0.21	0.23	0.22	0.26	0.10	**0.71**	0.20	0.76

* — No data. Values chosen for our nitrogen budgets, which best represent current agricultural management practices in the USA, are in boldface. If a more general ammonia emission factor was reported, we used the value for "cattle" for beef, dairy, and young, and we used the value for "poultry" fro layers, broilers, and turkeys.

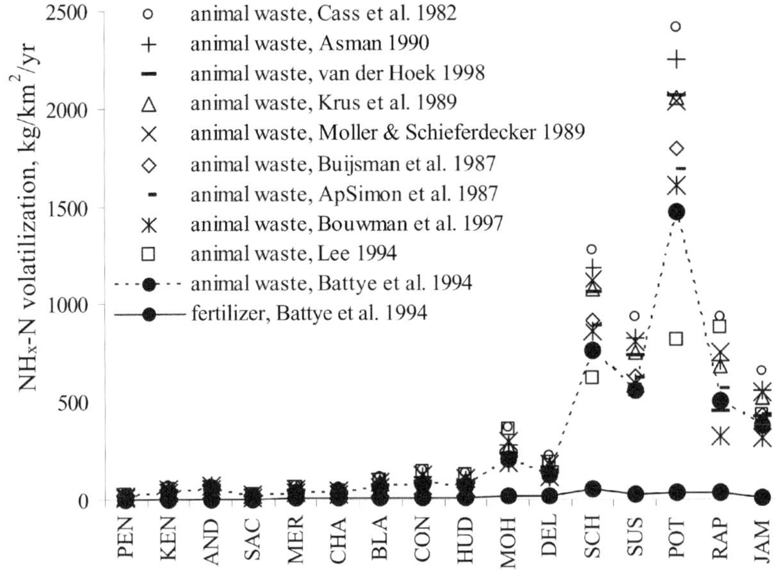

Figure 3. Estimates of NH_x volatilization from animal waste based on 10 published sets of animal emission factors. Values reported by Battye et al. 1994, which are recommended to describe U.S. agricultural practices, were used in our budget calculations. Estimates of NH_x volatilization from fertilizers are much lower in magnitude than the emissions from animal populations.

nitrate), 0.1% of anhydrous ammonia, and 2% of other combined fertilizers (Battye 1994).

An uncertain amount of reduced N redeposits within the same region from which it volatilized, but the existing wet deposition monitoring network may miss much of this redeposited N. NADP monitoring sites are generally located in forested areas rather than immediately downwind of farms and animal feedlots, and so NADP measurements likely capture long-range transport of NH_x rather than local sources. If we had detailed information on NH_x deposition downwind of these agricultural regions, we could simply have subtracted the NH_x volatilization emissions from the complete NH_x depositional inputs to estimate net NH_x inputs to each catchment. As a first approximation, we assumed that 75% of the ammonia volatilization that occurs from animal wastes and fertilizers is re-deposited locally, and thus is not an output from the catchment but rather represents a recycling within its boundaries. We assumed that the remaining 25% of the NH_x emissions from agricultural sources are transported long-range; this is treated as a volatilization output of NH_x-N from each catchment. Net NH_x gains or losses were then calculated as the difference between the depositional input and this volatilization output for each catchment.

Net organic nitrogen input

In addition to depositional inputs of inorganic N species, inputs of atmospheric organic nitrogen (AON) also can be substantial (see review by Neff et al. this volume). Measured fluxes of bulk DON deposition in New England ranged from 60 to 190 kg km^{-2} yr^{-1} (Campbell et al. 2000; Currie et al. 1996). Compiling data measurements from 41 different environments, Neff et al. found that the percentage contribution of organic nitrogen to total deposition is consistently around 30%. For our N budgets, we must consider how much of this AON is a new input to the region. Some of this organic N reflects recycling within the boundaries of our catchments, as it comes from natural biological sources (e.g. pollen) and from agricultural sources. However, AON is also formed as reaction byproducts between NO_y and hydrocarbons and is transported regionally (Neff et al. 2002). Modeled estimates of the long-range transport of AON (that is, new inputs of AON) predict about 44 kg km^{-2} yr^{-1} of AON deposition to the NE, largely derived from the Midwest (Neff et al. this volume). Given both measured and modeled estimates, we know that a significant fraction of the AON is a new input to the region. We assume that half of the AON (or 15% of total atmospheric N) is a new N input that is transported to each catchment (J. Neff, personal communication).

Input: Nitrogenous fertilizer use

The United States produces and consumes large amounts of fertilizers. In 1990, over 10 million metric tons of nitrogenous fertilizers were used in the U.S. (Battaglin & Goolsby 1994). We obtained digital data from the USGS on the N content of fertilizers sold within each county in 1991 (Battaglin & Goolsby 1994). These spatial maps are based on state data from the U.S. EPA, which were disaggregated to the county level by the Tennessee Valley Authority National Fertilizer and Environmental Research Center. Fertilizer sales estimates are available for each county, broken down by form: ammonium nitrate, anhydrous ammonia, nitrogen solutions, urea, and miscellaneous forms. We assumed that fertilizers are applied in the same county in which they are sold, which is a potential source of error. We aggregated the county level fertilizer data to the catchment level using GIS software by weighting each county estimate by the percent of county covered by each catchment.

Input: Net N import in human food and animal feed

Both humans and animals require food and feed, and these demands are met both by local agricultural production and by imports from other regions. Transfers of agricultural products can be important sources of N to a region. For example, Howarth et al. (1996) estimate that the net import of N in food and feed was 28% of total N inputs to the NE U.S. region as a whole. For each of the 16 NE catchments, we used the general method put forth by Jordan and Weller (1996), who quantified the net import of N as the balance between production of N in crops and animal products and the consumption of N by both humans and animals. We obtained crop and animal production data from the U.S. Department of Agriculture, National Agricultural Statistics Service (USDA/NASS). An agricultural census is conducted in the U.S. every 5 years (USDA/NASS 1992). We used county-level data from the 1992 U.S. Agriculture Census on the number of cows (beef and dairy), horses, pigs, sheep, chickens (layers and broilers), and turkeys, as well as information on pasture acreage and the annual production of crops typically used as food and feed, such as corn grain, corn silage, wheat, barley, oats, soybeans, and hay. These animal census and crop data were aggregated from the county to the catchment level by weighting the numbers reported for each county by the percent of county covered by each catchment.

Food and feed consumption
Human consumption of N in food was estimated by multiplying population density for each catchment (see Table 1) by a per capita intake of 5.0 kg N per year, a value typical of western populations on a high protein diet (Garrow et al. 2000). Animals are usually fed according to relatively straightforward dietary prescriptions designed for maintaining or gaining weight (van Horn et al. 1996). We estimated total demand for N in animal feed by multiplying per animal annual N requirements by the animal inventories from the 1992 agricultural census. We compared four sets of published values for the typical feed intake, or consumption, of N per animal type (Table 3). For our budgets, we chose the values reported by van Horn (1998) because they were developed based on current U.S. agricultural practices.

Crop production
To determine how much of the demands for N in consumption of food and feed can be satisfied by crop production, we calculated the N content of the entire crop harvest in each catchment. Crop production data were obtained from the 1992 agricultural census (USDA/NASS 1992), and nutrient contents for each crop type were assigned based on conversion factors reported by Lander and Moffitt (1996). Although most crops are produced for animal

Table 3. N consumption and waste production rates from animals*, kg N animal^{-1} yr^{-1}

Animal	Consumption: N intake rates				Waste production: N excretion rates					
	Thomas & Gilliam 1977	Bleken & Bakken 1997	van Horn 1998	van der Hoek & Bouwman 1999	Thomas & Gilliam 1977	Bleken & Bakken 1997	van Horn 1998	van der Hoek & Bouwman 1999	Smil 1999	SCS 1992
Beef cattle	56.00	74.60	**66.75**	51.30	44.00	66.60	**58.51**	40.70	50.00	41.72
Dairy cattle	166.00	126.70	**156.00**	177.00	120.00	93.90	**121.00**	140.50	80.00	79.47
Young cattle	—	57.70	—	74.10	—	36.45	—	67.70	—	36.26
Pigs & hogs	8.70	5.85	**8.51**	14.93	6.10	4.34	**5.84**	10.46	10.00	19.70
Sheep	—	9.76	**5.97**	27.60	—	6.70	**5.00**	25.00	5.00	13.04
Goats	—	13.90	**5.97**	23.70	—	12.30	**5.00**	19.90	—	—
Horses	—	—	**44.80**	—	—	50.00	**40.00**	45.00	35.00	27.81
Chickens (layers)	1.10	0.94	**0.84**	1.17	0.83	0.61	**0.55**	0.81	0.30	0.21
Chickens (broilers)	0.82	0.08	**0.13**	1.03	0.39	0.03	**0.07**	0.57	0.30	0.55
Turkeys	2.12	1.29	**0.62**	—	1.29	0.34	**0.39**	0.50	0.30	0.37

* — No data. Values chosen for our nitrogen budgets, which best represent current agricultural management practices in the USA, are in boldface.

consumption, a small fraction is grown for humans. To partition crop yields, we followed the distribution given by Jordan and Weller (1996), and assumed that 4% of corn, 61% of wheat, 6% of oats, 3% of barley, 17% of rye, 2% of soybeans, and 100% of potatoes were for human consumption. The remaining percentages of those crops went to feed animals, as did 100% of sorghum, hay, and pasture production. Following Jordan and Weller (1996), we assumed that pests, spoilage, and processing caused a 10% loss of all crops but hay and silage.

Animal production

Humans consume both animal and plant products. We quantified animal N production (i.e. meat, milk, and eggs) as the difference between animal feed consumption (intake) and animal excretion (waste production). Estimates of typical per animal feed intake and waste production vary significantly between studies (Table 3), because they depend on animal weights and effi- ciencies (for example, with the amount of milk a dairy cow produces) and on agricultural management practices. We compare values for N intake and excretion reported in the literature for Norway from Bleken and Bakken (1997), for the Netherlands and Europe from van der Hoek and Bouwman (1999), for the U.S. from Thomas and Gilliam (1977) and for the U.S. from van Horn et al. (1998). For our budgets, we chose the values reported by van Horn (1998), as these are the most current values that we could find that are based on U.S. agricultural practices. We assumed that spoilage and inedible components caused a 10% loss of animal products available for consumption.

Net import in food and feed

We estimated the net import of N in food and feed using a mass balance of needs versus production. We assumed that N import in feed equaled the differ- ence between animal N demands and N produced in crops grown for animal consumption, and that N import in food equaled the difference between human N demands and N produced in food for humans. Imports were assumed to have come from regions outside of each catchment boundary. Thus: *[net import in food and feed = human consumption + animal consump- tion – crop production for animal consumption – crop production for human consumption – animal production for human consumption].* In some cases the balances were negative, with crop and animal production exceeding human and animal demands; this indicates a net export of N in food and feed.

Calculations of net N import in food and feed were sensitive to the coeffi- cients used to describe rates of animal intake and excretion (Figure 4(a)). For our N budgets, we chose the rates reported by van Horn (1998), which are based on current agricultural practices in the U.S. Considering just imports in feed, we checked our estimate of the net import of N in feed based

152

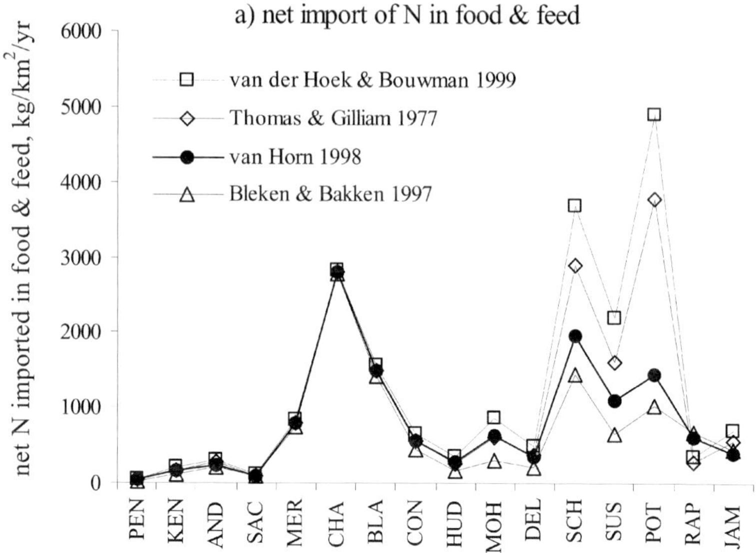

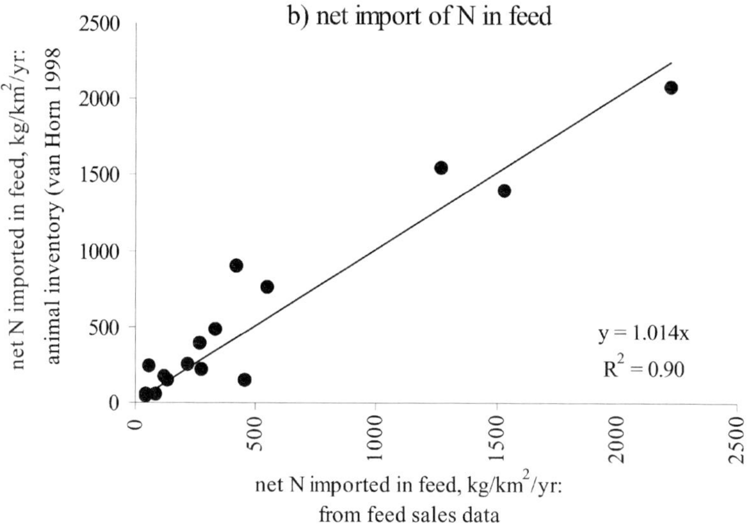

Figure 4. (a) We compare 4 estimates of N inputs to each catchment from net imports of food and feed, each based on a different set of coefficients describing the consumption and the waste production of N by animals. All estimates indicate that import of food is significant in the highly populated Charles and Blackstone catchments, while import of feed is significant in the highly agricultural Susquehanna, Potomac, and Rappahannock catchments. For our N budgets, we chose the rates reported by van Horn (1998), which are based on U.S. agricultural practices. (b) We checked our estimate of net N import in animal feed based on animal inventory data against an independent estimate based on feed expenditures. This nearly 1:1 relationship provides support that the coefficients of van Horn (1998) are appropriate to describe animal production in the northeastern U.S.

on animal inventory data against an independent estimate based on feed expenditure data (Figure 4(b)). Net N imports in feed were quantified from the animal inventory data using the coefficients for intake and excretion by van Horn 1998 that we used in our N budgets. County-level data on feed expenditures are reported in the 1992 Census of Agriculture. Feed expenditures were converted to estimates of N in feed by multiplying by average feed costs for the northeast region of approximately 3.0 kg dry matter per U.S. dollar, or 0.15 kg N per dollar (USDA/NASS 1997). The nearly 1:1 relationship between the net N import in feed calculated from the animal inventory data versus those calculated from feed expenditures within each catchment provide support that the coefficients of van Horn (1998) are appropriate to describe animal production in the northeastern U.S.

Input: Nitrogen fixation

Agricultural land
Alfalfa is the major N-fixing crop grown in the catchments, accounting for 23% of the N in total crop production, and 47% of the total N fixed in agricultural land. We estimated agricultural N fixation rates by multiplying the area of N-fixing species by literature-derived N fixation rates. We obtained data on the area of soybeans, alfalfa, hay, pasture, and snap beans from the 1992 agricultural census (USDA/NASS 1992). We used a fixation rate of 9600 kg km^{-2} yr^{-1} for soybeans (average of reported values by Rennie et al. 1978; Ham and Caldwell 1978; Deibert et al. 1979) and 22400 kg N km^{-2} yr^{-1} for alfalfa (Heichel et al. 1984); both rates are based on measurements in U.S. agroecosystems by the ^{15}N isotope dilution technique. We calculated the area of non-alfalfa leguminous hay by subtracting the reported area of grass hays (which do not fix N) and the area of alfalfa hay from the total area of hay. We assume that the remaining leguminous hay contains largely clovers and vetches, and assume a fixation rate of 11700 kg N km^{-2} yr^{-1}, based on averages of reported values for clover varieties (Brink 1990; Labandera et al. 1988; Rice 1980). Pasture was assumed to fix 1500 kg N km^{-2} yr^{-1} (after Jordan & Weller 1996). Snap beans grown in the catchments were assigned a rate of 90 kg N ha^{-1} yr^{-1} (Westerman et al. 1981). Other leguminous crops (such as other beans, peas, seeds, and peanuts) were not grown in significant quantities in the boundaries of our catchments. Although N-fixation rates for crops can vary significantly, the values we use for our N budgets are within the range of those considered by Smil (1999) to be most reliable.

Forested land
Both symbiotic and non-symbiotic N fixation can occur in eastern U.S. forests. Studies using the acetylene-reduction technique consistently report

154

extremely low rates of N fixation by free-living microbes in soils and woody litter, with a range of 0.2–200 kg km^{-2} yr^{-1} (Tjepkema 1979; Roskowski 1980; Grant & Binkley 1987; Hendrickson 1990; Barkmann & Schwintzer 1999; Cleveland et al. 1999). We assumed a constant value of 40 kg km^{-2} yr^{-1} for non-symbiotic N fixation in all forests in the region. Rates of N fixation in eastern species with symbiotic associations can be quite high (~3,000–7,500 kg km^{-2} yr^{-1}; Boring & Swank 1984), but both N fixation rates and the spatial distribution of these species are extremely variable and not well quantified. Black locust (*Robinia pseudoacacia*), a legume, occurs largely in southern Appalachia, where it ranges from 71% of the basal area in young stands to <1% in old, uneven-aged stands (Boring & Swank 1984). We obtained information on forest types from the USDA Forest Service's Forest Inventory and Analysis program (Hansen et al. 1992). For lack of additional information, we assumed that locust made up an average of 10% of oak-hickory stands, and that it fixed N at a mean rate of 5,000 kg km^{-2} yr^{-1} when in a pure stand (Boring & Swank 1984). Speckled alder (*Alnus incana* spp. rugosa) and common or hazel alder (*A. serrulata*) are actinorhizal N-fixing shrubs that occur in wet soils across the region. We assumed that alder covered 10% of wetland areas, and that it fixed N at a mean rate of 4,000 kg km^{-2} yr^{-1} when in a pure stand, as observed in the Adirondack Mountains, New York (Hurd et al. 2001).

Output: Riverine export

N export, or loading, in streamflow was quantified by multiplying stream N concentration by the flow rate of water. We obtained daily discharge values for the stream gaging stations located at the outlet of each catchment (see Table 1) from the USGS National Water Information System (USGS 2000). Stream N concentrations (of total N, including NO_3^-, NH_4^+ and dissolved organic nitrogen) were sampled at each of these sites approximately monthly, and were obtained from the databases of the USGS water quality monitoring networks (Alexander et al.1998). Stream N export was calculated from these data using the USGS 'estimator' software, which is a regression-based approach that allows flow-weighted interpolation of the discrete measurements of concentration, and has bias corrections (Cohn et al. 1992). We report average values of the annual N exports (or mass loadings) computed over the years 1988–1993.

Results

Atmospheric deposition was an important pathway by which N entered the catchments (Table 4). Total new inputs of N in deposition ranged from 575 kg km^{-2} yr^{-1} in the Penobscot to 1212 kg km^{-2} yr^{-1} in the Delaware (Table 4). Over the combined area of the catchments, NO_y, NH_x, and AON accounted for 69%, 15%, and 16% of the net N inputs, respectively, and new inputs of N in atmospheric deposition accounted for 31% of the total N inputs to the landscape.

Total atmospheric inputs from deposition of NO_y were lowest in the northern Maine catchments and highest in the Mid-Atlantic region, ranging from 362 kg km^{-2} yr^{-1} in the Penobscot to 885 kg km^{-2} yr^{-1} in the Schuylkill. Total net NH_x deposition, representing the fraction of total NH_x that is a new input to each catchment, averaged 120 kg km^{-2} yr^{-1}. Ammonia volatilization from fertilizer ranged from 1 kg km^{-2} yr^{-1} in the Saco to 58 kg km^{-2} yr^{-1} in the Schuylkill, while ammonia volatilization from animal waste ranged from 13 kg km^{-2} yr^{-1} in the Penobscot to 1459 kg km^{-2} yr^{-1} in the Potomac (see Figure 3). Net NH_x input was negative in the heavily agricultural Potomac basin (-119 kg km^{-2} yr^{-1}), with net volatilization outputs exceeding net depositional inputs due largely to the large number of poultry in the catchment and their associated emissions. Total net AON input, representing the fraction of AON that is a new input to each catchment, ranged from 88 kg km^{-2} yr^{-1} in the Potomac to 205 kg km^{-2} yr^{-1} in the Schuylkill.

Annual fertilizer N inputs increased with the percent of catchment area in agriculture, from less than 60 kg km^{-2} yr^{-1} on the forested catchments of the Saco and Kennebec to over 1000 kg km^{-2} yr^{-1} on the Schuylkill, Potomac and Rappahannock, where approximately 35% of the land area is in agriculture. Across the 16 catchments, N inputs from nitrogenous fertilizers averaged 474 kg km^{-2} yr^{-1} or 15% of the total N inputs. Fertilizer was applied mostly in compound forms (56%), although significant quantities of mixed urea and ammonium nitrate solutions (24%) and urea (14%) were applied. Much smaller quantities were applied in the form of ammonium nitrate (3%) and anhydrous ammonia (3%). The fractions containing urea are most susceptible to ammonia volatilization.

Human N consumption is proportional to the distribution of population, which is greatest in the Charles, Blackstone, Schuylkill, and Merrimack basins (Table 1). Animal N consumption was greatest on the Potomac, Schuylkill, Susquehanna, and Rappahannock, driven by large cattle herds and poultry demands (Table 5). Human and animal dietary N demands are satisfied by consumption of crops (forage, grains, fruits, vegetables) and animal products (meat, milk, and eggs). Across the 16 catchments, hay dominated the crop N production (38%), followed by pasture forages (23%), corn grain

Table 4. Atmospheric nitrogen deposition*, kg N km^2 yr^{-1}

Watershed	Total NO_y dep.	Total NH_x dep.	Total inorganic NO_y+NH_x dep.	NH_x vol. from fertilizer	NH_x vol. from an Waste	Total NH_x vol.	Net NH_x dep.	Net atm. N dep.	Total net N dep.
Penobscot	362	129	491	3	13	17	125	88	575
Kennebec	428	154	582	2	38	40	144	105	677
Androscoggin	495	176	671	3	48	51	163	121	779
Saco	566	187	753	1	15	16	183	136	885
Merrimack	606	184	790	5	40	45	173	142	921
Charles	674	178	852	8	31	39	168	153	996
Blackstone	707	190	897	12	66	77	171	162	1040
Connecticut	631	204	835	12	84	95	181	150	962
Hudson	658	234	893	8	73	81	214	161	1033
Mohawk	708	250	958	16	207	223	195	172	1075
Delaware	811	248	1059	22	127	149	211	191	1212
Schuylkill	885	253	1138	58	740	798	53	205	1143
Susquehanna	816	269	1085	28	540	569	127	195	1138
Potomac	714	255	969	35	1459	1494	−119	174	769
Rappahannock	615	256	871	33	506	538	121	157	893
James	652	237	889	12	373	384	141	160	953
Area-wght avg.	*677*	*228*	*904*	*19*	*412*	*431*	*120*	*163*	*959*

*See text for description of the net depositional terms. Boldface values were used in our budgets.

Table 5. Nitrogen transfers in food and feed, kg N km^{-2} yr^{-1}

Watershed	Animal N consumption	Human N consumption	Crop production N for animals	Crop production N for humans	Animal N products for humans	Net N import in feed for animals	Net N import in food for humans	**Net N import in food and feed**
Penobscot	122	38	67	32	24	55	−18	36
Kennebec	293	46	122	2	61	171	−17	154
Androscoggin	373	84	126	9	85	247	−10	237
Saco	123	82	74	4	23	49	55	104
Merrimack	335	713	185	2	64	150	647	797
Charles	201	2781	138	1	36	62	2745	2807
Blackstone	501	1380	284	2	99	217	1279	1495
Connecticut	764	324	367	8	149	398	167	565
Hudson	687	161	436	6	135	251	20	271
Mohawk	1977	272	1219	14	391	758	−134	624
Delaware	1001	426	846	32	197	155	197	352
Schuylkill	3523	1464	2122	102	810	1401	551	1952
Susquehanna	3097	270	1543	50	679	1554	−459	1095
Potomac	3822	313	1737	53	892	2085	−633	1452
Rappahannock	2673	121	1775	34	379	898	−291	607
James	1299	120	812	6	205	487	−91	395
Area-wght. Avg.	*1847*	*289*	*960*	*31*	*397*	*887*	*−139*	*748*
Sum	*20791*	*8595*	*11855*	*358*	*4230*	*8936*	*4008*	*12944*

(15%) and corn silage (12%). Wheat, barley, and oats each contributed less than 2% of total N in crop production; soybeans contributed an average of 6% with peak production in the Schuylkill catchment. Overall, N produced by crop and animal production was used to satisfy human and animal dietary needs. Of the total N produced in food and feed within the catchments, 69% was from crops that were fed to animals, 2% was from crops that were fed to humans, and 29% was from animals that were fed to humans. In catchments where the dietary demands for N could not be met by local crop and animal production, N was imported in food and feed (Table 5). Import of N in animal feed was most significant in the largely agricultural basins of the mid-Atlantic: the Potomac (2085 kg km^{-2} yr^{-1}), Susquehanna (1554 kg km^{-2} yr^{-1}), Schuylkill (1401 kg km^{-2} yr^{-1}), and the Rappahannock (898 kg km^{-2} yr^{-1}). All catchments had some import of N in feed, and averaged 887 kg km^{-2} yr^{-1} for animal feed. Import of N in human food was largest in the Charles (2745 kg km^{-2} yr^{-1}), Blackstone (1279 kg km^{-2} yr^{-1}), Schuylkill (551 kg km^{-2} yr^{-1}), and Merrimack (647 kg km^{-2} yr^{-1}), which are the 4 basins with the highest population densities (see Table 1). Many of the basins with large agricultural production or low population densities export foods to other regions.

Another important source of N was agricultural N fixation. N fixation from leguminous crop species averaged 740 kg km^{-2} yr^{-1} and varied directly with total land area in agriculture. Fixation rates ranged from 74 kg km^{-2} yr^{-1} on the Saco to 1439 kg km^{-2} yr^{-1} on the Rappahannock (Table 6). Alfalfa and other leguminous (e.g. clover) hays accounted for the vast majority of total agricultural fixation inputs (47% and 41%, respectively), with smaller inputs from fixation occurring in the small areas on which soybeans were harvested (6%), from eastern pasture (5%), and from other crops (<1%).

Forest fixation was a small N source to each catchment (Table 6). These estimates were largely dependent on assumptions regarding the abundance and mean N fixation rate of black locust, the main N-fixing tree species in the region. Estimated total forest N fixation rates per catchment ranged from 50 kg km^{-2} yr^{-1} on the Kennebec to 361 kg km^{-2} yr^{-1} on the James, with a mean of 167 kg km^{-2} yr^{-1}. Heterotrophic N fixation accounted for over half of the small amount of N (50–70 kg km^{-2} yr^{-1}) fixed in the northern Maine catchments, whereas symbiotic N fixation associated with black locust contributed over 70% of the N fixed in the Massachusetts, Pennsylvania, and Virginia catchments. N fixation associated with alder was most important in northern Maine, but never exceeded 30 kg km^{-2} yr^{-1}.

Total nitrogen budgets were established by aggregating the N input terms (net deposition, fertilizer use, fixation, and net inputs in food & feed) for each catchment (Table 6). Total new inputs of N to each catchment ranged from

Table 6. Overall nitrogen budget, kg N km^{-2} yr^{-1}

	Net Atmos. pheric N Dep.	Nitro- genous Fertilizer use	N fixation in forest lands	N fixation in agricult. lands	Net N import in food & feed	Total nitrogen inputs	Stream- flow N export*	% of N inputs exported in stream flow	% of N inputs stored or lost in basin
Penobscot	575	91	58	74	36	835	317	38	62
Kennebec	677	54	50	164	154	1099	333	30	70
Androscoggin	779	80	69	146	237	1310	404	31	69
Saco	885	42	107	96	104	1233	389	32	68
Merrimack	921	147	151	213	797	2228	499	22	78
Charles	996	197	218	187	2087	4406	1756	40	60
Blackstone	1040	307	260	305	1495	3407	1140	33	67
Connecticut	962	274	102	360	565	2262	538	24	76
Hudson	1033	204	103	374	271	1985	502	25	75
Mohawk	1075	411	70	1239	624	3420	795	23	77
Delaware	1212	527	201	675	352	2967	961	32	68
Schuylkill	1143	1207	190	1225	1952	5717	1755	31	69
Susquehanna	1138	615	179	1147	1095	4173	977	23	77
Potomac	769	1024	271	1173	1452	4689	897	19	81
Rappahannock	893	1030	277	1439	607	4246	470	11	89
James	953	361	361	703	395	2773	314	11	89
Area-wght. Avg.	*959*	*474*	*167*	*740*	*748*	*3088*	*718*	*25*	*75*

– no data. *Export of N in streamflow for the Charles River was 644 kg N km^{-2} yr^{-1}. The value shown includes 1112 kg N km^{-2} yr^{-1} of wastewater that originated within the Charles watershed but is diverted out of the basin boundary.

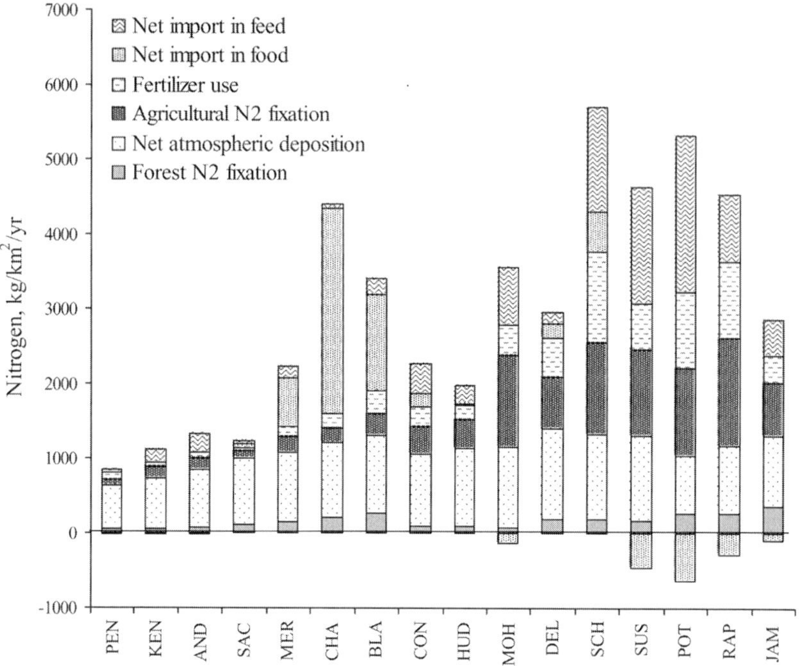

Figure 5. The relative importance of the N sources varies widely by catchment. Net atmospheric deposition was the largest N source (>60%) to the forested basins of northern New England (e.g. Penobscot & Kennebec); net import of N in food was the largest source of N to the more populated regions of southern New England (e.g. Charles & Blackstone); and agricultural inputs were the dominant N sources in the Mid-Atlantic region (e.g. Schuylkill & Potomac).

835 kg km^{-2} yr^{-1} (Penobscot) in the forested northeast to 5717 kg km^{-2} yr^{-1} (Schuylkill) in the agricultural southeast. The magnitude and relative importance of the N inputs vary widely between the catchments (Figure 5). Losses of N in riverine export ranged from 314 kg km^{-2} yr^{-1} on the James to 1,755 kg N km^{-2} yr^{-1} on the Schuylkill, with an average of 747 kg N km^{-2} yr^{-1} (Table 6). R^2 values describing the amount of variance in streamflow N export that can be explained by the variance in the individual N input terms with simple, pairwise regressions were 0.83 for net food & feed imports, 0.33 for atmospheric deposition, 0.21 for fertilizer, and 0.10 for fixation. N loss in streamflow was strongly related to the total N inputs ($R^2 = 0.62$, Figure 6). However, only 25% of the total N inputs are represented by N losses in streamflow export.

A correlation matrix provides further insight into the relationships between streamflow N loading (of nitrate-N and total N) and individual N inputs (Table 7). Atmospheric deposition terms are more closely correlated

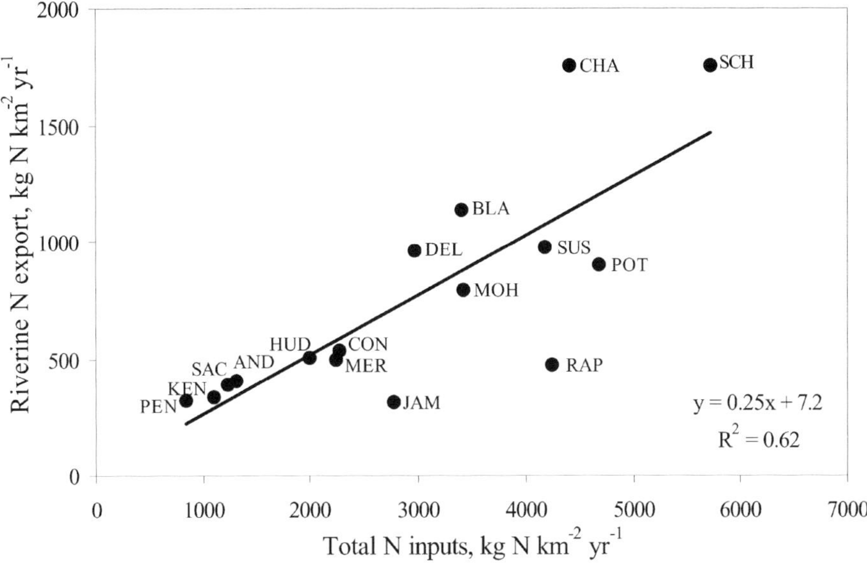

Figure 6. N export in streamflow is strongly related to total new inputs of nitrogen to each catchment.

with nitrate-N loss from the stream than with total-N loss in the stream. NO_y-N deposition is the most highly correlated input term with NO_3^-N export in streamflow ($R = 0.82$). In contrast, net import of N in food and feed is the most highly correlated input term with total-N export in streamflow ($R = 0.91$), yet this is less strongly correlated with nitrate-N export in streamflow ($R = 0.56$).

Discussion

We quantified nitrogen budgets for 16 catchments in the northeastern U.S., comparing new inputs of N from deposition, fertilizer, fixation, and inputs in food & feed to outputs of N transported in streamflow. The importance of the relative N sources varies widely by catchment and related strongly to land use (see Figure 5). These results emphasize that landscape management plans need to be developed on a watershed-by-watershed basis. Given the extreme variability among basins in both the sources of nutrients and controls on their transport, average values of the importance of the individual input terms across broad areas tend to be inappropriate descriptors of the individual catchments. For example, net atmospheric deposition was the most important N source (>60%) to the forested basins of northern New England (e.g. Penobscot and Kennebec); net import of N in food was the largest source of N to the

Table 7. Pearson correlation coefficients (R) relating streamflow N export (as NO_3-N and total-N) to watershed N inputs

	NO_3-N export in streamflow	Total N export in streamflow
NO_y-N dep	0.82	0.70
NH_x-N dep	0.56	0.29
Net AON dep	0.78	0.63
Total net N dep	0.66	0.57
Net import in feed	0.55	0.32
Net import in food	0.11	0.63
Food & feed N import	0.56	0.91
Fertilizer N use	0.70	0.46
Agric. N fixation	0.57	0.30
Forest N fixation	0.22	0.31
Total Agric. N inputs	0.74	0.78
Total N inputs	0.76	0.79
NO_3-N export in streamflow	1.00	0.78
Total N export in streamflow	0.78	1.00

more populated basins in southern New England (e.g. Charles & Blackstone); and agricultural inputs were the dominant N sources to the basins in the Mid-Atlantic (e.g. Schuylkill & Potomac). Over the region covering all of the NE catchments, net atmospheric deposition was the largest single source input (31%), followed by net imports of N in food and feed (25%), fixation in agricultural lands (24%), fertilizer use (15%), and fixation in forests (5%). The combined effect of fertilizer use, fixation in agricultural lands, and food & feed imports (64%) makes agriculture the largest overall source of N to the region. However, it is important to highlight that the net import of N in food and feed accounts for, among other things in the food production cycle, human and animal waste.

Although all 16 of the catchments are predominantly forested, shifts in land use to include even relatively small percentages of agricultural or urban land (see Table 1) have profound impacts on the annual N budgets. Total N inputs have a strong negative correlation with the fraction of land area in forest (Figure 7(a), $R^2 = 0.77$). N inputs increase directly with the fraction of land area in agriculture (Figure 7(b), $R^2 = 0.70$). The two outliers (in Figure 7(b), $\sim$10% agricultural land) are the highly urbanized catchments of the Charles (with a population density of 556 persons per km^2, 8.4% agricultural land use, and 22.2% urban areas) and Blackstone (with a population density of 276 persons per km^2, 8.1% agricultural land use, and 17.6% urban areas). This indicates the importance of urbanization, in addition to agriculture, as a large human-derived source of N to the region. Taking the sum of agricultural and urban lands, there is a direct and strong relationship between these disturbed landscapes and total N loading (Figure 7(c), $R^2 = 0.96$), highlighting the effects of anthropogenic manipulations.

Over the combined area of the catchments, 44% of food and feed requirements had to be supplied from imports from outside of the catchment boundaries. Animal demands exceeded crop production in all of the basins, and import of feed was necessary (see Table 5). Food was imported for human consumption in the New England and northern Mid-Atlantic region (including the Saco, Merrimack, Charles, Blackstone, Connecticut, Hudson, Delaware, and Schuylkill catchments). The more heavily agricultural catchments of the southern Mid-Atlantic region exported food, possibly supporting the demands to the north and to the urban centers below the points of watershed delineation for these analyses. These transfers of N illustrate the de-coupling of production and consumption inherent in many contemporary agricultural ecosystems, requiring the transfer of large quantities of N in food and feed across large distances (e.g. Jordan & Weller 1996). Because the outlets of our 16 catchments are located above many of the large population centers along the east coast, it is expected that the net import of N in food and feed is probably even more important to these larger drainage basins than our results would suggest.

Total N inputs greatly exceeded losses of N in riverine export. The fraction of N inputs represented by riverine export ranged from 11% to 40% and averaged 25%. This result is consistent with the findings of other studies; only a small fraction of N inputs to the landscape are explained by export in streamflow, whether considered at the scale of small catchments (e.g. Campbell et al. 2000), large river basins (Jaworski et al. 1997; Castro et al. 2000), or continents (Howarth et al. 1996). Questions remain about the fate of N that was attenuated by the catchment; i.e. converted to gaseous forms through denitrification, and/or stored in biomass, groundwater, or soils of the landscape. Because most of the nitrogen added to regions through human activity is stored within the region or denitrified, it is critical to understand the other major controls over loss and storage of this N. The role of these processes in the 16 catchments is evaluated in an accompanying manuscript (see Van Breemen et al. 2002).

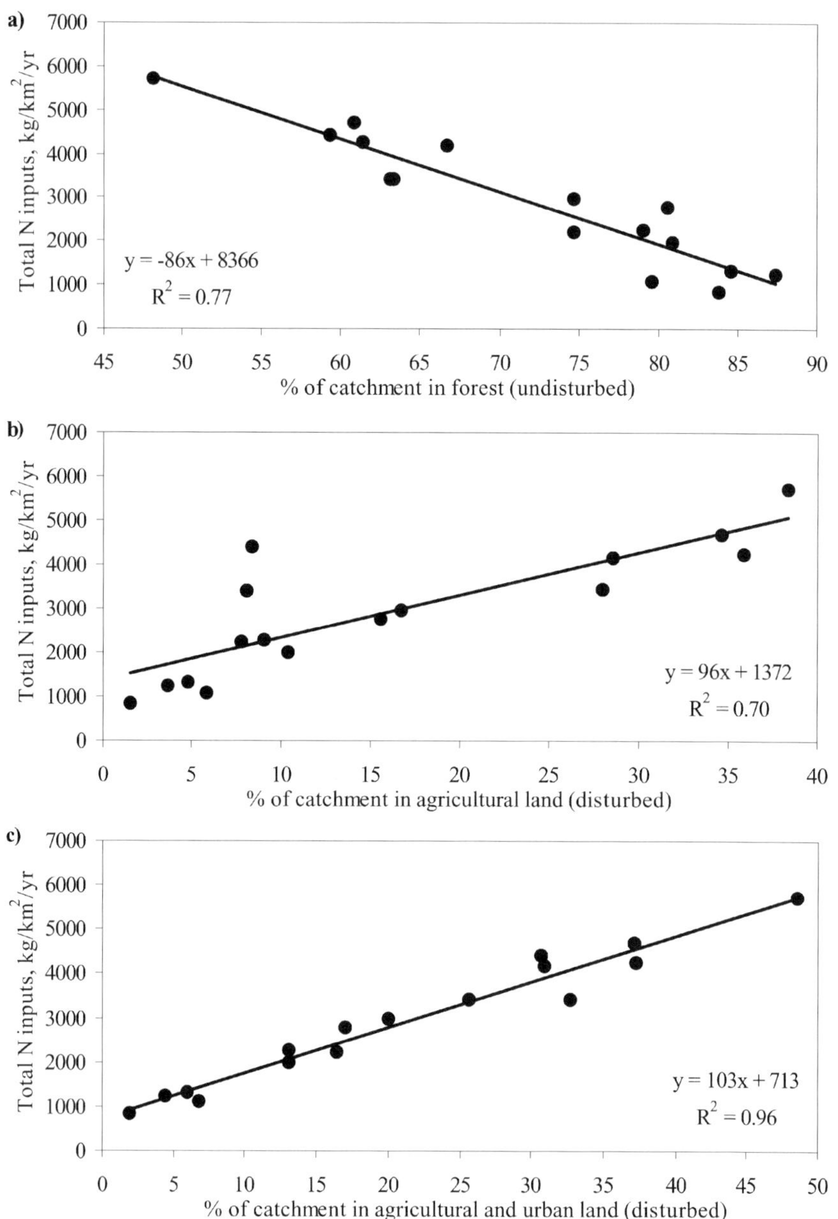

Figure 7. Nitrogen inputs to each catchment are related to land use, having: a negative correlation with land in forest (a); a positive correlation with land in agriculture (b); and a strong positive correlation with urbanized and agricultural lands (c). This highlights the effects of anthropogenic manipulations of the landscape.

Acknowledgements

This work was initiated as part of the International SCOPE Nitrogen Project, which received support from both the Mellon Foundation and from the National Center for Ecological Analysis and Synthesis. Thanks to T. Kittel & N. Rosenblum for sharing the VEMAP II climate data. We are grateful to Douwe Van Dam for literature regarding agricultural N fixation rates. We thank the following for their helpful discussions: Richard Alexander, Frank Dentener, Jim Galloway, Leo Hetling, Kate Lajtha, Gary Lovett, Jason Neff, Sybil Seitzinger, Nico van Breemen, Douwe Van Dam, and Hal Walker. The thoughtful comments of 4 reviewers – Jill Baron, Walter Boynton, Max Holmes, and Tom Fischer – substantially improved the manuscript.

References

Alexander RB, Slack JR, Ludtke AS, Fitzgerald KK & Schertz TL (1998) Data from selected U.S. Geological Survey national stream water quality monitoring networks. Water Resources Research 34(9): 2401–2405

ApSimon HM, Kruse M & Bell JNB (1987) Ammonia emissions and their role in acid deposition. Atmospheric Environment 21: 1939–1946

Barkmann J & Schwintzer CR (1999) Rapid N_2 fixation in pines? Results of a Maine field study. Ecol. 79(4): 1453–1457

Battye R, Battye W, Overcash C & Fudge S (1994) Development and selection of ammonia emission factors. Final Report prepared by EC/R Incorporated for EPA Atmospheric Research and Exposure Assessment Lab, EPA Contract Number 68-D3-0034

Battaglin WA & Goolsby DA (1994) Spatial Data in Geographic Information System Format on Agricultural Chemical Use, Land Use, and Cropping Practices in the United States. USGS Water-Resources Investigations Report 94-4176 Report available on-line at http://water.usgs.gov/pubs/wri944176/. Data and metadata available on-line at http://water.usgs.gov/GIS/metadata/usgswrd/nit91.html

Bleken MA & Bakken LR (1997) The nitrogen cost of food production: Norwegian society. Ambio 26(3): 134–142

Boring LR & Swank WT (1984) The role of black locust (Robinia pseudo-acacia) in forest succession. J. Ecol. 72: 749–766

Bouwman AF, Lee DS, Asman WAH, Dentener FJ, Van Der Hoek KW & Olivier GJ (1997) A global high-resolution emission inventory for ammonia. Global biogeochemical cycles 11(4): 561–587

Bricker SB, Clement CG, Pirhalla DE, Orland SP & Farrow DGG (1999) National Estuarine Eutrophication Assessment: A Summary of Conditions, Historical Trends, and Future Outlook. National Ocean Service, National Oceanic and Atmospheric Administration, Silver Springs MD

Brink GE (1990) Seasonal dry matter, nitrogen and dinitrogen fixation patterns of crimson and subterranean clovers. Crop Sci. 30: 1115–1118

Buijsman E, Maas HFM & Asman WAH (1987) Anthropogenic NH_3 emissions in Europe. Atmospheric Environment 21: 1009–1022

Campbell JL, Hornbeck JW, McDowell JW, Buso DC, Shanley JB & Likens GE (2000) Dissolved organic nitrogen budgets for upland, forested ecosystems in New England. Biogeochemistry 42: 123–142

Caraco NF & Cole JJ (1999) Human impact on nitrate export: an analysis using major world rivers. Ambio 28(2): 167–170

Castro MS, Driscoll CT, Jordan TE, Reay WG, Boynton WR, Seitzinger SP, Styles RV & Cable JE (2001) Contribution of atmospheric deposition to the total nitrogen loads to thirty-four estuaries on the Atlantic and Gulf coasts of the United States. In: Valigura RA et al. (Eds) Nitrogen Loading in Coastal Water Bodies: an Atmospheric Perspective (pp 77–106). AGU

Cleveland CC, Townsend AR, Schimel DS, Fisher H, Howarth RW, Hedin LO, Perakis SS, Latty EF, Von Fischer JC, Elseroad A & Wasson MF (1999) Global patterns of terrestrial biological nitrogen fixation in natural systems. Global Biogeochem. Cycles 13: 623–645

Cohn TA et al. (1992) The validity of a simple statistical model for estimating fluvial constitute loads: an empirical study involving loads entering the Chesapeake Bay. Water Resour. Res. 28: 2353–2363

Currie WS, Aber JD, McDowell WH, Boone RD & Magill AH (1996) Vertical transport of dissolved organic C and N under long-term N amendments in pine and hardwood forests. Biogeochemistry 35: 471–505

Deibert EJ, Bijeriego M & Olson RA (1979) Utilization of 15N fertilizer by nodulating and non-nodulating soybean isolines. Agron. J. 71: 717–723

Dentener FJ & Crutzen PJ (1994) A three-dimensional model of the global ammonia cycle. Journal of Atmospheric Chemistry 19: 331–369

EMEP (2001) Co-operative Programme for Monitoring and Evaluation of the Long-Range Transmission of Air Pollutants in Europe (EMEP), source-receptor matrices and import/export budgets for 1998. [online] URL: http://www.emep.int/index.html

Fangmeier A, Hadwiger-Fangmeier A, Van der Eerden L & Jager HJ (1994) Effects of Atmospheric Ammonia on Vegetation-A Review. Environmental Pollution

Galloway JN, Schlesinger WH, Levy H II, Michaels A & Schnoor JL (1995) Nitrogen fixation: anthropogenic enhancement-environmental response. Global Biogeochem. Cycles 9: 235–252

Garrow JS, James WPT & Ralph A (Eds) (2000) Human nutrition and dietetics 10th ed. Churchill Livingstone, Edinburgh, 900 pp

Galperin MV & MA Sofiev (1998) The long-range transport of ammonia and ammonium in the northern hemisphere. Atmospheric Environment, 32(3): 373–380

Goolsby DA, Battaglin WA, Lawrence GB, Artz RS, Aulenbach BT & Hooper RP (1999) Gulf of Mexico hypoxia assessment, Topic #3, Flux and sources of nutrient in the Mississippi-Atchafalaya River basin. Committee on Environmental and Natural Resources, Hypoxia Work Group for the Mississippi River/Gulf of Mexico Catchment Nutrient Task Force

Grant D & Binkley D (1987) Rates of free-living nitrogen fixation in some Piedmont forest types. Forest Sci. 33(2): 548–551

Ham GE & Caldwell AC (1978) Fertilizer placement effects on soybean seed yield, N_2 fixation and ^{33}P uptake. Agron. J. 70: 779–783

Heichel GH, Barnes DK, Vance CP & Henjum KI (1984) N_2 fixation, and N and dry matter partitioning during a 4-year alfalfa stand. Crop Sci. 24: 811–815

Hendrickson QQ (1990) Asymbiotic nitrogen fixation and soil metabolism in three Ontario forests. Soil Biol. Biochem. 22: 967–971

Holland EA, Dentener FJ, Braswell BH & Sulzman JM (1999) Contemporary and pre-industrial global reactive nitrogen budgets. Biogeochem. 46: 7–43

Howarth RW, Billen G, Swaney D, Townsend A, Jaworski N, Lajtha K, Downing JA, Elmgren R, Caraco N, Jordan T, Berendse F, Freney J, Kudeyarov V, Murdoch P & Zhao-Liang Z (1996) Regional nitrogen budgets and riverine N & P fluxes for the drainages to the North Atlantic Ocean: natural and human influences. Biogeochem. 35: 75–139

Hurd TM, Raynal DJ & Schwintzer CR (2001) Symbiotic N_2 fixation of Alnus incana spp. rugosa in shrub wetlands of the Adirondack Mountains, New York, U.S.A. Oecologia 126: 94–103

Jaworski NA, Howarth RW & Hetling LJ (1997) Atmospheric deposition of nitrogen oxides onto the landscape contributes to coastal eutrophication in the Northeast United States. Environ. Sci. Technol. 31: 1995–2004

Johnson DW & Lindberg SE (Eds) (1991) Atmospheric deposition and forest nutrient cycling. Ecological Series 91. Springer-Verlag, NY

Jordan TE & Weller DE (1996) Human contributions to terrestrial nitrogen flux. BioScience 46: 655–664

Kittel TGF, Royle JA, Daly C, Rosenbloom NA, Gibson WP, Fisher HH, Schimel DS, Berliner LM & VEMAP2 Participants (1997) A gridded historical (1895–1993) bioclimate dataset for the conterminous United States. In: Reno NV (Ed) Proceedings of the 10th Conference on Applied Climatology, 20–24 October 1997 (pp 219–222). American Meteorological Society, Boston

Labandera C, Danso SKA, Pastorini D, Curbelo S & Martin V (1988) Nitrogen fixation in a white clover-fescue pasture using three methods of nitrogen-15 application and residual nitrogen-15 uptake. Agron. J. 80: 265–268

Lander CH & Moffitt D (1996) 1996 Nutrient Use in Cropland Agriculture (Commercial Fertilizers and Manure): Nitrogen and Phosphorus. Working Paper 14, RCAIII, NRCS, United States Department of Agriculture

Lovett GM & Lindberg SE (1993) Atmospheric deposition and canopy interactions of nitrogen in forests. Can. J. For. Res. 23: 1603–1616

Lovett GM & Rueth H (1999) Soil nitrogen transformations in beech and maple stands along a nitrogen deposition gradient. Ecol. Appl. 9(4): 1330–1344

Möller D & Schieferdecker (1989) Ammonia emission and deposition of NH_x in the G.D.R. Atmospheric Environment 23(6). 1187–1193

MRLC (1995) Multi-Resolution Land Characteristics (MRLC) Consortium Documentation Notebook; national land cover database. [online] URL: http://www.epa.gov/mrlc/

NADP; National Atmospheric Deposition Program/National Trends Network (2000) NADP Program Office, Illinois State Water Survey, 2204 Griffith Dr., Champaign, IL 61820. [online] URL: http://nadp.sws.uiuc.edu/nadpdata

Neff JC, Holland EA, Dentener FJ, McDowell WH & Russell KM (2002) The origin, composition and rates of organic nitrogen deposition: A missing piece of the nitrogen cycle? Biochemistry 57/58: 99–136

NRC; National Research Council (2000) Clean Coastal Waters: Understanding and Reducing the Effects of Nutrient Pollution. National Academy Press, Washington, DC

Ollinger SV, Aber JD, Lovett GM, Millham SE, Lathrop RG & Ellis JM (1993) A spatial model of atmospheric deposition for the northeastern U.S. Ecol. Appl. 3: 459–472

Prospero JM, Barrett K, Church T, Dentener F, Duce RA, Galloway JN, Levy H II, Moody J & Quinn P (1996) Atmospheric deposition of nutrients to the North Atlantic Basin. Biogeochemistry 35: 27–73

Puckett LJ (1995) Identifying the major sources of nutrient water pollution: a national catchment-based analysis connects nonpoint and point sources of nitrogen and phosphorus with regional land use and other factors. Environ. Sci. Technol. 29: 408–414

Rennie RJ, Rennie DA & Fried M (1978) Concepts of 15N usage in dinitrogen fixation. In: Isotopes in biological dinitrogen fixation (pp 107–113) International Atomic Energy Agency, Vienna

Rice WA (1980) Seasonal patterns of nitrogen fixation and dry matter production by clovers grown in the Peace River region. Can. J. Plant Sci. 60: 847–858

Roskoski JP (1980) Nitrogen fixation in hardwood forests of the northeastern United States. Plant Soil 54: 33–44

(SCS) Soil Conservation Service (1992) Agricultural Waste Management Field Handbook, Chapter 4, U.S. Government Printing Office, Washington, DC

Schlesinger W & Hartley AE (1992) A global budget for atmospheric NH_3. Biogeochemistry 15: 191–211

Smil V (1997) Global population and the nitrogen cycle. Scientific American, July: 76–81

Smil V (1999) Nitrogen in crop production: an account of global flows. Global Biogeochemical Cycles 13(2): 647–662

Smith RA, Schwarz GE & Alexander RB (1997) Regional interpretation of water-quality monitoring data. Wat. Resour. Res. 33(12): 2781–2798

Stoddard JL (1991) Trends in Catskill stream water quality: evidence from historical data. Water Resour. Res. 27: 2855–2864

Thomas GW & Gilliam JW (1977) Agro-ecosystems in the U.S.A. AgroEcosystems 4: 182–239

Tjepkema J (1979) Nitrogen fixation in forests of central Massachusetts. Can. J. Bot. 57: 11–16

Turner RE & Rabalais NN (1991) Changes in Mississippi River water quality this century. Biosci. 41: 140–147

USDA/NASS, U.S. Department of Agriculture, National Agricultural Statistics Service (1992) 1992 Census of Agriculture. [online] URL: http://www.nass.usda.gov/census/

USDA/NASS, U.S. Department of Agriculture, National Agricultural Statistics Service (1997) Virginia Agricultural Statistical Bulletin, 1997. Richmond, VA: United States Department of Agriculture, 1997

U.S. Department of Commerce (1990) 1990 Census of Population: General population characteristics, United States. 1990-CP-1-1, Bureau of the Census. [online] URL: http://www.census.gov/main/www/cen1990.html

USGS, United States Geological Survey (2000) National Water Information System Data Retrieval [online] URL: http://waterdata.usgs.gov/nwis-w/US/

Van Breemen N, Boyer EW, Goodale CL, Jaworski NA, Seitzinger S, Paustian K, Hetling L, Lajtha K, Eve M, Mayer B, Van Dam D, Howarth RW, Nadelhoffer KJ & Billen G (2002) Where did all the nitrogen go? Fate of nitrogen inputs to large watersheds in the northeastern U.S.A. Biochemistry 57/58: 267–293

Van der Hoek KW (1998) Estimating ammonia emission factors in Europe: summary of the work of the UNECE ammonia expert panel. Atmospheric Environment 32(3): 315–316

Van der Hoek KW & Bouwman AF (1999) Upscaling of nutrient budgets from agroecological niche to global scale. In: Smaling EMA, O Oenema & LO Fresco (Eds) Nutrient Disequilibria in Agroecosystems (pp 57–73) CAB International

Van Horn HH, Newton GL & Kunkle WE (1996) Ruminant nutrition from an environmental perspective: factors affecting whole-farm nutrient balance. J. Animal Sci. 74: 3082–3102

Van Horn HH (1998) Factors affecting manure quantity, quality, and use. Proceedings of the Mid-South Ruminant Nutrition Conference, Dallas-Ft. Worth, May 7–8, 1998. Texas Animal Nutrition Council, pp 9–20

Vitousek PM, Aber JD, Howarth RW, Likens GE, Matson PA, Schindler DW, Schlesinger WH & Tilman DG (1997) Human alteration of the global nitrogen cycle: sources and consequences. Ecol. Appl. 7(3): 737–750

Westerman DT, Kleinkopf GE, Porter LK & Leggett GE (1981) Nitrogen sources for bean seed production. Agronomy Journal 73: 660–664

Biogeochemistry **57/58**: 171–197, 2002.
© 2002 *Kluwer Academic Publishers. Printed in the Netherlands.*

Sources of nitrate in rivers draining sixteen watersheds in the northeastern U.S.: Isotopic constraints

BERNHARD MAYER[1,*], ELIZABETH W. BOYER[2], CHRISTINE GOODALE[3], NORBERT A. JAWORSKI[4], NICO VAN BREEMEN[5], ROBERT W. HOWARTH[6], SYBIL SEITZINGER[7], GILLES BILLEN[8], KATE LAJTHA[9], KNUTE NADELHOFFER[10], DOUWE VAN DAM[5], LEO J. HETLING[11], MILOSLAV NOSAL[12] & KEITH PAUSTIAN[13]

[1]*University of Calgary, Departments of Geology & Geophysics and Physics & Astronomy, 2500 University Drive NW, Calgary, Alberta, Canada T2N 1N4;* [2]*State University of New York, College of Environmental Science and Forestry, 1 Forestry Drive, Syracuse, NY 13210, U.S.A.;* [3]*Carnegie Institution of Washington, Department of Plant Biology, 260 Panama St., Stanford, CA 94305, U.S.A.;* [4]*USEPA (retired), 2004 S. Magnolia Ave., Sanford, FL 32771, U.S.A.;* [5]*Laboratory of Soil Science and Geology and Wageningen Institute for Environment and Climate Research, Wageningen University, Wageningen, the Netherlands;* [6]*Cornell University, Department of Ecology & Environmental Biology, Corson Hall, Ithaca, NY 14853, U.S.A.;* [7]*Rutgers University, Institute of Marine and Coastal Sciences, Rutgers/NOAA CMER Program, 71 Dudley Road, New Brunswick, NJ 08901, U.S.A.;* [8]*UMR Sisyphe, University of Paris VI, 4 Place Jussieu, 75005 Paris, France;* [9]*Department of Botany and Plant Pathology, Oregon State University, Corvallis OR 97331, U.S.A.;* [10]*Marine Biological Laboratory, The Ecosystems Center, Woods Hole, MA 02543, U.S.A.;* [11]*Department of Energy and Environmental Engineering, Rensselaer Polytechnic Institute, Troy, New York, U.S.A.;* [12]*University of Calgary, Department of Mathematics & Statistics, 2500 University Drive NW, Calgary, Alberta, Canada T2N 1N4;* [13]*Natural Resource Ecology Laboratory, Colorado State University, Fort Collins, Colorado 80523, U.S.A.*
(*author for correspondence, e-mail: bernhard@earth.geo.ucalgary.ca)*

Key words: denitrification, nitrate, nitrate sources, rivers, stable isotopes, $\delta^{15}N_{nitrate}$, $\delta^{18}O_{nitrate}$

Abstract. The feasibility of using nitrogen and oxygen isotope ratios of nitrate (NO_3^-) for elucidating sources and transformations of riverine nitrate was evaluated in a comparative study of 16 watersheds in the northeastern U.S.A. Stream water was sampled repeatedly at the outlets of the watersheds between January and December 1999 for determining concentrations, $\delta^{15}N$ values, and $\delta^{18}O$ values of riverine nitrate.

In conjunction with information about land use and nitrogen fluxes, $\delta^{15}N_{nitrate}$ and $\delta^{18}O_{nitrate}$ values provided mainly information about sources of riverine nitrate. In predominantly forested watersheds, riverine nitrate had mean concentrations of less than 0.4 mg NO_3^--N L^{-1}, $\delta^{15}N_{nitrate}$ values of less than +5‰, and $\delta^{18}O_{nitrate}$ values between +12 and +19‰. This indicates that riverine nitrate was almost exclusively derived from soil nitrification processes with potentially minor nitrate contributions from atmospheric deposition in some catchments. In watersheds with significant agricultural and urban land use, concentrations of riverine nitrate were as high as 2.6 mg NO_3^--N L^{-1} with $\delta^{15}N_{nitrate}$ values between +5 and +8‰ and $\delta^{18}O_{nitrate}$ values generally below +15‰. Correlations between nitrate concentrations, $\delta^{15}N_{nitrate}$ values, and N fluxes suggest that nitrate in waste water constituted a major, and nitrate in manure a minor additional source of riverine nitrate. Atmospheric nitrate deposition or nitrate-containing fertilizers were not a significant source of riverine nitrate in watersheds with significant agricultural and urban land use. Although complementary studies indicate that in-stream denitrification was significant in all rivers, the isotopic composition of riverine nitrate sampled at the outlet of the 16 watersheds did not provide evidence for denitrification in the form of elevated $\delta^{15}N_{nitrate}$ and $\delta^{18}O_{nitrate}$ values. Relatively low isotopic enrichment factors for nitrogen and oxygen during in-stream denitrification and continuous admixture of nitrate from the above-described sources are thought to be responsible for this finding.

Introduction

Human activity has greatly altered the nitrogen (N) cycle in terrestrial and aquatic ecosystems (e.g. Kinzing & Socolow 1994; Vitousek et al. 1997) causing increased nitrogen loads in many rivers (e.g. Paces 1982; Turner & Rabalais 1991; Jaworski & Hetling 1996; Goolsby 2000). According to mass balances, less than 30% of the anthropogenic N inputs to large watersheds are exported to the oceans with surface runoff in rivers and streams (Howarth et al. 1996; Boyer et al. 2002). Consequently, more than 70% of human-controlled N inputs are stored, denitrified, or volatilized in the watersheds. Because of their spatial and temporal variations, the relative importance of these N retention and transformation mechanisms is difficult to quantify on a watershed scale (Van Breemen et al. 2002). It also is difficult to determine the origin of nitrate that is exported from catchments, although there is evidence that different anthropogenic N inputs are differentially retained in large watersheds (Howarth et al. 1996).

Isotopic techniques have been successfully used in numerous case studies to identify nitrogen sources and to describe nitrogen transformations in terrestrial and aquatic ecosystems (e.g. Letolle 1980; Hübner 1986; Nadelhoffer & Fry 1994; Kendall 1998). Nitrogen isotope ratios have proven useful in quantifying the extent of point and non-point nitrogen sources to rivers (e.g. Fogg et al. 1998; Harrington et al. 1998). Nitrate derived from manure or sewage is usually characterized by $\delta^{15}N$ values between +7 and more than +20‰ (Kreitler & Jones 1975; Gormly & Spalding 1979; Kreitler 1979; Kreitler & Browning 1983; Aravena et al. 1993; Wassenaar 1995; Aravena & Robertson 1998). It is therefore isotopically distinct from N in atmospheric deposition (–10 to +8‰), from N in most synthetic fertilizers (0 $\pm$ 3‰), from natural soil organic N (–3 to +5‰) and nitrate generated therein by

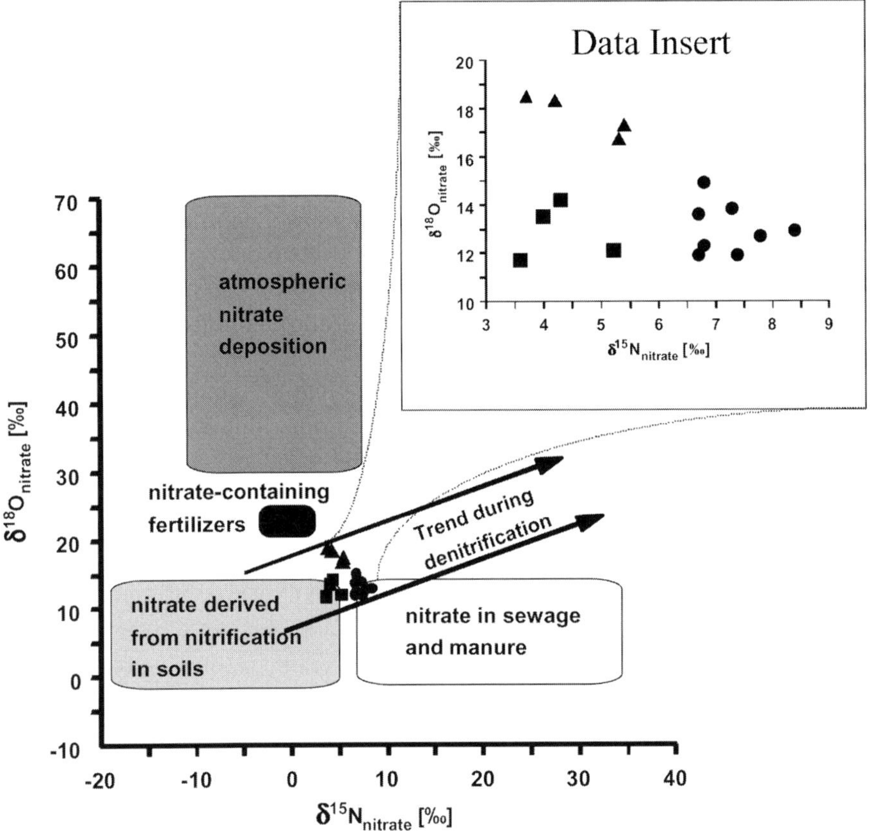

Figure 1. Mean δ^{15}N and mean δ^{18}O values of riverine nitrate collected at the outlet of 16 watersheds located in the Mid-Atlantic and New England states of the U.S.A. (cluster 1 = squares; cluster 2 = triangles; cluster 3 = circles; explanations see text). Ranges of isotopic compositions for four major nitrate sources are indicated by boxes: (a) atmospheric nitrate deposition, (b) nitrate-containing fertilizers, (c) nitrate derived from nitrification e.g. in soils, and (d) nitrate in manure and/or sewage. Also shown is the expected trend for the isotopic composition of residual nitrate undergoing microbial denitrification, assuming that the initial nitrate was derived from soil nitrification processes.

microbial nitrification (e.g. Kendall 1998). Usually, the latter three sources can not be differentiated by nitrogen isotope ratios alone, because of their wide and overlapping ranges of δ^{15}N values (Figure 1).

Recent advances in analytical methodology now allow the measurement of oxygen isotope ratios of nitrate (Amberger & Schmidt 1987; Voerkelius 1990; Wassenaar 1995; Revesz et al. 1997; Chang et al. 1999; Bräuer & Strauch 2000; Silva et al. 2000). Nitrate in atmospheric deposition has positive δ^{18}O values ranging from +25 to more than +70‰ (e.g. Voerkelius 1990; Durka

174

et al. 1994; Kendall 1998; Mayer et al. 2001). Nitrate-containing synthetic fertilizers typically have $\delta^{18}O$ values near $+22 \pm 3\%$ (Amberger & Schmidt 1987; Voerkelius 1990; Wassenaar 1995). Nitrate derived from microbial soil nitrification processes has $\delta^{18}O$ values between less than 0 and $+14\%$ depending on the nitrification pathway and the oxygen isotope ratios of the ambient water and O_2 at the site of nitrate formation (Mayer et al. 2001). The few available oxygen isotope ratio determinations for nitrate from manure (Wassenaar 1995) and sewage (Aravena et al. 1993) indicate comparatively low $\delta^{18}O$ values of less than $+15\%$ for these sources. Hence, the combined analysis of both $\delta^{15}N$ and $\delta^{18}O$ values of nitrate provides a tool for distinguishing between four major nitrate sources: (a) atmospheric deposition of nitrate, (b) nitrate-containing fertilizers, (c) nitrate derived from nitrification e.g. in soils, and (d) nitrate in manure and/or sewage (Figure 1).

The isotopic composition of nitrate is not only a powerful tool to determine its origin, but can also provide clues about nitrogen transformation processes such as ammonia volatilization and denitrification. Volatilization is typically accompanied by isotopic fractionation enriching the lighter isotope ^{14}N in the product ammonia gas (Hübner 1986) causing the remaining nitrogen-bearing compounds such as ammonium and subsequently nitrate to become isotopically enriched in ^{15}N. This has been frequently observed in farmlands after urea and manure applications (Heaton 1986), in sewage treatment plants, and septic systems (Aravena et al. 1993; McClelland et al. 1997; McClelland & Valiela 1998). Another process capable of causing significant alterations to the isotopic composition of nitrate is microbial denitrification, during which the lighter isotopes ^{14}N and ^{16}O are preferentially metabolized by microorganisms and are converted to N_2 and N_2O, causing an enrichment of the heavy isotopes ^{15}N and ^{18}O in the remaining nitrate through kinetic isotope effects (e.g. Blackmer & Bremner 1977; Mariotti et al. 1982; Mariotti et al. 1988; Böttcher et al. 1990). In single source closed system scenarios, microbial denitrification results in progressively increasing $\delta^{15}N_{nitrate}$ and $\delta^{18}O_{nitrate}$ values as nitrate concentrations decrease. The extent of nitrogen isotope fractionation can be variable and is influenced by several factors including temperature and concentration of the substrate (Mariotti et al. 1982). The increase in $\delta^{15}N_{nitrate}$ values due to microbial denitrification appears to be between 1.5 and 2.0 times that of $\delta^{18}O_{nitrate}$ values in groundwater systems (Böttcher et al. 1990; Aravena & Robertson 1998) and riparian zones (Cey et al. 1999; Mengis et al. 1999). Hence, the remaining nitrate eventually obtains elevated $\delta^{15}N$ and $\delta^{18}O$ values, which are unique for nitrate that has undergone denitrification under closed system conditions (Figure 1). Microbial denitrification may occur in soils, in aquifers, in riparian and hyporheic zones, in river water and sediments, and in sewage treatment systems, provided that organic carbon or reduced inorganic compounds are available as electron donors and that the appropriate redox conditions are achieved (Knowles 1982).

Current and expected future human alterations to the N balances in catchments make it desirable that we understand how N is cycled through watersheds. For assessing the consequences of increasing anthropogenic nitrogen inputs on issues such as acidification of terrestrial and aquatic ecosystems, nitrate concentrations in aquifers, and eutrophication of surface waters and coastal oceans, a detailed understanding of the fate of N from individual watershed sources is required. These N sources include (a) atmospheric deposition, (b) fertilizers, (c) N fixation in forests and crops, (d) net import of N in food and feedstocks, (e) mineralization of soil organic matter, (f) animal manure, and (g) municipal and industrial waste water (note that a-d represent new N inputs to catchments, whereas e-g can be considered a recycling of watershed-internal N). Nitrogen isotope techniques have proven useful for obtaining information about sources of nitrate or N transformation processes in several compartments of watersheds (e.g. Knowles & Blackburn 1993; Macko & Ostrom 1994; Nadelhoffer & Fry 1994). New analytical capabilities allowing the precise and accurate determination of both δ^{15}N and δ^{18}O values of dissolved nitrate have added a promising tool for elucidating the nitrogen cycle in watersheds.

Here we evaluate the usefulness of the isotopic composition of riverine nitrate in describing N cycling in a comparative study of 16 watersheds in the northeastern U.S.A. Our specific goals were to determine whether the isotopic composition of riverine nitrate can be used to identify its watershed sources and to test whether the isotopic composition of riverine nitrate provides evidence for microbial denitrification in the watersheds. Hence, the objective was to evaluate whether the isotopic composition of riverine nitrate provides source or process information, or a combination of both.

Methods

Sixteen watersheds in the mid-Atlantic and New England states of the U.S.A. with well-constrained N budgets were selected for this study (Figure 2). Related papers in this issue detail watershed characteristics (Boyer et al. 2002), develop nitrogen budgets (Boyer et al. 2002), and quantify nitrogen storage and sinks (Van Breemen et al. 2002) for these basins. The reader is referred to these companion papers for a detailed description of the methods used for assessing nitrogen and nitrate fluxes.

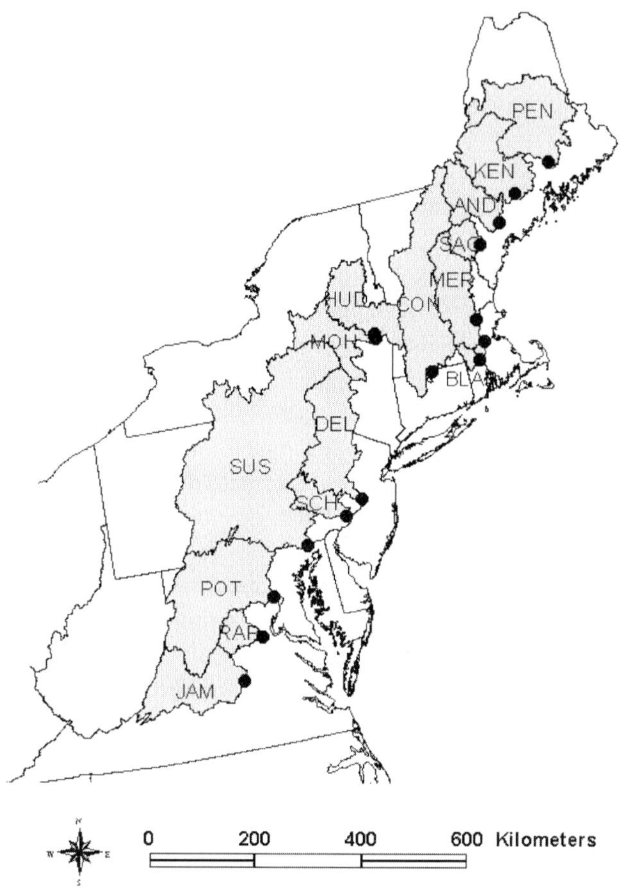

Figure 2. Location of the following 16 watersheds in the in the Mid-Atlantic and New England states of the U.S.A.: Penobscot (PEN), Kennebec (KEN), Androscoggin (AND), Saco (SAC), Merrimack (MER), Charles, Blackstone (BLA), Connecticut (CON), Hudson (HUD), Mohawk (MOH), Delaware (DEL), Schuylkill (SCH), Susquehanna (SUS), Potomac (POT), Rappahannock (RAP), James (JAM). The watershed boundaries shown are delineated upstream of the following USGS stations (black dots), from which streamflow data and water samples for concentration and isotope analyses were obtained: Penobscot River at Eddington, ME; Kennebec River at North Sidney, ME; Androscoggin River near Auburn, ME; Saco River at Cornish, ME; Merrimack River below Concord River at Lowell, MA; Charles River at Dover, MA; Blackstone River at Manville, RI; Connecticut River at Thompsonville, CT; Hudson River above lock 1 near Waterford, NY; Mohawk River at Cohoes NY; Delaware River at Trenton, NJ; Schuylkill River at Philadelphia, PA; Susquehanna River at Conowingo, MD; Potomac River near Washington, DC Lower Falls Pump Station; Rappahannock River near Fredericksburg, VA; James River at Cartersville, VA.

Sample collection

Water from rivers draining the 16 watersheds was sampled up to 7 times between January and December 1999 at USGS stream gaging stations located at the outlet of each basin (Figure 2). Typically, three liters of water were sampled by USGS personnel and shipped in cooled containers by overnight courier to the Isotope Science Laboratory at the University of Calgary.

Nitrate concentrations and fluxes

Nitrate and nitrite concentrations in stream water at the gaging stations, which define the outlet of the 16 watersheds, were determined monthly by various northeastern offices of the USGS Water Resources Division. Daily flows from these gaging stations were obtained from the USGS national watershed information system (USGS 2000).

Riverine fluxes of oxidized inorganic N (defined as the dominant oxidized forms nitrate plus nitrite) were calculated for water year 1999 (October 1, 1998 to September 30, 1999) using an automated implementation of the Beale ratio estimator (Richards & Holloway 1987). This method uses a flow-based stratification to estimate daily and annual loads from the infrequent concentration measurements and daily flow values. For five basins, for which 1999 fluxes could not be calculated due to a lack of concentration or flow data, average annual flux estimates reported for the period 1988–1993 (Boyer et al. this issue) were used as a surrogate. Mean annual concentrations of oxidized inorganic N in each river draining the 16 watersheds were estimated by dividing riverine oxidized inorganic N fluxes by the mean annual discharge (the average of the daily flows observed throughout the water year).

Analysis of nitrogen and oxygen isotope ratios of nitrate

Within two days of collection, water samples were passed through anion exchange resins to retain and store nitrate for subsequent isotopic analyses in the Isotope Science Laboratory at the University of Calgary, Canada. The nitrate was later eluted and converted to $AgNO_3$ using a modified version of a technique described by Silva et al. (2000).

Water samples were passed through a cation exchange resin (2 mL of 50W-X4, H^+-form, Bio-Rad) at a rate of 5 mL min^{-1} to exchange cations with H^+ and simultaneously remove HCO_3^- through acidification. Subsequently, nitrate, sulfate, and phosphate were retained quantitatively on an anion exchange resin (2 mL AG 1-X8 resin, Cl^--form, Bio-Rad). After rinsing with 10 mL deionized water, the anion exchange resins were stored at 5 °C in darkness until further processing.

Nitrate, sulfate, and phosphate were eluted from the anion exchange resins into a beaker by passing 15 mL 3 M HCl through the columns. One mL of 0.2 M $BaCl_2$ solution was added to the HCl eluate to precipitate sulfate and phosphate as $BaSO_4$ and $Ba_3(PO_4)_2$, respectively. After 24 hours, $BaSO_4$ and $Ba_3(PO_4)_2$ were removed by filtration (0.45 μm membrane filter). Excess Ba^{2+} was removed by passing the sample through a cation exchange resin in H^+-form (2 mL 50W-X4 resin, Bio-Rad). The remaining, almost DOC-free solution containing HNO_3 and HCl was neutralized by adding approximately 7.5 g pure and pre-washed Ag_2O (Merck, Darmstadt, Germany). The resulting AgCl precipitate was removed by membrane filtration (0.45 μm) leaving only Ag^+ and NO_3^- in solution (equation 1).

$$Ag_2O + HNO_3 + HCl \rightarrow Ag^+ + NO_3^- + H_2O + AgCl \downarrow \qquad (1)$$

Thereafter, the solution was freeze-dried yielding pure $AgNO_3$. For oxygen isotope analyses on nitrate, 10 mg $AgNO_3$ was mixed with 2 mg pure graphite powder. This mixture was placed in a 6 mm quartz tube, which was evacuated and flame sealed. The mixture was thermally decomposed at 860 °C for 3 hours, followed by slow cooling to ensure complete conversion of the nitrate-oxygen to CO_2 (equation 2):

$$2AgNO_3 + 3C \rightarrow 2Ag + N_2 + 3CO_2 \qquad (2)$$

The resulting CO_2 was cryogenically purified and analyzed mass spectrometrically. Accuracy and precision of the measurements was assured by repeated analyses of laboratory internal and international reference materials. The mean $\delta^{18}O_{nitrate}$ value obtained for IAEA-NO-3 was with $+23.1 \pm 0.7\%_0$ ($n = 12$) within the range of previously reported $\delta^{18}O$ values for this reference material (e.g. Revesz et al. 1997; Bräuer & Strauch 2000). The reproducibility of nitrate extraction, gas preparation, and mass spectrometric measurement was found to be better than $\pm 1.0\%_0$ for $\delta^{18}O_{nitrate}$, as determined by duplicate analyses.

Nitrogen isotope ratios were determined on N_2 after thermal decomposition of $AgNO_3$ in an elemental analyzer (Carlo Erba NA 1500) and subsequent continuous-flow isotope ratio mass spectrometry (CF-IRMS). $\delta^{15}N$ values for all samples were calibrated against international reference materials (IAEA N1 and N2). The reproducibility of nitrate extraction, gas preparation, and mass spectrometric measurement was better than $\pm 0.3\%_0$ for $\delta^{15}N_{nitrate}$ determinations.

Oxygen isotope ratios for water samples were obtained using standard equilibration techniques (Epstein & Mayeda 1953) with a reproducibility of better than $\pm 0.2\%_0$. All stable isotope ratios are expressed in the usual delta per mil ($\%_0$) notation:

$$\delta_{sample}(\%_0) = [(R_{sample} - R_{standard})/R_{standard}] * 1000 \qquad (3)$$

where R is the $^{15}N/^{14}N$ or $^{18}O/^{16}O$ ratio of the sample and the standard, respectively. $\delta^{15}N$ values are reported with respect to AIR and $\delta^{18}O$ values with respect to Vienna Standard Mean Ocean Water (V-SMOW).

Results

Watershed and land use characteristics and mean annual N inputs and exports (1988–1993) for the 16 watersheds as described by Boyer et al. (2002) are summarized in Table 1. Because river gaging stations were generally located upstream from the major coastal population centers, urban land use constituted less than 5% of the catchment area in most watersheds. Only the watersheds of the Merrimack River (9%), the Schuylkill River (10%), the Blackstone River (18%), and of the Charles River (22%) had significant portions of urban land use. Agricultural land use varied between 2 and 38% among the catchments. Between 48 and 87% of the area of the 16 watersheds was forested, with the remainder being wetlands and water surfaces.

Annual new nitrogen inputs to the 16 watersheds from four major sources – atmospheric NO_y deposition, fertilizers, N fixation in forests and crops, and net import of N in food and feedstocks – ranged from less than 1000 kg N km^{-2} a^{-1} in the predominantly forested watersheds to more than 3000 kg N km^{-2} a^{-1} in catchments with significant agricultural and urban land use (Table 1). Mean annual export of total nitrogen in streamflow ranged from more than 40% of the N input in some predominantly forested watersheds to less than 20% of the total N input in some catchments with significant agricultural and urban land use. On average, 28% of the N inputs to the watersheds were exported in riverine flows (Boyer et al. 2002).

Selected mean annual nitrogen fluxes for the 16 watersheds such as N inputs with atmospheric deposition and fertilizers, and N transfers with human wastewater flows and with animal waste (manure) production, are summarized in Table 2. These data were determined for the period 1988 to 1993 (Boyer et al. 2002), but are believed to be representative for the latter part of the 90's. Nitrate-N inputs with wet and dry deposition varied between less than 400 kg N km^{-2} a^{-1} in the northernmost forested catchment to more than 800 kg N km^{-2} a^{-1} in watersheds receiving significant amounts of industrial emissions. Nitrogen inputs with fertilizers varied markedly from

Table 1. Area, land use, population density, total nitrogen input, and total nitrogen export with riverine flow (1988–1993) for the 16 watersheds in the mid-Atlantic and New England states, U.S.A. (c.f. Boyer et al. 2002)

River/ Watershed	Area [km^2]	Land Area Forested [%]	Land Area Agricultural [%]	Land Area Urban [%]	Population [km^{-2}]	Total N Inputs [kg km^{-2} yr^{-1}]	River N Export [% of N Input]
Penobscot	20109	83.8	1.5	0.4	8	835	38
Kennebec	13994	79.6	5.9	0.9	9	1099	30
Androscoggin	8451	84.6	4.8	1.1	17	1310	31
Saco	3349	87.4	3.6	0.8	16	1233	32
Merrimack	12005	74.7	7.8	8.7	143	2228	22
Charles	475	59.3	8.4	22.2	556	4406	40
Blackstone	1115	63.3	8.1	17.6	276	3407	33
Connecticut	25019	79.0	9.0	4.0	65	2262	24
Hudson	11942	80.8	10.4	2.7	32	1985	25
Mohawk	8935	63.1	28.0	4.7	54	3420	23
Delaware	17560	74.7	16.7	3.3	85	2967	32
Schuylkill	4903	48.1	38.4	10.2	293	5717	31
Susquehanna	70189	66.7	28.5	2.4	54	4173	23
Potomac	29940	60.8	34.6	2.6	63	4689	19
Rappahannock	4134	61.3	35.9	1.4	24	4246	11
James	16206	80.6	15.6	1.4	24	2773	11

Table 2. Selected mean annual N fluxes for the 16 watersheds averaged for the years 1988 to 1993 (c.f. Boyer et al. 2002)

River/ Watershed	Wet & Dry NO_3^--N Deposition [kg km^{-2} yr^{-1}]	Fertilizer N [kg km^{-2} yr^{-1}]	Waste Water N [kg km^{-2} yr^{-1}]	Animal Waste N [kg km^{-2} yr^{-1}]
Penobscot	362	91	25	78
Kennebec	428	54	56	187
Androscoggin	495	80	62	237
Saco	566	42	44	82
Merrimack	606	147	171	219
Charles	674	197	1372	143
Blackstone	707	307	885	333
Connecticut	631	274	123	488
Hudson	658	204	60	439
Mohawk	708	411	110	1261
Delaware	811	527	170	651
Schuylkill	885	1207	618	2147
Susquehanna	816	615	178	1909
Potomac	714	1024	57	2583
Rappahannock	615	1030	35	2234
James	652	361	145	1096

less than 100 kg N km^{-2} a^{-1} in predominantly forested watersheds to more than 1000 kg N km^{-2} a^{-1} in watersheds with intensive agricultural land use. Transfers of N with human wastewater and animal waste were not considered new inputs of N to each watershed, but rather are a recycling of watershed-internal N (Boyer et al. 2002). N transfers in wastewater were correlated with population densities and thus urban land use (Table 1). Waste water inputs ranged from 25 to 178 kg N km^{-2} a^{-1} in watersheds with less than 9% urban land, and from 618 to 1372 kg N km^{-2} a^{-1} in watersheds with more than 10% urban land. Nitrogen transfers in animal waste (manure) ranged from 78 kg N km^{-2} a^{-1} in predominantly forested catchments to 2583 kg N km^{-2} a^{-1} in watersheds with significant agricultural land use.

Average concentrations, fluxes, and isotopic compositions for riverine nitrate determined at the outlet of each of the 16 watersheds are summarized in Table 3. Mean (discharge weighted) annual nitrate concentrations ranged

Table 3. Mean nitrate concentrations, average NO_3^--N fluxes, and mean isotopic composition of nitrate and water-oxygen for the water year 1999 in rivers draining 16 watersheds in the mid-Atlantic and New England states, U.S.A.

River/ Watershed	$[NO_3^-$-N] [mg L^{-1}]	# of $[NO_3^-]$ analyses	# of discharge measurements	NO_3^--N Export [kg km^{-2} yr^{-1}]	$\delta^{15}N_{nitrate}$ [‰]$_{AIR}$	n	$\delta^{18}O_{nitrate}$ [‰]$_{SMOW}$	n	$\delta^{18}O_{water}$ [‰]$_{SMOW}$	n
Penobscot	0.11*	0	0	66*	3.7 ± 2.9	5	18.5	1	−9.5 ± 0.7	5
Kennebec	0.15*	0	0	87*	4.2 ± 1.5	4	18.3	1	−9.5 ± 0.5	4
Androscoggin	0.18*	0	365	112*	4.0 ± 0.9	4	13.5 ± 2.6	3	−9.6 ± 0.6	5
Saco	0.12*	0	365	81*	5.4 ± 1.1	3	17.3 ± 3.2	2	−9.1 ± 0.7	5
Merrimack	0.21	13	365	100 ± 29	6.8 ± 2.4	3	12.3 ± 0.8	2	−7.6 ± 1.2	5
Charles	0.54	19	365	260 ± 18	6.8 ± 2.7	4	14.9 ± 2.0	4	−6.1 ± 1.0	5
Blackstone	0.68	9	365	336 ± 217	7.8 ± 2.2	4	12.7 ± 4.4	4	−6.7 ± 0.8	6
Connecticut	0.31	10	365	153 ± 26	5.3 ± 1.4	6	16.7 ± 1.9	4	−9.3 ± 1.0	7
Hudson	0.36*	0	365	222*	3.6 ± 1.0	2	11.7 ± 0.8	2	−10.9 ± 0.3	2
Mohawk	0.62	12	365	249 ± 16	5.2 ± 0.9	2	12.1 ± 0.8	2	−11.4 ± 0.2	2
Delaware	0.87	13	365	341 ± 41	7.3 ± 1.7	3	13.8 ± 4.1	4	−7.9 ± 1.1	4
Schuylkill	2.57	20	365	1025 ± 64	8.4 ± 2.0	3	12.9 ± 3.8	4	−7.1 ± 0.8	4
Susquehanna	1.10	16	365	319 ± 47	7.4 ± 1.7	5	11.9 ± 2.4	5	−9.0 ± 1.0	5
Potomac	1.07	68	365	162 ± 9	6.7 ± 1.1	5	11.9 ± 2.5	6	−7.4 ± 0.8	7
Rappahannock	0.49	32	365	69 ± 5	6.7 ± 0.5	3	13.6 ± 1.6	2	−6.0 ± 1.2	3
James	0.21	31	365	41 ± 4	4.3 ± 0.3	3	14.2 ± 2.3	3	−7.1 ± 1.0	4

*Data derived from 1988–1993 estimates (Jaworswki, personal communication).

from as low as 0.11 mg NO_3^--N L^{-1} in rivers draining predominantly forested watersheds to 2.57 mg NO_3^--N L^{-1} in the Schuylkill River, which drains considerable areas of urban and agricultural land. Mean annual NO_3^--N fluxes ranged from less than 50 to approximately 225 kg km^{-2} a^{-1} in watersheds with less than 20% combined agricultural and urban land use, and between 100 and 1025 kg km^{-2} a^{-1} in watersheds with more than 20% combined agricultural and urban land use. Mean $\delta^{15}N$ values for riverine nitrate varied from less than +5‰ in predominantly forested watersheds to more than +7‰ in catchments with high percentages of agricultural and/or urban land use. The oxygen isotope ratios of nitrate in 12 of the 16 rivers were uniform with an average $\delta^{18}O_{nitrate}$ value of +13.0 $\pm$ 1.0‰. Only in four mainly forested catchments $\delta^{18}O_{nitrate}$ values were markedly higher: between +16.7 and +18.5‰. The oxygen isotope ratios of the river water ranged between –11.4 and –6.0‰ during the observation period. No significant linear relation between the oxygen isotope ratios of water and nitrate was observed for the individual samples from the sixteen watersheds (r^2 = 0.10, p = 0.517, n = 49).

Discussion

The following discussion is based on mean annual concentrations and mean isotopic compositions of riverine nitrate (estimated as described above) from 16 different watersheds. The current database is insufficient to interpret seasonal or flow-dependent variations of nitrate concentrations and isotopic compositions for individual rivers or changes of these parameters along the flowpath within each watershed. Such investigations were beyond the scope of this study, but clearly deserve further attention.

Although the overall range of $\delta^{15}N$ and $\delta^{18}O$ values was comparatively narrow, the isotopic composition of riverine nitrate appeared to be grouped in three different clusters (see data insert Figure 1). At the outlet of the watersheds Androscoggin, Hudson, Mohawk, and James (cluster 1: squares in Figure 1), mean $\delta^{15}N$ values of riverine nitrate were below +5‰ and mean $\delta^{18}O_{nitrate}$ values below +14‰. Samples from Penobscot, Kennebec, Saco, and Connecticut (cluster 2: triangles in Figure 1) had similarly low $\delta^{15}N_{nitrate}$ values, but slightly elevated $\delta^{18}O_{nitrate}$ values (+16 to +19‰). Riverine nitrate from the remaining eight watersheds (cluster 3: circles in Figure 1) had $\delta^{18}O$ values generally below +15‰ and $\delta^{15}N$ values of more than +6‰. The isotopic composition of nitrate in cluster 1 suggests that nitrification processes in soils were the major source of riverine nitrate (Figure 1). The isotopic composition of riverine nitrate in clusters 2 and 3 could be explained by mixing of nitrate from soil nitrification and from other sources (e.g.

atmospheric deposition, fertilizers, sewage and/or manure). Alternately, denitrification of nitrate that was initially formed by nitrification processes in soils, could also result in an isotopic composition similar to that of riverine nitrate in clusters 2 and 3 (see arrows in Figure 1). Combined evaluation of concentration, flux, and isotope data was pursued to gain a better understanding of the somewhat ambiguous information provided by the isotopic composition of riverine nitrate in 12 of the 16 watersheds (clusters 2 and 3).

Mixing of nitrate from various sources or denitrification?

Plotting nitrogen isotope ratios versus nitrate concentrations often reveals whether denitrification or mixing of nitrate from various sources is responsible for increasing $\delta^{15}N_{nitrate}$ values in a given aquatic system. Microbial denitrification typically results in progressively increasing $\delta^{15}N_{nitrate}$ values as nitrate concentrations decrease, whereas mixing of nitrate from two or more sources can result in patterns of increasing $\delta^{15}N$ and concentration values (see schematic inserts in Figure 3). Although we compared data from 16 different rivers, a clear trend of increasing $\delta^{15}N_{nitrate}$ values with increasing nitrate concentrations was evident (Figure 3) with a 2nd degree polynomial regression yielding a r^2 value of 0.78 ($p < 0.001$; $n = 16$). This suggest that riverine nitrate in most watersheds contained contributions from at least two different sources: one nitrate source generating low concentrations and $\delta^{15}N_{nitrate}$ values below +4‰, and another nitrate source with variable but generally high concentrations and a $\delta^{15}N_{nitrate}$ value above +8‰. It appears plausible that the variability of $\delta^{15}N$ values of riverine nitrate at the outlet of the 16 watersheds was governed primarily by mixing of two or more sources of nitrate, rather than by microbial denitrification within the watershed, although the latter can not be excluded based on the presented data set.

Effect of land use

Figure 4 displays a significant positive linear relation ($r^2 = 0.75$, $p = 0.001$, $n = 16$) between increasing $\delta^{15}N$ values of riverine nitrate and percentages of agricultural plus urban land. In watersheds with less than 15% agricultural and urban land, or more than 80% forest cover, $\delta^{15}N_{nitrate}$ values varied between +3.5 and +5.5‰. In watersheds with more than 15% agricultural and urban area, $\delta^{15}N$ values of riverine nitrate were typically above +6‰. Hence, nitrogen isotope ratios of riverine nitrate appear to be directly related to land use practices with higher percentages of urban and/or agricultural land in the watershed causing elevated $\delta^{15}N_{nitrate}$ values.

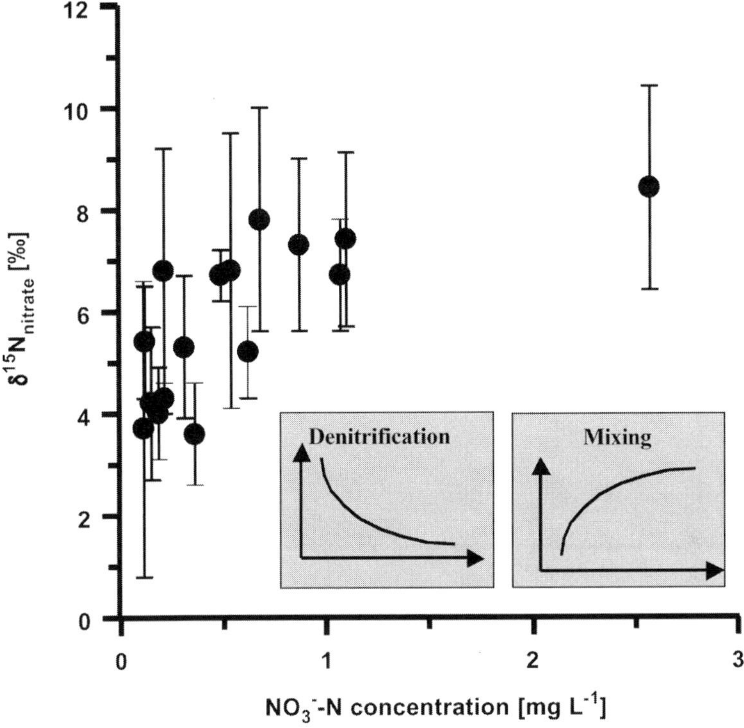

Figure 3. Mean δ^{15}N values of riverine nitrate versus NO_3^--N concentrations. A 2nd degree polynomial regression yielded a positive relation with a r^2 value of 0.78 ($p < 0.001$; $n = 16$). The left insert shows the expected trend of increasing $\delta^{15}N_{nitrate}$ values with decreasing nitrate concentration typical for denitrification in a single source closed system scenario (no nitrate from other sources added). The right insert displays a trend of increasing $\delta^{15}N_{nitrate}$ values with increasing nitrate concentrations. Such a scenario could be explained as a result of mixing of nitrate from two sources: one with low nitrate concentrations and $\delta^{15}N_{nitrate}$ values and another one with high nitrate concentrations and $\delta^{15}N_{nitrate}$ values.

Nitrogen sources

Figures 3 and 4 provide evidence that the source generating nitrate with low concentrations and δ^{15}N values of less than +4‰ is located in forested areas. We suggest that this source is mineralization of soil organic matter, which typically causes low nitrate concentrations in seepage water and surface runoff from forested catchments (Sollins & McCorison 1981; Stottlemyer & Troendle 1992; Hedin et al. 1995; Vanderbilt & Lajtha 2000) and $\delta^{15}N_{nitrate}$ values of less than +4‰ (e.g. Durka et al. 1994; Nadelhoffer & Fry 1994). This is presumably true for most first and second order streams in the 16

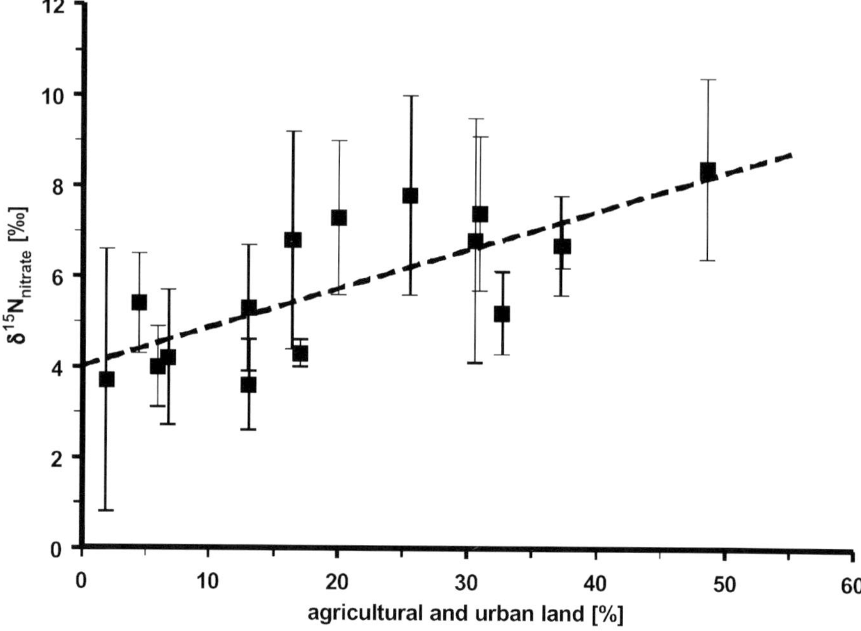

Figure 4. Mean δ^{15}N values of riverine nitrate versus percentage of agricultural plus urban land in the watersheds. A significant positive linear relation with $r^2 = 0.75$ ($p = 0.001$, $n = 16$) was observed.

watersheds, which all have forested headwater portions (Boyer, personal communication).

Combined evaluation of Figures 3 and 4 also suggests that in watersheds with more than 15% agricultural plus urban land use, concentrations and nitrogen isotope ratios of riverine nitrate were markedly higher than in predominantly forested areas. This can in principle result from three different mechanisms: (1) admixture of nitrate with runoff from manured agricultural areas (e.g. Kreitler & Jones 1975; Kreitler 1979); (2) admixture of nitrate from sewage treatment plants or septic systems (e.g. Aravena & Robertson 1998); (3) microbial denitrification (e.g. Farrell et al. 1996; Ostrom et al. 1998). The first two mechanisms typically entail increasing concentrations and δ^{15}N values of nitrate in aquatic systems, since nitrate derived from manure or sewage has usually δ^{15}N values higher than +7‰ (Fogg et al. 1998). The latter process, in contrast, is generally characterized by increasing δ^{15}N$_{nitrate}$ values and decreasing nitrate concentrations (Böttcher et al. 1990; Aravena et al. 1993; Farrell et al. 1996). Comparison of riverine nitrate from 16 different watersheds revealed that increasing δ^{15}N$_{nitrate}$ values were accompanied by increasing nitrate concentrations (Figure 3). Also, δ^{15}N$_{nitrate}$ values

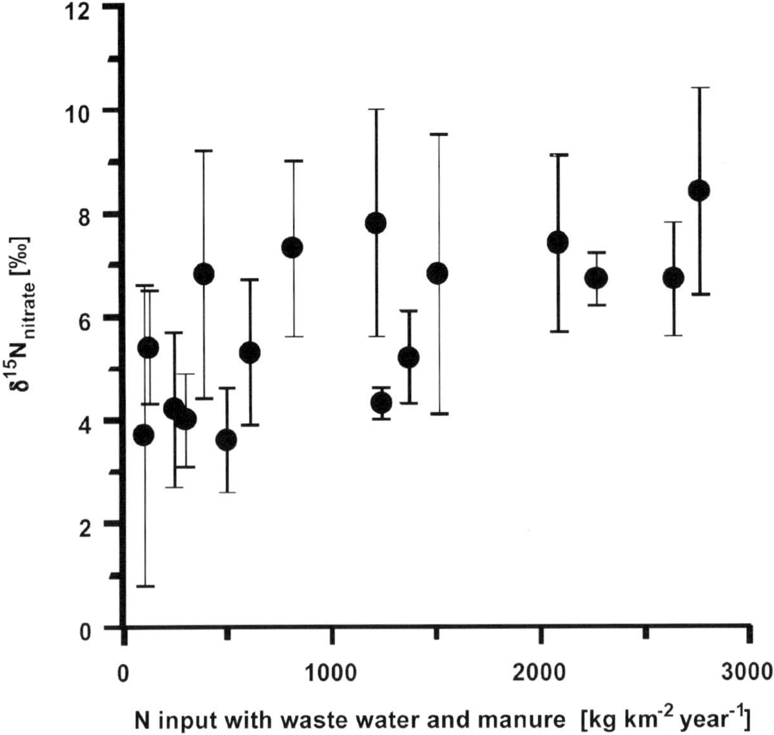

Figure 5. Mean $\delta^{15}N$ values of riverine nitrate versus combined N input with waste water and animal waste (manure) in the 16 watersheds. A 2nd degree polynomial regression yielded a significant positive relation with a r^2 value of 0.68 ($p < 0.006$; $n = 16$).

and the annual N fluxes with waste water and manure in the respective watersheds (Figure 5) displayed a significant positive relation with a r^2 value of 0.68 ($p < 0.006$; $n = 16$). We propose, therefore, that elevated $\delta^{15}N_{nitrate}$ values in rivers draining watersheds with significant urban and agricultural land use were caused by nitrate from sewage and/or manure. Since nitrate derived from both of these sources is typically characterized by $\delta^{15}N$ values higher than +7‰, we used the Best Regression Algorithm to test a variety of bivariate polynomial regression models to describe the effect of waste water and animal waste on the $\delta^{15}N$ values of riverine nitrate. A highly significant polynomial regression of the 2nd degree ($r^2 = 0.70$; $p = 0.002$; $n = 16$) between mean $\delta^{15}N_{nitrate}$ values and the N fluxes with sewage and manure in the watersheds suggests that about 50% of the variability in the $\delta^{15}N$ values of riverine nitrate could be attributed to N fluxes in sewage, whereas approximately 20% was caused by manure applied to the agriculturally used portions of the watersheds. Riverine nitrate in watersheds with waste water

N fluxes above 150 kg N km^{-2} a^{-1} had $\delta^{15}N$ values of more than +6.5‰ (Tables 2, 3). In contrast, watersheds with wastewater N fluxes of less than 150 kg N km^{-2} a^{-1} had typically $\delta^{15}N_{nitrate}$ values below +5.5‰. The two exceptions – the Potomac and the Rappahannock River having sewage N fluxes of less than 100 kg km^{-2} a^{-1} in their catchments – had $\delta^{15}N_{nitrate}$ values of +6.7‰. These comparatively high nitrogen isotope ratios were in all likelihood caused by the very high manure applications ($>$2000 kg N km^{-2} a^{-1}) in these watersheds (Table 2). Sewage, and to a lesser extent manure were probably responsible for the high concentrations and nitrogen isotope ratios of riverine nitrate in watersheds with significant agricultural and urban land use. Lack of information about the spatial and seasonal variation of $\delta^{15}N_{nitrate}$ values of sewage and manure precludes a more quantitative assessment of the contributions of these anthropogenic N fluxes. Nevertheless, the good positive correlation between $\delta^{15}N$ values and (a) concentrations of riverine nitrate and (b) N inputs with manure and sewage indicate that denitrification was not the major cause for the elevated $\delta^{15}N$ values observed in riverine nitrate at the outlet of some of the 16 watersheds.

Since most of the variability in the $\delta^{15}N$ values of riverine nitrate from the 16 different watersheds could be sufficiently explained by mixing of nitrate from various sources, the variability of the oxygen isotope ratios of riverine nitrate was also evaluated in terms of multiple source mixing. A significant linear relation (r^2 = 0.76; p = 0.001; n = 16) between mean $\delta^{18}O$ values of riverine nitrate and inverse nitrate concentrations (Figure 6) suggests two source mixing. One source of nitrate causing variable but high concentrations had a $\delta^{18}O$ value of +12‰ (= y intercept in Figure 6). Such oxygen isotope ratios are typical for nitrate derived from nitrification processes in manure, sewage, or soils (Figure 1). The second source responsible for low nitrate concentrations was apparently associated with higher $\delta^{18}O_{nitrate}$ values of more than +17‰. $\delta^{18}O_{nitrate}$ values higher than +17‰ can be caused by denitrification or by mixing of nitrification nitrate ($\delta^{18}O_{nitrate}$ $<$ +14‰) with fertilizer nitrate ($\delta^{18}O_{nitrate}$ $\sim$ +22‰) or atmospherically deposited nitrate ($\delta^{18}O_{nitrate}$ $>$ +25‰). To evaluate the influence of the latter two nitrate sources on the oxygen isotope ratios of riverine nitrate, mean $\delta^{18}O_{nitrate}$ values were plotted versus the N inputs with atmospheric nitrate deposition (Figure 7(a)) and synthetic fertilizers (Figure 7(b)) to the 16 watersheds expressed as percentage of total N input.

Mean $\delta^{18}O_{nitrate}$ values showed a weak positive correlation (r^2 = 0.63, p = 0.01; n = 16) with the percentage of atmospheric NO_3^--N deposition of the total N inputs (Figure 7(a)). A tendency to increasing $\delta^{18}O_{nitrate}$ values was particularly obvious when atmospheric NO_3^--N deposition represented more than 40% of the entire N inputs to the watersheds, as in the Kennebec, Penob-

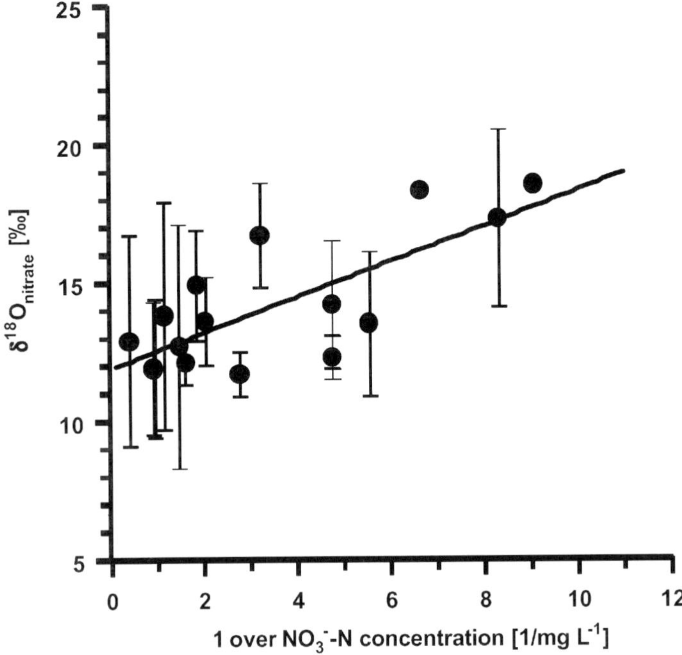

Figure 6. Mean $\delta^{18}O$ values of riverine nitrate versus inverse nitrate concentration ($r^2 = 0.76$, $p = 0.001$, $n = 16$).

scot, and Saco catchments, which yielded $\delta^{18}O_{nitrate}$ values between +16 and +19‰. The comparatively high $\delta^{18}O$ values suggest that not all nitrate in the surface runoff from these predominantly forested ($>79\%$) watersheds was derived from nitrification ($\delta^{18}O_{nitrate} < +14\%$). Since $\delta^{18}O$ values of atmospheric nitrate vary typically between +25 and more than +70‰ (e.g. Durka et al. 1994; Kendall 1998), $\delta^{18}O$ values of up to +19‰ potentially indicate that a small proportion of riverine nitrate might have been derived from atmospheric NO_3^- deposition. This was only observable in predominantly forested watersheds with nitrate deposition as the dominant N input and at low concentrations of riverine nitrate. Oxygen isotope ratios of less than +15‰ in riverine nitrate in most other watersheds revealed no direct contribution of atmospheric nitrate to surface runoff. Apparently, nitrate from atmospheric deposition is intensively cycled through the organic N pool in all watersheds. Its origin is no longer isotopically recognizable after one immobilization/mineralization cycle, since nitrate derived from re-mineralization of the organic nitrogen compounds acquires $\delta^{18}O$ values of less than +15‰ (Mayer et al. 2001).

190

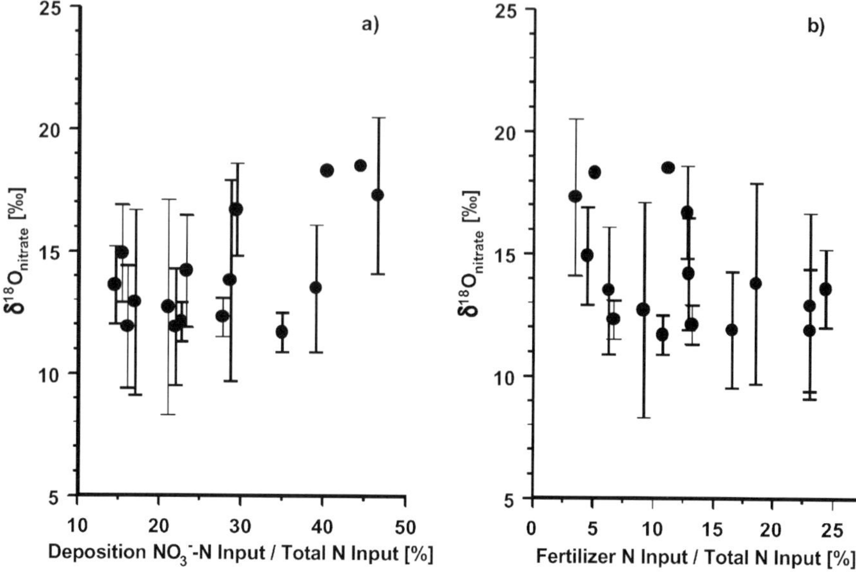

Figure 7. Mean $\delta^{18}O$ values of riverine nitrate versus N inputs with (a) atmospheric nitrate deposition and (b) with synthetic fertilizers to the 16 watersheds expressed as percentage of total N input.

$\delta^{18}O_{nitrate}$ values did not vary with the percentage of fertilizer N on the total N input (Figure 7(b)) indicating that nitrate-containing fertilizers ($\delta^{18}O \sim +22\%_0$) were not a major source of riverine nitrate. The lowest $\delta^{18}O$ values of riverine nitrate were observed in watersheds with the highest fertilizer N inputs. The absence of a trend of increasing $\delta^{18}O_{nitrate}$ values with increasing application of commercial fertilizers has in all likelihood two reasons. Uptake of fertilizer-nitrate by crops and microbes followed by re-mineralization and nitrification produces nitrate with $\delta^{18}O$ values below $+15\%_0$ (Mayer et al. 2001), thereby changing the original oxygen isotope ratio of the fertilizer-nitrate. Also, less than 50% of the fertilizers applied in agricultural areas of the northeastern U.S.A. contain nitrate (c.f. Boyer et al. 2002). Nitrogen from ammonium or urea-containing fertilizers must be nitrified before it can contribute to riverine nitrate. In analogy to what was described above, nitrate derived from nitrification of ammonium or urea-based fertilizers is typically characterized by $\delta^{18}O$ values of less than $+15\%_0$. This may explain the observed tendency of decreasing $\delta^{18}O$ values in riverine nitrate with increasing fertilizer loads (Figure 7(b)). Most ammonia and urea-containing fertilizers have $\delta^{15}N$ values around $0\%_0$. Hence, the isotopic composition of nitrate derived from nitrification of these fertilizers can not be distinguished isotopically from that generated by nitrification of organic soil N.

$\delta^{18}O_{nitrate}$ data suggest that nitrate-containing synthetic fertilizers do not directly contribute to riverine nitrate, but may do so after cycling through the organic N pool of the watersheds.

Denitrification

Microbial denitrification, the reduction of nitrate to N_2O and N_2 when oxygen is limited and degradable organic carbon is available (Knowles 1982), constitutes another mechanism, which increases $\delta^{18}O_{nitrate}$ values while decreasing nitrate concentrations (Figure 6). Therefore, this process deserves attention even though most of the observed variability in the isotope composition of riverine nitrate can be explained by mixing of nitrate from various sources.

Denitrification is potentially important in various compartments of large watersheds. These include soils (e.g. Ostrom et al. 1998), aquifers (Fustec et al. 1991; Aravena & Robertson 1998), riperian zones (Warwick & Hill 1988; Lowrance et al. 1995; Hill 1996), hyporheic zones (Duff & Triska 1990), and stream sediments (Cooper 1990; Seitzinger et al. 2002). The extent of denitrification and its influence on N budgets is difficult to assess on a watershed scale, because of the spatial and temporal variability of this process.

Kinetic isotope effects during microbial denitrification progressively enrich the remaining nitrate in ^{15}N and ^{18}O as concentrations decrease. In closed systems, this isotopic enrichment obeys the kinetics of a Rayleigh process, with isotopic enrichment factors for nitrogen varying from less than 10‰ to more than 30‰ depending on environmental conditions (e.g. Mariotti et al. 1981; Mariotti et al. 1988; Fustec et al. 1991). Oxygen isotope fractionation during microbial denitrification varies typically between 8 and 15‰ (Böttcher et al. 1990; Aravena & Robertson 1998; Cey et al. 1999; Mengis et al. 1999). Closed system conditions are often prevalent in anaerobic soil compartments, along the groundwater flowpath in aquifers and riparian zones, and occasionally in river sediments (e.g. Kellman & Hillaire-Marcel 1998). Nitrate that has undergone partial denitrification in these watershed compartments should, therefore, have elevated $\delta^{15}N$ and $\delta^{18}O$ values and lowered concentrations compared to input values. Nitrate entering such compartments has usually undergone at least one immobilization/mineralization cycle resulting in $\delta^{18}O$ values of less than +15‰ (Mayer et al. 2001).

Of the N removal processes within rivers (uptake by biota, burial in sediments, leakage into underlying aquifers, microbial denitrification below the water-sediment interface), only denitrification is accompanied by significant isotope fractionation enriching the remaining nitrate in ^{15}N (Kellman & Hillaire-Marcel 1998). Therefore, changes in the concentration and the isotopic composition of riverine nitrate might be expected, if in-stream denitrification is occurring. Since our samples were taken at the outlet of the 16

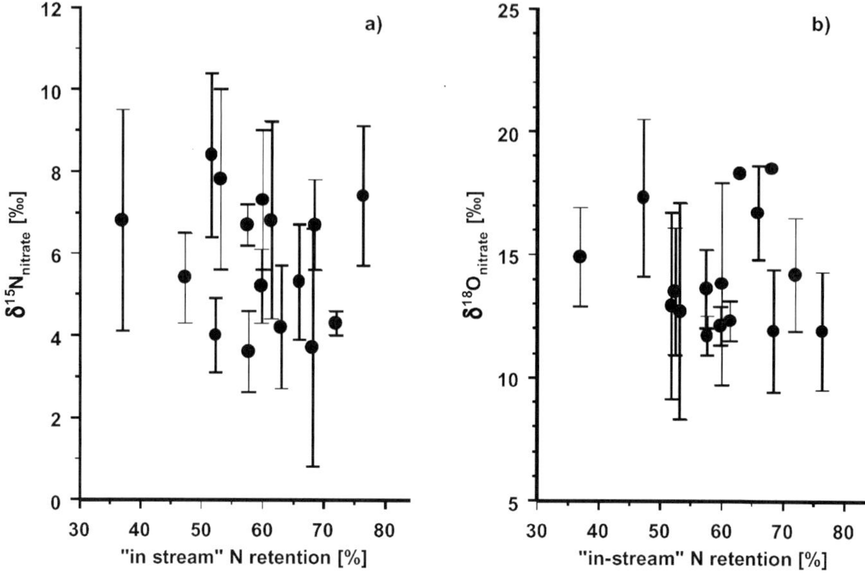

Figure 8. Mean δ^{15}N values (a) and δ^{18}O values (b) of riverine nitrate versus the extent of 'in-stream' N retention in the 16 rivers as determined by Seitzinger et al. (2002).

watersheds, and between 37% and 72% of the nitrogen entering the rivers was removed by in-stream processes (Seitzinger et al. 2002), in-stream denitrification was a potential candidate for influencing the isotopic composition of riverine nitrate.

Figures 8a and 8b show mean δ^{15}N and δ^{18}O values of riverine nitrate plotted versus in-stream N retention estimated by Seitzinger et al. (2002). Although these authors pointed to denitrification as the major in-stream N removal process, we found no statistically significant trend of increasing nitrogen or oxygen isotope ratios of riverine nitrate with increasing in-stream N retention. There are at least two possible explanations for this apparent contradiction. Microbial denitrification below the water-sediment interface may be limited by the diffusion of nitrate from the well-mixed aerobic water column to the anaerobic sediments, a process, which does not discriminate isotopically (Sebilo et al. forthcoming). This would result in little nitrogen and oxygen isotope fractionation during microbial denitrification in benthic sediments. Furthermore, while denitrification decreases nitrate concentrations, concomitant admixture of nitrate from other sources such as waste water and/or animal waste tends to increase riverine nitrate concentrations. Hence, even if denitrification would generate increasing δ^{15}N and δ^{18}O values in riverine nitrate, the simultaneous addition of nitrate from sewage or manure in watersheds with significant urban and agricultural land use would readily mask any isotopic denitrification signal. This is true also for partially denitrified nitrate entering rivers from soils, aquifers, riparian and hyporheic zones. Because in-stream denitrification is not a single source closed system process, occasional analysis of riverine nitrate at the outlet of large watersheds is clearly inappropriate for determining the location and the extent of microbial denitrification in large watersheds. Detailed monitoring of the evolution of both the concentration and the isotopic composition of dissolved nitrate along a river holds more promise for a conclusive evaluation of the role of in-stream denitrification during riverine N removal (c.f. Kellman & Hillaire-Marcel 1998).

Conclusions

Detailed nitrogen budgets for 16 watersheds with significant variability of nitrogen inputs enabled us to evaluate the usefulness of the isotopic composition of riverine nitrate as a potential tracer for N sources and N transformations in large watersheds. Our data indicate that the isotopic composition of riverine nitrate collected at the outlet of a watershed provides mainly information about nitrate sources.

In predominantly forested watersheds, nitrate was mainly derived from nitrification processes in soils, resulting in comparatively low concentrations of riverine nitrate and $\delta^{15}N_{nitrate}$ values of less than +5‰. In watersheds with significant agricultural and urban land use, $\delta^{15}N$ of riverine nitrate increased to values between +6 and +9‰. Elevated $\delta^{15}N_{nitrate}$ values in some rivers were predominantly caused by admixture of sewage- and manure-derived nitrate. Directly introduced into rivers, waste water caused marked increases in $\delta^{15}N$ of riverine nitrate at fluxes as low as 150 kg N km^{-2} a^{-1}. By contrast, animal waste increased $\delta^{15}N_{nitrate}$ values only at inputs exceeding 1000 kg N km^{-2} a^{-1}. Since increasing $\delta^{15}N$ values were positively correlated with increasing nitrate concentrations, we conclude that sewage was the primary and manure a secondary source of nitrate responsible for increasing concentrations of riverine nitrate, and thus increasing N export from the watersheds. Hence, nitrogen isotope ratios of riverine nitrate appeared to reflect land use practices, with increasing percentages of urban or agricultural land causing increasing $\delta^{15}N$ values of riverine nitrate. Oxygen isotope analyses of riverine nitrate revealed that nitrate in atmospheric deposition and from commercial fertilizers was not a major source of riverine nitrate in the 16 watersheds. Despite significant in-stream denitrification (Seitzinger et al. 2002), this process was not revealed by the isotopic composition of riverine nitrate sampled at the outlet of the 16 watersheds. We suggest that comparatively low isotopic enrichment factors for nitrogen and oxygen

194

during diffusion-controlled denitrification in river sediments in concert with concomitant admixture of nitrate from waste water and manure is responsible for this finding.

We conclude that a combined evaluation of concentrations and isotope compositions of riverine nitrate can provide information about nitrate sources, whereas detection of N removal processes such as in-stream denitrification was not possible using isotope techniques and the here employed sampling protocol.

Acknowledgements

This work was initiated as part of the International SCOPE Nitrogen project, which received support from both the Mellon Foundation and from the National Center for Ecological Analysis and Synthesis (NCEAS) in Santa Barbara, California. Special thanks to the USGS personnel who collected water samples at the gaging stations associated with each of the 16 watersheds.

References

Amberger A & Schmidt HL (1987) Natürliche Isotopengehalte von Nitrat als Indikatoren für dessen Herkunft. Geochim. Cosmochim. Acta. 51: 2699–2705

Aravena R, Evans ML & Cherry JA (1993) Stable isotopes of oxygen and nitrogen in source identification of nitrate from septic systems. Ground Water 31: 180–186

Aravena R & Robertson WD (1998) Use of multiple isotope tracers to evaluate denitrification in ground water: study of nitrate from a large-flux septic system plume. Ground Water 36: 975–982

Blackmer AM & Bremner JM (1977) Nitrogen isotope discrimination in denitrification of nitrate in soils. Soil Biol. Biochem. 9: 73–77

Böttcher J, Strebel O, Voerkelius S & Schmidt H-L (1990) Using isotope fractionation of nitrate-nitrogen and nitrate-oxygen for evaluation of microbial denitrification in a sandy aquifer. J. Hydrol. 114: 413–424

Boyer EW, Goodale CL, Jaworski NA & Howarth RW (2002) Anthropogenic nitrogen sources and relationships to riverine nitrogen export in the northeastern U.S.A. Biochemistry 57/58: 137–169

Bräuer K & Strauch G (2000) An alternative procedure for the ^{18}O measurement of nitrate oxygen. Chem. Geol. 168: 283–290

Cey EE, Rudolph DL, Aravena R & Parkin G (1999) Role of the riparian zone in controlling the distribution and fate of agricultural nitrogen near a small stream in southern Ontario. J. Contaminant Hydrol. 37: 45–67

Chang CCY, Langston J, Riggs M, Campbell DH, Silva SR & Kendall C (1999) A method for nitrate collection for δ^{15}N and δ^{18}O analysis from waters with low nitrate concentrations. Can. J. Fish. Aquat. Sci. 56: 1856–1864

Cooper AB (1990) Nitrate depletion in the riparian zone and stream channel of a small headwater catchment. Hydrobiologia 202: 13–26

Duff JH & Triska FJ (1990) Denitrification in sediments from the hyporheic zone adjacent to a small forested stream. Can. J. Fish. Aquat. Sci. 47: 1140–1147

Durka W, Schulze E-D, Gebauer G & Voerkelius S (1994) Effects of forest decline on uptake and leaching of deposited nitrate determined from ^{15}N and ^{18}O measurements. Nature 372: 765–767

Epstein S & Mayeda T (1953) Variation of O–18 content of waters from natural sources. Geochim. Cosmochim. Acta. 4: 213–224

Farrell RE, Sandercock PJ, Pennock DJ & Van Kessel C (1996) Landscape-scale variations in leached nitrate: relationship to denitrification and natural nitrogen-15 abundance. Soil Sci. Soc. Am. J. 60: 1410–1415

Fogg GE, Rolston DE, Decker DL, Louie DT & Grismer ME (1998) Spatial variation in nitrogen isotope values beneath nitrate contamination sources. Ground Water 36: 418–426

Fustec E, Mariotti A, Grillo X & Sajus J (1991) Nitrate removal by denitrification in alluvial ground water: role of a former channel. J. Hydrol. 123: 337–354

Goolsby DA (2000) Mississippi basin nitrogen flux believed to cause gulf hypoxia. EOS 81: 321–327

Gormly JR & Spalding RF (1979) Sources and concentrations of nitrate-nitrogen in ground water of the Central Platte Region, Nebraska. Ground Water 17: 291–301

Harrington RR, Kennedy BP, Chamberlain CP, Blum JD & Folt CL (1998) ^{15}N enrichment in agricultural catchments: field patterns and applications to tracking Atlantic salmon (*Salmo salar*). Chem. Geol. 147: 281–294

Heaton THE (1986) Isotopic studies of nitrogen pollution in the hydrosphere and atmosphere: a review. Chem. Geol. 5: 87–102

Hedin LO, Armesto JJ & Johnson AH (1995) Patterns of nutrient loss from unpolluted, old-growth temperate forests: evaluation of biogeochemical theory. Ecology 76: 493–509

Hill AR (1996) Nitrate removal in stream riparian zones. J. Environ. Qual. 25: 743–755

Howarth RW, Billen G, Swaney D, Townsend A, Jaworski N, Lajtha K, Downing JA, Elmgren R, Caraco N, Jordan T, Berendse F, Freney J, Kudeyarov V, Murdoch P & Zhao-Liang (1996) Regional nitrogen budgets and riverine N & P fluxes for the drainages to the North Atlantic Ocean: natural and human influences. Biogeochem. 35: 75–139

Hübner H (1986) Isotope effects of nitrogen in the soil and biosphere. In: Fritz P & Fontes JC (Eds) Handbook of Environmental Isotope Geochemistry: The Terrestrial Environment (pp 361–425). Elsevier, Amsterdam

Jaworski NA & Hetling LJ (1996) Water quality trends of Mid-Atlantic and northeast watersheds over the past 100 years. Presented at Watershed's 96, Baltimore, MD

Kellman L & Hillaire-Marcel C (1998) Nitrate cycling in streams: using natural abundances of $NO_3^- -\delta^{15}N$ to measure *in-situ* denitrification. Biogeochemistry 43: 273–292

Kendall C (1998) Tracing nitrogen sources and cycling in catchments. In: Kendall C & McDonnell JJ (Eds) Isotope Tracers in Catchment Hydrology (pp 521–576). Elsevier, Amsterdam

Kinzing AP & Socolow RH (1994) Human impacts on the nitrogen cycle. Physics Today (November 1994): 24–31

Knowles R (1982) Denitrification. Microbiol. Rev. 46: 43–70

Knowles R & Blackburn TH (1993) Nitrogen Isotope Techniques. Academic Press, San Diego, 311 pp

Kreitler CW (1979) Nitrogen-isotope ratio studies of soils and groundwater nitrate from alluvial fan aquifers in Texas. J. Hydrol. 42: 147–170

Kreitler CW & Browning LA (1983) Nitrogen-isotope analysis of groundwater nitrate in carbonate aquifers: natural sources versus human pollution. J. Hydrol 61: 285–301

Kreitler CW & Jones DC (1975) Natural soil nitrate: the cause of the nitrate contamination of ground water in Runnels County, Texas. Ground Water 13: 53–61

Letolle R (1980) Nitrogen-15 in the natural environment. In: Fritz P & Fontes JC (Eds) Handbook of Environmental Isotope Geochemistry: The Terrestrial Environment (pp 407–433). Elsevier, Amsterdam

Lowrance R, Vellidis G & Hubbard RK (1995) Denitrification in a restored riparian forest wetland. J. Environ. Qual. 24: 808–815

Macko SA & Ostrom NE (1994) Pollution studies using stable isotopes. In: Lajtha K & Michener RH (Eds) Stable Isotopes in Ecology and Environmental Science (pp 45–62). Blackwell Scientific Publications, Oxford

Mariotti A, Germon JC, Hubert P, Kaiser P, Letolle R, Tardieux A & Tardieux P (1981) Experimental determination of nitrogen kinetic isotope fractionation: some principles; illustration for the denitrification and nitrification processes. Plant and Soil 62: 413–430

Mariotti A, Germon JC & Leclerc A (1982) Nitrogen isotope fractionation associated with the $NO_2 \rightarrow N_2O$ step of denitrification in soils. Can. J. Soil Sci. 62: 227–241

Mariotti A, Landreau A & Simon B (1988) ^{15}N isotope biogeochemistry and natural denitrification process in ground water: application to the chalk aquifer in northern France. Geochim. Cosmochim. Acta. 52: 1869–1878

Mayer B, Bollwerk SM, Mansfeldt T, Hütter B & Veizer J (2001) The oxygen isotope composition of nitrate generated by nitrification in acid forest floors. Geochim. Cosmochim. Acta. 65: 2743–2756

McClelland JW & Valiela I (1998) Linking nitrogen in estuarine producers to land derived sources. Limnol. Oceanogr. 43: 577–585

McClelland JW, Valiela I & Michener RH (1997) Nitrogen-stable isotope signatures in estuarine food webs: a record of increasing urbanization in coastal watersheds. Limnol. Oceanogr. 42: 930–937

Mengis M, Schiff SL, Harris M, English MC, Aravena R, Elgood RJ & MacLean A (1999) Multiple geochemical and isotopic approaches for assessing ground water NO_3^- elimination in a riparian zone. Ground Water 37: 448–457

Nadelhoffer KJ & Fry B (1994) Nitrogen isotope studies in forest ecosystems. In: Lajtha K & Michener RM (Eds) Stable Isotopes in Ecology and Environmental Science (pp 22–44). Blackwell Scientific Publishers, Oxford

Ostrom NE, Knoke KE, Hedin LO, Robertson GP & Smucker AJM (1998) Temporal trends in nitrogen isotope values of nitrate leaching from an agricultural soil. Chem. Geol. 146: 219–227

Paces T (1982) Natural and anthropogenic fluxes of major elements from Central Europe. Ambio. 11: 206–208

Revesz K, Böhlke JK & Yoshinari T (1997) Determination of $\delta^{18}O$ and $\delta^{15}N$ in nitrate. Anal. Chem. 69: 4375–4380

Richards RP & Holloway J (1987) Monte carlo studies of sampling strategies for estimating tributary loads. Water Resources Research 23: 1939–1948

Sebilo M, Billen G, Grably M & Mariotti A (in review) Isotopic composition of nitrate-nitrogen as a marker of riparian and benthic denitrification at the scale of the whole Seine River system. Biogeochemistry, forthcoming

Seitzinger PS, Styles RV, Boyer E, Alexander RB, Billen G, Howarth RW, Mayer B & Van Breemen N (2002) Nitrogen retention in rivers: model development and application to watersheds in the northeastern U.S.A. Biogeochemistry 57/58: 199–237

Silva SR, Kendall C, Wilkinson DH, Ziegler AC, Chang CCY & Avanzino RJ (2000) A new method for collection of nitrate from fresh water and the analysis of nitrogen and oxygen isotope ratios. J. Hydrol. 228: 22–36

Sollins P & McCorison FM (1981) Nitrogen and carbon solution chemistry of an old growth coniferous forest watershed before and after cutting. Water Resources Research 17: 1409–1418

Stottlemyer R & Troendle CA (1992) Nutrient concentration patterns in streams draining alpine and subalpine catchments, Fraser Experimental Forest, Colorado. J. Hydrol. 140: 179–208

Turner RE & Rabalais NN (1991) Changes in Mississippi River water quality this century. BioSci. 41: 140–147

USGS (2000) National water information system data retrieval. http://waterdata.usgs.gov/nwis-w/US/. USGS

Van Breemen N, Boyer EW, Goodale CL, Jaworski NA, Seitzinger S, Paustian K, Hetling L, Lajtha K, Eve M, Mayer B, Van Dam D, Howarth RW, Nadelhoffer KJ & Billen G (2002) Where did all the nitrogen go? Fate of nitrogen inputs to large watersheds in the northeastern U.S.A. Biogeochemistry 57/58: 267–293

Vanderbilt KL & Lajtha K (2002) Annual and seasonal patterns of nitrogen dynamics at the H. J. Andrews Experimental Forest, Oregon. Biogeochemistry, in review

Vitousek PM, Aber JD, Howarth RW, Likens GE, Matson PA, Schindler DW, Schlesinger WH & Tilman DG (1997) Human alteration of the global nitrogen cycle: sources and consequences. Ecol. Appl. 7: 737–750

Voerkelius S (1990) Isotopendiskriminierungen bei der Nitrifikation und Denitrifikation: Grundlagen und Anwendungen der Herkunfts-Zuordnung von Nitrat und Distickstoffmonoxid. PhD thesis TU Munich, Munich, 119 pp

Warwick J & Hill AR (1988) Nitrate depletion in the riparian zone of a small woodland stream. Hydrobiologia 157: 231–240

Wassenaar L I (1995) Evaluation of the origin and fate of nitrate in the Abbotsford Aquifer using the isotopes of ^{15}N and ^{18}O in NO_3^-. Appl. Geochem. 10: 391–405

Biogeochemistry **57/58:** 199–237, 2002.

Nitrogen retention in rivers: model development and application to watersheds in the northeastern U.S.A.

SYBIL P. SEITZINGER[1]*, RENÉE V. STYLES[1], ELIZABETH W. BOYER[2], RICHARD B. ALEXANDER[3], GILLES BILLEN[4], ROBERT W. HOWARTH[5], BERNHARD MAYER[6] & NICO VAN BREEMEN[7]
[1]*Rutgers University, Rutgers/NOAA CMER Program, Institute of Marine and Coastal Sciences, 71 Dudley Rd, New Brunswick, NJ 08901, USA;* [2]*State University of New York, College of Environmental Science and Forestry, 1 Forestry Drive, Syracuse, NY 13210, USA;* [3]*U.S. Geological Survey, 413 National Center, 12201 Sunrise Valley Drive, Reston, VA 20192, USA;* [4]*UMR Sisyphe, Université Pierre et Marie Curie, 4 place Jussieu, 75005 Paris, France;* [5]*Department Ecology & Environmental Biology, Corson Hall, Cornell University, Ithaca, NY 14853, USA;* [6]*University of Calgary, Departments of Physics and Astronomy and Geology and Geophysics, 2500 University Drive NW, Calgary, Alberta, Canada T2N 1N4;* [7]*Laboratory of Soil Science and Geology and Wageningen Institute for Environment and Climate Research, Wageningen University, P.O. Box 37, 6700 AA Wageningen, the Netherlands (*author for correspondence, e-mail: sybil@imcs.rutgers.edu)*

Key words: budgets, denitrification, model, nitrogen, rivers, watersheds

Abstract. A regression model (RivR-N) was developed that predicts the proportion of N removed from streams and reservoirs as an inverse function of the water displacement time of the water body (ratio of water body depth to water time of travel). When applied to 16 drainage networks in the eastern U.S., the RivR-N model predicted that 37% to 76% of N input to these rivers is removed during transport through the river networks. Approximately half of that is removed in 1st through 4th order streams which account for 90% of the total stream length. The other half is removed in 5th order and higher rivers which account for only about 10% of the total stream length. Most N removed in these higher orders is predicted to originate from watershed loading to small and intermediate sized streams. The proportion of N removed from all streams in the watersheds (37–76%) is considerably higher than the proportion of N input to an individual reach that is removed in that reach (generally <20%) because of the cumulative effect of continued nitrogen removal along the entire flow path in downstream reaches. This generally has not been recognized in previous studies, but is critical to an evaluation of the total amount of N removed within a river network. At the river network scale, reservoirs were predicted to have a minimal effect on N removal. A fairly modest decrease (<10 percentage points) in the N removed at the river network scale was predicted when a third of the direct watershed loading was to the two highest orders compared to a uniform loading.

Introduction

Human activities have markedly increased nitrogen (N) inputs to watersheds (Vitousek et al. 1997). A portion of this N enters river networks, degrading river water quality and increasing nutrient input and subsequent eutrophication of coastal marine ecosystems (National Academy of Sciences 2000). However, rivers do not act as "inert pipelines", transporting N from the land to the coastal zone. Processes such as denitrification, organic matter burial in sediments, sediment sorption, and plant and microbial uptake can remove N from the river network, and thus affect the amount of N that is transported by rivers to coastal ecosystems (Billen et al. 1991).

A number of studies have examined factors controlling N retention and/or denitrification rates in rivers including nitrate concentration, sediment properties, flow rate, and land use (e.g., Robinson et al. 1979; Cooke & White 1987; Christensen & Sørensen 1988; Jansson et al. 1994; Howarth et al. 1996; Mulholland et al. 2000; Peterson et al. 2001). In general studies have addressed N removal in short sections of a river, but give no direct information about the amounts of N removed throughout the complex network of rivers draining a watershed. A river network approach is needed for a watershed-wide quantification of the proportion of the N input that is removed during riverine transport. Such an approach, used in some recent watershed mass-balance studies (Smith et al. 1997; Alexander et al. 2000), is also needed to understand how the amount of N transported to the watershed outlet is affected by such factors as number and size of reservoirs in the network, location of N inputs within the network relative to the river mouth, size of watershed, and network configuration.

In this paper we: (1) use published data to develop a predictive model relating the physical and hydraulic properties of rivers and lakes to the proportion of the N input that is removed in these water bodies, (2) apply that model in sixteen drainage networks in the eastern U.S. to estimate riverine N removal, and (3) explore the effect of various watershed and river properties on N removal in the drainage network including size of watershed, presence of reservoirs and their location in the river network, distribution of N input to the river network within the watershed, and the scale of the river network data.

Model

Model development

Estimates of nitrogen removal in streams and rivers (expressed as a proportion of the external input) were obtained from published data. The studies were conducted in river reaches that ranged from first order headwater rivers to the tidal freshwater portion of major rivers (Table 1 plus Swank & Caskey 1982; Billen et al. 1985; Jacobs & Gilliam 1985; Cook & White 1987; Chesterikoff et al. 1992; Mulholland 1992; Jansson et al. 1994; Sjodin et al. 1997 and others). Land use in the watersheds included agricultural, urban, forested and mixed. No studies were found in which N removal was measured throughout an entire river network. Rather, measurements were usually made in only a short section, or reach, of a river (e.g., first order rivers, tidal freshwater portion of a river) which had fairly uniform characteristics (e.g., depth, water residence time, N loading, etc.). Combining the results of studies from sites with a wide range of characteristics is desirable to develop relationships between N removal and river characteristics that are widely applicable both within a drainage network and across different watersheds.

N removal in the river studies was quantified using one of two basic approaches: (1) mass balance calculations based on N concentration and water discharge measurements at an upstream and downstream location, or (2) denitrification measurements on sediment cores. There are many variations on the two approaches, and depending on the approach a slightly different set of N removal mechanisms may be captured. In some mass balance studies, calculations were based on nitrate or dissolved inorganic N (DIN = nitrate, nitrite plus ammonia) only; in others total N was used. Measurements of denitrification in sediment cores included a range of techniques. N removal in hyporheic zones, as well as uptake by benthic biota and burial in sediment generally would be captured in mass balance studies but not in denitrification measurements using core incubations of surface sediments. Some studies were conducted during a portion of the year, others over one or more annual cycles; whenever possible, we used annual studies. Generally, N input from upstream as well as direct watershed input (groundwater, point and non-point sources) for the study reach were used to compare with N removal within the study reach; however, in some cases only upstream N input was reported. We included the results from studies using all approaches in the model development, at the same time recognizing the limitations and uncertainties imposed by this approach. In the following, we use the term "N removal" to indicate all N removal/retention processes. However, denitrification, the only permanent N removal process in rivers, is expected to be the dominant loss process reflected in the literature measurements. Uptake by

Table 1. Studies used to develop model equation for estimating proportion of N input to a river reach that is removed as a function of depth/TOT

River	Watershed land use	Strahler order	Stream study length (km)	Depth (m)	Time of Travel (TOT) (hr)	Depth/TOT ($m\ y^{-1}$)	%N removed	Period of study	Approach	Reference
Delaware (tidal freshwater), USA	Urban, Agriculture		120	6.8	967	62	20 (DIN)	Aug. 1984	Denitrification in cores	Seitzinger 1988a,b flow from Cole et al. 1993
Duffin Creek, ON, Canada	50% Agriculture 20% Forest 20% Abandoned farmland	6	26	0.35	16	190	6 $(TN)^{a,b}$	May–Oct. 1973–75; Nov.–March 1978	Mass balance and denitrification in cores	Hill 1979, 1981, 1983
Gelbaek, Denmark	Agriculture	1	1	0.40	1.4	2500	1 $(NO_3)^a$	Spring 1993– Summer 1994	Denitrification in cores	Christensen & Sorensen 1988; Christensen et al. 1990
Neversink, NY, USA	Forest	1 3	0.6 1.0	0.2 0.5	0.8 1.4	2190 3130	11$(NO_3)^c$ 12$(NO_3)^c$	April, June, July, Sept. 1992	Mass balance	Burns 1998

Table 1. Continued

River	Watershed land use	Strahler order	Stream study length (km)	Depth (m)	Time of Travel (TOT) (hr)	Depth/TOT $(m\,y^{-1})$	%N removed	Period of study	Approach	Reference
Potomac (tidal freshwater), USA	Urban, Forest			5.0	3264	13	35 (TN)	Sept. 1985	Denitrification in cores	Seitzinger 1988a, 1991
Purukohukohu, New Zealand	Pasture Forest	1 1	0.08 0.08	0.03 0.03	1.0 1.3	270 200	14(NO$_3$)[b] 17(NO$_3$)[b]	Aug., Nov. 1982; June 1983	Mass balance	Cooper & Cooke 1984
River Dorn Oxfordshire, England	Agriculture		35	0.5	65	67	15(NO$_3$)[d]	June 1985	Denitrification in cores	Cooke & White 1987
Swifts Brook, ON, Canada	Forest, Unused pasture	1	2	0.11	19	51	20(TN)[a,b]	1975-1976	Mass balance	Robinson et al. 1979; Kaushik & Robinson 1976

[a] Annual estimate.
[b] Removal attributed to denitrification.
[c] Average % N removal over 4 sampling periods. Includes removal due to both biological uptake and denitrification.
[d] Under summer, baseflow conditions.

204

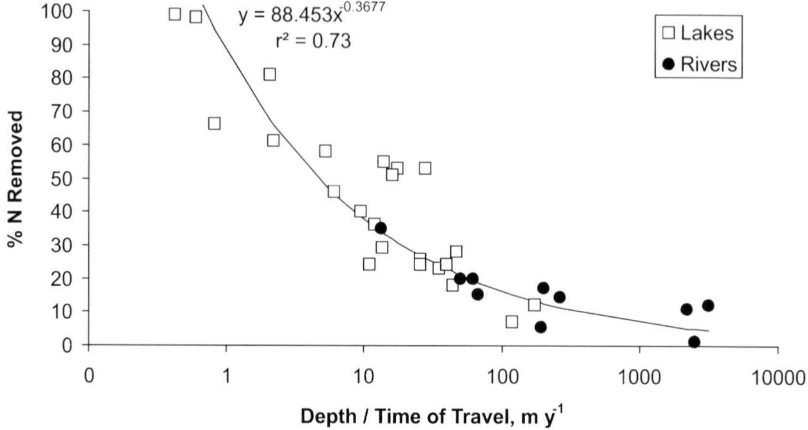

Figure 1. Empirical relationship between % N removed in a study area and depth divided by time of travel (water residence time for lakes) based on studies in river reaches and in lakes. Data sources and details for rivers are in Table 1. Data sources for lakes include: Garnier et al. 1999; Kelley et al. 1987; Andersen 1977; Calderoni et al. 1978; Ayers 1970 cited by Schelske 1975; Dillon and Molot 1990. All lake studies were conducted over at least a 1-year period.

biota and to some extent sedimentation only temporarily retain nitrogen, and generally would not be expected to influence the mean estimates of loss from mass balance studies of annual conditions. The studies used to develop the model do not include N removal in aquifers or wetlands in a watershed.

There is a wide range (1 to 80%) in the efficiency with which N is removed in river reaches (Table 1 plus Swank & Caskey 1982; Billen et al. 1985; Jacobs & Gilliam 1985; Cook & White 1987; Chesterikoff et al. 1992; Mulholland 1992; Jansson et al. 1994; Sjodin et al. 1997, others). Relationships between N removal and the following variables were examined: river order, river discharge, land use, N-loading, water residence time, and water displacement (ratio of water depth to water residence time). Nitrogen removal (R; expressed as a percentage of N input) was best described (i.e., based on r^2) as an inverse relation with water displacement according to:

$$R = 74.61 \left(\frac{D}{T} \right)^{-0.344} \tag{1}$$

($r^2 = 0.43$; n = 10) where depth (D) is in meters and time of travel (T) is in years. Time of travel was calculated from length of study reach divided by velocity.

A separate model fit that combined data from the 10 river observations with data from 23 lakes and reservoirs (see Figure 1 for data sources) was found to have similar coefficients as the model using only the river data (eq. 1):

$$R = 88.45 \left(\frac{D}{T} \right)^{-0.3677} \tag{2}$$

($r^2 = 0.73$) where T (in years) reflects the water residence time for lakes and reservoirs and the water time of travel for rivers (Figure 1). Therefore, equation (2), based on the combined data provides a generally consistent description of nitrogen loss in streams, reservoirs and lakes. A relationship which is applicable to both river reaches and reservoirs within a river is useful in evaluating N removal in a river network. This model is similar to the theoretical relationship for river reaches, lakes and reservoirs presented in Howarth et al. (1996) and Billen et al. (1998), which was based on the extension of a lake mass-balance model of nitrate removal quantifying the effects of benthic denitrification in oxic well-mixed shallow lakes (Kelly et al. 1987).

A log-linear form of the model (eq. 2) was used to estimate the model parameters by applying a natural log transformation to both the response and explanatory variables. The estimates of %N loss were not adjusted to correct for the log retransformation bias associated with the application of a log-linear model (Duan 1983). The river N removal studies which contained information on depth and time of travel and which, therefore, formed the basis of the final model are listed in Table 1. The range of stream sizes covered by the relation gives reasonable coverage of the stream sizes of interest in the northeastern U.S. watersheds. Most of the studies are for small streams, although the Potomac and Delaware extend the coverage to higher order streams relevant to the larger northeastern U.S. streams. An examination of the residuals of this model shows that there is relatively constant variance over the range of the data and that they are approximately normally distributed. The data (and residuals) do display greater scatter for larger values of the ratio (generally above 100 which includes many of the river observations), suggesting greater model uncertainty in this range.

The model in equation (2) empirically quantifies the fraction of nitrogen removed as a function of the rate of water displacement in streams, reservoirs/lakes (the ratio of depth to water travel time represents the height of the water column that is annually displaced from the water body). This hydraulic property provides a measure of the time for dissolved and particulate N to react with the benthic sediments. Variations in water contact time may occur as a result of the natural properties of streams related to channel size (i.e., water depth, velocity) as well as attributes intrinsic to the measurements of a particular study (i.e., reach length, velocity during study period). These characteristics may influence the removal of nitrogen from the water column by or to the benthic sediment, by such processes as denitrification, uptake by biota, and particulate settling (Wollheim et al. 2001). For example, increases in water residence times in streams and reservoirs/lakes generally increase particulate settling as well as increase the amount of nitrate (per unit volume of water) that can diffuse into the sediments, leading to the removal of a larger fraction of nitrogen. This model is consistent with a relationship between the fraction of N input that is removed and water residence time in estuaries (Nixon et al. 1996). It is also consistent with the notion that the fraction of nitrogen removed per unit of water travel time (i.e., first-order rate of nitrogen loss) varies inversely with stream channel depth (Smith et al. 1997; Alexander et al. 2000); this could reflect the smaller volume of water that is processed by a unit area of benthic sediment in shallow streams (Peterson et al. 2001).

Model application

Most of the river studies used to develop equation 2 were conducted over fairly short reaches of a river (i.e., sections of rivers with relatively uniform time of travel and depth). This has important implications when using this equation to estimate N retention throughout an entire river network. The equation must be applied to the individual reaches of the river network, with exports from a reach routed downstream to subsequent reaches for processing.

N removal in river networks was calculated for sixteen watersheds in the eastern U.S. between Maine and Virginia (Figure 2; Table 2). The most downstream reach of each watershed coincided with the location of a USGS station with a high density of water quality data (Boyer et al. this volume); thus the most downstream reach of each watershed area was not necessarily at the fall line or coast. The study watersheds varied considerably in physiographic and hydrographic characteristics such as watershed area, total river length and discharge (Table 2). A complete description of the watersheds is presented in Boyer et al. (this volume) and van Breemen et al. (this volume). We applied equation 2 using EPA-USGS reach network files for the sixteen watersheds. Subsets of the sixteen watersheds were chosen for more detailed analysis of the model output.

Base case scenario

The river network configuration was based on the enhanced EPA-USGS Reach File Version 1 (RF1; Alexander et al. 1999), which we supplemented with the higher resolution EPA-USGS National Hydrography Dataset – NHD (as described below). RF1 is a spatial data file that contains 60,000 major river and stream reaches in the conterminous U.S. and has been used in predictive

Table 2. Characteristics of the sixteen watershed river networks, and base case ($RF1_m$ + NHD') and reduced scale (RF1 only) model predictions of proportion of N inputs to the river network that are removed by in-river processes

Watershed[1]	Watershed area[2] km^2	Discharge[2] $m^3 s^{-1}$	River export[2] kg TN km^{-2} y^{-1}	Number of reservoirs	RF1 total reach length km	NHD total reach length km	Number of reaches	% N input removed $RF1_m$ + NHD' Base case	% N input removed RF1 only Reduced scale
Penobscot	20,109	375	317	22	4,344	19,798	5,917	68	59
Kennebec	13,994	251	333	14	2,716	12,182	3,651	63	52
Androscoggin	8,451	171	404	8	1,076	5,737	1,938	52	44
Saco	3,349	71	389	5	644	2,896	880	47	33
Merrimack	12,005	224	499	13	1.070	12,784	3,509	61	38
Charles	475	9	644	0	78	626	205	37	15
Blackstone	1,115	23	1,140	0	108	1,148	317	53	22
Connecticut	25,019	509	538	18	4,748	23,173	7,380	66	55
Upper Hudson	11,942	236	502	5	2,213	8,420	2,384	58	50
Mohawk	8,935	155	795	6	1,677	8,471	2,704	60	47
Delaware	17,560	304	961	13	2,612	14,228	5,011	60	48
Schuylkill	4,903	76	1,755	2	550	3,625	966	52	31
Susquehanna	70,189	1,084	977	21	9,612	69,248	17,499	76	63
Potomac	29,940	312	897	3	5,089	25,105	8,624	68	60
Rappahannock	4,134	47	470	0	844	3,041	806	57	42
James	16,206	209	314	3	3,322	15,375	4,885	72	61

[1] see Figure 1 for watershed delineations.
[2] From Boyer et al. (2002).

Figure 2. Delineation of the 16 watershed regions used in the current analysis. Black filled circles denote watershed outlets at the following locations: Penobscot River at Eddington, ME; Kennebec River at North Sidney, ME; Androscoggin River near Auburn, ME; Saco River at Cornish, ME; Merrimack River below Concord River at Lowell, MA; Charles River at Dover, MA; Blackstone River at Manville, RI; Connecticut River at Thompsonville, CT; Hudson River above lock 1 near Waterford, NY; Mohawk River at Cohoes NY; Delaware River at Trenton, NJ; Schuylkill River at Philadelphia, PA; Susquehanna River at Conowingo, MD; Potomac River near Washington, DC Lower Falls Pump Station; Rappahannock River near Fredericksburg, VA; James River at Cartersville, VA. (From Boyer et al. 2002)

water quality models of stream nutrient transport (Smith et al. 1997; Alexander et al. 2000). RF1 is approximately the resolution of the "blue-line" drainages that show up on a USGS 1:500,000 scale map. Reach characteristics used from the RF1 file included reach length, Strahler river order, reach time of travel (TOT), reach watershed area, reach mean flow, reservoir time of travel, and reach and reservoir node identification. Mean depth of each reach was calculated from mean flow (Q) based on the following relationship (Alexander et al. 2000):

$$D = 26.12 Q^{-0.3966} \qquad (3)$$

where reach depth is in meters and flow is in $m^3\ sec^{-1}$ ($r^2 = 0.83$). This relationship was based on stream morphology and hydraulic data for 112 rivers in the U.S. (Leopold & Maddock 1953). Mean reservoir depth was computed as the ratio of the normal capacity to surface area (Ruddy & Hitt 1990). The reach network configuration and flow paths were constructed based on the node identifiers.

The effect of map scale on defining existing rivers in a watershed is well known. A more comprehensive representation of the river network in each watershed (EPA-USGS National Hydrography Dataset – NHD – see <http://nhd.usgs.gov>) was used to supplement the information in RF1. NHD is a newly-released reach file available in beta form, and depicts approximately the resolution of the "blue-line" drainages that show up on a USGS 1:100,000 scale map (7.5 degree quadrangle). A comparison of the drainage density for these two map scales is shown for the Saco watershed (Figure 3A, B). The relatively coarse scale of RF1 generally does not capture true 1st, 2nd and 3rd order rivers. While NHD gives much better information on total length and reach lengths of rivers in each watershed, it does not contain information on the flow, depth and TOT of stream reaches and reservoirs. Therefore, for the additional river lengths not contained in RF1 (termed NHD'; calculated as NHD total reach lengths minus RF1 total reach lengths), reach depth and TOT were estimated using established relationships between hydraulic and physiographic factors which were tailored to each watershed based on the characteristics documented in RF1 (see Appendix for detailed description).

The total watershed area did not change when the NHD' reaches were added to the RF1 file. However, obviously, the total length of the river reaches did increase. Therefore, the direct watershed area to RF1 reaches was reapportioned to account for the stream lengths added with the inclusion of NHD'. The direct watershed area was uniformly apportioned per meter of stream length (RF1+NHD') throughout the watershed. We refer to this modified RF1 file in which the direct watershed area to RF1 reaches has been adjusted as $RF1_m$.

There were very few reaches in RF1 for the two smallest watersheds, the Blackstone and Charles. Therefore, there was increased uncertainty in extrapolating depth and TOT for NHD' reaches (Appendix) in those two watersheds. As such, our model estimates of N removal in those two watersheds are likely to be less certain than for the other watersheds in this analysis.

A. RF1

B. NHD

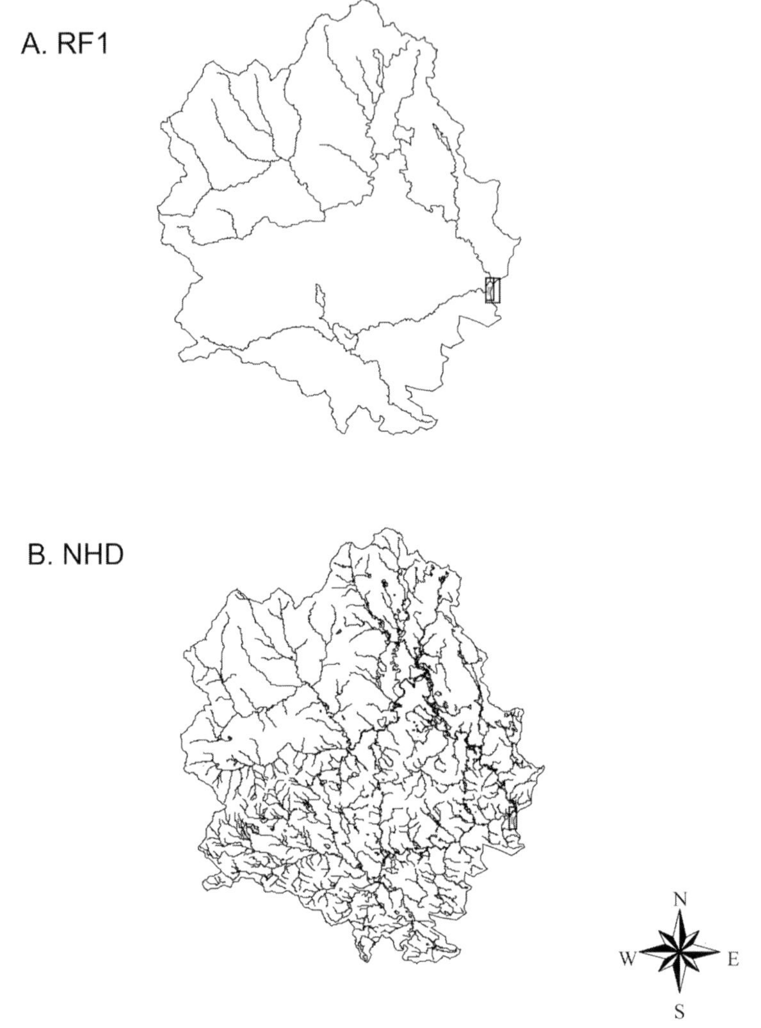

Figure 3. Map of river reach network in the Saco basin as defined by: (A) RF1 and (B) NHD. The scale of the reach networks used in the base case scenario for the RivR-N model is approximately that shown in (B), while the reduced scale scenario used only RF1 files.

The proportion of N loading to the river that was removed in the river network was calculated for each watershed using the $RF1_m + NHD'$ network configuration. N entered each reach from direct watershed loading and from N exported from the adjacent upstream reach (Figure 4). Direct watershed loading refers to all N inputs to the river from the direct watershed area for a reach and would include groundwater, point and non-point source inputs. Whole-watershed N inputs and N inputs to the rivers were estimated by Boyer

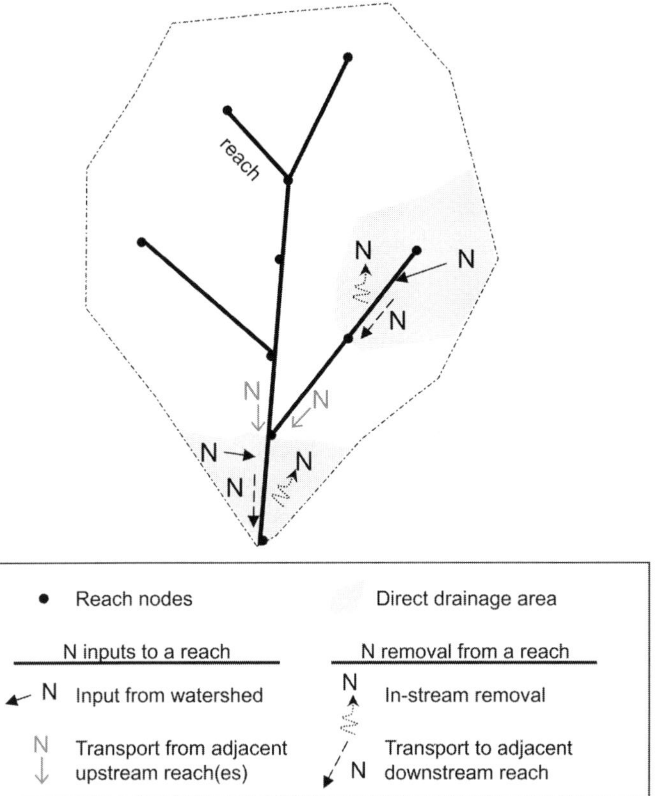

Figure 4. Simplified schematic of a river network. The network was assembled using reach node identifiers. N enters a reach from the direct drainage area for that reach plus transport from the adjacent upstream reach or reaches. N is removed from a reach by both in-stream processes (e.g., denitrification, burial, biotic uptake) and transport to the adjacent downstream reach. The direct drainage area and the N input/output detail are shown for 2 reaches.

et al. (2002) and van Breemen et al. (2002). However, the final estimates were not available when the final RivR-N model runs were conducted; in addition, those estimates did not maintain information in the spatial distribution of N inputs within a watershed. Therefore, in the RivR-N model base case scenario, land use throughout a watershed was assumed uniform, with direct watershed N loading to a reach (kg N per km^2 of direct watershed per year) uniform throughout the watershed. We choose 100 kg N per km^2 of direct watershed per year as a loading rate, which is similar to, although on the low end, of N loading to surface waters from forested areas (summarized in Howarth et al. 1996). The model calculates the proportion of the N input that is removed, which does not vary as a function of the rate of N loading; therefore, the model can be run without knowing the exact N loading rate to the river. To summarize, the base case scenario used the combined $RF1_m$ + NHD' network configuration and a uniform direct watershed N loading.

Alternate scenarios
Additional model runs were conducted to explore the following: (1) effect of reservoirs on N removal, (2) scale of river network, and (3) distribution of N loading to the river within the watershed. The amount of N removed in reservoirs was examined by removing reservoirs from each watershed file and substituting the pre-reservoir TOT and depth information for those reaches (also available in RF1). The effects of using only the lower resolution RF1 network files were examined by calculating the proportion of N removed using only the RF1 network files; we refer to this scenario as the reduced scale scenario. The effect of a non-uniform distribution of N loading within the watershed on N removal in the river network was examined by allocating a third of the total N loading in a watershed to reaches of the two highest orders.

Statistical analysis of model uncertainty
Statistical uncertainties in model coefficients and predictions were quantified using bootstrapping techniques (Efron 1982). Estimation of coefficient errors was based on a resampling with replacement from the original set of lake and stream observations (sample size = 45) and fitting of separate regression models to the resampled data for a total of 200 iterations. Confidence intervals were developed from the resulting statistical distributions for the model coefficients. Errors in model predictions of the proportion of nitrogen removed for each of the sixteen watersheds included uncertainties in both the model coefficients and the observed data (i.e., regression residuals). Contributions from the latter term were based on a random sampling of the model residuals associated with each of the bootstrap regressions. For each of the 200 iterations (i.e., bootstrap regression), a model residual was randomly assigned to each reach or reservoir prediction of the proportion of nitrogen removed. The resulting proportion was applied to the reach or reservoir mass flux, which was added to the cumulative mass of stored or denitrified nitrogen for each watershed. The residual error contributed only to the estimates of the quantities of nitrogen removed from individual reaches or reservoirs, and was not applied to the nitrogen mass transported to downstream segments. Errors in equation (2) are assumed to be multiplicative because of the log-linear form of the model. Inclusion of the regression residual errors serves to correct for the log retransformation bias associated with the application of a log-linear model. The uncertainty estimates were only computed for losses in the RF1 streams and reservoirs (reduced scale scenario).

Results and discussion

Base case results

N removal at the river network scale
A substantial proportion of the N input to rivers is removed during transport through the river network according to the model (hereafter referred to as the RivR-N model; **Riv**er **R**emoval of **N**itrogen). Between 37% and 76% of the N inputs were removed in the16 river networks for the $RF1_m$ + NHD' scenario (Figure 5A; Table 2); on average 60% (+/– 10%; S.D.) of the N inputs were removed. The smallest proportion removed was in the Charles and the largest proportion removed was in the Susquehanna. The proportion removed at the river network scale is considerably higher than the proportion of N input to an individual reach that is removed in that reach (reach-specific removal). Each river network in our suite of 16 watersheds consists of between $\sim$200 and $\sim$17,500 reaches (Table 2) with reach lengths ranging from less than 1 km to approximately 100 km in length. Approximately 85–90% of the reaches have depth/TOT characteristics and consequently reach-specific N removals (based on an analysis of the 5 scenario watersheds) within the range of literature values for river reaches that were used to develop the empirical model equation (1% to 35%; Table 1; Figure 1). At the river network scale the proportion of total N input that is removed is considerably larger than for an individual reach because of the cumulative effect of continued nitrogen removal along the entire flow path in downstream reaches (Figure 4). This has not generally been recognized in previous studies, but is critical to an evaluation of the total amount of N removed within a river network.

The amount of N removed in the respective river networks is not only a substantial portion of the total N input to the rivers (Figure 5A), but also a substantial portion of the total N input to these 16 watersheds (Figure 5B). The RivR-N model predicts the proportion of N input to a river network that is removed within the river. It does not require knowing the amount of N input to the river, which is difficult to quantify given the multiple sources and pathways of N inputs throughout a river network. However, both the amount of N removed in a river network ($totN_{rem}$) and the input of N to a river network (N_{input}) can be estimated by combining the RivR-N model estimate of the proportion of N input to a river network that is removed ($fractionN_{rem}$) (Figure 5A) with the river export of N at the mouth of the defined watershed (N_{export}) as follows:

$$N_{input} = N_{export}/(1 - fractionN_{rem}) \tag{4}$$

$$totN_{rem} = N_{input} * fractionN_{rem} \tag{5}$$

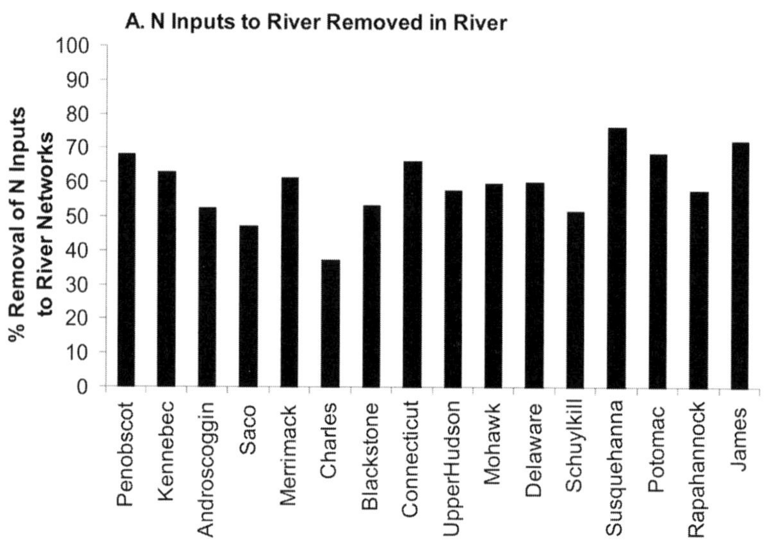

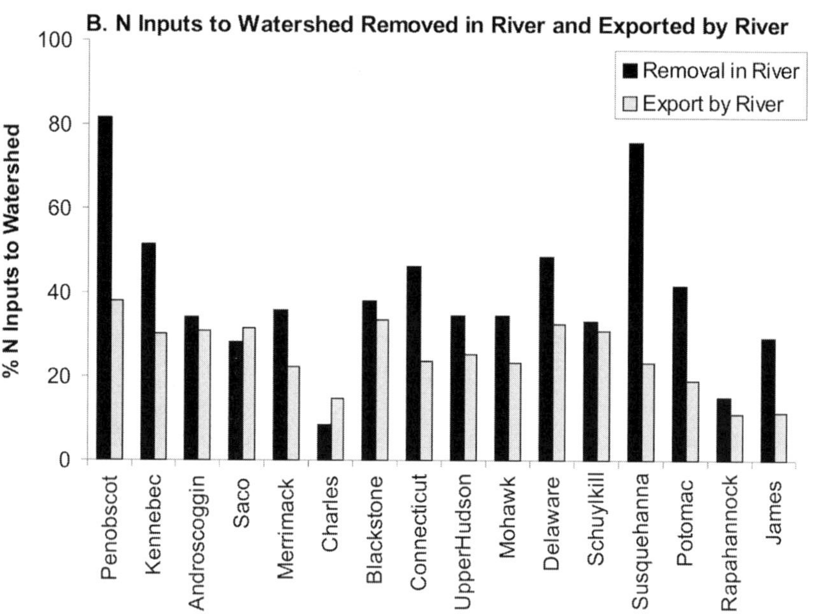

Figure 5. (A) Model-predicted proportion of N input to the river networks that is removed by in-river processes, and (B) proportion of N input to the watershed that is removed by in-river processes and river export for the 16 river networks (shown in Figure 2) for the $RF1_m + NHD'$ scenario (base case).

We estimated totN$_{rem}$ by the river using N$_{export}$ for each river network calculated from USGS monitoring data by Boyer et al. (2002) and fractionN$_{rem}$ from the RivR-N model output (Figure 5A). We then estimated the proportion of total watershed inputs that were removed by the river (Figure 5B) by comparing this value to total N inputs to these watersheds from atmospheric deposition (wet and dry NOy), fertilizer use, N$_2$-fixation in forests and agriculture, and net import of food and feed (Boyer et al. 2002; van Breemen et al. 2002).

On average about half (48% +/– 19) of the total N inputs to these watersheds was removed within the river, based on the RF1$_m$ + NHD' scenario (base case). There was considerable variation among rivers with approximately 10% (Charles River) to 80% (Penobscot River) of the total known N inputs to the watershed removed within the associated river network (Figure 5B). On average, N export by the rivers was equal to 25% (+/– 8) of the total N inputs to the watersheds (Boyer et al. 2002). The combined removal of N by in-river processes (RivR-N model estimates) plus river export accounts for an average of 65% (+/– 24) of the total N input to the watershed.

There are numerous processes within the watershed that remove N, in addition to river export and denitrification. Comprehensive N budgets were developed for these sixteen watersheds by van Breemen et al. (2002) that included all known input terms (noted above; as developed by Boyer et al. 2002) and N removal and storage terms (soil denitrification in forests, agriculture and urban land; forest harvest; storage in wood and soils; waste volatilisation; river N export; denitrification in rivers). Overall, the three largest sinks for N were denitrification in soils and rivers and river export. On average, there was considerable agreement between the estimated watershed total N inputs and total N storage and losses (van Breemen et al. 2002). However, there are considerable uncertainties in a number of budget components, including denitrification in soils and in rivers, and the watershed N balance depends to a considerable extent on the choice of values used for these terms, as they discuss. Uncertainties in the river denitrification term, as estimated by the RivR-N model, are discussed in more detail below (Model Uncertainty).

Calculation of N removal in a river network with the RivR-N model requires detailed reach scale information on depth, time of travel, and river network configuration. Comparison of the model-estimated N removal at the river network scale (Figure 5A) with more readily available river network scale parameters suggests that both watershed area and total length of all reaches in a watershed are reasonably good predictors of the proportion of N input to rivers that is removed in the river network of a watershed, under

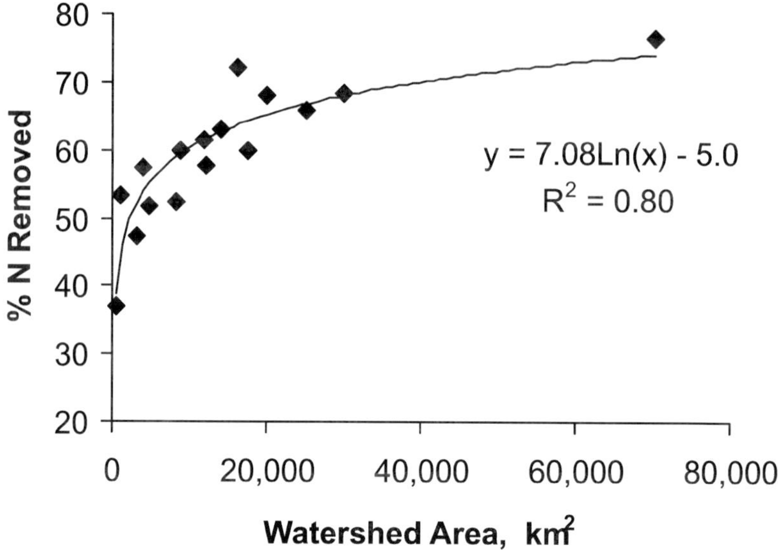

Figure 6. Relationship between watershed area and the proportion (as a percentage) of N input to the river network that is removed by in-river processes according to the RivR-N model ($RF1_m$ + NHD' scenario, base case).

the base case scenario (constant N loading per unit area of watershed). The RivR-N model predicted proportion of N removed is related to the log of the watershed area (r^2 = 0.80) (Figure 6) or log of the total length of all reaches (r^2 = 0.84; %N Removed = 7.5*Ln(total reach length, m) – 59). These relationships may provide approaches for estimating N removal in watersheds where detailed river network information is not available, at least in watersheds with hydrological and physiographic characteristics similar to the watersheds used in this study.

N removal patterns within the river networks
There are a number of perspectives from which to examine the pattern of N removal within a river network. N removal varies as a function of stream size, both at the reach scale and at the river network scale. Reaches within a river network range from small, shallow headwater streams to wider, deeper reaches of rivers. There are a number of methods to classify stream systems. In the Strahler channel-ordering system, a first order stream has no tributaries; where two 1st order channels join, a stream of 2nd order is formed, and so on (Strahler 1952 modified from Horton 1945). Strahler order is identified for each reach in the RF1 file, and therefore, provides one perspective from which to examine the contribution of different stream sizes to N removal in the16 river networks.

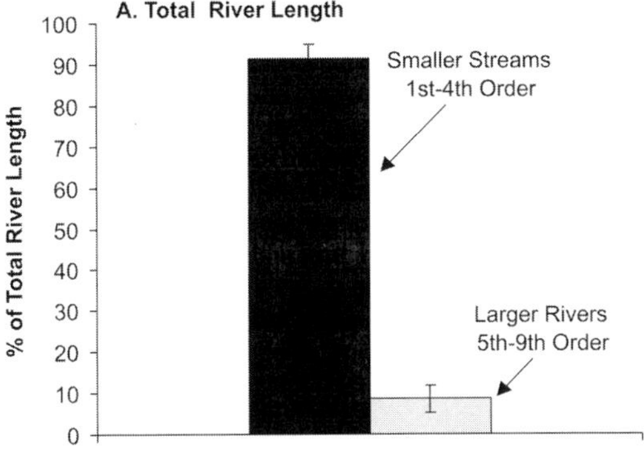

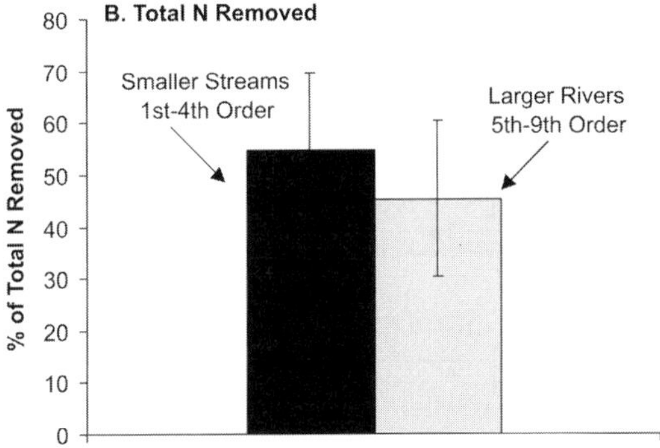

Figure 7. Proportion of: (A) total river length, and (B) model predicted percent of total N removed in the river network, for small to intermediate (1st through 4th order) and larger (5th order and greater) streams/rivers ($RF1_m$ + NHD' scenario). Average $\pm$ S.D. for the 16 study watersheds.

The vast majority of the total stream length in a river network is in relatively small streams (Leopold & Maddock 1953). In the 16 study watersheds, reaches identified as 1st through 4th order account for an average of 91% (+/− 3%) of the total stream length in the river network, while reaches that are 5th order and greater account for only 9% (+/− 3%) of the total stream length (Figure 7A). There is considerable variation among reaches within a stream order in reach-specific N removal (proportion of N input to a reach that is removed in that reach). However, overall, the model suggests that reach-specific N removal generally decreases with increasing stream order, as illustrated by the Penobscot, Connecticut and Kennebec rivers (Figure 8A). In these three rivers, approximately 20–25% of the N input to first order reaches is removed within the reach, while in 7th and 8th order reaches only about

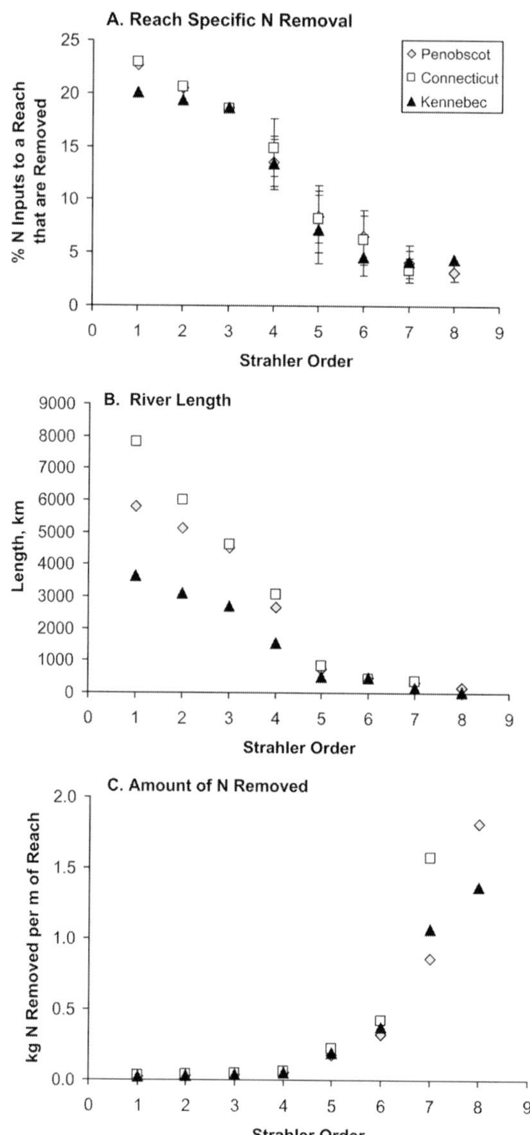

Figure 8. Relationship to Strahler order for the: (A) proportion of N input to a reach that is removed within that reach (average $\pm$ S.D.), (B) amount (km) of the total river length in a watershed, and (C) average amount of N (kg) removed per meter of stream/river for the $RF1_m$ + NHD' scenario with uniform direct watershed N loading to river (100 kg N km^{-2} watershed y^{-1}). Results for the Penobscot, Connecticut and Kennebec river networks are shown.

5% of the N input (direct watershed input plus input from upstream reaches) are removed within the reach. This general pattern is similar across the 16 watersheds and is related to the depth/TOT characteristics which, on average, increase with increasing channel size. Both the water depth and velocity increase with increasing channel size. However, depth increases relatively faster than velocity, resulting in an increase in the ratio of depth to time of travel, and thus a decrease in the model predicted proportion of N input that is removed in the reach.

The contribution of streams of different sizes (i.e., orders) to the total N removed in a river network depends on a number of factors related to stream size according to the RivR-N model. These factors include the: (1) contribution of streams of different sizes to total river length, (2) depth and time of travel characteristics of streams of various sizes, and (3) magnitude of N input to streams of different sizes (from both upstream reaches and direct watershed input). The contribution of each stream order to total stream length in a river network decreases with increasing stream order (Figure 8B). In addition, reach-specific N removal is greatest in smaller streams (Figure 8A). However, the model suggests that the total mass of nitrogen removed per meter of stream length increases with increasing stream order (Figure 8C). The reason for this seemingly contradictory pattern is the effect of stream morphology on N inputs and losses, specifically the dendritic pattern displayed in stream channels. Higher order streams and rivers have many more interconnected segments arranged in a linear series that extend over long distances as compared to the more parallel, and hydrologically independent, arrangement of segments that is reflected in lower order streams (see Figures 3, 4). This, in combination with N export from upstream lower order streams, leads to a greater cumulative effect on the absolute quantities of nitrogen that are input and removed in the individual reaches of higher ordered streams. Therefore, while a smaller proportion of N entering a reach is removed in higher order reaches, larger amounts (mass) of N are input in these reaches resulting in a larger amount of N removed per meter of stream length in higher order reaches (Figure 8C). At the river network scale this results in approximately half (45 +/– 15%) of the total N being removed in streams of 5th order and higher, and half (55 +/– 15%) in 1st through 4th order streams, according for the RivR-N model (Figure 7B). Thus, the larger river channels as well as small to intermediate streams are important in total N removal at the river network scale, although the larger channels account for only ~10% of the total length of river channels.

The RivR-N model output was used to explore how much of the river N input from the watershed to streams of various sizes is removed before reaching the watershed outlet. A detailed examination of results from the

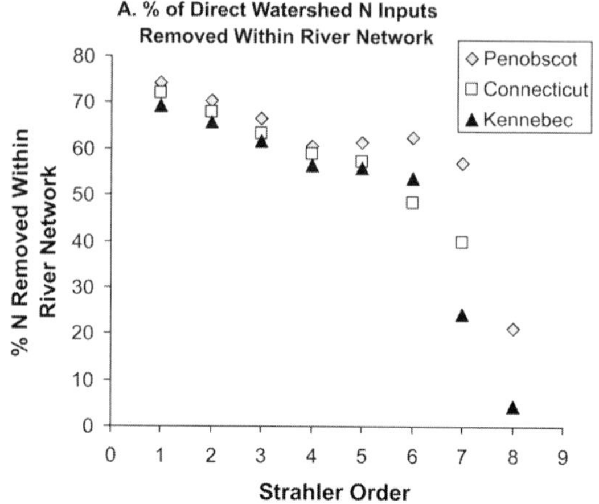

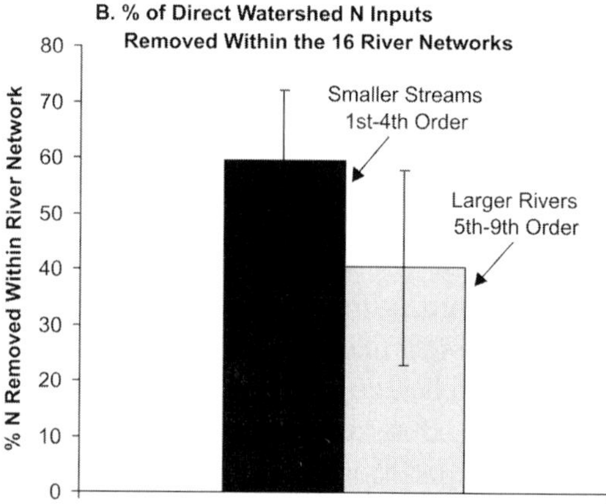

Figure 9. Model predicted percent of direct watershed input to a particular Strahler order that is removed within the river network for (A) the Penobscot, Connecticut and Kennebec river networks and (B) all 16 river networks (average ± S.D.). For example, (A) approximately 60% of the direct watershed N input to 4th order streams is removed within streams of order four and greater.

Penobscot, Connecticut and Kennebec river networks suggests that about 60 to 70% of the direct watershed N input to 1st through 5th order rivers is removed within the river network (Figure 9A). Conversely, 30 to 40% of that N input is exported by the river to the watershed outlet. In rivers of 6th order and higher a decreasing proportion of the direct watershed N input is removed within the river network, which reflects the decreased in-river processing time as well as lower reach-specific removal in higher order rivers (Figure 8A). In the 16 river networks, an average of 60% (±13) of the river

N inputs from the watershed areas draining 1st through 4th order streams is removed before reaching the watershed outlet, according to the RivR-N model (Figure 9B). A more variable and smaller proportion (40% ± 18) of the N entering larger river channels (≥5th order) from the surrounding watershed is removed within the river network. These results are consistent with the SPARROW (Spatially-Referenced Regression On Watershed attributes) model analysis of the Mississippi River basin and demonstrate the importance of distance traveled through the river network for efficient removal of N (Alexander et al. 2000). SPARROW is a statistical watershed model (see Smith et al. 1997) that provides empirical estimates of nitrogen loss in streams and rivers.

Finally, the RivR-N model output was used to explore where N is removed that originates from watershed areas draining 1st, 2nd, etc. order streams. Direct watershed N loading to 1st order streams is removed throughout the river network, with a significant amount removed even in the highest order rivers, as illustrated by the Penobscot river results (black bars, Figure 10). The largest amount of the direct watershed N loading to 1st order streams is removed within the 1st order streams compared to any other single downstream order. Similarly, 2nd order streams remove more of the direct watershed N loading to 2nd order streams than is removed in any single downstream order, although all downstream orders contribute to additional removal. The same is true of direct watershed N loading to all orders. Examination of the total N removed by river order shows that, in general, the highest order rivers remove more N than any other single lower order (Figure 10), although they account for only a small percentage of the total river length (Figure 7A and 8B). Furthermore, most of the N removed in these highest orders originates from the smaller (1st through 4th) order rivers under the base case scenario of uniform direct watershed N loading.

Scenario results

Effect of reservoirs
The effect of human modification of river hydrology on nutrient transport is the subject of considerable interest (Garnier et al. 2000; Vörösmarty et al. 1997). Reservoirs are one such modification. The model equation (eq. 2) for predicting reach-specific N removal was based on both river and lake studies and thus was considered applicable to reservoirs. We used the RivR-N model to explore the potential effect of reservoirs on N removal at the network scale. The number of reservoirs in each of the study watersheds ranged from 0 (Blackstone and Charles) to over 20 (Susquehanna and Penobscot) (Table 2).

At the river network scale, the effect of reservoirs on N removal was minimal. Removing all reservoirs from a watershed river network and substi-

222

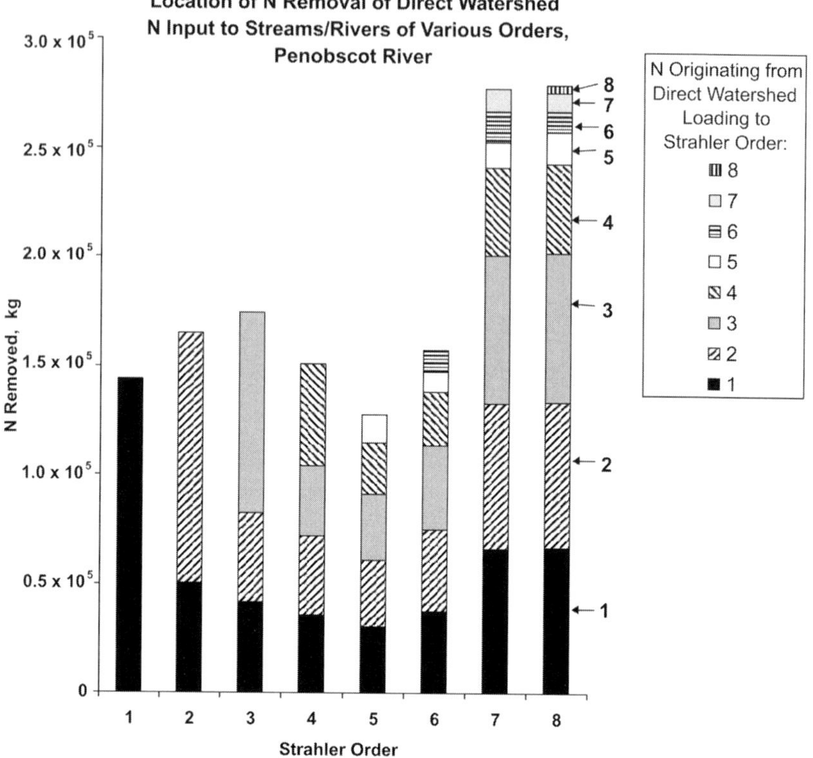

Figure 10. The location of N removal of direct watershed N input to streams/rivers of various Strahler orders for the Penobscot river network according to the RivR-N model. For example, the total amount of N removed in each Strahler order that originates from direct watershed N loading to 1st order reaches is represented by the black portion of the bars. The amount of N removed (kg) is determined using the RF1 + NHD' scenario with uniform direct watershed N loading (100 kg N km^{-2} watershed y^{-1}).

tuting pre-reservoir depth and TOT information for those reaches had little effect on the proportion of N removed (Figure 11A). The difference in the proportion of N removed with and without reservoirs was less than 2 percentage points, and in some cases removal of all reservoirs slightly increased N removal.

Examination of individual reservoirs on N removal provides insight into why the total contribution of reservoirs within the river network of a watershed is often predicted to be small by the RivR-N model. While reservoirs increase both depth and water residence time in a section of river, it is the ratio of these two characteristics that determines the proportion of N removed according to the model (eq. 2). The proportion of N input to a reservoir that is removed can be greater, less than, or essentially the same as the propor-

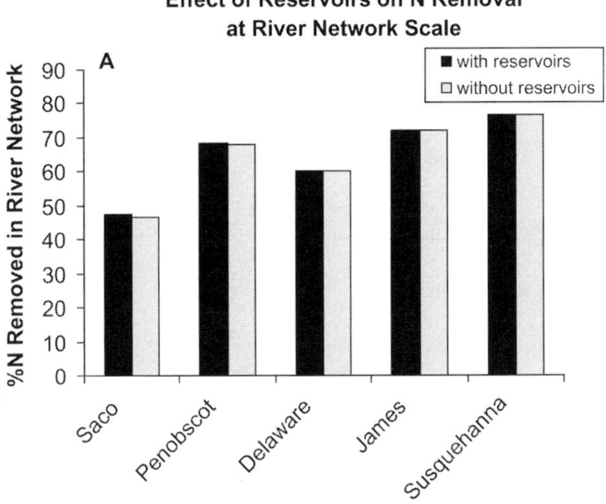

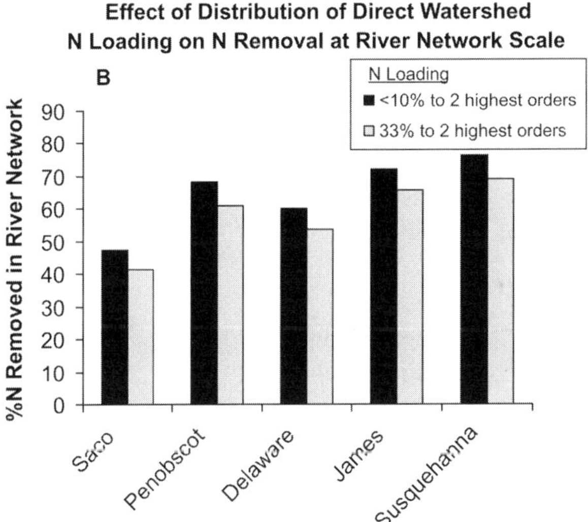

Figure 11. River network scale comparison of the proportion (as a percentage) of N input that is removed: (A) with reservoirs (base case) and without reservoirs, and (B) with $\leq$ 10% and 33% of the total direct watershed N load entering rivers of the 2 highest orders.

tion of N input removed in that river section under pre-reservoir depth and TOT (Figure 12). For example, the depth (8.9 m) and TOT (1.06 years) for Schoodic Lake (designated #1078 in the RF1 file) in the Penobscot basin results in a considerably higher (40.5%) proportion of N input to the reservoir that is removed compared to the pre-reservoir removal (14%) which had a

224

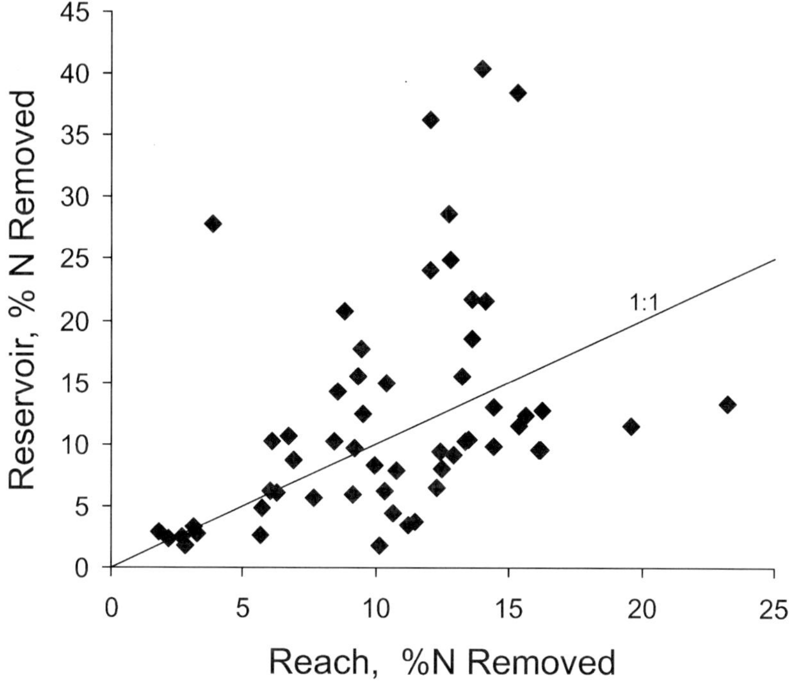

Figure 12. Reach scale comparison of the proportion (as a percentage) of N input that is removed in a reservoir and in the same reach under pre-reservoir depth and time of travel characteristics, according to the RivR-N model.

depth of 0.23 m and a TOT of 0.0015 y. In contrast, the Toronto reservoir (designated #1683 in the RF1 file) in the Delaware basin with a depth of 89.6 m and TOT of 0.42 y yielded a 12.4% removal of N input to the reservoir, which is slightly less than the proportion of N input that was removed in the reach (15.6%) before the reservoir (0.34 m depth; 0.003 y TOT), according to the RivR-N model.

The position of a reservoir within the river network, in addition to the ratio of the depth to the TOT of water in the reservoir, is also important in determining the contribution of a reservoir to N removal at the river network scale. For example, a reservoir with optimal depth and TOT (shallow with long residence time) placed near the outlet of a watershed would be relatively more effective at reducing N export by a river than the same reservoir placed in the upper reaches of a watershed where relatively less of the total N loading would pass through the reservoir. To explore this, we artificially replaced a 4th order and a 7th order reach of the Delaware with a theoretical reservoir with a depth of 16 m, a 1 y TOT, and a model predicted N removal of 32% of the N input to the reservoir. These were the characteristics of reservoir #1070 (Rangeley Lake; RF1 file) in the Androscoggin watershed; the proportion of N input to this reservoir that is removed is one of the highest in the 5 scenario watersheds. Placement of this reservoir in a 4th order reach (0.42 m depth, 0.00318 y TOT) of the Delaware river network increased the N removal within that reach from 14.7% to 32% of the N input to that reach. However, at the river network scale the N removal increased by only 0.1 percentage points (from 60.0% to 60.1%). Placement of this reservoir in the most downstream reach (depth 2.7 m, TOT of 0.00099 y) of the Delaware river network where it is a 7th order river increased the N removal within that reach from 4.8% to 32% of the input to the reach. The total N removal in the Delaware river network increased from 60% to 70%. The discharge of the Delaware River at this location (365 m^3 s^{-1}) would necessitate a very large reservoir (719 km^2) to maintain the theoretical reservoir depth and TOT. However, the exercise illustrates the relatively small additional N removal that a reservoir even with "optimal" depth, TOT and location is predicted to have in a watershed under the base case scenario with uniform N loading from the direct watershed.

Scale of river network information
Two reach files with different scales were combined to apply the RivR-N removal model to the study watersheds in the base case scenario (as detailed above in Model application: Base case scenario). RF1 and NHD depict approximately the resolution of the "blue-line" drainages that show up on USGS 1:500,000 and 1:100,000 scale maps, respectively (Figure 3A, B). The effect of the spatial scale of the reach network file used to estimate N removal in a watershed was examined by comparing the RivR-N model output for the base case (approximately the scale of NHD) with the reduced scale (RF1 data only). The reduced scale scenario eliminated all 1st, 2nd, 3rd and in some cases 4th order rivers in the base case scenario. The total watershed area and total watershed direct N loading was the same as in the base case, however, the watershed area draining directly into a specific reach in RF1 was greater than under the base case scenario, because of the greatly reduced total reach length in a watershed. Direct watershed area for each reach in RF1 was available in the RF1 file. The predicted proportion of N removed in a river network in the reduced scale scenario was less than with the higher resolution base case (Table 2). This was expected given the greatly reduced total reach length, and thus reduced cumulative N removal, in the RF1 only scenario. However, the decrease in the proportion of N removed under the reduced scale scenario averaged only 14 percentage points less than the higher resolution base case (S.D. +/– 7; range ~8 to 31 percentage points), although the total reach length was approximately 80% less. The somewhat smaller than

expected difference in the predicted N removal between the $RF1_m$ + NHD' and RF1 scenario is due to the larger direct watershed area for reaches in RF1 in the RF1 scenario compared to those same reaches in the $RF1_m$ + NHD' scenario. This resulted in a larger N input to the RF1 reaches (and thus larger N removal) in the RF1 scenario compared to the $RF1_m$ + NHD' scenario. This effect is also illustrated by comparison of Figure 7B which shows that in the $RF1_m$ + NHD' scenario approximately 50% of the total N was removed in 1st through 4th order streams (Figure 7B) which are primarily the NHD' reaches, with the finding that the difference in N removed in the RF1 scenario was only 14 percentage points, on average, less than in the $RF1_m$ + NHD' scenario.

Distribution of N loading within the watershed
The varying distribution of source inputs across the landscape in a watershed is expected to affect the N removed in a river network (Alexander et al. 2000). Total N loadings and N inputs to the rivers for the 16 study watersheds were estimated using GIS land-use distributions coupled with N loading information that was developed for specific land uses (Boyer et al. 2002; van Breemen et al. 2002). However, the geographic distribution of N loading within each watershed was not maintained in those budgets. So, as a first approximation of the effect of varying degrees of source inputs across the landscape on N removal in rivers, we explored several theoretical patterns of the N distribution. In the base case scenario, N loading to the rivers per unit area of watershed (referred to as direct watershed N loading) was held constant (100 kg N per km^2 per year). Such a scenario could be considered to represent a uniform distribution of land use in a watershed, such as a completely forested watershed. Uniform land use is not generally the case today, given the mosaic of human development/alteration in watersheds, including agriculture and urban land uses.

It is not uncommon for major cities to be located at the base of a watershed, on the higher order river segments, resulting in a high N input from wastewater treatment facilities, urban runoff, etc. near the base of a watershed. To explore the effect of the distribution of N loading (land use) in a watershed on the amount of N removed during river transport, we allocated a third of the total direct watershed N loading to reaches of the two highest orders (these were generally between 6th and 9th orders); lower order reaches received the remaining N at a constant rate. The total direct watershed N loading to the river network remained the same as in the base case scenario. The five test watersheds were used for this scenario run.

In the base case, approximately 90% of the total direct watershed N loading entered 1st through 4th order reaches. This was the combined result of 1st through 4th order reaches generally accounting for ~90% of the total

Table 3. Regression estimates for nitrogen loss model. CI = Confidence interval

Model parameters	Parametric coefficient	Bootstrap coefficient	Lower 90% CI	Upper 90% CI	Standard deviation
Intercept	88.45	90.19	74.85	110.06	14.5
Ratio of depth to water residence time	0.3677	–0.3722	–0.2930	–0.4665	0.064
R-squared	0.73				
Mean square error (log units)	0.2403				

river length (Figure 7B), a constant drainage density (km^2/river length) in the watershed, and an assumed even N loading per unit area of watershed. The two highest orders account for a total of between 1% and 10% of the total river length, and, therefore, 1% to 10% of the total direct watershed N loading. Re-allocating a third of the direct watershed loading to the two highest orders resulted in a fairly modest decrease in the proportion of N removed in the river network: 6 to 7 percentage points (Figure 11B). These results are generally consistent with the findings of a SPARROW-based analysis of non-uniformities in the spatial distribution of sources in these watersheds and their effects on in-stream N loss (see Alexander et al. 2002). The decrease in the N removal compared to the base case is partly the result of a larger amount of N entering the higher order reaches which are subject to less additional removal in downstream reaches (shortened flow path to watershed outlet). In addition, these higher order reaches tend to have a lower proportion of reach-specific N removal than lower order reaches due to their depth and water residence time characteristics (Figure 8A).

Model uncertainty

There are several sources of model uncertainty, including: (1) statistical uncertainties in model coefficients and predictions, and (2) uncertainties in the river reach files (RF1 and NHD) used in the model application.

Statistical uncertainties in model coefficients (Table 3) and predictions (Table 4) were quantified using bootstrapping techniques (Efron 1982), as described above (Model application: Statistical analysis of model uncertainty). A number of factors contribute to the statistical uncertainties in the model coefficients (Table 3). These include the small number of river studies that were available for use in the development of the model equation, as well as uncertainties associated with each study in the determination, or our extra-

Table 4. Evaluations of equation 2 (RivR-N Model) for estimating N loss in river networks using only the RF1 data files (reduced scale scenario) for the 16 watersheds. Results from the SPARROW Model (Alexander et al. 2000) also using the RF1 data files are shown for comparison

Watershed	Equation 2 parametric[a]	Bootstrap % boot mean[b]	Bootstrap % w/model error[c]	Lower 90%[d]	Upper 90%[e]	SPARROW
Penoboscot	59	61	67	51	82	47
Kennebec	52	53	57	42	70	46
Androscoggin	44	47	51	37	64	40
Saco	33	35	38	27	49	35
Merrimack	38	45	50	35	63	34
Charles	15	15	15	7	24	30
Blackstone	22	24	26	16	35	31
Connecticut	55	60	66	49	80	47
Upper Hudson	50	48	52	39	65	45
Mohawk	47	51	55	39	69	40
Delaware	48	53	59	42	74	38
Schuylkill	31	30	33	22	42	28
Susquehanna	63	67	73	57	87	45
Potomac	60	59	64	49	78	59
Rappahannock	42	41	45	33	56	59
James	61	57	62	46	74	58
25th Percentile	37	40	43			35
Median	48	50	54			43
75th Percentile	56	58	62			47

[a]RF1 loss based on equation 2 %N = 88.45* Ratio**–0.3677 with parametric estimates.
[b]RF1 loss – based on bootstrap mean coefficients for equation 2 : %N = 90.19* Ratio**–0.3722.
[c]RF1 loss – based on bootstrap mean coefficients for equation 2 as above with model error (regression coefficients & residuals).
[d]Lower 90% confidence interval on bootstrap %N based on inclusion of model error.
[e]Upper 90% confidence interval on bootstrap %N based on inclusion of model error.

polation, of the proportion of TN removed on an annual basis in each of those studies. In addition, variables not currently included in the model, but which are important in controlling the proportion of TN removed in a river reach, would contribute to the uncertainty in the model coefficients. Future studies quantifying N removal in rivers are needed for independent comparison with the model predictions and to potentially improve the model equations.

Errors in model predictions of the proportion of nitrogen removed for each of the sixteen watersheds included uncertainties in both the model coefficients and the observed data (i.e., regression residuals). Errors in model predictions at the river network scale were computed utilizing the uncertainties associated with the regression residuals (see Model application: Statistical analysis of model uncertainty). These computations were only done for the reduced scale scenario for the RivR-N model. Therefore, they include errors in model predictions of N losses in the RF1 streams and reservoirs, but not losses in NHD' reaches. The prediction uncertainties for the 90% confidence interval in the proportion of nitrogen removed at the river network scale (Table 4) range from +/− 17 to +/− 32 percentage points with a median of +/− 28. For example, in the Delaware basin the 90% confidence interval on the mean nitrogen loss of 59% ranges from 42 to 74% or +/− 32 percentage points. Expressed as a proportion of the mean nitrogen loss in the river networks, the 90% confidence intervals range from 39 to 113 percent (median = 54 percent). The greatest uncertainty is generally displayed in the smallest watersheds, such as the Charles (113 percent) and Blackstone (73 percent), where the least amount of averaging of errors would be expected to occur.

The in-stream losses calculated for the sixteen eastern U.S. watersheds using the RivR-N model are very similar to those using the SPARROW model (see Alexander et al. 2000) (Table 4), based on a comparison of estimates for both models using the RF1 scale reaches (reduced scale scenario of RivR-N model) (see Alexander et al this volume for additional comparisons). SPARROW estimates of the rates of nitrogen loss decline rapidly with channel size from 0.45 per day of water travel time in small streams to 0.005 per day in large rivers (Alexander et al. 2000). The SPARROW estimates show general agreement with the pattern in the RivR-N model (Figure 8A) and with literature estimates of loss, including those for streams used in the calibration of the RivR-N model.

While the river reach files used in the model application are the most comprehensive available, they both have shortcomings. As previously noted, approximately 80% of the total channel length (true 1st, 2nd and 3rd order rivers) in a network are not contained in the RF1 files because of their lower spatial resolution. The higher resolution NHD files contain more accurate total length of reaches in the rivers in each watershed but do not contain information on the flow, depth and TOT of stream reaches. We estimated these parameters for NHD' for each watershed based on extrapolations using characteristics of the higher order RF1 reaches for the associated watershed (detailed in Appendix). The errors incurred by combining of NHD' and RF1 and with estimating reach characteristics for NHD' are not known. Within the next few years it will be possible to rerun the RivR-N model when the fully annotated NHD reach files are available (R. Pierce, U.S. Geological Survey, pers. comm. 2000).

Summary

Many coastal ecosystems are receiving increased N inputs due to anthropogenic N inputs to their watersheds. Management of nutrient inputs to coastal ecosystems requires not only knowing the sources and magnitude of N to their watersheds, but also knowing where and how much of that N is removed during its movement through the terrestrial and freshwater ecosystems. A substantial amount of nitrogen can be removed within the network of streams and rivers draining watersheds as shown by studies in a wide range of stream/river sizes and geographic locations. However, most studies have examined N removal in short sections of a river (reach), such as a first order stream or a larger channel of the river and the proportion of N removed at the river reach scale varies widely (1% to >80%) (e.g., Table 1; Swank & Caskey 1982; Chesterikoff et al. 1992; Jansson et al. 1994). Previous studies generally have not addressed how to scale-up those results to the whole river network, with the exception of Smith et al. (1997) and Alexander et al. (2000). The current analysis suggests that the water displacement of streams and reservoirs (ratio of depth to time of travel for a river reach) can explain a considerable amount of the variation in the reported N removal efficiencies. In order to evaluate the N removed in a river network, the relationship must be applied to reaches throughout a river network, and any N not removed in a reach must be routed to downstream reaches where it is subject to additional removal. As a result of this cumulative removal, the proportion of the N inputs to a river network are considerably larger than the proportion removed in any one reach or section of river, as shown by the RivR-N model developed in this study. Application of the RivR-N model to sixteen river networks in the eastern U.S. indicates that in-river processes significantly reduce N transport to coastal ecosystems. Between 37% and 76% of the total N input to these river networks was removed during transport. In the context of the total N inputs to these watersheds, in-river processes removed an average of 40% ($\pm$ 19) of the inputs.

The relationship developed for the RivR-N model describing the proportion of N input that is removed in a river reach as an inverse function of "reaction time", in this case water displacement (depth/water time of travel), is consistent with findings across a wide range of aquatic ecosystems. This includes studies within an individual river reach where N removal decreased as flow increased (Jansson et al. 1994; Pind et al. 1997). It is also consistent with the watershed mass-balance model, SPARROW, in which the proportion of N removed decreases with increasing stream channel depth (Smith et al. 1997; Alexander et al. 2000), with a relationship developed for estuaries in which the proportion of N removed decreases with decreasing water residence time (Nixon et al. 1996), and with a relationship developed for shallow, well-mixed lakes in which the proportion of nitrate removed decreases with increasing nitrate concentration and the ratio of depth/water residence time (Kelly et al. 1987). Modifications to the RivR-N model would likely be necessary for rivers with characteristics widely different than those used to develop the model, such as rivers with extensive hyporheic flow (Sjodin et al. 1997). The model does not include processes in riparian areas where a substantial amount of N also can be removed (Peterjohn & Correll 1984; Lowrance et al. 1984; Billen & Garnier 1999).

Analysis of the RivR-N model output was used to explore where in the river network most of the N removal is occurring and what factors contribute to those patterns, based on the $RF1_m + NHD'$ scenario. Larger river channels generally have the lowest reach-specific N removal (proportion of total N input to a reach that is removed in that reach). However, per meter of stream, the greatest amount of N removed is in the larger river channels, because of the large amount of N passing through them due to export from upstream reaches. Approximately half of the N in the sixteen eastern U.S. river networks is removed in larger rivers ($\geq$5th order) and half in small to medium size streams ($\leq$4th order), although streams of $\geq$5th order account for only about 10% of the total stream length. In the 16 river networks, an average of 60% ($\pm$ 13) of the river N inputs from the direct watershed areas draining 1st through 4th order streams is removed before reaching the watershed outlet. A more variable and smaller proportion (40% $\pm$ 18) of the N entering larger river channels ($\geq$5th order) from the directly surrounding watershed is removed within the river network. Direct watershed N loading to, for example, 1st order streams is removed throughout the downstream river network, with a significant amount removed even in the highest order rivers. Most of the N removed in the highest orders originates from the smaller (1st through 4th) order rivers. The RivR-N model indicates that reservoirs account for a small proportion (less than 2 percentage points) of the total N removed throughout a river network, although within any one reach a reservoir can substantially increase or decrease N removal. Varying the distribution of N loading within the river network indicated that the N removal decreased by less than 10 percentage points when the direct watershed N input to the two highest orders increased from less than 10% to 33%. Clearly, N removal in river networks is important in determining the amount of N delivered from watersheds to coastal marine systems. Future field studies and annotation of high resolution river reach files will be important in improving our understanding of this important sink for N.

Appendix 1
Development of depth and TOT characteristics for NHD' reaches

The numbers of 1st, 2nd, 3rd, and in some cases 4th order rivers in NHD' in each watershed were estimated using the average length of 1st, 2nd, 3rd and 4th order rivers (1609 m, 3701 m, 8528 m, 19308 m, respectively) and a bifurcation ratio of approximately 3.5, both derived from many samples of basins in the U.S. (Leopold et al. 1964). Bifurcation ratios normally range between 2 and 4; the ratio varied between 2.6 and 3.7 in our analysis to accommodate the various NHD' lengths in the different watersheds while maintaining the length of nth order rivers constant.

Reach depths for NHD' were estimated as a function of reach discharge using eq. 3 (main text). Time of travel (TOT) for NHD' reaches was estimated from river length divided by velocity. Velocity is a power function of discharge (Leopold & Miller 1956). The relationship between velocity and discharge (Figure A1) for each watershed was calculated using RF1 data; the relationship varied within a watershed, depending on discharge. Therefore, only data from reaches with discharges less than 1–2 m^3/s were used to develop the relationship for NHD' velocities (which have low discharges). NHD' velocities were then used with NHD' reach lengths to calculate the time of travel for reaches in NHD'.

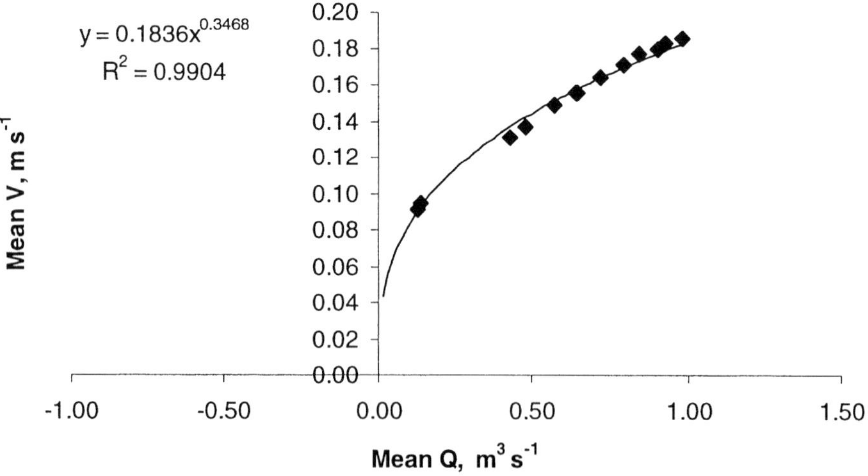

Figure A1. Graph of Q vs V for Delaware (all reaches with Q less than 1).

Discharge is needed in both equations 2 and 3 (main text). Within a basin, river order generally is related to the log of the discharge (Leopold & Miller 1956). There was a significant relationship between log of discharge and river order in RF1 (0.001 level of significance) for each of the study watersheds. Therefore, the relationship between river order and discharge in RF1 for each watershed was used to estimate the discharge of river orders in NHD' (i.e., 1st, 2nd, and 3rd order rivers) for each watershed (Figure A2). There are multiple rivers of nth order in a river network. Within RF1 often there are multiple contiguous reaches of the same order within a river. Only data for the most downstream reach within each river of order n were used to derive the relationship between river order and discharge.

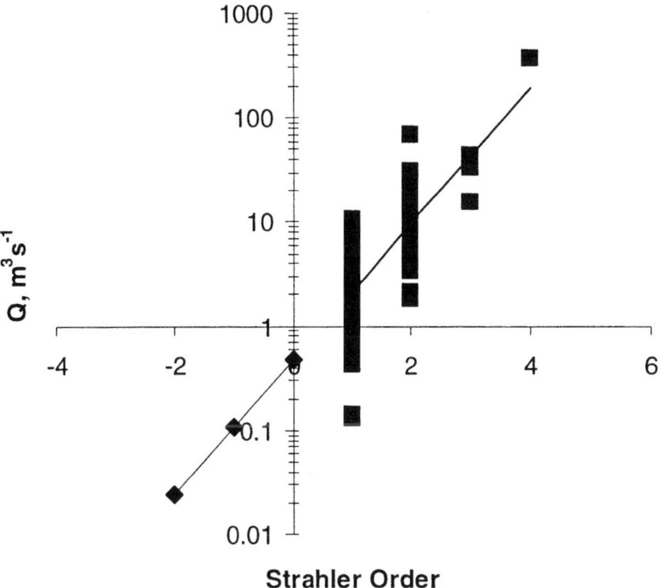

Figure A2. Example graph of Q vs Strahler order for Delaware [regression equation: Q = (0.651 ∗ Strahler #) – 0.318]. Predicted Q for NHD' Strahler orders (here denoted as 0, –1 and –2; in the RF1$_m$ + NHD' scenario all orders were renumbered such that in the current example the final orders would be 1 through 6).

NHD' reaches were connected to the RF1 network by assuming that half of all rivers of order n discharge to rivers of n+1 order (Leopold et al. 1964). The remaining rivers of order n were then connected to rivers of greater than n+1, equally per unit length of river.

234

Acknowledgements

This work was initiated as part of the NOAA Atmospheric Assessment Project and continued under the International SCOPE Nitrogen Project that received support from both the Mellon Foundation and from the National Center for Ecological Analysis and Synthesis. This work was supported, in part, by funding (to SPS) from NOAA NJ Sea Grant (NJSG # 99–427) and the NOAA Coastal Ocean Program. The views expressed herein are those of the authors and do not necessarily reflect the views of NOAA or any of its sub-agencies. We would like to thank Hilairy Hartnett and Tracy Wiegner for useful discussions during the development of this project and three anonymous reviewers of the manuscript who provided insightful comments.

References

Alexander RB, Johnes PJ, Boyer EA & Smith RA (2002) A comparison of methods for estimating the riverine export of nitrogen from large watersheds. Biogeochemistry 57/58: 295–339

Alexander RB, Brakebill JW, Brew RE, & Smith RA (1999) ERF1 – Enhanced River Reach File 1.2, U.S. Geological Survey Open-File Report 99–457. Reston, Virginia

Alexander RB, Smith RA & Schwarz GE (2000) Effect of stream channel size on the delivery of nitrogen to the Gulf of Mexico. Nature 403: 758–761

Andersen JM (1977) Rates of denitrification of undisturbed sediment for six lakes as a function of nitrate concentration, oxygen, and temperature. Arch. Hydrobiol. 80: 147–159

Ayers JC (1970) Lake Michigan environmental survey. University of Michigan, Great Lakes Research Division Special Report 49

Billen G, Lancelot C & Meybeck M (1991) N, P and Si retention along the aquatic continuum from land to ocean. In: Mantoura RFC, Martin JM & Wollast R (Eds) Ocean Margin Processes in Global Change (pp 19–44). Dahlem Workshop Reports, Wiley

Billen G, Somville M, De Becker E & Servais P (1985) A nitrogen budget of the Scheldt hydrographical basin. Netherlands Journal of Sea Research 19(3/4): 223–230

Billen G, Garnier J & Meybeck M (1998) Les sels nutritifs: l'ouverture des cycles. In: Meybeck, de Marsily & Fustec (Eds) "La Seine en son bassin: fonctionnement écologique l'un système fluvial anthropisé". Elsevier

Billen G & Garnier J (1999) Nitrogen transfers through the Seine drainage network: a budget based on the application of the "Riverstrahler" model. Hydrobiologia 410: 139–150

Boyer EW, Goodale CL, Jaworski NA & Howarth RW (2002) Anthropogenic nitrogen sources and relationships to riverine nitrogen export in the northeastern U.S.A. Biochemistry 57/58: 137–169

Burns DA (1998) Retention of NO_3^- in an upland stream environment: A mass balance approach. Biogeochemistry 40: 73–96

Calderoni A, Mosello R & Tartari G (1978) P, N and Si budget in Lago Mergozzo. Verh. Int. Verein. Limnol. 20: 1033–1037

Chesterikoff A, Garban B, Billen G & Poulin M (1992) Inorganic nitrogen dynamics in the River Seine downstream from Paris (France). Biogeochemistry 17: 147–164

Christensen PB, Nielsen LP, Sørensen J & Revsbech NP (1990) Denitrification in nitrate-rich streams: Diurnal and seasonal variation related to benthic oxygen metabolism. Limnology Oceanography 35(3): 640–651

Christensen PB & Sørensen J (1988) Denitrification in sediment of lowland streams: Regional and seasonal variation in Gelbaek and Rabis Baek, Denmark. FEMS Microbiology Ecology 53: 335–344

Cole JJ, Peierls BL, Caraco NF & Pace ML (1993) Nitrogen loading of rivers as a human-driven process. In: McDonnel MJ (Ed) Humans as Components of Ecosystems (pp 141–157). Springer-Verlag, New York

Cooke JG & White RE (1987) The effect of nitrate in stream water on the relationship between denitrification and nitrification in a stream-sediment microcosm. Freshwater Biology 18: 213–226

Cooper AB & Cooke JG (1984) Nitrate loss and transformation in 2 vegetated headwater streams. New Zealand Journal of Marine and Freshwater Research 18: 441–450

Dillon PJ & Molot LA (1990) The role of ammonium and nitrate retention in the acidification of lakes and forested catchment. Biogeochem. 11: 23–44

Duan N (1983) Smearing estimate: A nonparametric retransformation method. J. Am. Stat. Assoc. 78: 605–610

Efron B (1982) The jackknife, the bootstrap and other resampling plans, in CBMS-NSF Regional Conference Series in Applied Mathematics, Soc for Ind. And Appl. Math., Philadelphia, PA

Garnier J, Leporcq B, Sanchez N & Philippon X (1999) Biogeochemical mass-balances (C, N, P, Si) in three large reservoirs of the Seine Basin (France). Biogeochemistry 47: 119–146

Garnier J, Billen G, Sanchez N & Leporcq B (2000) Ecological functioning of the Marne reservoir (upper Seine basin, France). Regulated Rivers-Research & Management 16(1): 51–71

Hill AR (1979) Denitrification in the nitrogen budget of a river ecosystem. Nature 27: 291–292

Hill AR (1981) Nitrate-nitrogen flux and utilization in a stream ecosystem during low summer flows. Canadian Geographer 25: 225–239

Hill AR (1983) Nitrate-nitrogen mass balances for two Ontario Rivers. In: III Fontaine (Ed) Dynamics of Lotic Ecosystems (pp 457–477). Ann Arbor Sciences Publishers, Ann Arbor

Horton RE (1945) Erosional development of streams and their drainage basins. Geol. Soc. America Bull. 56: 275–370

Howarth RW, Billen G, Swaney D, Townsend A, Jaworski N, Lajtha K, Downing JA, Elmgren R, Caraco N, Jordan T, Berendse F, Freney J, Kudeyarov V, Murdoch P & Zhao-Liang Z (1996) Regional nitrogen budgets and riverine N & P fluxes for the drainages to the North Atlantic Ocean: Natural and human influences. Biogeochemistry 35: 75–139

Jacobs TC & Gilliam JW (1985) Headwater stream losses of nitrogen from two coastal plain watersheds. J. Environ. Qual. 14(4): 467–472

Jansson M, Leonardson L & Fejes J (1994) Denitrification and Nitrogen Retention in a Farmland Stream in Southern Sweden. Ambio 23: 326–331

Kaushik NK & Robinson JB (1976) Preliminary observations on nitrogen transport during summer in a small spring-fed Ontario stream. Hydrobiologia 49(1): 59–63

Kelly CA, Rudd JWM, Hesslein RH, Schindler DW, Dillon PJ, Driscoll CT, Gherini SA & Hecky RE (1987) Prediction of biological acid neutralization in acid-sensitive lakes. Biogeochemistry 3: 129–140

Leopold & Maddock (1953) The hydraulic geometry of stream channels and some physiographic implications. Geological Survey Professional Paper #252, 57 pp, US Government Printing Office

Leopold LB & Miller JP (1956) Ephemeral streams – hydraulic factors and their relation to the drainage net. U.S. Geol. Survey Prof. Paper 282-A: 16–24

Leopold LB, Wolman MG & Miller JP (1964) The drainage basin as a geomorphic unit. In: Gilluly J & Woodford AO (Eds) Fluvial Processes in Geomorphology. San Francisco

Lowrance R, Todd R & Fail J (1984) Riparian forests as nutrient filters in agricultural watersheds. BioScience 34: 374–377

Mulholland PJ (1992) Regulation of nutrient concentrations in a temperate forest stream: Roles of upland, riparian, and instream processes. Limnology Oceanography 37(7): 1512–1526

Mulholland PJ, Tank JL, Sanzone DM, Wollheim WM, Peterson BJ, Webster JR & Meyer JL (2000) Nitrogen cycling in a forest stream determined by a ^{15}N tracer addition. Ecological Monographs 70: 471–493

National Academy of Sciences (2000) Clean Coastal Waters: Understanding and Reducing the Effects of Nutrient Pollution. Committee on the Causes and Management of Eutrophication, Ocean Studies Board, Water Science and Technology Board, National Research Council, 428 pp

Nixon SW, Atkinson LP, Berounsky VM, Billen G, Boicourt WC, Boynton WR, Church TM, Ditoro DM, Elmgren R, Garber JH, Giblin AE, Jahnke RA, Owens NJP, Pilson MEQ & Seitzinger SP (1996) The fate of nitrogen and phosphorus at the land-sea margin of the North Atlantic Ocean. Biogeochemistry 35: 141–180

Peterjohn WT and Correll DL (1984) Nutrient dynamics in an agricultural watershed: observations on the role of a riparian forest. Ecology 65: 1466–1475

Peterson BJ, Wollheim WM, Mulholland PJ, Webster JR, Meyer JL, Tank JL, Marti E, Bowden WB, Valett HM, Hershey AE, McDowell WH, Dodds WK, Hamilton SK, Gregory S & Morrall DD (2001) Control of nitrogen export from watersheds by headwater streams. Science 292 (5514): 86–90

Pind A, Risgaard-Petersen N & Revsbech NP (1997) Denitrification and microphytobenthic NO_3^- consumption in a Danish lowland stream: Diurnal and seasonal variation. Aquatic Microbial Ecology 12: 275–284

Robinson JB, Whiteley HR, Stammers W, Kaushik NK & Sain P (1979) The fate of nitrate in small streams and its management implications. In: Lohr RC (Ed) Best Management Practices for Agriculture & Silviculture (pp 247–259). Ann Arbor Science Publications, Ann Arbor

Ruddy BC & Hitt KJ (1990) Summary of selected characteristics of large reservoirs in the United States and Puerto Rico 1988, U.S. Geological Survey Open File Report 90–163, Reston, Virginia

Schelske CL (1975) Silica and nitrate depletion as related to rate of eutrophication in lakes Michigan, Huron, and Superior. In: Hasler AD (Ed) Coupling of Land and Water Systems (pp. 277–298). Springer-Verlag, New York

Seitzinger SP (1988a) Denitrification in freshwater and coastal marine ecosystems: ecological and geochemical importance. Limnology Oceanography 33: 702–724

Seitzinger SP (1988b) Benthic nutrient cycling and oxygen consumption in the Delaware Estuary. In: Majumdar SK (Ed) The Ecology and Restoration of the Delaware River Basin (pp 132–147). PA Acad. Sci.

Seitzinger SP (1991) The effect of pH on the release of phosphorus from Potomac Estuary sediments: Implications for blue-green algal blooms. Estuarine, Coastal and Shelf Science 33: 409–418

Sjodin AL, Lewis J & Saunders III (1997) Denitrification as a component of the nitrogen budget for a large plains river. Biogeochemistry 39: 327–342

Smith RA, Schwarz GE & Alexander RB (1997) Regional interpretation of water-quality monitoring data. Water Resources Research 33(12): 2781–2798

Strahler AN (1952) Hyposometric (area-altitude) analysis of erosional topography. Geol. Soc. America Bull. 63: 1117–1142

Swank WT & Caskey WH (1982) Nitrate depletion in a second-order mountain stream. J. Environ. Qual. 11(4): 581–584

Van Breemen N, Boyer EW, Goodale CL, Jaworski NA, Paustian K, Seitzinger SP, Lajtha K, Mayer B, van Dam D, Howarth RW, Nadelhoffer KJ, Eve M & Billen G (2002) Where did all the nitrogen go? Fate of nitrogen inputs to large watersheds in the northeastern U.S.A. Biogeochemistry 57/58: 267–293

Vitousek PM, Aber JD, Howarth RW, Likens GE, Matson PA, Schindler DW, Schlesinger WH & Tilman DG (1997) Human alteration of the global nitrogen cycle: sources and consequences. Ecological Applications 7: 737–750

Vörösmarty CJ, Sharma KP, Fekete BM, Copeland AH, Holden J, Marble J et Al. (1997) The storage and aging of continental runoff in large reservoir systems of the world. Ambio 26(4): 210–219

Wollheim WM, Peterson BJ, Deegan LA, Hobbie JE, Hooker B, Bowden WB, Edwardson KJ, Arscott DB, Hershey AE & Finlay J (2001) Influence of stream size on ammonium and suspended particulate nitrogen processing. Limnology and Oceanography 46(1): 1–13

Biogeochemistry **57/58**: 239–266, 2002.
© 2002 *Kluwer Academic Publishers. Printed in the Netherlands.*

Forest nitrogen sinks in large eastern U.S. watersheds: estimates from forest inventory and an ecosystem model

CHRISTINE L. GOODALE[1,*], KATE LAJTHA[2], KNUTE J. NADELHOFFER[3], ELIZABETH W. BOYER[4] & NORBERT A. JAWORSKI[5]

[1] *Carnegie Institution of Washington, Department of Plant Biology, Stanford CA 94305. Current address: Woods Hole Research Center, Woods Hole, MA 02543;* [2] *Oregon State University, Department of Botany and Plant Pathology, Oregon State University, Corvallis, OR 97331-2902;* [3] *Marine Biological Laboratory, The Ecosystems Center, Woods Hole, MA 02543;* [4] *State University of New York, College of Environmental Science and Forestry, Syracuse, NY 13210;* [5] *Retired (US Environmental Protection Agency), 2004 S Magnolia Ave, Sanford, FL 32771 (*author for correspondence, e-mail: cgoodale@whrc.org)*

Key words: forest growth, forest inventories, nitrogen budget, nitrogen sinks, nitrogen uptake, wood production

Abstract. The eastern U.S. receives elevated rates of N deposition compared to preindustrial times, yet relatively little of this N is exported in drainage waters. Net uptake of N into forest biomass and soils could account for a substantial portion of the difference between N deposition and solution exports. We quantified forest N sinks in biomass accumulation and harvest export for 16 large river basins in the eastern U.S. with two separate approaches: (1) using growth data from the USDA Forest Service's Forest Inventory and Analysis (FIA) program, and (2) using a model of forest nitrogen cycling (PnET-CN) linked to FIA information on forest age-class structure. The model was also used to quantify N sinks in soil and dead wood, and nitrate losses below the rooting zone. Both methods agreed that net growth rates were highest in the relatively young forests on the Schuylkill watershed, and lowest in the cool forests of northern Maine. Across the 16 watersheds, wood export removed an average of 2.7 kg N ha^{-1} yr^{-1} (range: 1–5 kg N ha^{-1} yr^{-1}), and standing stocks increased by 4.0 kg N ha^{-1} yr^{-1} (–3 to 8 kg N ha^{-1} yr^{-1}). Together, these sinks for N in woody biomass amounted to a mean of 6.7 kg N ha^{-1} yr^{-1} (2–9 kg N ha^{-1} yr^{-1}), or 73% (15–115%) of atmospheric N deposition. Modeled rates of net N sinks in dead wood and soil were small; soils were only a significant net sink for N during simulations of reforestation of degraded agricultural sites. Predicted losses of nitrate depended on the combined effects of N deposition, and both short- and long-term effects of disturbance. Linking the model with forest inventory information on age-class structure provided a useful step toward incorporating realistic patterns of forest disturbance status across the landscape.

240

Introduction

Human activity has dramatically increased the emission of fixed nitrogen compounds to the atmosphere and the deposition of fixed N across large regions in the temperate zone, particularly in western Europe, eastern Asia, and the eastern U.S. (Galloway et al. 1995; Holland et al. 1999; Vitousek 1994). Watershed-scale mass balance studies repeatedly show that riverine N export can account for only about 20–30% of the total N input to terrestrial systems; the fate of the remaining 70–80% is not well quantified, but likely consists of a combination of storage in biomass, soils, and groundwater, and loss to the atmosphere through denitrification (Galloway et al. 1995; Howarth et al. 1996). One potentially large N sink in terrestrial systems is incorporation of N into forest biomass and soils. Aggrading forests in the eastern U.S. currently act as a considerable sink for atmospheric carbon (Birdsey & Heath 1995; Turner et al. 1995). In this paper, we used forest inventory data and an ecosystem process model (PnET-CN; Aber et al. 1997) to quantify the N sinks in forest biomass, harvested material, dead wood, and soil in the eastern U.S. We focused on forested lands in 16 large watersheds ranging from Maine to Virginia for which other components of N budgets have been quantified (see Boyer et al. and van Breemen et al. 2002). Together, these watersheds occupy 248,000 km^2, and forests cover 72% of this area (Figure 1).

Nutrient uptake into forest biomass is not measured directly in the field, but rather is inferred from measurements of the production of wood and litter, and the nutrient concentrations of those components (e.g. Whittaker et al. 1979; Woodwell et al. 1975). Most of the N that trees take up annually is allocated to foliage and fine roots, but these components provide only a transient sink for N, as fine litter annually returns much of this N to the forest floor. In contrast, N accumulation in wood of aggrading forests represents a long-term N sink. Woody tissue has a low N concentration, but it makes up about 95% of total forest biomass (Whittaker et al. 1974, 1979). Here, we focused on N accumulation in woody tissues and export in harvested material.

Forest standing stocks and growth are measured regularly across the U.S. as a part of the USDA Forest Service's Forest Inventory and Analysis (FIA) program (Hansen et al. 1992; Woudenberg & Farrenkopf 1995). We used these data as the basis for estimating sinks of N into wood and harvested material, but the other N fluxes needed to calculate net forest N retention – such as nitrate losses or changes in woody debris or soil N stocks – are not presently measured throughout the inventory program. Ecosystem process models were designed to be able to estimate the importance and temporal variation of several of these unmeasured fluxes. However, biogeochemical models are just beginning to consider the effects of forest harvests and other

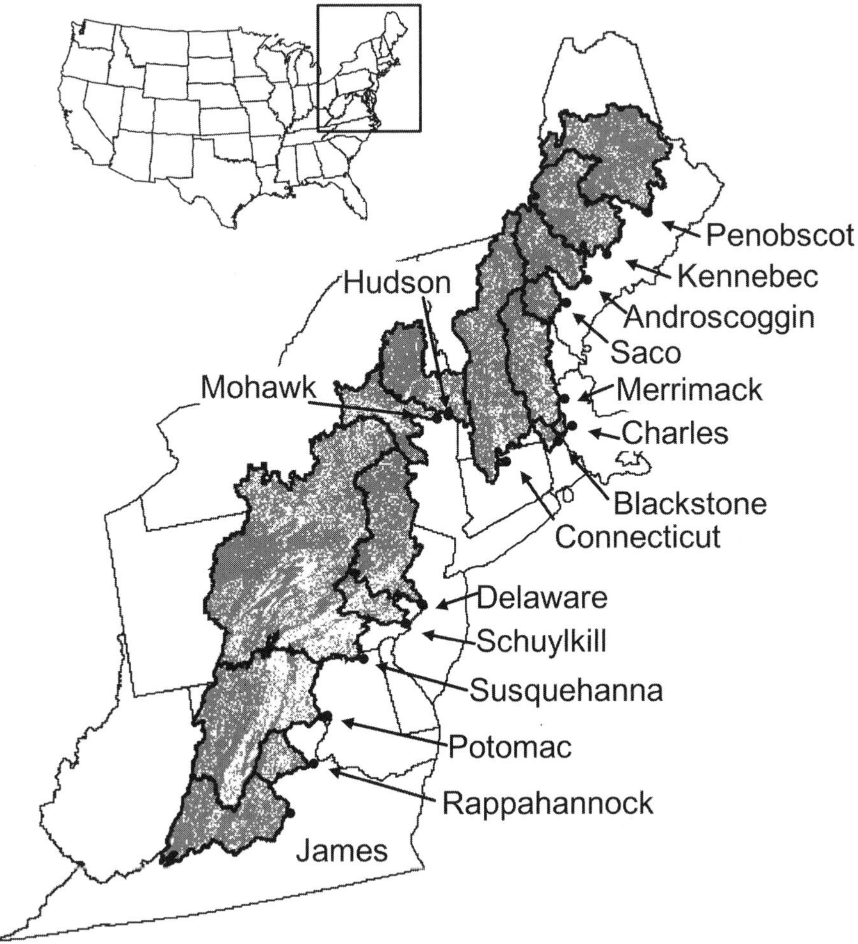

Figure 1. Boundaries of the focus watersheds and MRLC (1995) forest cover.

disturbances in complex landscapes (e.g. McGuire et al. 2001; Schimel et al. 1997, 2000). Disturbances have major impacts on forest N cycles that cannot be ignored when simulating N retention at large scales. Although aggrading forests retain N in biomass and soil, disturbed systems often experience large pulses of nitrate loss, due to reduced plant uptake and continued or enhanced N mineralization (e.g. Likens et al. 1970; Vitousek et al. 1979; Vitousek & Matson 1984). For this modeling effort, we used FIA data on forest age structure to simulate forests in a variety of stages of growth, disturbance, and recovery within the 16 watersheds. This linkage represents a first step toward consideration of both natural and human-induced disturbances in the ecosystem process model at the landscape scale.

The FIA age class information was combined with two contrasting scenarios of land-use history (long-term forestry and agricultural abandonment) to bracket the range of known short-term and potential long-term effects of disturbance on forest N retention. These disturbances, combined with input data on monthly climate and N deposition, allowed model predictions of forest N retention that incorporated the net effects of watershed differences in age-class structure, N deposition, and spatial and temporal variations in climate. On all 16 watersheds, both FIA data and PnET-CN were used to estimate forest growth, mortality, harvest export, and standing stocks. The model was also used to estimate N sinks in dead wood and soil, and losses of nitrate below the rooting zone.

Methods

Forest N sinks were quantified for 16 river basins: the Penobscot, Kennebec, Androscoggin, Saco, Merrimack, and Connecticut Rivers in northern New England; the Charles and Blackstone Rivers in southern New England; the Hudson, Mohawk, Delaware, Schuylkill, and Susquehanna Rivers in the Mid-Atlantic states; and the Potomac, Rappahannock, and James Rivers largely in Virginia. (Figure 1, Table 1). Boyer et al. (2002) describe characteristics of the 16 river basins in detail. This paper focused on only the forested portion of the basins, and all stocks and fluxes were expressed per unit forest area.

Our modeling efforts focused on predictions for the years of 1988–1992 in order to be consistent with river flux and N deposition records from this period (Boyer et al. 2002). The forest inventory data reflect the varied dates of surveys in different states, completed 1983–1995 and covering 6–20 years.

Forest inventory

County-level estimates of forest volume, area, age-class structure, and annual net growth, mortality, and harvests were obtained for Maine through Virginia from the FIA program, available on-line at: http://www.srsfia. usfs.msstate.edu/scripts/ew. htm (Hansen et al. 1992). Growth, mortality and harvest data were not available from this source for Vermont and New Hampshire; limited net growth data for these states were obtained from FIA's Forest Inventory Tablemaker, version 1.0 (Jan. 1, 1997, USDA Forest Service). County-level data were converted to watershed-level estimates by weighting forest biomass or N in each county by the percent of that county within each watershed boundary, as obtained by GIS overlays of county and watershed boundaries using Arc/Info (Redlands, CA).

Table 1. Watershed, forest, and timberland area (km^2), and percent of timberland area by type and by age class. Forest type in bold indicates the types used in PnET-CN simulations

Watershed	Area (km^2)			[2]Forest Type (%)				[2]Forest Age Class (%)						
	Total	[1]Forest land	[2]Timber-land	Pine	Spruce-Fir	Mixed Oak	North. Hardwd.	1–10	11–20	21–40	41–60	61–80	>80	Mixed
Penobscot	20,109	16,854	16,488	5	**41**	1	**53**	9	9	3	9	18	29	23
Kennebec	13,994	11,155	12,884	4	**33**	1	**61**	9	10	5	7	14	16	38
Androscoggin	8,451	7,142	7,092	10	**23**	5	**61**	2	7	4	14	15	12	47
Saco	3,349	2,932	2,764	**22**	10	13	**55**	1	2	3	8	11	6	69
Merrimack	12,005	8,853	8,854	**31**	4	29	**35**	<1	0	3	13	14	2	69
Charles	475	275	218	**23**	0	**56**	22	0	0	2	28	34	8	27
Blackstone	1,115	691	686	**20**	0	**65**	15	0	0	2	28	44	3	24
Connecticut	25,019	20,802	20,193	**17**	14	13	**56**	1	0	3	11	10	3	72
Hudson	11,942	9,662	6,336	**20**	5	15	**60**	1	0	8	19	26	27	19
Mohawk	8,935	5,751	4,491	**21**	2	11	**65**	4	0	20	22	24	23	8
Delaware	17,560	13,098	10,166	9	1	43	**46**	1	2	32	30	18	14	3
Schuylkill	4,903	2,370	1,565	4	1	**70**	23	3	15	53	24	1	3	0
Susquehanna	70,189	46,782	39,801	8	1	45	**45**	3	7	39	32	8	9	1
Potomac	29,940	18,200	15,296	7	0	**80**	12	3	7	17	21	19	20	14
Rappahannock	4,134	2,546	1,943	**15**	0	**82**	2	6	5	23	30	24	11	0
James	16,206	12,876	11,497	**16**	0	**81**	2	10	9	14	21	26	20	<1
Total/Wtd. Ave.	248,326	179,989	160,275	12	11	34	43	4	6	17	20	15	14	24

[1]From GIS overlays of watershed boundaries and the National Land Cover Database (NLCD) of the Multi-Resolution Land Characteristics Interagency Consortium (MRLC).
[2]From weighted averages of county-level FIA timberland data.

The FIA data were reported in terms of growing-stock volume by forest type and size- or age-class. Growing stock consists of the merchantable portion of commercial tree species ≥ 5 inches diameter at breast height (12.7 cm dbh) growing on land classified as timberland. Timberland is land capable of producing at least 20 ft^3 acre^{-1} yr^{-1} (1.4 m^3 ha^{-1} yr^{-1}) of commercial wood, on which harvesting is not prohibited by law or regulation. Non-merchantable tree parts (e.g. branches, coarse roots) and small, poor-quality, and noncommercial trees are excluded in these estimates, as are lands such as the state-owned 'forever wild' forests in New York's Adirondack Park.

The FIA growth data are reported as net growth of growing stock. Net growth represents the total production of growing stock (gross growth) after losses due to natural mortality.

(1)   Net growth = gross growth – natural mortality

When converted to units of forest biomass, gross growth is essentially net primary production of wood (woody NPP), excluding woody litter production other than whole-tree mortality (e.g. branch drop). Subtracting harvest losses from net growth yields the forest's net change in standing stock.

(2)   Net change = net growth – harvest mortality

Net growth represents the forest's rate of accumulation of wood into live trees; net change in standing stock represents the net balance of wood production and consumption across the landscape. After harvest, a portion of the harvested biomass is removed from the forest. The rest (woody roots, tops, etc.) remains as harvest debris that decomposes and eventually releases its N. In the northeastern U.S., 55–67% of harvested biomass (including roots) is removed (Birdsey 1996); we used a default of 60% for all watersheds.

(3)   Harvest export = harvest mortality – logging debris

For estimating watershed-level sinks for N in wood, we focused on two terms representing long-term net N sinks: net change in standing stock, and harvest export. Using FIA data to estimate woody N sinks required three steps: (1) converting FIA estimates of growing-stock volume (m^3/ha) to estimates of total forest biomass (Mg/ha); (2) calculating the N in this biomass; and (3) extending these estimates to forest land not classified as timberland.

Two approaches have been developed to convert U.S. growing-stock volume data to estimates of dry biomass of the whole forest, methods developed largely in order to quantify forest carbon balance. In the first approach (Method 1), Birdsey (1992) and Turner et al. (1995) multiplied growing-stock volume by forest-type specific wood density values and two additional factors: the ratio of all tree species to commercial species; and the ratio of all woody tree parts (above- and belowground) to bolewood (Appendix 1, Table 1). Alternatively, (Method 2) Schroeder et al. (1997), Brown and Schroeder (1999), and Brown et al. (1999) developed a method for predicting biomass expansion factors (BEF) or empirical conversion factors that vary with both forest type and growing-stock volume. Low-volume forests are predicted to have high BEF values, roughly compensating for the FIA practice of including only trees ≤ 12.7 cm dbh in growing stock statistics. Although the conversion factors for Method 1 were developed to quantify both above- and belowground woody biomass, the Method 2 equations predict aboveground biomass, only. Coarse root biomass generally amounts to 15–25% of aboveground biomass (Birdsey 1992; Cairns et al. 1997; Johnson & Lindberg 1992; Whittaker et al. 1974). We added 20% to values derived from Method 2 to account for belowground woody biomass. We applied both conversion approaches to the same set of growing-stock volume data to obtain two sets of FIA-based estimates for each watershed. To obtain a single set of final values, we used the mean of these two biomass conversion methods and present their range as an additional component of uncertainty.

For both sets of biomass estimates, mass of N was calculated by multiplying woody biomass by its mean percent N as derived from the literature (Table 2). Nitrogen concentrations are quite low in all woody tissues, but concentrations in branches and woody roots are two to three times those of the bole (Table 2). We used the mean (± 1 SD) weighted average N concentration of all woody tissues for softwoods (0.19% $\pm$ 0.08%) and for hardwoods (0.26% $\pm$ 0.06%). Concentrations varied considerably, and we assessed the potential variability of our FIA-based estimates of woody N sinks by varying %N by the standard deviation of the estimates from the literature (Table 2).

For lack of additional information, we assumed that forests growing on land designated as timberland were representative of those growing on other forested land. Total N stocks of all forests observed on the watersheds were estimated by multiplying mean values for timberland by the total area of forest land, as determined by GIS overlays of land cover data and watershed boundaries. The land cover data were obtained from the Multi-Resolution Land Characteristics (MRLC) consortium's National Land Cover Data Base (http://www.epa.gov/mrlc) (MRLC 1995). Our approach of weighting county-level FIA data by fractional watershed coverage assumed a fairly even distribution of forest coverage within each county. Comparison of the area of land classified as timberland by the Forest Service with the MRLC estimates of forest area (Table 1) indicated that this approach did

Table 2. Weighted average nitrogen concentrations of wood in branches, boles, and roots in eastern North American forests (% of dry weight)

Location	Forest type	Woody tissue N concentration (%)			
		Branch	Bole	Root	Wtd. Ave.
Hardwoods					
[1]Brookhaven, NY	Mixed oak	0.32	0.18	0.23	0.22
[2]Hubbard Brook, NH	N. hardwood	0.46	0.16	0.70	0.33
[3]Oak Ridge, TN	Yellow poplar	0.29	0.18	0.34	0.24
[3]Walker Branch, TN	Poplar/hardwoods	0.26	0.16	0.39	0.23
[3]Walker Branch, TN	Oak-hickory	0.50	0.19	0.39	0.29
[3]Walker Branch, TN	Chestnut oak	0.46	0.18	0.37	0.27
[3]Coweeta, NC	Oak-hickory	0.45	0.18	0.83	0.40
[4]Berlin, NH	N. hardwood				*0.20
[5]Oak Ridge, TN	Mixed oak	0.24	0.17	0.20	0.19
[6]Cockaponset, CT	Ctrl. hardwood	0.30	0.13	0.35	0.20
[7]Coweeta, NC	S. hardwood	0.23	0.20	0.83	0.33
[7]Smoky Mtns., TN	American beech	0.14	0.14	0.80	0.22
[7]Huntington Forest, NY	N. hardwood	0.10		0.25	0.21
[7]Turkey Lakes, ON	N. hardwood	0.35	0.16	0.15	0.20
[8]Hubbard Brook W5, NH	N. hardwood; 6 yr.		0.33		0.33
[9]Nantahala NF, NC	Mixed oak				*0.22
[10]Harvard Forest, MA	Mixed oak		0.16		
Average		0.32	0.19	0.46	0.26
Standard Deviation		0.13	0.05	0.25	0.06
Softwoods					
[1]Brookhaven, NY	Pitch pine	0.22	0.11	0.17	0.15
[3]Oak Ridge, TN	Shortleaf pine	0.24	0.11	0.34	0.19
[7]Coweeta, NC	White pine	0.39	0.13	0.70	0.31
[7]Duke Forest, NC	Loblolly pine	0.23	0.11	0.11	0.13
[7]Gainesville, FL	Slash pine	0.24	0.13	0.41	0.20
[7]Oak Ridge, TN	Loblolly pine	0.14	0.09		0.10
[7]Howland Forest, ME	Red spruce	0.31	0.15	0.22	0.19
[7]Smoky Mtns., NC	Red spruce	0.13	0.07	0.27	0.13
[7]Smoky Mtns., NC	Red spruce	0.16	0.06	0.47	0.14
[7]Whiteface, NY	Spruce-fir	0.75	0.12	0.53	0.36
[9]Nantahala NF, NC	Mixed pine				*0.18
[10]Harvard Forest, MA	Red pine		0.06		
Average		0.28	0.10	0.36	0.19
Standard Deviation		0.18	0.03	0.19	0.08

*mixed bole and branch.
From: [1]Woodwell et al. 1975; [2]Whittaker et al. 1979; [3]Cole & Rapp 1981; [4]Hornbeck & Kropelin 1982; [5]Johnson et al. 1982; [6]Tritton et al. 1987; [7]Johnson & Lindberg 1992; [8]Mou et al. 1993; [9]Vose & Swank 1993; [10]Nadelhoffer et al. 1999a.
N. hardwood = northern hardwood.

not introduce large biases. However, county-based estimates of timberland area occasionally exceeded estimates from MRLC overlays (the Kennebec and Merrimack), presumably due to slightly less forest cover observed within watershed boundaries than occurred within the county as a whole. The reverse case (MRLC estimates of forest area exceeded county-weighted estimates of timberland) more likely indicated watersheds that contained non-timberland forest, reaching 22–34% of the forest area on the Hudson, Mohawk, Delaware, Schuylkill, and Rappahannock. Some of this land is not classified as timberland because of low productivity, and our approach likely overestimated growth rates on these areas. However, most of the non-timberland forest was within watersheds where the timberland designation was prevented by administrative rule rather than growth potential (e.g. portions of the Adirondack Park, NY or Shenandoah Park, VA).

Forest ecosystem modeling

Although nitrogen sinks in wood production and harvest are likely most reliably quantified with forest inventory data, an alternative approach is needed to estimate N sinks in dead wood and soil and nitrate losses below the rooting zone. We used an ecosystem process model, PnET-CN, to estimate these fluxes, and we used forest inventory information on age-class structure to better represent patterns of disturbance within the model.

Model description
The PnET-CN model simulates the cycles of carbon, water, and N through forest ecosystems. The model uses inputs of monthly climate and N deposition to predict photosynthesis (Pn), evapotranspiration (ET), and N cycling on a monthly time-step for several forest types. The model's photosynthetic routine (PnET-Day) has been tested against C flux data from the eddy covariance tower at the Harvard Forest, MA (Aber et al. 1996); its water balance (PnET-II), against measured stream flow at Hubbard Brook, NH (Aber & Federer 1992) and elsewhere in the northeastern U.S. (Ollinger et al. 1998). The N cycling routine was added in Aber et al. (1997), and tested against measurements of production, N cycling, and nitrate loss at Harvard Forest and Hubbard Brook (Aber et al. 1997; Aber & Driscoll 1997). Currie and Aber (1997) and Currie et al. (1999) developed more detailed soil models to describe fluxes of dissolved organic matter (DOCMOD) and gross N cycling in litter and humus (TRACE). For our simulations, we used the simpler version described in Aber et al. (1997). Both PnET-CN and TRACE use the same formulation for net N mineralization from humus, although TRACE distinguishes humus in the forest floor and the mineral soil and includes a separate routine to describe litter decomposition. PnET-CN mineralizes N

248

along with C from a single soil pool that contains both non-woody litter and soil humus. This pool represents the top, active layers of litter and soil organic matter, and does not include 'passive' carbon. The pool's turnover (0.075 yr^{-1}) is modified by empirical temperature and moisture relationships (Aber et al. 1997; Aber & Driscoll 1997). A fraction of the mineralized N is re-immobilized in the soil pool, as determined by soil C:N ratio; no other direct or indirect mechanisms of soil N retention are considered. After net mineralization, N is available for plant uptake, and excess N can nitrify and leach below the rooting zone. Predicted rates of N cycling and nitrate losses can vary greatly from year to year due to climatic variation and its relative effects on plant growth and decomposition (Aber & Driscoll 1997).

Modeled plant N uptake, N concentrations in foliage and wood, and net nitrification are all controlled by an index of plant N limitation, which is the ratio of mobile, unallocated plant N to a specified maximum. Increasing the foliar N concentration drives an increase in photosynthesis and production as long as light, water, N and temperature limitations permit. Net primary production and available N are allocated first to foliage, then fine roots, and finally to wood. Modeled dead wood pools result from logging residue and from annual production of woody litter, determined as a constant fraction of woody biomass (1.7% yr^{-1}). This term includes both whole-tree mortality and branch turnover. FIA measurements indicate that whole-tree mortality in these forests averages $\sim$0.8–1.0% yr^{-1} of biomass (Brown & Schroeder 1999). Data from Whittaker et al. (1974, 1979) suggest that dropped branches and other woody litter production not associated with whole-tree mortality amounts to an additional 0.7% yr^{-1}. Dead wood turns over at 10% yr^{-1} (Arthur et al. 1993); 80% of the mass lost is respired and the residual 20% enters the soil pool along with 100% of the turned-over N, effectively decreasing the C:N ratio of the decayed woody material transferred to the soil pool (Aber et al. 1991).

Input data

PnET-CN requires monthly climate data, N deposition values, and scenarios describing past disturbances and historical removals of C and N. As described below, the model was run for each watershed using its 1–2 most dominant vegetation types (Table 1), a single climate time-series, a range of age classes, and two historical disturbance scenarios.

We developed spatially averaged climate time-series for each watershed from the climate data set developed for the VEMAP II modeling comparison. For VEMAP II, Kittel et al. (1997) reconstructed historical (1895–1993) monthly precipitation, solar radiation, and monthly averaged maximum and minimum daily temperature on a half-degree grid. We weighted climate values in each grid cell by the fraction of watershed area covered by that grid cell, as determined by GIS overlays of grid cells and watershed boundaries. Simulations of years prior to 1895 used average monthly climate values.

Wet and dry deposition of inorganic N was estimated for each watershed using spatial regression equations developed by Ollinger et al. (1993), which we modified by using dry deposition coefficients from the review in Lovett and Rueth (1999). The Potomac, Rappahannock, and James watersheds were outside of the range in which the regression equations applied, and total inorganic N deposition on these watersheds was estimated by doubling wet N deposition, as measured by the National Atmospheric Deposition Program/National Trends Network (NADP/NTN). See Boyer et al. (2002) for additional details. Preindustrial N deposition was set at 1 kg ha^{-1} yr^{-1} (Galloway et al. 1995, Holland et al. 1999), increased linearly from 1880 to 1980, and was held constant through 1993.

Forest inventory data were used to partition forests on each watershed into six age classes and a seventh mixed-age class (Table 1). Forests in each age class likely resulted either from even-aged forest harvest or from agricultural abandonment. We performed sets of simulations for each of these scenarios, with model predictions weighted by the fraction of forest area in each age class (Figure 2). In the first, continuous-forest (forest-only) scenario, forests were assumed to be on 80-year cycles of harvest since the mid- to late-1700s. Growth and N losses vary greatly within the first decade after disturbance, and so the 1-10 year age class was split into smaller increments. On each watershed, harvests were simulated 0, 2, 5, and 8 years before 1990 for the 1–10 year age class, and 15, 30, 50, 70, and 90 years before 1990 for the 11–20, 21–40, 41–60, 61–80, and >80 year old forests, respectively. Harvest intensity was assumed to increase over time from affecting 40% of stand biomass in the late 1700s to 90% for harvests after 1950, and the fraction of harvested material removed increased from 40% in the late 1700s to 60% after 1970. Mixed-aged forests potentially reflect a wide range of management histories. A subset of FIA data suggested that these forests had approximately the same net growth rate as 41–60 year-old forests, and so, for lack of additional disturbance information, they were assumed to also have similar N cycling rates. FIA harvest data suggested that in addition to stand-replacing harvests, logging in older or mixed-aged forests occurred at a rate of approximately 1.0% yr^{-1}; we used this value in the model for all watersheds, for both the forest-only and old-field scenarios.

In the second, post-agricultural (old-field) scenario, forests were assumed to have regenerated on old fields abandoned 5, 15, 30, 50, 70, and 90 years before 1990, after a century of farming with minimal N inputs. Farming in the eastern U.S. peaked around 1880 (Figure 2), and has declined steadily

250

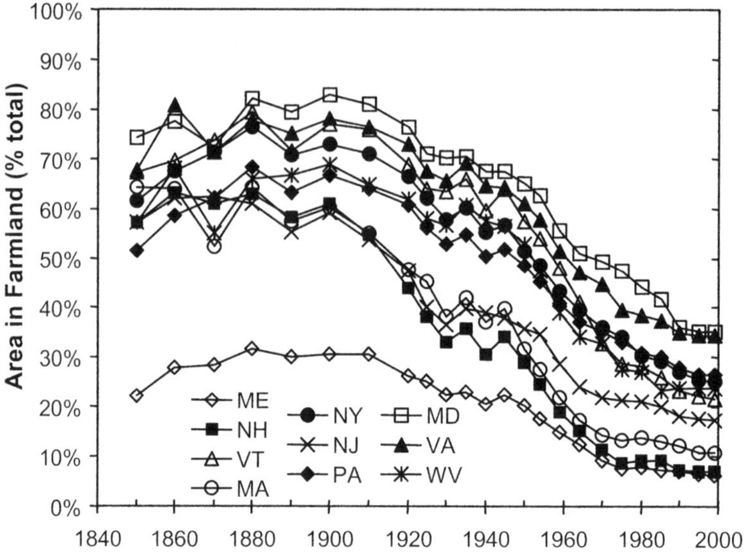

Figure 2. Percent of eastern U.S. land area in farmland, 1850 to 1999, by state. Data for 1850–1969 from U.S. Bureau of Census (1977); 1970–1999 from the National Agricultural Statistics Service (http://www.nass.usda.gov:81/ipedb/).

since then (Ramankutty & Foley 2001; U.S. Bureau of the Census 1977). We roughly estimated the relative importance of agricultural abandonment as a land-use history on each watershed using state-level agricultural statistics for 1850–1969 (U.S. Bureau of the Census 1977) and 1970–1995 (http://www.nass.usda.gov:81/ipedb/) (Figure 2). These data suggested that 25% of Maine consisted of former farm land, with higher fractions for the Mid-Atlantic states (37–49%), Vermont and New Hampshire (both 56%) and southern New England (52–72%). This approach ignores all spatial variation of land-use history within states, and assumes no bias of whether former farmland reverted to forest or to other land uses. The true disturbance histories of these watersheds likely consisted of forest harvest, agriculture, and other land-uses of varied intensity; this set of simulations offers a preliminary consideration of disturbance on these watersheds.

Results

We first present results of forest biomass, N uptake and harvest export as derived from forest inventory data, and then compare these with PnET-CN estimates of these terms. We then present PnET-CN estimates of N sinks in pools that cannot be estimated from forest inventory data.

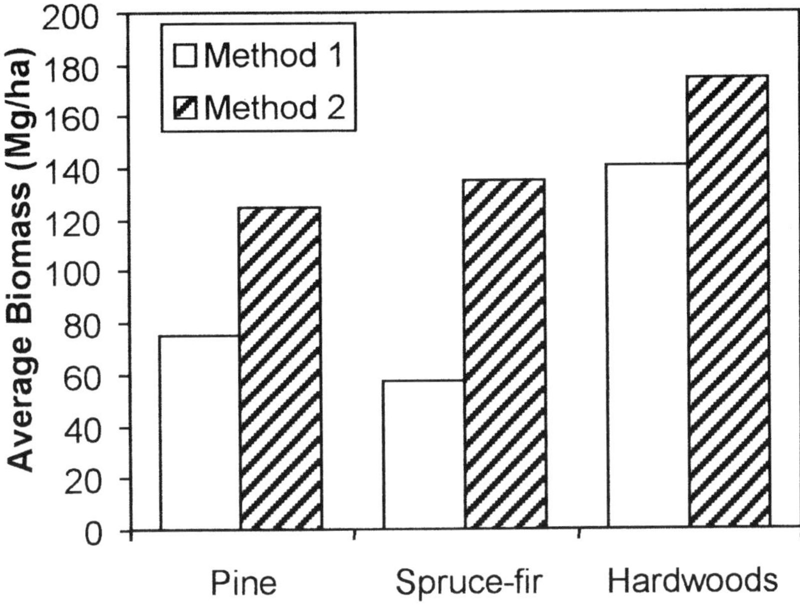

Figure 3. Comparison of two methods of converting growing stock volume data to estimates of total (above- and belowground) biomass for pine, spruce-fir, and hardwood forests. Method 1 used conversion factors from Birdsey (1992) and Turner et al. (1995); Method 2 used an approach described in Brown et al. (1999).

Forest inventory data

The two methods used to convert growing stock volume to biomass produced different estimates of forest biomass, growth and harvest. Above- and below-ground forest biomass averaged 126 Mg ha^{-1} using Method 1 (Birdsey 1992; Turner et al. 1995), and 164 Mg ha^{-1} using the Method 2 (Brown & Schroeder 1999 + 20% for belowground biomass). Corresponding N stocks in woody biomass were calculated using mean $\pm$ 1 SD of %N values in Table 2, and averaged 310 $\pm$ 84 and 400 $\pm$ 110 kg ha^{-1}, respectively. Overall, Method 2 produced biomass estimates 30% greater than Method 1, with larger differences in the Penobscot (57%) and Kennebec (46%) watersheds. These two watersheds had large areas of spruce-fir forests (Table 1), and conversion differences were far greater for estimates of spruce-fir biomass than for pine or hardwoods (Figure 3). Both conversion approaches agreed that forests on the Rappahannock watershed had the greatest biomass, and those on the Penobscot and Kennebec had the least.

Uptake of N in net growth varied across the region (Figure 4). Our county-level calculations of aboveground biomass production using Method 2 closely matched those reported by Brown and Schroeder (2000). Slower growth occurred in the cool forests of northern Maine and along the Appalachian Mountain range in Pennsylvania and Virginia. The fastest net growth occurred on the Schuylkill, Rappahannock, and James, and the slowest, on the Penobscot and Kennebec watersheds in northern Maine. Across all watersheds, net growth averaged 2.9 Mg ha^{-1} yr^{-1} using Method 1, and 3.8 Mg ha^{-1} yr^{-1} using Method 2. These net growth rates corresponded to rates of N uptake into live wood of 7.1 $\pm$ 1.9 kg ha^{-1} yr^{-1} and 9.2 $\pm$ 2.5 kg ha^{-1} yr^{-1}. Nitrogen uptake into net wood growth approximately balanced N deposition (Figure 5). Net wood growth correlated with N deposition (Figure 5; $R = 0.70$ for Method 1; $R = 0.65$ for Method 2), but this does not necessarily indicate that N deposition stimulated forest production, as N deposition co-varied with mean annual temperature ($R = 0.73$) and correlations of wood growth and temperature ($R = 0.65$ for Method 1 and 0.62 for Method 2) are nearly as strong as correlations between wood growth an N deposition.

Not all N taken up in net growth remains in long-term storage, as harvests allow some of this N to be recycled through production and decomposition of logging debris. The main forms of N sinks that we considered here were N in harvest exports and N in accumulated wood. Harvest rates could not be estimated for the Saco, Merrimack, or Connecticut River basins because FIA harvest data were not available for New Hampshire or Vermont. On the remaining 13 watersheds, harvest mortality was highest in Maine, on the Kennebec and Penobscot, followed by the Rappahannock and James in Virginia. Harvest rates were lowest on the Mohawk and Delaware. Harvest mortality on the Penobscot and Kennebec between 1979 and 1995 exceeded net growth during this period, causing declines in forest biomass (negative rates of forest increment) (Figure 6). On the other watersheds, net growth exceeded harvest mortality and allowed forest biomass to accumulate, particularly on the Mohawk and the Delaware. Across the 13 watersheds with harvest data, harvest mortality rates averaged 1.3 and 1.8 Mg ha^{-1} yr^{-1}, containing 3.8 $\pm$ 1.0 and 5.1 $\pm$ 1.4 kg N ha^{-1} yr^{-1}, using the Method 1 and Method 2 conversion approaches, respectively. However, not all harvested material was removed from the watershed; woody roots, stumps, and other slash were presumably left behind to decompose. Assuming that 60% of harvested biomass was removed from the forest, we estimated that export of N in harvested wood averaged 2.3 $\pm$ 0.6 (Method 1) and 3.0 $\pm$ 0.8 kg N ha^{-1} yr^{-1} (Method 2). Harvest export was the dominant N sink for the northern Maine watersheds, while N accumulation in living tree biomass was the dominant N sink on the other watersheds (Figure 6).

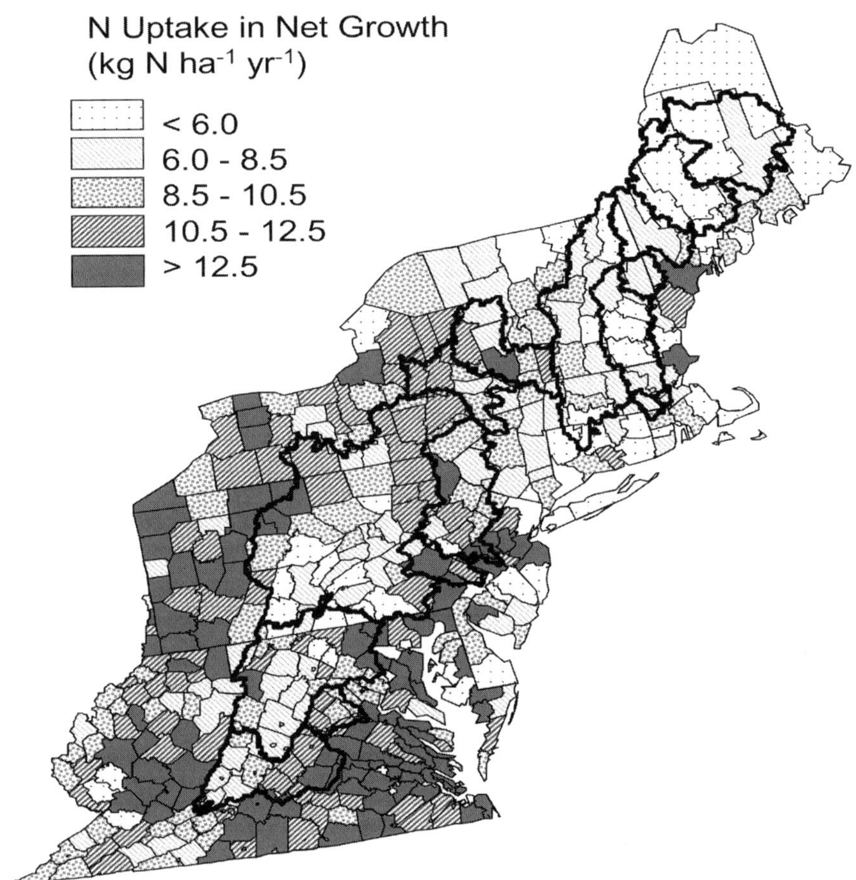

Figure 4. Uptake of N in net growth of wood (kg N ha^{-1} yr^{-1}) across the eastern U.S., estimated from FIA data using the mean of two methods of biomass conversion and %N values derived from a literature review (Table 2).

Model-inventory comparison

Comparing inventory and model-based estimates of forest growth, harvest, and standing stocks allows for evaluation of model performance. This testing is important for determining whether the use of forest age classes and assumptions about past land-use history fit with observed patterns of biomass and growth, and it helps identify potential strengths and weaknesses in model predictions of unmeasured fluxes.

Standing biomass is particularly challenging to predict, as it depends on accurately representing both growth and mortality (natural and harvested) correctly over time. Modeled estimates of stand biomass compared rather

254

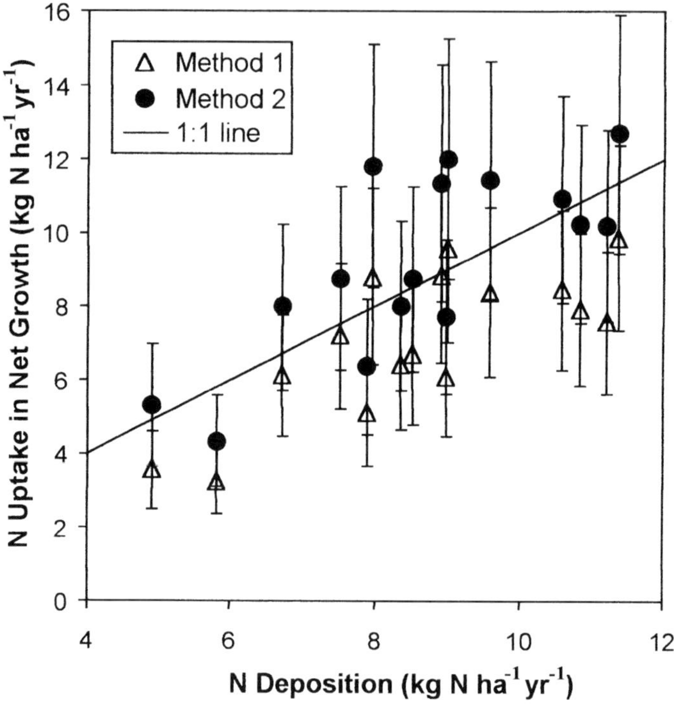

Figure 5. Comparison between N deposition and estimated N uptake in net wood growth for 16 watersheds, using two methods of biomass conversion. Error bars represent ± 1 SD of %N values from a literature review.

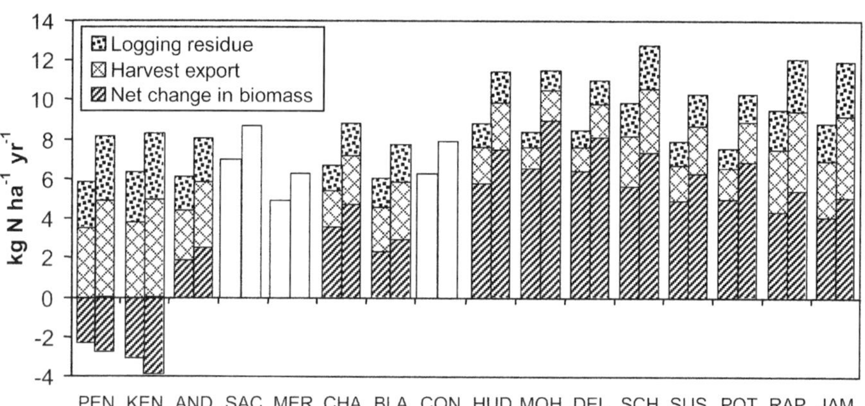

Figure 6. Nitrogen sinks in aggrading biomass and harvest export (kg N ha^{-1} yr^{-1}). Nitrogen in harvest residue is retained in the forest and eventually recycled. Together, these three terms amount to N uptake in net growth. Harvest data were not available for the Saco, Merrimack, and Connecticut, and only net growth is shown. Paired bars represent different methods of biomass conversion, with Method 1 on the left and Method 2 on the right.

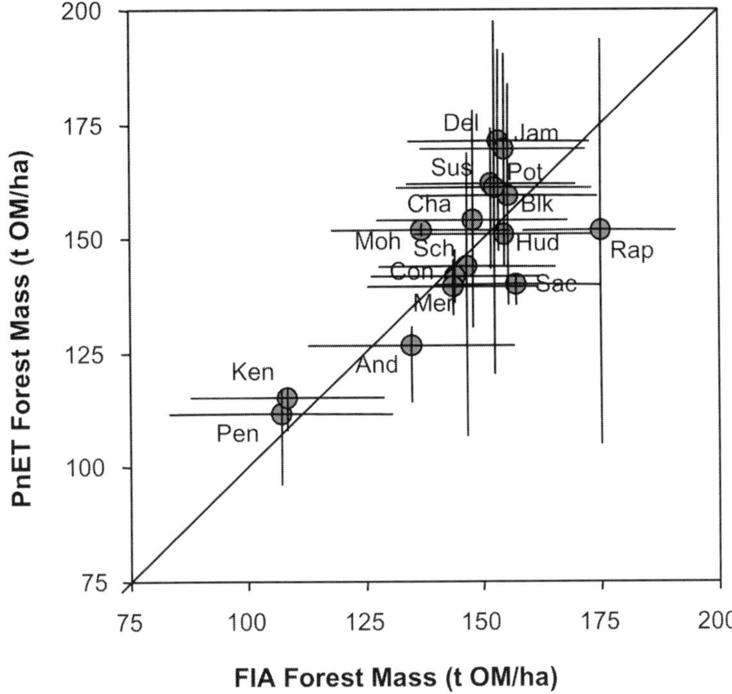

Figure 7. Comparison of modeled and FIA-based estimates of stand biomass by watershed. FIA values indicate the mean and range of two methods of biomass conversion, with Method 1 on the low end of the range and Method 2 on the high. Modeled values represent the weighted mean and range of contrasting scenarios of past land-use history, with forestry scenarios on the high end of the range and old-field scenarios on the low end.

well with estimates based on FIA data, although there were large uncertainties for both approaches (Figure 7). FIA-based estimates are presented as the mean and range of the two biomass conversion methods, while PnET-based estimates are presented as the weighted mean and range from simulations using the two land-use histories. Averaged across all watersheds, PnET-CN estimates of standing biomass were lower under old-field simulations (136 Mg/ha) than simulations assuming only harvest disturbances (158 Mg/ha). Differences due to land-use history were pronounced in the mixed-oak dominated stands of the southern watersheds. Comparisons of stand N stocks were similar to comparisons of stand biomass, although there is additional uncertainty in the FIA-based estimates due to potential variability of wood %N (Table 2).

Climate, age-class structure, and land-use history all affected PnET-CN's predictions of net growth and N uptake. Predicted net growth rates were highest during the first two decades after harvest, when woody NPP was

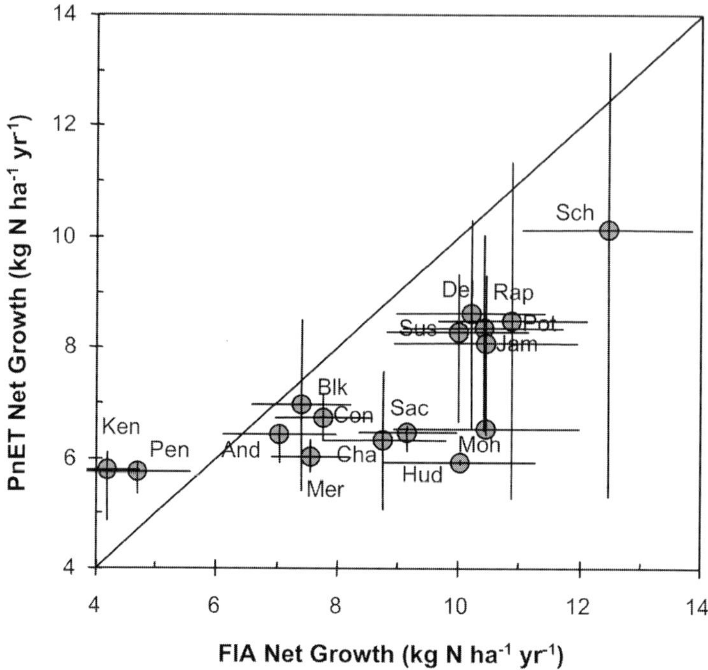

Figure 8. Comparison of modeled and FIA-based estimates of uptake of N in net growth of wood. Error bars represent ranges as described for Figure 7. Additional variation in FIA-based values due to variability of wood %N is not included in this figure.

greatest and mortality was low. The model does not have any prescribed effects of age on growth rate; these changes resulted from predicted changes in C and N pool sizes and fluxes. Net growth rates declined in later years due to increases in natural mortality and woody litter production. High net growth rates were predicted for watersheds with mild climates and large fractions of forest area made up of young forests. Consistent with the FIA data, the model predicted the lowest net growth rates for forests on the cool northern Maine watersheds and the highest uptake rate for forests on the Schuylkill (Figure 8), the watershed with the greatest fraction of forest area in the 11–20 and 21–40 year age classes (Table 1). The greatest discrepancy between modeled and FIA-based net growth occurred on the Mohawk and Hudson watersheds, where the model underestimated net growth by about 4 kg N ha⁻¹ yr⁻¹. Forest-harvest simulations for the other watersheds matched the mean FIA-based estimates of N accumulation in net growth rather well (Figure 8). Across the watersheds, N uptake in net growth averaged 8.1 kg N ha⁻¹ yr⁻¹ using the forest-harvest scenario. Slower net growth rates were predicted for forests on former agricultural sites, averaging 6.2 kg N ha⁻¹ yr⁻¹ in wood production. As for biomass, differences due to land-use history were particularly large on the mixed oak-dominated southern watersheds. Weighting these land-use history scenarios by their approximate historical importance as suggested by state-level farmland data (Figure 2), we underestimated mean FIA-based estimates of net growth by 1.6 kg N ha^{-1} yr^{-1} (Figure 8).

PnET-CN simulated harvests through both stand-replacing clearcuts and chronic, low-intensity logging of older stands. Stand-replacing clearcuts were assigned based on the land area in 1–10 year age class (Table 1) and selective cutting of older stands was set at 1.0% yr^{-1} of stand biomass, as indicated by a subset of FIA data. Hence, modeled and FIA estimates of harvest mortality were not wholly independent. Simulated N fluxes in harvest-induced mortality were naturally much higher under the scenario that the forest age structure was determined by stand-replacing harvests (6.0 kg N ha^{-1} yr^{-1}) rather than through patterns of agricultural abandonment (3.1 kg N ha^{-1} yr^{-1}). Weighting PnET-CN simulations by roughly estimated land-use histories, predicted harvest mortality was within 2 kg N ha^{-1} yr^{-1} of the FIA-derived values for all watersheds but the Kennebec and Androscoggin, where the model underestimated harvest mortality by 2.8 and 3.0 kg N ha^{-1} yr^{-1}, respectively. Harvest mortality was underestimated on these two watersheds even under the forest-only scenario alone. Harvest mortality was partitioned into harvest export and logging debris using the same method for both model- and FIA-based estimates, and so model-FIA comparisons for harvest export and residue had patterns similar to the comparison for harvest mortality. Mean estimates of harvest export were about 0.8 kg N ha^{-1} yr^{-1} lower using land-use history-weighted PnET simulations (2.3 kg N ha^{-1} yr^{-1}) than using mean FIA-based estimates (3.1 kg N ha^{-1} yr^{-1}), with larger underestimates (about 2.0 kg N ha^{-1} yr^{-1}) for the Maine watersheds.

Model-based estimates of net changes in standing stock depended on accurately predicting both net growth and harvest rates, and, like the results for net growth, model-based estimates of the net change in standing stock were on average 1.5 kg N ha^{-1} yr^{-1} lower than FIA-based estimates. The model-based approach overestimated net increment on the Maine watersheds where harvest rates were underestimated, and the model underestimated net increment on the Mohawk due to an underestimate of net growth and a slight overestimate in harvest rates.

Simulated N sinks

Both FIA data and PnET-CN can provide estimates of forest N sinks in accumulating biomass or exported wood, but the ecosystem process model is needed to estimate net N sinks in dead wood, green plant tissues (foliage, fine roots, internal N stores), and soil, and nitrate losses below the rooting zone. The model's ability to generally reproduce patterns of wood growth supports its ability to estimate N sinks in the unmeasured pools.

Dead wood

Dead wood is produced by both natural mortality and harvest debris. Dead wood pools might be expected to be increasing due to increasing woody litter production by aggrading eastern forests. Harvest slash contributed variable amounts of dead wood on the different watersheds, depending on the amount and biomass of recently cut forest. In forest harvest scenarios, dead wood pools were largest immediately following harvests, and they decreased in size as decomposition of harvest slash exceeded woody litter production by the recovering forests. PnET-CN suggested that on average, dead wood was an extremely small net sink for N in the harvest scenario (0.1 kg N ha^{-1} yr^{-1}), and only a slightly larger sink in the old-field scenario (0.6 kg N ha^{-1} yr^{-1}). In the forest harvest scenario, inputs of N in woody litter and harvest debris were largely balanced by decomposition and release from older dead wood. On the Penobscot, Schuylkill, and Susquehanna, decomposition of dead wood produced by large harvests 10–30 years ago exceeded new production of dead wood, so that the dead wood pool was a small source of N (Figure 9a). In the old-field scenario, dead wood pools were small at agricultural abandonment, and formed a consistent but small sink across all watersheds as forests aggraded.

Soil

PnET-CN predicted variations in N pools in litter and soil organic matter due to differences in land-use history, forest age-class structure, and spatial and interannual variability in climate. Losses were predicted to occur on watersheds with relatively large areas of recent harvest, due to reduced inputs of litter and continued decomposition and net N mineralization. Older, never-farmed forests were predicted to have small changes in soil organic matter and N that varied from year-to-year due to variability in temperature and moisture conditions. In the continuous-forest scenario, changes in soil N stocks varied by watershed due to local climate conditions experienced during 1988–1992 – or even during one year with particularly large effects. For example, the three Virginia watersheds had climate conditions in 1989 that were unusually favorable toward net N mineralization, and the predicted soil N losses from this year dominated the 1988–1992 period (Figure 9a). On the Mohawk and Hudson, 1989 and 1991 had climate conditions particularly conducive to litter production and N immobilization relative to decomposition, driving net increases in soil N for the 1988–1992 period (Figure 9a). Averaged over all watersheds, PnET-CN estimated that litter and soils were a small net source

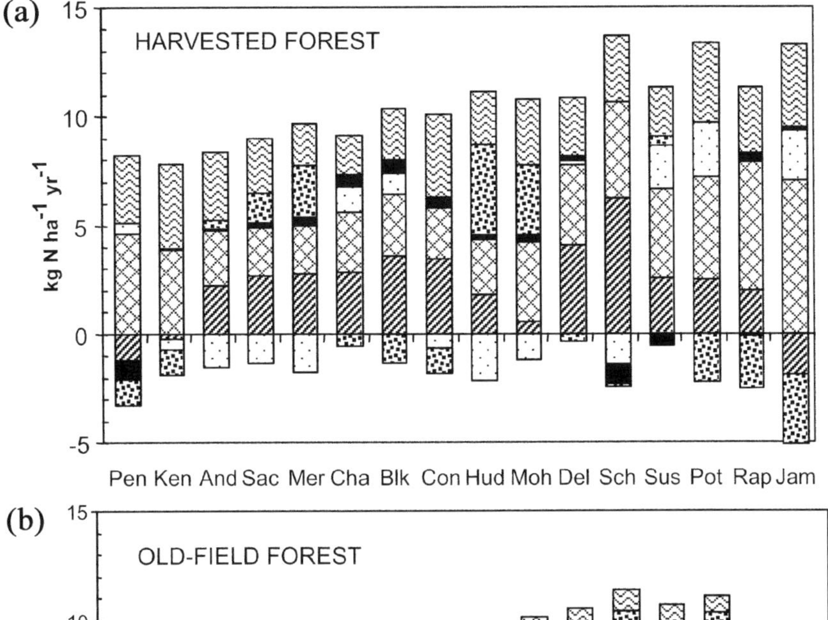

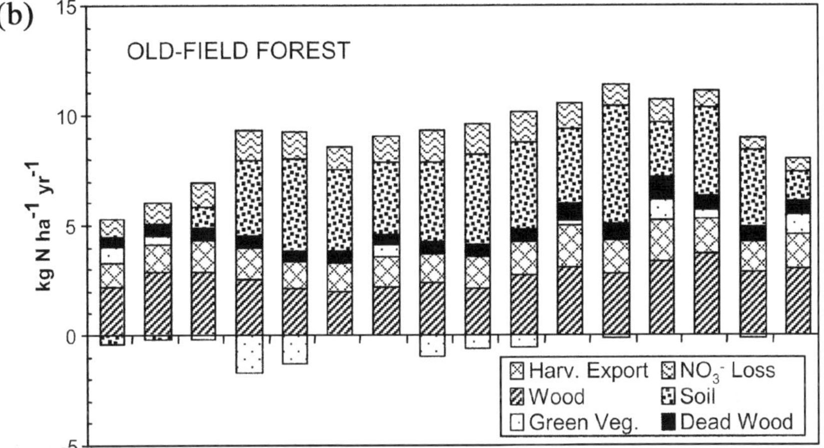

Figure 9. Modeled N balance for all watersheds using FIA forest age class information and scenarios assuming that forests of different ages derived wholly form forest harvest (top) or abandoned fields (bottom). Harvest export and NO$_3^-$ loss represent net losses from the system; all other terms are the net sink (or source) of N into wood, green tissues (foliage, roots or internal stores), soil and litter, and dead wood.

of N (0.1 kg N ha^{-1} yr^{-1}) under the forest harvest scenario, and a substantial sink for N (2.5 kg N ha^{-1} yr^{-1}) under the old-field scenario (Figure 9). In these simulations, harvests induced relatively small losses of C and N from soil pools, thereby allowing rapid recovery. In the agricultural scenarios, 100 years of farming with minimal inputs severely depleted the soil C and N pools, which became net sinks for N as forests regrew.

260

Green vegetation
Foliage, fine roots, and internal N stores do not form significant N sinks over the long term, because they turn over so quickly. However, N in these pools can fluctuate from year to year in response to interannual variability in climate conditions, and transient changes in these pools predicted for 1988–1992 can cause short-term sources or sinks of N. Short-term changes in these pools generally counteracted changes in the soil pool, such that N stores in green vegetation pools increased in years that climate conditions favored net N mineralization over N accumulation in forest litter and soils (Figure 9).

Nitrate loss
Predicted nitrate losses did not correlate with N deposition across the watersheds, but instead varied with short- and long-term disturbance patterns. In PnET-CN, net mineralization from the soil is available for uptake by plants; the excess can nitrify and leach below the rooting zone with drainage water. Large nitrate losses were predicted to occur during the first decade after harvest, when soil N pools are large and forest uptake is low. This pattern was damped in the old-field scenario, where more N was immobilized in the N-poor soil than was leached below the rooting zone. In the forest harvest scenario, watersheds with relatively large amounts of forest area in the 1–10 year age class (the Penobscot, Kennebec, and James) were predicted to have relatively high rates of nitrate loss, despite low N deposition rates (Figure 9a). These differences in harvesting regimes obscured any direct correlations between N deposition and predicted nitrate leaching ($R < 0.02$). Land-use history had a large effect on predicted nitrate losses. Forests growing on degraded agricultural sites, where soils were large N sinks, had much lower mean predicted rates of nitrate loss (1.1 kg N ha^{-1} yr^{-1}) than forests growing on never-farmed land (2.9 kg N ha^{-1} yr^{-1}) (Figure 9).

Discussion

Forests of the eastern U.S. are aggrading, forming sinks for atmospheric carbon (Birdsey & Heath 1995; Turner et al. 1995), and as suggested here, for nitrogen. Much of the observed C sink is due to regrowth of previously cleared lands, with harvest rates lagging rates of regrowth (Birdsey & Heath 1995). On these watersheds and elsewhere in the eastern U.S., the potential for continued and future accumulation of C and N in living biomass is largely constrained by the extent to which historically disturbed forests have recovered. Current forests contain about half the biomass observed in some old-aged northern hardwood forests, suggesting that, if left unharvested, these forests could continue to sequester large amounts of C (and N) (Brown et al. 1997). Harvesting is likely to continue, however, and forest harvest can remove significant quantities of N through export of harvested wood (e.g. Hornbeck & Kropelin 1982; Johnson et al. 1982; Tritton et al. 1987). For the basins studied here, export of N in harvested wood amounted to 18–88% of the N received in atmospheric deposition. Increasing forest harvest rates is not an advised means of managing excess N deposition, however, as harvests in regions of elevated N deposition cause local episodic pulses of nitrate in streams (e.g. Hornbeck & Kropelin 1982; Likens et al. 1970; Martin & Pierce 1980), and export of other nutrients in wood, particularly calcium, may lead to future nutrient limitations (Federer et al. 1989). Observed harvest rates on the Penobscot and Kennebec already exceeded growth rates, and hence were not sustainable.

Local patterns of disturbance and recovery across the landscape cause substantial variation in forest C and N accumulation. This variation can lead to highly variable nitrate export within regions of relatively homogeneous N deposition (e.g. Lovett et al. 2000; Goodale et al. 2000; Williard et al. 1997), and can obscure direct correlations between nitrate export and deposition across large gradients (e.g. Dise et al. 1998). Linking the model with forest inventory information on age-class structure provided a useful step toward incorporating realistic patterns of forest disturbance status across the landscape.

Both predicted (PnET-CN) and observed (FIA-based) rates of N uptake into wood are large enough to nearly balance the N input from deposition, allowing for little net accumulation of N in soil. Similarly, Johnson (1992) observed that across a range of intensively studied forested sites, N taken up in aggrading vegetation was of the same magnitude as N received in atmospheric deposition. However, these observations do not necessarily indicate that vegetation takes up N deposition directly, nor that increased deposition will directly increase vegetation growth and N uptake. Studies using ^{15}N tracers indicate that when ^{15}N is added to forested sites, little of it enters the vegetation; most is rapidly incorporated into forest floor and mineral soil organic mater, for at least the first 1–3 years after application (Nadelhoffer et al. 1999a, b). These field studies of ^{15}N tracers and related simulations (Currie et al. 1999; Currie & Nadelhoffer 1999) suggest that the N taken up and accumulated in forest vegetation does not derive directly from deposition, but from N redistributed from litter and soils. Although the ^{15}N studies indicate that soils are the immediate direct sink for added N, it is less clear whether this N is later mineralized from soil organic matter, and if so, how quickly. Soils of the eastern U.S. have been exposed to elevated N deposition for many decades, and these long-term additions may have altered current soil N availability. The mass balance of net sequestration of N in plant biomass requires a significant source of N. If new additions of N from atmospheric deposition are largely retained in the soil, then there must be a source of older or otherwise available soil N for plant uptake. Understanding controls on soil N retention remains a vitally important question for determining the rate at which N becomes available for plant uptake. Aggrading forests can accumulate N in woody biomass, but the extent to which this process can truly offset N from deposition requires a better understanding of processes controlling soil N retention and turnover.

Acknowledgments

We are grateful to the Forest Service's FIA Program for making these data available on the Web, and to T. Kittel and N. Rosenblum for sharing the VEMAP II climate data sets. Thanks to Bill Currie, John Aber, and Rita Freuder for modeling advice and to Jennifer Jenkins, Sandra Brown, and Tom Frieswyk for assistance in interpreting the FIA data. This work was initiated as part of the International SCOPE N Project, which received support from the Mellon Foundation and the National Center for Ecological Analysis and Synthesis. CLG was supported by the Alexander Hollaender Postdoctoral Fellowship Program, which is sponsored by the Office of Biological and Environmental Research of the U.S. Department of Energy, and administered by the Oak Ridge Institute for Science and Education.

Appendix 1

Table 1: Multipliers used in converting growing-stock volume (m^3/ha) to total forest biomass (Mg/ha) for the northeastern U.S.; from Turner et al. (1995).

Forest type	Wood Density (Mg/m^3)	Noncommercial species	Biomass ratio (total wood: bole)
Pine	0.378	1.01	1.61
Spruce-fir	0.369	1.01	1.69
Oak-hickory	0.636	1.14	1.75
Northern hardwood	0.600	1.14	2.08
Bottomland hardwood	0.580	1.14	1.64

*Specific gravity of wood, ratio of all trees to commercial species, and the ratio of total wood (bolewood plus roots, stumps, branches, and cull trees) to merchantable bolewood.

References

Aber JD & Federer CA (1992) A generalized, lumped-parameter model of photosynthesis, evapotranspiration and net primary production in temperate and boreal forest ecosystems. Oecologia 92: 463–474

Aber JD & Driscoll CT (1997) Effects of land use, climate variation and N deposition on N cycling and C storage in northern hardwood forests. Global Biogeochem. Cycles 11(4): 639–648

Aber JD, Melillo JM, Nadelhoffer KJ, Pastor J & Boone RD (1991) Factors controlling nitrogen cycling and nitrogen saturation in northern temperate forest ecosystems. Ecol. Appl. 1: 303–315

Aber JD, Reich PB & Goulden ML (1996) Extrapolating leaf CO_2 exchange to the canopy: a generalized model of forest photosynthesis compared with measurements by eddy correlation. Oecologia 106: 257–265

Aber JD, Ollinger SV & Driscoll CT (1997) Modeling nitrogen saturation in forest ecosystems in response to land use and atmospheric deposition. Ecol. Model. 101: 61–78

Arthur MA, Tritton LM & Fahey TJ (1993) Dead bole mass and nutrients remaining 23 years after clear-felling of a northern hardwood forest. Can. J. For. Res. 23: 1298–1305

Birdsey RA (1992) Carbon storage and accumulation in United States forest ecosystems. Gen. Tech. Rep. WO-59. USDA Forest Service, Washington, DC

Birdsey RA (1996) Carbon storage for major forest types and regions in the conterminous United States. In: Sampson RN & Hair D (Eds) Forests and Global Change. Vol. 2. (pp 1–25). American Forests, Washington, DC

Birdsey RA & Heath LS (1995) Carbon changes in U.S. forests. In: Joyce LA (Ed) Productivity of America's forests and climate change. Gen. Tech. Rep. RM-271. USDA Forest Service, Rocky Mountain Forest and Range Experiment Station, Fort Collins, Colorado, U.S.A. pp 56–70

Boyer EA, Goodale CL, Jaworski NA & Howarth RW (2002) Anthropogenic nitrogen sources and relationships to riverine nitrogen export in the northeastern U.S.A. Biogeochemistry 57/58: 137–169

Brown SL & Schroeder PE (1999) Spatial patterns of aboveground production and mortality of woody biomass for eastern U.S. forests. Ecol. Appl. 9(3): 968–980

Brown SL & Schroeder PE (2000) Spatial patterns of aboveground production and mortality of woody biomass for eastern U.S. forests: Erratum. Ecol. Appl. 10: 937

Brown SL, Schroeder PE & Birdsey R (1997) Aboveground biomass distribution of US eastern hardwood forests and the use of large trees as an indicator of forest development. For. Ecol. and Mange. 96: 37–47

Brown SL, Schroeder PE & Kern JS (1999) Spatial distribution of biomass in forests of the eastern U.S.A. For. Ecol. Manage. 123: 82–90

Cairns MA, Brown S, Helmer EH & Baumgardner GA (1997) Root biomass allocation in the world's upland forests. Oecologia 111: 1–11

Cole DW & Rapp M (1981) Elemental cycling in forest ecosystems. In: Reichle DW (Ed) Dynamic Properties of Forest Ecosystems. IBP 23. Cambridge University Press, Cambridge

Currie WS & Aber JD (1997) Modeling leaching as a decomposition process in humid montane forests. Ecol. 78(6): 1844–1860

Currie WS & Nadelhoffer KJ (1999) Dynamic redistribution of isotopically labeled cohorts of nitrogen inputs in two temperate forests. Ecosystems 2: 4–18

Currie WS, Nadelhoffer KJ & Aber JD (1999) Soil detrital processes controlling the movement of ^{15}N tracers to forest vegetation. Ecol. Appl. 9(1): 87–102

Dise NB, Matzner E & Gundersen P (1998) Synthesis of nitrogen pools and fluxes from European forest ecosystems. Water Air Soil Pollut. 105: 143–154

Federer CA, Hornbeck JW, Tritton LM, Martin CW, Pierce RS & Smith CT (1989) Long-term depletion of calcium and other nutrients in eastern U.S. forests. Environ. Mange. 13: 593–601

Galloway JN, Schlesinger WH, Levy H II, Michaels A & Schnoor JL (1995) Nitrogen fixation: anthropogenic enhancement-environmental response. Global Biogeochem. Cycles 9: 235–252

Goodale CL, Aber JD & McDowell WH (2000) The long-term effects of disturbance on organic and inorganic nitrogen export in the White Mountains, New Hampshire. Ecosystems 3: 433–450

Hansen MH, Frieswyk T, Glover FF & Kelly JF (1992) The Eastwide forest inventory data base: users manual. Gen. Tech. Rep. NC-151. St. Paul, MN: U.S. Department of Agriculture, Forest Service, North Central Forest Experiment Station. 48 pp

Holland EA, Dentener FJ, Braswell BH & Sulzman JM (1999) Contemporary and pre-industrial global reactive nitrogen budgets. Biogeochem. 46: 7–43

Hornbeck JS & Kropelin W (1982) Nutrient removal and leaching from a whole-tree harvest of northern hardwoods. J. Environ. Qual 11(2): 309–316

Howarth RW, Billen G, Swaney D, Townsend A, Jaworski N, Lajtha K, Downing JA, Elmgren R, Caraco N, Jordan T, Berendse F, Freney J, Kudeyarov V, Murdoch P & Zhao-Liang Z (1996) Regional nitrogen budgets and riverine N & P fluxes for the drainages to the North Atlantic Ocean: natural and human influences. Biogeochem. 35: 75–139

Johnson DW (1992) Nitrogen retention in forest soils. J. Environ. Qual. 21: 1–12

Johnson DW & Lindberg SE (Eds) (1992) Atmospheric deposition and forest nutrient cycling. Ecological Studies vol. 91. Springer-Verlag, New York

Johnson DW, West DC, Todd DE & Mann LK (1982) Effects of sawlog vs. whole-tree harvesting on the nitrogen, phosphorous, potassium, and calcium budgets of an upland mixed oak forest. Soil Sci. Soc. Am. J. 46: 1304–1309

Kittel TGF, Royle JA, Daly C, Rosenbloom NA, Gibson WP, Fisher HH, Schimel DS, Berliner LM & VEMAP2 Participants (1997) A gridded historical (1895–1993) bioclimate dataset for the conterminous United States. In: Proceedings of the 10th Conference on Applied Climatology, 20–24 October 1997, Reno, NV. American Meteorological Society, Boston. pp 219–222

Likens GE, Bormann FH, Johnson NM, Fisher DW & Pierce RS (1970) Effects of forest cutting and herbicide treatment on nutrient budgets in the Hubbard Brook watershed-ecosystem. Ecol. Monogr. 40(1): 23–47

Lovett GL & Rueth H (1999) Soil nitrogen transformations in beech and maple stands along a nitrogen deposition gradient. Ecol Appl 9(4): 1330–1334

Lovett GL, Weathers KC & Sobczak WV (2000) Nitrogen saturation and retention in forested watersheds of the Catskill Mountains, New York. Ecol. Appl. 10(1): 73–84

Martin CW & Pierce RS (1980) Clearcutting patterns affect nitrate and calcium in streams in New Hampshire. J. For. 78: 268–272

McGuire AD, Sitch S, Clein JC, Dargaville R, Esser G, Foley J, Heimann M, Joos F, Kaplan J, Kicklighter DW, Meirer RA, Melillo JM, Moore B III, Prentice IC, Ramankutty N, Reichenau T, Schloss A, Tian H, Williams LJ & Wilttenberg U (2001) Carbon balance of the terrestrial biosphere in the twentieth century: analyses of CO_2, climate and land use effects with four process-based ecosystem models. Global Biogeochem. Cycles 15(1): 183–206

Mou P, Fahey TJ & Hughes J (1993) Effects of soil disturbance on vegetation recovery and nutrient accumulation following whole-tree harvest of a northern hardwood ecosystem. J. Appl. Ecol. 30: 661–675

MRLC (1995) Multi-Resolution Land Characteristics (MRLC) Consortium Documentation Notebook; national land cover database. [online] URL: http://www.epa.gov/mrlc/

Nadelhoffer KJ, Downs MR & Fry B (1999a) Sinks for ^{15}N-enriched additions to an oak forest and a red pine plantation. Ecol. Appl. 9(1): 72–86

Nadelhoffer KJ, Emmett BA, Gundersen P, Kjonaas OJ, Koopmans CJ, Schleppi P, Tietema A & Wright RG (1999b) Nitrogen deposition makes a minor contribution to carbon sequestration in temperate forests. Nature 398: 145–148

Ollinger SV, Aber JD, Lovett GM, Millham SE, Lathrop RG & Ellis JM (1993) A spatial model of atmospheric deposition for the northeastern U.S. Ecol. Appl. 3: 459–472

Ollinger SV, Aber JD & Federer CA (1998) Estimating regional forest productivity and water yield using an ecosystem model linked to a GIS. Landscape Ecol. 13: 323–334

Ramankutty N & Foley JA (1999) Estimating historical changes in global land cover: croplands from 1700 to 1992. Global Biogeochem. Cycles 13(4): 997–1027

Schimel DS, VEMAP Participants & Braswell BH (1997) Continental scale variability in ecosystem processes: models, data, and the role of disturbance. Ecol. Monogr. 67(2): 251–271

Schimel DS, Melillo J, Tian H, McGuire AD, Kicklighter D, Kittel T, Rosenbloom N, Running S, Thornton P, Ojima D, Parton W, Kelly R, Sykes M, Neilson R & Rizzo B (2000) Contribution of increasing CO2 and climate ot carbon storage by ecosystems in the United States. Science 287: 2004–2006

Tritton LM, Martin CW, Hornbeck JW & Pierce RS (1987) Biomass and nutrient removals from commercial thinning and whole-tree clearcutting. Environ. Manage. 11: 659–666

Turner DP, Koerper GJ, Harmon ME & Lee JJ (1995) A carbon budget for forests of the conterminous United States. Ecol. Appl. 5(2): 421–436

United States Bureau of the Census (1977) Historical Statistics of the United States from Colonial Times to 1970. Series K 17–81. US Bureau of the Census, Washington, DC

Van Breemen N, Boyer EW, Goodale CL, Jaworski NA, Seitzinger S, Paustian K, Hetling L, Lajtha K, Eve M, Mayer B, Van Dam D, Howarth RW, Nadelhoffer KJ & Billen G (2002) Where did all the nitrogen go? Fate of nitrogen inputs to large watersheds in the northeastern U.S.A. Biogeochemistry 57/58: 267–293

Vitousek PM (1994) Beyond global warming: Ecology and global change. Ecol. 75: 1861–1876

Vitousek PM & Matson PA (1984) Disturbance, nitrogen availability, and nitrogen losses in an intensively managed loblolly pine plantation. Ecol. 66(4): 1360–1376

Vitousek PM, Gosz JR, Grier CC, Melillo JM, Reiners WA & Todd RL (1979) Nitrate losses from disturbed ecosystems. Science 204: 469–475

Vose JM & Swank WT (1993) Site preparation burning to improve southern Appalachian pine-hardwood stands: aboveground biomass, forest floor mass, and nitrogen and carbon pools. Can J For Res 23: 2255–2262

Whittaker RH, Bormann FH, Likens GE & Siccama TG (1974) The Hubbard Brook ecosystem study: forest biomass and production. Ecol. Monog. 44: 233–254

Whittaker RH, Likens GE, Bormann FH, Eaton JS & Siccama TG (1979) The Hubbard Brook ecosystem study: forest nutrient cycling and element behavior. Ecol. 60: 203–220

Williard KW, DeWalle DR, Edwards PJ & Schnabel RR (1997) Indicators of nitrate expot from forested watersheds of the mid-Appalachians, United States of America. Global Biogeochem. Cycles 11(4): 649–656
Woodwell GM, Whittaker RH & Houghton RA (1975) Nutrient concentrations in plants in the Brookhaven oak-pine forest. Ecol 56: 318–332
Woudenberg SW & Farrenkopf TO (1995) The Westwide forest inventory data base: users manual. Gen. Tech. Rep. Int-GTR-317. USDA Forest Service, Intermountain Research Station, Ogden, Utah. 67 pp

Biogeochemistry **57/58**: 267–293, 2002.

Where did all the nitrogen go? Fate of nitrogen inputs to large watersheds in the northeastern U.S.A.

N. VAN BREEMEN[1]*, E.W. BOYER[2], C.L. GOODALE[3],
N.A. JAWORSKI[4], K. PAUSTIAN[5], S.P. SEITZINGER[6], K. LAJTHA[7],
B. MAYER[8], D. VAN DAM[1], R.W. HOWARTH[9], K.J. NADELHOFFER[10],
M. EVE[6] & G. BILLEN[11]
[1]*Wageningen University, Laboratory of Soil Science & Geology and Wageningen Institute for Environment and Climate Research, POB 37, 6700 AA Wageningen, the Netherlands;*
[2]*State University of New York, College of Environmental Science and Forestry, Syracuse, NY;*
[3]*Carnegie Institution of Washington, Department of Plant Biology, Stanford, CA;*
[4]*Retired, US EPA, Wakefield, RI;* [5]*Colorado State University, Natural Resource Ecology Laboratory, Fort Collins, CO;* [6]*Rutgers University, Institute of Marine and Coastal Sciences, New Brunswick, NJ;* [7]*Oregon State University, Department of Botany & Plant Pathology, Corvallis OR;* [8]*University of Calgary, Department of Geology & Geophysics, Canada;*
[9]*Cornell University, Department of Ecology & Evolutionary Biology, Ithaca, NY;*
[10]*Marine Biological Laboratory, The Ecosystems Center, Woods Hole, MA;*
[11]*Free University of Brussels, Belgium*
(*author for correspondence, e-mail: Nico.vanBreemen@bodeco.beng.wau.nl)*

Abstract. To assess the fate of the large amounts of nitrogen (N) brought into the environment by human activities, we constructed N budgets for sixteen large watersheds (475 to 70,189 km^2) in the northeastern U.S.A. These watersheds are mainly forested (48–87%), but vary widely with respect to land use and population density. We combined published data and empirical and process models to set up a complete N budget for these sixteen watersheds. Atmospheric deposition, fertilizer application, net feed and food inputs, biological fixation, river discharge, wood accumulation and export, changes in soil N, and denitrification losses in the landscape and in rivers were considered for the period 1988 to 1992. For the whole area, on average 3420 kg of N is imported annually per km^2 of land. Atmospheric N deposition, N$_2$ fixation by plants, and N imported in commercial products (fertilizers, food and feed) contributed to the input in roughly equal contributions. We quantified the fate of these inputs by independent estimates of storage and loss terms, except for denitrification from land, which was estimated from the difference between all inputs and all other storage and loss terms. Of the total storage and losses in the watersheds, about half of the N is lost in gaseous form (51%, largely by denitrification). Additional N is lost in riverine export (20%), in food exports (6%), and in wood exports (5%). Change in storage of N in the watersheds in soil organic matter (9%) and wood (9%) accounts for the remainder of the sinks. The presence of appreciable changes in total N storage on land, which we probably under-rather than overestimated, shows that the N budget is not in steady state, so that drainage and denitrification exports of N may well increase further in the future.

Introduction

Through processes such as manufacturing fertilizers, burning fossil fuels, and cultivating crops that fix nitrogen (N) symbiotically, humans have greatly accelerated the fixation of atmospheric N to plant-available forms (Galloway et al. 1995; Vitousek et al. 1997). As a result, the amount of reactive N that enters the biosphere each year worldwide has roughly doubled since pre-industrial times (Galloway et al. 1995; Howarth et al. 1996; Mosier et al. 2001). Because N is the second-most abundant plant nutrient (after CO_2) and limits primary production in many terrestrial, freshwater and near-coastal marine ecosystems, these anthropogenic activities have major environmental consequences (Vitousek et al. 1997).

Although the amounts of N fixed from natural and human activities have received much attention, the fate of this reactive N is poorly understood. Several recent budget studies have presented estimates of all known new N inputs to terrestrial landscapes. For example, Boyer et al. (2002) quantified N inputs to individual watersheds that drain to the northeast (NE) coast of the U.S.A., following the methodology put forth by Howarth et al. (1996) who estimated N inputs to all of the large regions that drain to the North Atlantic Ocean, including one value for the NE region as a whole. Despite the large differences in scale, both studies indicate that streamflow exports account for only about 25% of N inputs to the landscape (Figure 1). Regardless of how the input terms and boundary conditions are defined, nearly all such studies conclude that only 20–60% of N inputs are explained by N export in streamflow, whether considered at the scale of small watersheds (e.g. Campbell et al. 2000; Dise et al. 1998), large river basins (e.g. Jaworski et al. 1997; Castro et al. 2001), or regional drainage areas (e.g. Howarth et al. 1996). The N input in excess of riverine export has been termed the 'missing nitrogen', highlighting the uncertainty in the scientific community of how to quantify the fate and transport of N in the landscape. The fraction of N 'consumed' by the landscape and not delivered to streamflow is partly stored in pools with residence times exceeding decades to centuries (soil, wood, groundwater) and partly returned to the pool of highly inert atmospheric N_2 by denitrification. The relative sizes of these storage and loss terms of the N budget are highly uncertain. The aim of this paper is to increase our understanding of the amounts of N that are stored in these different slow but potentially reactive pools and that are lost to the atmosphere due to N transformation processes.

We chose 16 watersheds in the northeastern U.S.A. (Figure 2, taken from Boyer et al. 2002) for our analyses, a region where N cycling is of particular importance due to problems in coastal waters caused by over-enrichment of N (Bricker et al. 1999). Further, the availability of high-quality, long-term

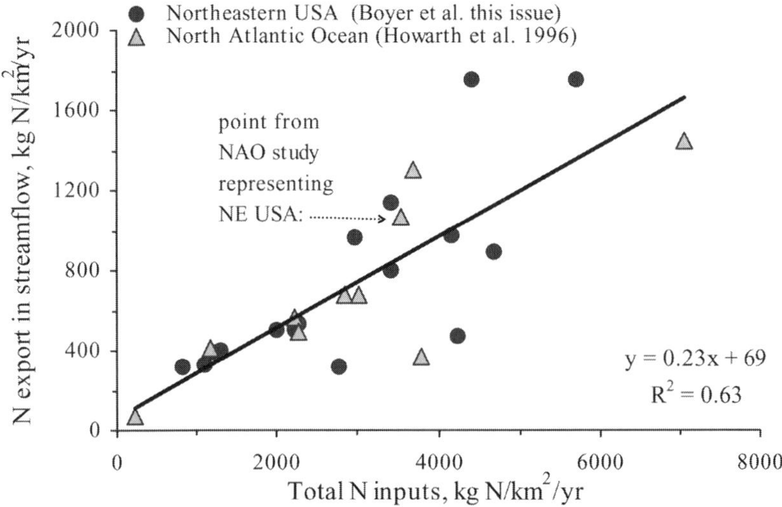

Figure 1. Nitrogen budget studies for large watersheds in the northeast (NE) U.S.A. (Boyer et al. 2002) and for regions in north America and in western Europe that drain to the north Atlantic Ocean (NAO, Howarth et al. 1996). Total N inputs are the net anthropogenic inputs from atmospheric deposition, fertilizer inputs, nitrogen fixation, and the net import of N in food & feed. The NAO regional analysis represents the entire NE U.S.A. as one region.

monitoring data in the northeast allowed us to quantify, with good confidence, N inputs to these regions as a starting point for investigating storage and losses of N in the landscape. The details of our calculations of N inputs and riverine exports for each watershed are presented in a companion paper (see Boyer et al. 2002). Nitrogen budgets were established by quantifying total annual inputs of N to each catchment. Most of the N inputs are derived from human activities, and include atmospheric deposition, fertilizer use, net imports (or exports) in food & feed, and biological fixation in agricultural areas and in forests. As shown in Figure 1, riverine export of N was well correlated with N inputs, but represented only a fraction (11–40%) of the total N inputs. The unresolved N inputs in excess of streamflow export are either stored (e.g. in vegetation, soil, or groundwater) or lost (e.g. denitrified, volatilized, or exported in transfers of commodities) in the watershed. In this paper, we attempt to close input-output N budgets for these watersheds by explaining the fate of the N inputs. We estimate storage and loss terms from a combination of statistical and process models and with data on land use change.

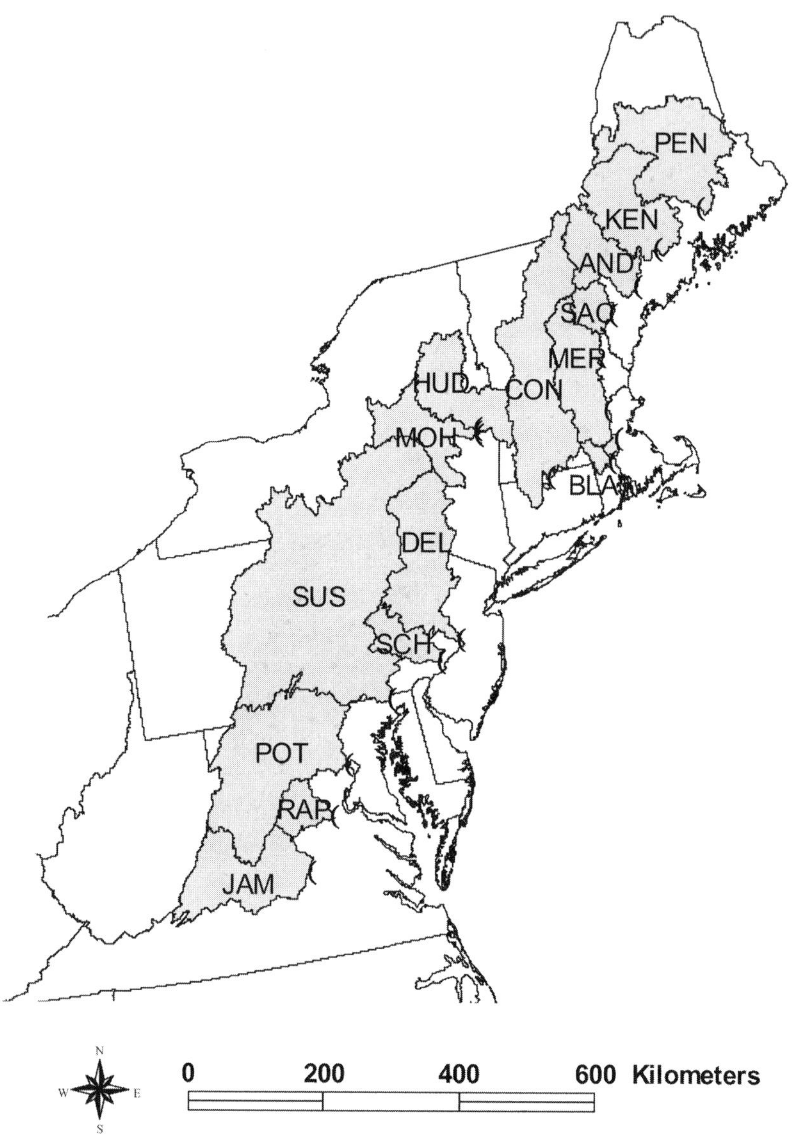

Figure 2. (from Boyer et al. 2002). Location of 16 watersheds draining to the northeast US coast. Watershed boundaries are delineated upstream of USGS stations (denoted with black circles) where streamflow and water quality characteristics were measured.

Methods

We begin with N inputs to 16 watersheds in the northeastern U.S.A., presented by Boyer et al. (2002), and refer to that paper for a detailed description of watershed characteristics, data sources, and methods used to estimate N inputs and riverine export. In brief, the watersheds provide the major drainage ways to the northeast coast, and are located in a latitudinal profile from Maine to Virginia. The basins range in size from 475 km^2 to 70,189 km^2. The combined total area of all watersheds is largely forested (72%) with some agricultural land (19%) and a small fraction of urban land (3%) (Table 1). Because we delineated the watershed boundaries upstream of suitable gaging stations (from which riverine streamflow and water quality data were obtained), most major coastal population centers are excluded. Data presented in this paper are representative of the early 1990's and, where possible, reflect average values over the period 1988–1993.

Sources of N to the 16 watersheds include atmospheric N deposition, nitrogenous fertilizer use, import of N in food & feed, and biological N fixation in crops and in forests (Table 2, after Boyer et al. 2002). The goal of this paper is to explain the fate of these N inputs. We take a mass balance approach to establish a complete N budget for each watershed, where inputs are balanced by outputs or changes in storage in the watershed. We utilize a conceptual model whereby N inputs are routed through one of the 4 major land use 'reservoirs' comprising the watershed: forested, agricultural, sub/urban, and water ecosystems (Figure 3). We establish a mass balance of N for each of the individual ecosystems, and aggregate their inputs and outputs to determine the total storages and losses from the watershed. In some cases, outputs from one ecosystem are inputs to another, transferring N internally but having no net effect on watershed output. In addition to the N exported in rivers, outputs of N from each watershed include removal in harvested wood, exports of food, ammonia volatilization losses, and gaseous losses due to denitrification (from sewage and waste water, from soil solutions in transit from soils to rivers, and within the water column). Changes in storage include the net accumulation of N in vegetation, in soil, and in groundwater.

Storage and losses in forest lands

Inputs of N to forested lands include atmospheric deposition and fixation, while losses include removal in harvested wood and denitrification (see Figure 3). Changes in storage include net accumulation in woody biomass, in dead wood and in green plant tissues (foliage & fine roots), and in forest soils. While not a loss when considered at the scale of the watershed, the internal transfer of N from the forest ecosystem to subsurface water reservoirs (e.g.

Table 1. Watershed Characteristics*

Watershed	Abbreviation	Area km^2	Sewer. Popul. #/km^2	Unsew. Popul. #/km^2	Temper. °C	Precip. mm yr^{-1}	Runoff yr^{-1}	Land Forest %	Land Agric. %	Land Urban %	Land Wat & Wetl. %	Land Other %
Penobscot	PEN	20109	3.0	4.5	4.3	1075	588	84	1	0	11	3
Kennebec	KEN	13994	4.4	4.8	4.3	1085	566	80	6	1	10	4
Androscoggin	AND	8451	8.2	8.6	4.6	1151	640	85	5	1	8	1
Saco	SAC	3349	7.6	8.7	5.8	1218	672	87	4	1	7	1
Merrimack	MER	12005	94.7	47.9	7.4	1148	589	75	8	9	8	1
Charles	CHA	475	428.9	127.4	9.7	1207	583	59	8	22	10	0
Blackstone	BLA	1115	217.5	58.5	9.0	1260	651	63	8	18	10	1
Connecticut	CON	25019	44.0	20.8	6.3	1160	642	79	9	4	7	1
Hudson	HUD	11942	19.7	12.6	6.6	1126	622	81	10	3	6	0
Mohawk	MOH	8935	32.3	22.0	6.8	1142	548	63	28	5	4	0
Delaware	DEL	17560	55.1	30.0	8.7	1131	547	75	17	3	5	0
Schuylkill	SCH	4903	249.9	42.8	10.6	1134	488	48	38	10	2	1
Susquehanna	SUS	70189	33.1	20.9	8.9	1022	487	67	29	2	2	1
Potomac	POT	29940	46.7	15.9	11.3	985	328	61	35	3	1	1
Rappahannock	RAP	4134	10.1	14.2	12.6	1045	360	61	36	1	1	1
James	JAM	16206	10.1	14.0	10.1	934	407	81	61	1	*1*	1
Area-wt. avg	—	*32666*	*38.0*	*19.0*	*8.0*	*1068*	*515*	*72*	*19*	*3*	*5*	*1*

*From Boyer et al. 2002. Watersheds are arranged by latitude of their outlets, from north to south.

Table 2. Sources of N to watersheds* (kg N per km^2 per year)

Watershed	Total NO_y dep.	Total NH_x dep.[1]	Net Org. N dep.	N fertilizer use	Forest N fixation	Agricul. N fixation	Net N import in feed	Net N import in feed	Total Sources
PEN	362	129	88	91	58	74	55	0	857
KEN	428	154	105	54	50	164	171	0	1126
AND	495	176	121	80	69	146	247	0	1332
SAC	566	187	136	42	107	96	49	55	1237
MER	606	184	142	147	151	213	150	647	2240
CHA	674	178	153	197	218	187	62	2745	4415
BLA	707	190	162	307	260	305	217	1279	3426
CON	631	204	150	274	102	360	398	167	2286
HUD	658	234	161	204	103	374	251	20	2005
MOH	708	250	172	411	70	1239	758	0	3610
DEL	811	248	191	527	201	675	155	197	3005
SCH	885	253	205	1207	190	1225	1401	551	5917
SUS	816	269	195	615	179	1147	1554	0	4774
POT	714	255	174	1024	271	1173	2085	0	5696
RAP	615	256	157	1030	277	1439	898	0	4671
JAM	652	237	160	361	361	703	487	0	2961
Wt. Avg.	*677*	*288*	*163*	*474*	*167*	*740*	*887*	*86*	*3420*

*From Boyer et al. 2002. [1]Rather than using a net input term, wer treat depositional inputs of total NH_x and volatilization losses of NH_x separately.

groundwater) and to the river affects the input of N to these other ecosystems. We estimate all of the N fluxes in forested ecosystems (Table 3). The forest calculations for each watershed are presented in detail in a companion paper by Goodale et al. (2002). Forest inventory data were used to estimate changes in biomass and harvest export, and an ecosystem model was used to estimate changes in dead wood, green plant tissues, forest soils, and leaching of nitrate below the rooting zone.

Briefly, county-level data were obtained from the U.S. Forest Service's Forest Inventory and Analysis (FIA) program (Hansen et al. 1992) on the volume of wood growth, mortality, and harvests in eastern forests. These values were converted to estimates of biomass with two different approaches, and multiplied by literature-derived estimates of the N content in wood (0.19 + 0.08% for softwoods and 0.26 + 0.06% for hardwoods) to obtain estimates in terms of N. Values included in our study are the mean of the two biomass conversion approaches. Net N accumulation in biomass consists of growth

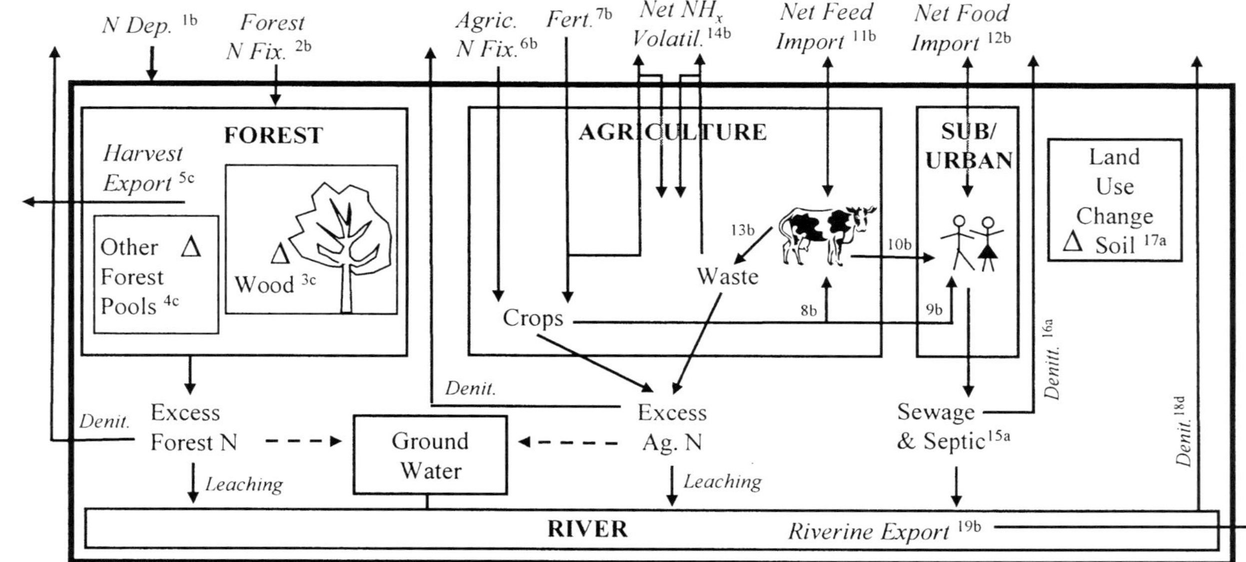

Figure 3. N fluxes to, from, and within each watershed: (1) NH_x, NO_y, and net organic N atmospheric deposition to whole watershed, (2) biological N fixation in forests, (3) increase in N stocks in woody biomass and (4) other forest pools (dead wood, forest soils, green tissues) (5) export of harvested wood, (6) biological nitrogen fixation in agricultural lands, (7) fertilizer use, (8) crop production for animal consumption and (9) for human consumption, (10) meat, milk and eggs, (11) Net import (or export) of animal feed and (12) human food, (13) production of animal waste, (14) Net NH_3 volatilization from animal waste and fertilizer, (15) sewage & septic waste and (16) denitrification during treatment, (17) increase in soil N due to land-use change, (18) in-stream denitrification, and (19) riverine export. Letters indicate the source of calculations: (a) this paper, (b) Boyer et al. (c) Goodale et al. (d) Seitzinger et al.

Table 3. N balance for forest land (kg N per km^2 of forest land per year)

Water-shed	% of land in forest	N Sources			N Storages and Losses							
		Atmospheric dep.	Forest N fixation	Total N inputs	Wood biomass storage[1]	Other forest storage[2]	Change in soil storage[3]	Harvest export[4]	NO$_3$ leaching	DON leaching	Total storage & loss	% inputs denitr. in soils[5]
PEN	84	579	70	649	−205	−6	−99	491	256	75	513	21
KEN	80	687	63	750	−220	−6	−96	524	312	75	589	21
AND	85	791	81	873	184	−103	55	313	259	75	783	10
SAC	87	889	123	1012	379	−113	191	181	218	75	931	8
MER	75	932	202	1134	267	−106	328	184	159	75	907	20
CHA	59	1006	368	1374	415	113	158	213	141	75	1115	19
BLA	63	1059	411	1470	263	131	98	257	174	75	997	32
CON	79	985	129	1114	385	−28	123	202	265	75	1021	8
HUD	81	1053	128	1181	426	−100	413	211	189	75	1212	−3
MOH	63	1131	111	1242	469	−40	357	130	222	75	1212	2
DEL	75	1250	269	1518	721	66	133	148	203	75	1346	11
SCH	48	1342	396	1738	649	−111	202	286	223	75	1325	24
SUS	67	1280	268	1548	550	167	123	215	179	75	1309	15
POT	61	1143	446	1589	570	182	69	193	231	75	1320	17
RAP	61	1027	451	1479	473	38	40	363	185	75	1173	21
JAM	81	1049	448	1497	392	200	−117	384	237	75	1171	22
Wt. Avg.	*72*	*1067*	*242*	*1309*	*395*	*66*	*102*	*259*	*218*	*75*	*1116*	*15*

1,2,3,4from Goodale et al. 2002. Change in storage in: [1]living wood biomas, [2]dead wood and live plant tissues, and [3]in forest land soils. [4]Removal of forest products from watershed. [5]In forest lands; estimated by difference of inputs − outputs.

minus losses due to natural mortality and harvests. Harvest export consists of only the harvested biomass removed from the forest; the rest of the harvested material is assumed to decompose within the watershed.

Accumulation of N in dead wood, green plant tissues, and soils was estimated with the forest ecosystem model PnET-CN (Aber et al. 1997; Aber & Driscoll 1997) as the balance between inputs from litter and logging slash, and losses from decomposition. Simulations were performed for a range of forest age classes as indicated by forest inventory data, and for two contrasting scenarios past land-use history: agriculture and forestry. Model-based N fluxes included here used the mean simulation results from these two land-use histories, weighted by the approximate prevalence of each history on each watershed.

Leaching of N from soils is significant in regions with excess N input, as in the northeastern U.S.A. Leaching losses of nitrate below the rooting zone were also estimated with the PnET-CN model. In the model structure, net mineralization from the soil is available for uptake by plants; the excess can nitrify and leach below the rooting zone with drainage water. Predicted losses of nitrate depended on the combined effects of N deposition, and both short- and long-term effects of disturbance (see Goodale et al. 2002). DON leaching losses from forested lands were estimated from literature values. In the northeastern U.S.A., DON losses to streams range from 30 to 240 kg km^{-2} yr^{-1}, and average about 75 kg km^{-2} yr^{-1} (Campbell et al. 2000; Lovett et al. 2000; Goodale et al. 2000). We assumed that the forested areas in each catchment leach 75 kg of DON km^{-2} yr^{-1} to streamwater. This conservative, constant estimate allows us to account for this important removal pathway while not biasing the pattern of residual uncertainties in our N budgets.

Some of the leached N is lost before the drainage water reaches rivers or groundwater, of which only part can be explained by plant uptake or retention in soils (Lajtha et al. 1995; Sollins & McCorison 1981). There is evidence for denitrification in the vadose zone or at the terrestrial-aquatic interface that is difficult to quantify (Montgomery et al. 1997; Valiela et al. 1997; Seely et al. 1998). Denitrification from well-drained, upland forest soils in the northeastern U.S.A. is generally very low (Groffman & Tiedje 1989; Bowden 1986). Overall, the forest ecosystem N balance was well constrained, with the estimated storage and loss terms accounting for approximately 85% of the inputs. The difference between N inputs and outputs, or the unresolved N, is attributed to denitrification or to changes in groundwater storage.

Storage and losses in sub/urban and agricultural lands

Sources of N to sub/urban lands include atmospheric deposition and food inputs. Food inputs come from two sources: from agricultural lands within the watershed (i.e. local crop and animal production of fruits, vegetables, meat, milk and eggs), and from imports of N from outside of the watershed (see Figure 3). Change in soil storage is potentially a significant sink for N. Watershed-scale losses of N from sub/urban lands include net export of N in food and denitrification of N during waste treatment.

Estimates of losses in food exports are presented by Boyer et al. (2002) in their discussion of the net import of food and feed to each watershed. About 1/2 of the watersheds produced more N in crop and animal products than could be consumed by the populations living there, and thus exported N in food sales to other regions (see Table 2). We calculated sewage treatment losses by comparing human consumption of N in food with data on N delivered to rivers in wastewater discharge. Data on populations with sewered waste were obtained from the 1990 U.S. Census (U.S. Dept. of Commerce 1990), and measurements of wastewater discharge and N concentration were obtained from a variety of agency reports (N Jaworski and L Hetling, personal communication). The correlation between the sewered population and wastewater discharge ($r^2 = 0.95$) yielded a mean per capita load of 3.1 kg N yr^{-1} per person (Figure 4). This is similar to the value of 3.3 kg N yr^{-1} per person reported by Meybeck et al. (1989). The difference between the per capita waste excretion and the per capita intake of 5 kg N yr^{-1} per person (Garrow et al. 2000), or 1.9 kg N yr^{-1} per person, is the amount of N that is either retained or, more likely, denitrified during sewage processing. We estimated total watershed losses from septic and sewage treatment by assuming the same per capita N loss from septic tanks as for the sewered population (1.9 kg N yr^{-1} per person). Although septic treatment is likely to retain N longer in the watershed than sewage treatment, the error introduced by this assumption is probably small because of the relatively low unsewered population (19%, see Table 1).

The northeastern U.S.A. has been experiencing rapid changes in land use, which could affect N storage in litter and soil organic matter stocks in all ecosystems (see Figure 3). Our analysis of data on land use change reveals that between 1982 and 1992 none of the watersheds gained land in agriculture, several gained small percentages of forest land, and all gained substantial amounts of sub/urban land (Figure 5). Therefore, we expected that changes in N in soil stores in sub/urban land uses to be significant factor in our N budgets, associated with the land use change itself (e.g. a shift from forest to urban land) or from land use legacies (e.g. land that had been previously fertilized that is not yet in steady state). Data on land use change were used to estimate net changes in soil organic N stocks using a modified version of the IPCC (Intergovernmental Panel on Climate Change) soil carbon (C) inventory procedure (IPCC 1997, Paustian et al. 1997a). The method employs

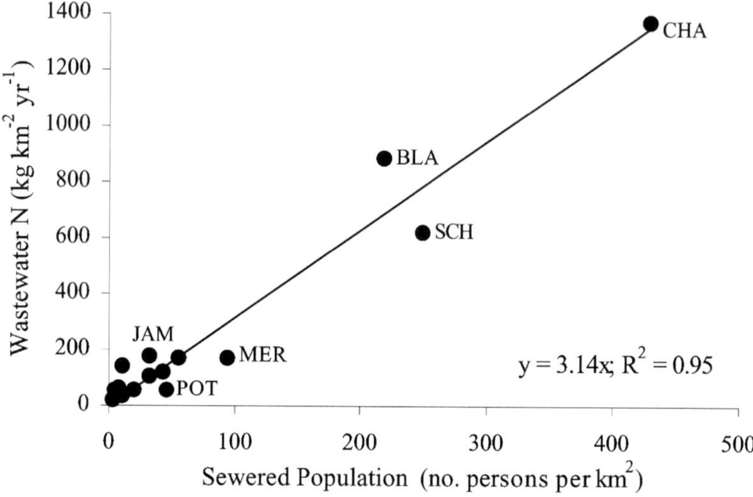

Figure 4. Relationship between sewered population and nitrogen fluxes due to sewage wastewater. The regression line indicates a per capita load in wastewater of 3.1 kg N yr^{-1} per person.

a net stock change approach that incorporates changes in areas according to multiple categories of land use and agricultural management systems, stratified by soil type and climate. Changes in soil C stocks in the upper 30 cm of the soil profile, which result from both land use change and management changes within continued agricultural use, were estimated using a series of coefficients based on climate, soil type, disturbance history, tillage intensity and C input rate (productivity).

We obtained data from the 1982 and 1992 National Resources Inventory (NRI), maintained by USDA's Natural Resources Conservation Service, to estimate land use and soil organic C and N changes. The NRI is a nationwide inventory of land cover and land use and management, comprised of >800,000 permanent sampling locations (Nusser & Goebel 1997). The inventory's statistical design includes an 'expansion factor' for each point location, which is used to estimate the total area represented by the point as specified by the hydrologic unit code (HUC) containing the point. A total of 55,289 NRI points representing 250,000 km^2 were located within the boundaries of the study watersheds.

Land use/land cover types included forest, agricultural, urban, rangeland, miscellaneous and non-cropland (includes abandoned, non-forested agricultural lands, non-forested wetlands). Within agricultural lands, crop rotations were grouped into several types of management systems (e.g. irrigated cropland, continuous row crops, row crop-small grain rotations, row crop-hay

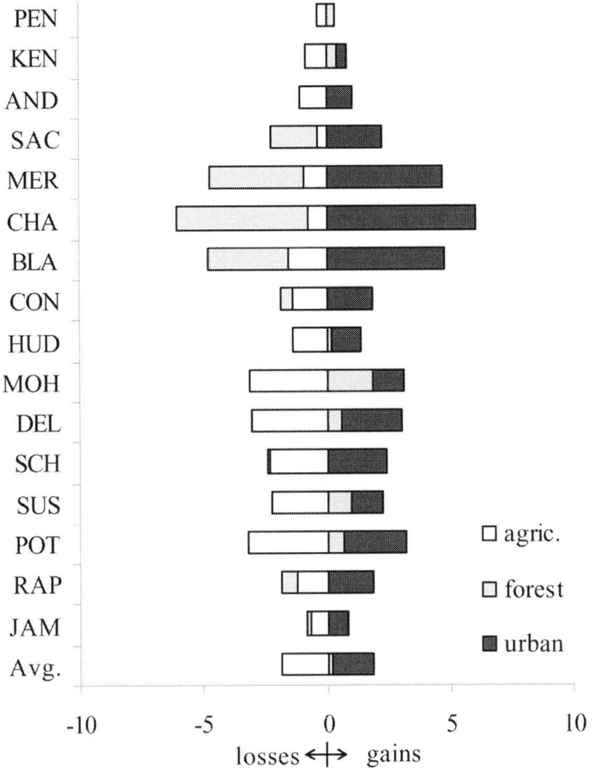

Figure 5. Changes in land use from 1982 to 1992 are shown as a percent of the land area of each watershed. Over the decade, all watersheds lost land in agriculture and gained urban land. Data are from the USDA-NRCS 1992 National Resources Inventory.

rotations, small grain hay rotations, and vegetable crops), based on the cropping history information in NRI. Crop rotations were then aggregated into three groups, i.e. high, medium and low input systems, based on the rates of production and return of organic residues to soil. Management systems were further stratified into three categories of tillage: conventional, reduced and no-till. Data from the CTIC (Conservation Tillage Information Center) (www.ctic.purdue.edu; Dan Towery, personal communication) were used to estimate the areas under the three tillage management scenarios in 1982 and 1992. Areas for each of the land use/management systems were then calculated by watershed, for 1982 and 1992. To account for our use of a 10 year inventory cycle instead of the default 20-year IPCC inventory period, we changed the coefficients to yield only 50% of the expected change in C stocks over a twenty year period. Changes in organic N stocks were estimated

280

from the C calculations, assuming a C:N ratio of 10 for agricultural and 15 for urban/suburban soils and early successional forests converted from agricultural use (Robertson & Vitousek 1981; Hamburg 1984; Zak et al. 1990).

Inputs of N to the agricultural ecosystem include atmospheric deposition, fixation, fertilizer, and net import (or export) in food & feed (see Figure 3). As in the sub/urban system, changes in soil N storage are potential N sinks. We calculate the change of N stored in agricultural soils due to land use change according to the IPCC methodology described above. Watershed-scale losses of N from agricultural lands include volatilization, feed exports, and denitrification. Internally, there are many important transfers. For example, a fraction of the crops produced and animal products (meat, milk & eggs) are outputs from the agricultural ecosystem but are input as food to the sub/urban ecosystem, thus having no net effect on the watershed-scale budget. We estimate both the internal and external cycling in agricultural lands (Table 4).

Estimates of net NH_3 volatilization losses are presented by Boyer et al. (2002) in their discussion of depositional inputs. Calculations of losses from each watershed in feed exports from agricultural lands are also presented by Boyer et al. (2002) in their discussion of net food & feed imports. None of the watersheds had net exports of N from agricultural lands in animal feed (see Table 2).

Leaching losses from applied N in agricultural fields to ground and surface water are difficult to quantify and depend on many factors. These losses increase with the amount of applied N and are generally higher in arable fields than in grasslands. Using literature values, we assumed that leaching losses from grassland equal $0.15 \times$ (N input $- 500$) kg km^{-2} yr^{-1} (Jordan et al. 1994; Scholefield et al. 1988; Magesan et al. 1996). Leaching from arable land was assumed to be $0.2 \times$ (N input $- 500$) kg km^{-2} yr^{-1} (Goss et al. 1988; Madramootoo et al. 1995; Watson et al. 1993; Wyland et al. 1996; Shephard 1992). N fertilizer inputs are normally somewhat higher on arable land than on pasture land, but varying the relative fertilizer application rates between 1 and 3 times higher on arable than on pasture land had no marked (<1%) effect on the estimated NO_3 leaching from agricultural land.

Denitrification in agricultural soils can be very significant, particularly in areas with high inputs of N fertilizers (Velthof et al. 1997). We estimated denitrification by difference between N total N inputs and the outputs from agricultural lands (soil N storage, NH_3 volatilization, food production removals, and NO_3 leaching). We attribute this budget discrepancy to denitrification, which appears to make up a significant fraction ($\sim$50%) of the N inputs to the agricultural ecosystem (Table 4).

Table 4. N balance for agricultural land (kg N per km^2 of agricultural land per year)

Water-shed	% of land in agric.	N Sources					N Storages and Losses						
		Atmospheric dep.	Fertilizer use	Agricul. N fixation	Net import in feed	Total N inputs	Crop products for food	Animal products for food	net NH$_x$ vol.[1]	NO$_3$ leaching[2]	Change in soil storage[1]	Total storage & loss	% inputs denitr. in soils[4]
PEN	2	579	4550	3694	2726	11549	1581	1347	208	1779	901	5816	50
KEN	6	687	900	2727	2852	7165	34	1131	166	1095	1493	3918	45
AND	5	791	1600	2913	4936	10241	172	1888	253	1583	1998	5894	42
SAC	4	889	1050	2389	1224	5551	104	633	103	782	2062	3683	34
MER	8	932	1896	2749	1930	7507	23	921	146	1171	902	3164	58
CHA	8	1006	2343	2229	740	6318	8	477	115	1044	1831	3474	45
BLA	8	1059	3791	3764	2675	11290	28	1363	238	1902	680	4212	63
CON	9	985	3041	3999	4412	12437	90	1838	265	1969	945	5107	59
HUD	10	1053	1958	3586	2410	9008	62	1439	194	1556	1462	4713	48
MOH	28	1131	1469	4431	2711	9742	52	1553	199	1767	1104	4674	52
DEL	17	1250	3155	4044	929	9377	190	1313	222	1678	1038	4442	53
SCH	38	1342	3144	3191	3650	11328	267	2346	520	2062	1757	6951	39
SUS	29	1280	2155	4018	5446	12899	175	2642	498	2360	1095	6771	48
POT	35	1143	2961	3391	6028	13522	154	2866	1080	2097	1302	7499	45
RAP	36	1027	2873	4014	2504	10418	95	1173	375	1949	993	4585	56
JAM	16	1049	2312	4503	3116	10981	38	1462	616	2066	1031	5213	53
Wt. Avg.	*19*	*1067*	*2504*	*3727*	*3979*	*11278*	*243*	*1954*	*431*	*1921*	*1167*	*5716*	*49*

[1] Volatilized NH$_x$ from fertilizers and animal waste that is transported outside of the watershed. [2] Subsurface leaching to ground water and to surface water. [3,4] In agricultural lands. [4] Estimated by difference of inputs − outputs.

Storage and losses in rivers

Inputs to the riverine ecosystem are made up of subsurface flow leached from agricultural, sub/urban, and forested landscapes, plus the very small fraction of atmospheric deposition that falls on the areas of water (see Figure 3). Watershed losses of N include riverine export and in-river denitrification losses. Calculations of N discharged from each watershed in riverine export are presented in Boyer et al. (2002).

The removal of N in the water column itself, which we attribute wholly to denitrification, is detailed in a companion paper by Seitzinger et al. (2002). Briefly, we estimated N loss in rivers using the robust inverse statistical relation observed with the ratio of water depth to water residence time in lakes or river stretches, integrated over whole river systems (Seitzinger et al. 2002). Both residence time of water and flow depth are surrogates to describe flow conditions, with high flows having higher depths of water and faster travel times. Under high flow conditions, there may be less settling of particulate N and less exchange with the subsurface sediments and hyporheic zone (Alexander et al. 2000). In general, the reduced contact times of N transported in streamflow that occur under high flow conditions result in less N loss to denitrification. To scale in-stream removal rates from short river stretches to whole watershed river systems, Seitzinger et al. (2002) used EPA-USGS reach network files for the sixteen watersheds, and their associated attributes describing depth and time of travel of each stream reach. The N loss values were calculated for each individual reach according to the inverse relationship between loss and flow conditions, then losses from all the reaches encompassing a river network were aggregated to provide a total loss estimate for each watershed. We estimate total edge-of-stream loading inputs to each river as the sum of the (insignificant) N deposition occurring on water areas plus the total N leaching losses calculated in the ecosystem budgets for agricultural, sub/urban (wastewater), and forest lands. We use the RivR-N model estimates of Seitzinger et al. (2002) based on the RF1 dataset, indicating the percentage of N inputs that are removed during transport through the river network, to estimate in-stream denitrification for each basin.

Results and discussion

Estimates of total watershed N inputs (Table 2) averaged 3420 kg N km^{-2} yr^{-1}. Atmospheric deposition was the largest single source input (31%), although the combination of N inputs from imports in food in feed (28%), by N$_2$ fixation in agricultural lands (22%), and from fertilizer inputs (14%) made agriculture the largest total source of N. Nitrogen fixation in forests contributed little (5%). (Note: Our presentation of the relative importance of the N input terms is slightly different than is reported for the same N sources in the accompanying manuscript by Boyer et al. They compare each input term to the total *net* anthropogenic N inputs. To facilitate our analysis of watershed N losses, we treat volatilization losses of ammonia and food exports as outputs from, rather than negative inputs to, each watershed, which accounts for the differences in the input terms between our papers). Total N inputs increase from around 1000 kg N km^{-2} yr^{-1} in the northern watersheds to between 2000 to 6000 kg N km^{-2} yr^{-1} in the southern watersheds. This increase is due mainly to increasing proportions of agricultural and urban land, at the expense of forested land, and the associated increase in N$_2$ fixation, and inputs of fertilizer and food and feed. In the three most densely populated, relatively small, watersheds (Charles, Blackstone & Schuylkill) net food and feed imports form the dominant N input category. Atmospheric deposition increased from north to south, but this increase in N inputs was small relative to that from agricultural and urban activities.

Aggregating the total N inputs to and outputs from the forested, agricultural, sub/urban, and riverine ecosystems yields a complete accounting of sources, storages, and losses occurring in the 16 watershed (Table 5). N inputs to each watershed are lost (as N removals or gaseous losses) or cause changes in storage within the system Loss terms make up, on average, 82% of the total sinks with the remaining 18% being stored in the watersheds.

Losses due to N removals

Riverine export
As we knew at the onset of this study, riverine export accounts for only a fraction of the total N inputs to each watershed, on average 20% of total N inputs. Our estimates of N inputs according to our riverine ecosystem N budget (see Figure 3) includes the (insignificant) fraction of N inputs from atmospheric deposition that lands on areas of water plus leaching of N from forest, agricultural, and sub/urban (wastewater) systems. These N inputs to the riverine ecosystem are well correlated with N losses in riverine export and with the sum of N discharged plus N removed in rivers due to denitrification (Figure 6). Although the rates of N drained from agricultural and forested areas are indirect model estimates, the N inputs to each river from wastewater discharge are based on independent measurements and are presumably reliable. Total N discharged from the watersheds in riverine export is about 84% of the N inputs to the river from N drainage from the landscape.

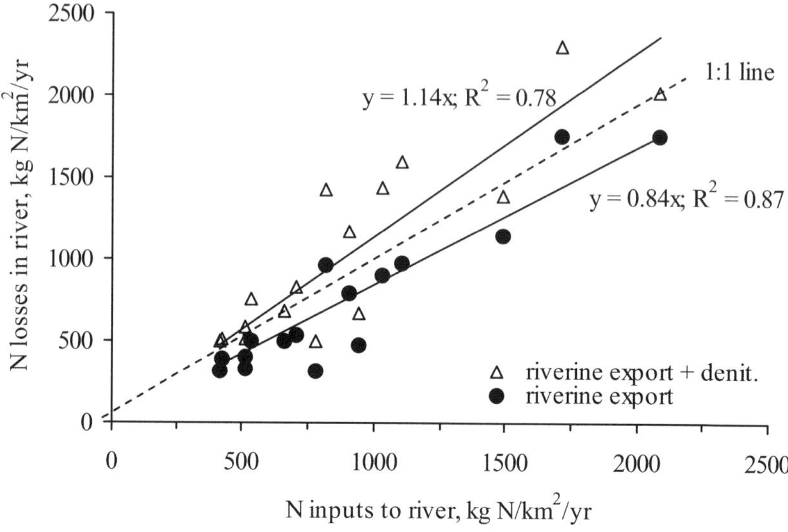

Figure 6. Total N inputs to the river from leaching of N from forested, agricultural, and urban (wastewater) landscapes, compared to the amount of N discharged from the watershed (export) and the sum of N discharged plus N denitrified in rivers (export + loss).

Export in commodities

N exports in commodities, including food products (6%, from removals in meat, milk & eggs) and forest products (5%, from removal of harvested timber) are small loss terms. However, these values do not adequately reflect the importance of commodity exports in individual watersheds (see Table 5). There were no net exports of N in food from most basins, and only in the Susquehanna, Rappahannock, and Potomac were the food exports significant fractions of outputs ($\sim$10%). In watersheds with the largest percentage of forested lands, harvest exports were among the dominant export terms. In the Penobscot, Kennebec, and Androscoggin watersheds, each supporting industrial timber production, exports of N in the forest harvest accounted for 44, 37, and 19% of the total outputs, respectively. We treated the entire wood harvest, which accounted for 5% of the losses on average, as an export term. However, some of the harvested wood will remain in the watershed for local use, and will decay at rates in the order of 0.1 to 0.005 yr^{-1} (Harmon et al. 1996), so this loss term may have been overestimated.

Gaseous losses

Denitrification in rivers
The relationship between N inputs to the river from leaching of N from forested, agricultural, and urban (wastewater) landscapes, and river N export (Figure 3) leaves room for denitrification inside the river system, but not as much as indicated by the analysis by Seitzinger et al. (2002). The sum of N discharged in riverine export plus N removed in the rivers via denitrification exceeds the estimated amount of N drained from the land: (N river discharge plus N denitrified in the rivers) = 1.14 * (N drained from the land), $R^2 = 0.78$. Given the uncertainties of both the land drainage and in-stream denitrification numbers, this good correlation is encouraging, and the overestimate of 14% is a reasonable error term. However, we used the 'low' estimate provided by Seitzinger et al. based on the RF1 dataset. Using their 'high' estimates of riverine denitrification based on the NHD dataset would yield an overestimate of N inputs by 38%: (N river discharge plus N denitrified in the rivers) = 1.38 * (N drained from the land), $R^2 = 0.92$. Assuming that the discharge values are correct, this would imply that either drainage from land has been underestimated or that in-river denitrification has been overestimated. If one equates the difference between N inputs to the river ecosystem and riverine export to in-stream denitrification, this estimate is in good agreement with the 'low' estimate of Seitzinger et al. Our best guess (according to the lower of the two Seitzinger et al. modeled estimates) indicates that, on average, in-stream denitrification accounts for 11% of the total N storage & loss terms.

Denitrification in the landscape
The estimates for denitrification on land must be considered as tentative, as they result from budget discrepancies of N, and therefore contain the accumulated uncertainties of the other estimates. Losses of N in septic and sewage treatment, like ammonia volatilization, accounted for only a small percentage (3%) of the overall N outputs from the watershed (3%). It was calculated in direct proportion to population, and was only significant as an output term in the basins with the highest percentages of urban populations: the Charles (23%), Blackstone (15%), and Merrimack (12%). Estimated soil denitrification was the dominant 'sink' for N inputs to the watersheds, accounting for 34% of the total storage and loss terms, on average. Though this estimate is very rough, it is believable given the vast amount of N sources to the agricultural landscape available to be denitrified. This estimate is similar to the to the value of 40% reported by Kroeze et al. (submitted) for the Netherlands where inputs and, therefore, losses are higher than in the watersheds considered here.

Table 5. Storages and losses of N in watersheds* (kg N per km^2 per year)

Water-shed	N losses							changes in N storage			total N source inputs[5]	total N storages & losses	% inputs unresolved[6]
	NH$_3$ vol.[1]	Denit. in soils	Denit. of human waste	Denit. in river	Riverine N export[2]	Wood export	Food export	Forest wood storage[3]	Forest soil storage	Other soil storage[4]			
PEN	4	233	14	187	317	411	18	−177	−83	17	857	943	−10
KEN	10	324	17	173	333	417	17	−180	−76	102	1126	1137	−1
AND	13	297	31	178	404	265	10	68	47	119	1332	1431	−7
SAC	4	153	30	128	389	158	0	233	167	91	1237	1353	−9
MER	11	506	264	190	499	138	0	120	245	149	2240	2121	5
CHA	10	393	1029	270	1756	126	0	313	93	561	4415	4552	−3
BLA	19	872	511	245	1140	162	0	249	62	174	3426	3436	0
CON	24	734	120	294	538	160	0	282	97	123	2286	2371	−4
HUD	20	422	60	251	502	171	0	263	334	191	2005	2213	−10
MOH	56	1437	100	374	795	82	134	270	225	361	3610	3834	−6
DEL	37	953	158	461	961	111	0	588	99	208	3005	3576	−19
SCH	199	1879	542	547	1755	138	0	259	97	853	5917	6270	−6
SUS	142	1907	100	616	977	143	459	478	82	338	4774	5244	−10
POT	374	2247	116	538	897	117	633	457	42	484	5696	5904	−4
RAP	135	2279	45	199	470	223	291	313	24	370	4671	4349	7
JAM	96	1163	45	191	314	309	91	477	−94	176	2961	2768	7
Wt. Avg.	*108*	*1253*	*107*	*397*	*718*	*192*	*224*	*313*	*71*	*259*	*3420*	*3641*	*−6*

[1]NH$_3$ Volatilization of fertilizers and animal waste that is transported outside of the watershed boundary. [2]Includes 1112 kg km^{-1} yr^{-1} from wastewater export from the Charles watershed area that is not drained on the Charles River. [3]In living biomass, dead wood, and live plant tissues. [4]In agricultural and urban land. [5]From table 2. [6]Estimated by difference of inputs − outputs.

NH$_3$ volatilization losses
N is also lost from the landscape in agricultural lands via the volatilization of fertilizers and animal waste. NH$_3$ volatilization losses in Table 5, which refer to net transport of NH$_3$ outside the watershed, are among the smallest loss terms, accounting on average for 3% of all outputs. Gross NH$_3$ volatilization (not shown) is about four times higher, but about 75% of that is assumed to be re-deposited within the watershed boundaries, and therefore does not figure in the input/output budget (Boyer et al. 2002). NH$_3$ volatilization losses increased from less than 10 kg km^{-2} yr^{-1} in the northernmost watersheds to over 100 kg km^{-2} yr^{-1} in several watersheds in the mid-Atlantic region. This trend reflects animal waste production.

Changes in storage

Storage in biomass
Averaged across all watersheds, changes in storage in forests due to net increment of living wood biomass and net accumulation of N in dead wood and in green plant tissues accounted for 9% of the total N storage and losses. Sinks for N in wood were greatest on those watersheds with both high growth rates per unit forest area and high fractions of watershed area in forest cover. Net N uptake rates were high on most watersheds, as many of these forests are aggrading after past harvests or agricultural abandonment. Forests across the eastern U.S. were cleared for agriculture several centuries ago, and forest regrowth stores large quantities of N in accumulating wood (Goodale et al. 2002). With the abandonment of marginal farmlands in New England, wood stocks increased by 50% between 1952 and 1992 (Birdsey & Heath 1995). This recovering biomass represents a large but finite N sink in old-field forests, although forest harvest will continue to export substantial quantities of N (e.g. Hornbeck & Kropelin 1982; Johnson 1992).

Storage in soils
Even though forested land acreage greatly exceeds agricultural and sub/urban acreage, N storage in non-forested land ($\sim$7%) is only slightly less than that in forest land (10%). This is mainly due to the growing sub/urban area, the land use category that increased most, by $\sim$3500 km^2, for the whole study area (Figure 5). Much of the landscape that was once under continuous row cropping was converted to urban/suburban expansion, especially in the watersheds in Maine and New York and southward; these systems are predicted to have higher soil organic matter stocks as a result. In the New England region, the predominant change was the conversion of forest to urban/suburban land use. Relatively little data is available on soil organic matter levels in urban lands. However, the management and high level of inputs (i.e. fertilizer and

irrigation) in urban forests and grasslands (i.e. lawns, parks, golf courses) are conducive to the buildup of soil organic matter and soil N. Groffman et al. (1995) reported 30% higher soil organic matter in urban forests compared to rural forests having the same species and soil types. Changes in land use resulted in a net increase of soil N stocks of almost 97,000 Mg N yr^{-1} for all watersheds combined. The Potomac and Susquehanna watersheds accounted for about 60% of the net soil organic N increase, because of significant areas of soils under continuous row cropping which were converted to urban lands, forests, and pasture. There was a significant increase in use of reduced till and no-till over the 10 year period which should increase soil organic C and N stocks on the land remaining in agriculture. Further, conversion of agricultural soils, which are relatively depleted in organic matter compared to native forests and grasslands (Paustian et al. 1997b) into grasslands also increased soil N. On average, the estimated increases in soil organic N stocks in forested, agricultural, and urban lands due to changes in land use and in land management were 330 kg km^{-2} yr^{-1}, accounting for 9% of the total N storages and losses.

Comparison of total sources and total storages & losses

Total N storage & losses for each watershed vary from 943 to 6270 kg km^{-2} yr^{-1}, with an area-weighted mean of 3641 kg km^{-2} yr^{-1} (Table 5). The percentage of inputs not accounted for by our estimates of storage & losses vary among the watersheds from −19% to +7%. On average, the budget discrepancy was small (−6%). Sources (inputs) and storages & losses (outputs) are very well correlated: inputs = 0.96 * outputs; R^2 = 0.98. Figure 7 shows our 'best guesses' for the N inputs to the watersheds and for the fate of these N inputs via storage and loss pathways.

Losses by denitrification in landscape soils are our most uncertain estimates, because they were calculated by difference between total inputs to and outputs from each ecosystem, and therefore contain accumulated errors from other estimates. This loss term by difference, which we ascribed to denitrification in forest, urban, or agricultural soils, may also reflect the change in N storage in groundwater. However, we assume that groundwater aquifers, though enriched with N in some areas in the northeast, are not gaining significant new N over the period of interest since fertilizer use rates have been relatively stable in the U.S.A. over the past decade. In-river denitrification estimates are also very uncertain (see multiple estimates presented in Seitzinger et al. 2002).

The presence of appreciable changes in total N storage on land (18% of total storage & losses) indicates that there is a non-steady state condition, presumably associated with increasing N inputs from commercial imports

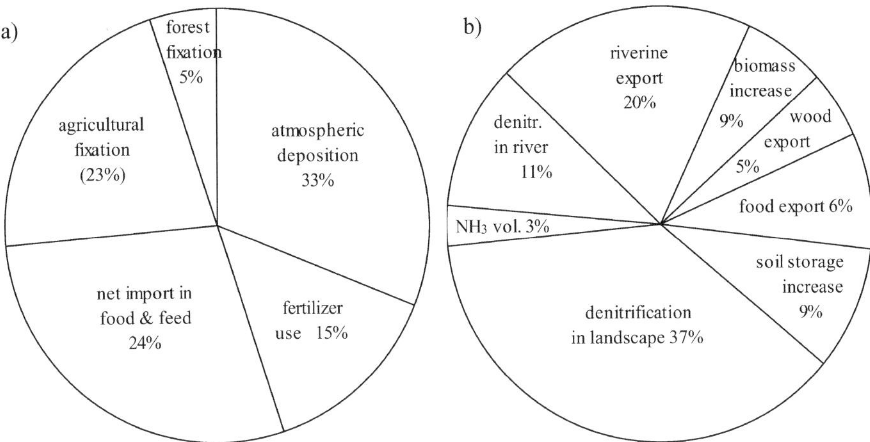

Figure 7. Best guess of (a) Nitrogen sources and (b) Nitrogen storages and losses. Values are the weighted average for the 16 watersheds.

and anthropogenically elevated atmospheric deposition. This in turn is ultimately due to the strongly increased industrial fixation of atmospheric N_2, and its application, mainly in agriculture. Increasing storage of N on land implies that drainage and denitrification exports of N are bound to increase further as storage terms reach a new steady state.

Understanding the sources and fate of N inputs to watersheds is necessary for mitigating N pollution problems in coastal and inland waters. Our estimates of N sources, storages, and losses are uncertain. Our ability to make these estimates is dependent on the availability a wide variety of statistical and spatial databases, and highlights the need for long-term monitoring. One element of uncertainty comes from the quality of these data themselves, and the methods used to scale information to the boundaries of our watersheds and to the timeframe of interest. For example, recent papers discuss challenges in estimating atmospheric N inputs (Meyers et al. 2001) and N loads in rivers (Brock 2001) based on incomplete and uncertain data from point monitoring networks. Another element of uncertainty comes from the empirical and process models from which we calculate storage and loss terms. Further research is needed to better understand the processes controlling N transport and transformations, and on how to best represent these processes in models that allow assessments at the scale of large regions.

Acknowledgements

This work was initiated as part of the International SCOPE Nitrogen Project, which received support from both the Mellon Foundation and from the National Center for Ecological Analysis and Synthesis. Thanks to Leo Hetling for his assistance with the estimates of N in wastewater. Thoughtful reviews by Doug Burns, Art Gold, and an anonymous reviewer substantially improved the manuscript.

References

Aber JD & Driscoll CT (1997) Effects of land use, climate variation and N deposition on N cycling and C storage in northern hardwood forests. Global Biogeochem. Cycles 11: 639–648

Aber JD, Ollinger SV & Driscoll CT (1997) Modeling nitrogen saturation in forest ecosystems in response to land use and atmospheric deposition. Ecol. Model. 101: 61–78

Alexander RB, Smith RA & Schwarz GE (2000) Effect of stream channel size on the delivery of nitrogen to the Gulf of Mexico. Nature 403: 758–761

Birdsey RA & Heath LS (1995) Carbon changes in U.S. forests. In: Joyce LA (Ed) Productivity of America's Forests and Climate Change, USDA Forest Service General Technical Report RM-271. United States Department of Agriculture

Bowden WB (1986) Gaseous nitrogen emissions from undisturbed terrestrial ecosystems: an assessment of their impacts on local and global nitrogen budgets. Biogeochem. 2: 249–279

Boyer EW, Goodale CL, Jaworski NA & Howarth RW (2002) Anthropogenic nitrogen sources and relationships to riverine nitrogen export in the northeastern U.S.A. Biogeochemistry 57/58: 137–169

Bricker SB, CG Clement, DE Pirhalla, SP Orland & DGG Farrow (1999) National Estuarine Eutrophication Assessment: A Summary of Conditions, Historical Trends, and Future Outlook. National Ocean Service, National Oceanic and Atmospheric Administration, Silver Springs, MD

Brock DA (2001) Uncertainties in individual estuary N-loading assessments. In: Valigura RA, Alexander RB, Castro MS, Meyers TP, Paerl HW, Stacey PE & Turner RE (Eds) Nitrogen Loading in Coastal Water Bodies: An Atmospheric Perspective (pp 171–185). American Geophysical Union, Washington, DC

Campbell JL, Hornbeck JW, McDowell WH, Buso DC, Shanley JB & Likens GE (2000) Dissolved organic nitrogen budgets for upland, forested ecosystems in New England. Biogeochem. 49: 123–142

Castro MS, Driscoll CT, Jordan TE, Ray WG, Boynton WR, Seitzinger SP, Styles RV & Cable JE (2001) Contribution of atmospheric deposition to the total nitrogen loads to thirty-four estuaries on the Atlantic and Gulf Coasts of the United States. In: Valigura RA, Alexander RB, Castro MS, Meyers TP, Paerl HW, Stacey PE & Turner RE (Eds) Nitrogen Loading in Coastal Water Bodies: An Atmospheric Perspective (pp 77–106). American Geophysical Union, Washington, DC

Dise NB, Matzner E & Gundersen P (1998) Synthesis of nitrogen pools and fluxes from European forest ecosystems. Water Air Soil Pollut. 105: 143–154

Galloway JN, Schlesinger HL, Michaels A & Schnoor JA (1995) Nitrogen fixation: atmospheric enhancement – environmental response. Global Biogeochemical Cycles 9: 235–252

Garrow JS, James WPT & Ralph A (2000) Human nutrition and dietetics, 0th ed. Churchill Livingstone, Edinburgh, 900 pp

Goodale CL, Aber JD & McDowell WH (2000) The effects of land-use history on organic and inorganic nitrogen losses in the White Mountains, NH. Ecosystems 3: 433–450

Goodale CJ, Lajtha K, Nadelhoffer KJ, Boyer EW & Jaworski NA (2002) Forest nitrogen sinks in large eastern US watersheds: estimates from forest inventory and an ecosystem model. Biogeochemistry 57/58: 239–266

Goss MJ, Colbourn P, Harris GL & Howse KR (1988) Leaching of nitrogen under autumn-sown crops and the effect of tillage. In: Jenkison DS & Smith KA (Eds) Nitrogen Efficiency in Agricultural Soils (pp 269–282). Elsevier Applied Science, London and New York

Groffman PM, Pouyat RV, McDonnell MJ, Pickett STA & Zipperer WC (1995) Carbon pools and trace gas fluxes in urban forest soils. In: Lal R, Kimble J, Levine E & Stewart BA (Eds) Soil Management and Greenhouse Effect. Advances in Soil Science (pp 147–158). CRC Press, Boca Raton

Groffman PM & Tiedje JM (1989) Denitrification in north temperate forest soils: Spatial and temporal patterns at the landscape and seasonal scales. Soil Biol. and Biochem. 21: 613–620

Hamburg SP (1984) Effects of forest growth on soil nitrogen and organic matter pools following release from subsistence agriculture. In: Stone EL (Ed) Forest soils and treatment impacts (pp 145–158). Univ. of Tennessee, Knoxville, TN

Hansen MH, Frieswyk T, Glover JF & Kelly JF (1992) The eastwide forest inventory database users manual. General Technical Report NC-151. St Paul, MN: North Central Forest Experimental Experiment Station, United States Department of Agriculture

Harmon ME, Harmon JM & Ferrell WK (1996) Modeling carbon stores in Oregon and Washington forest products: 1900–1992. Climate Change 33: 521–550

Howarth RW, Billen G, Swaney D, Townsend A, Jaworski N, Lajtha K, Downing JA, Elmgren R, Caraco N, Jordan T, Berendse F, Freney J, Kudeyarov V, Murdoch P & Zhu Zhao-Liang (1996) Regional nitrogen budgets and riverine N & P fluxes for the drainages to the North Atlantic Ocean: Natural and human influences. Biogeochemistry 35: 75–139

Hornbeck JS & Kropelin W (1982) Nutrient removal and leaching from a whole-tree harvest of northern hardwoods. J. Environ. Qual 11: 309–316

Johnson DW (1992) Nitrogen retention in forest soils. J. Environ. Qual. 21: 1–12

IPCC (1997) Vol 3: Revised 1996 IPCC Guidelines for National Greenhouse Gas Inventories. Reference Manual JT, Houghton JT, Meira Filho LG, Lim B, Treanton K, Mamaty I, Bonduki Y, Griggs DJ & Callander BA (Eds). Intergovernmental Panel on Climate Change

Jordan C, Milhalyfalvy E, Garret MK & Smith RV (1994) Modelling of nitrate leaching on a regional scale using a GIS. J. Environmental Management 42: 279–298

Kroeze C, Aerts R, van Breemen N, van Dam D, van der Hoek K, Hofschreuder P, Hoosbeek M, de Klein J, Kros H, van Oene H, Oenema O, Tietema A, van der Veeren R & de Vries W (submitted) Uncertainties in the fate of nitrogen. I: an overview of sources of uncertainty illustrated with a Dutch case study. Nutrient Cycling in Agroecosystems

Lajtha K, Seely B & Valiela I (1995) Retention and leaching losses of atmospherically-derived nitrogen in the aggrading coastal watershed of Waquoit Bay, MA. Biogeochemistry 28: 33–54

Lovett GM, Weathers KC & Sobczak WV (2000) Nitrogen saturation and retention in forested watersheds of the Catskill Mountains, New York. Ecological Applications 10: 73–84

Madramootoo CA, Wiyo KAW & Enright P (1995) Simulating tile drainage and nitrate leaching under a potato crop. Water Resources Bulletin 31: 463–473

Magesan GN, White RE & Scotter DR (1996) Nitrate leaching from a drained, sheep-grazed pasture. 1. Experimental results and environmental implications. Austr. J. Soil Res. 34: 55–67

Meybeck M, Chapman DV & Helmer R (1989) Global freshwater quality: a first assessment. World Health Organization/United Nations Environment Programme. Basil Blackwell, Inc., Cambridge, MA

Meyers T, Sickles J, Dennis R, Russell K, Galloway J & Church T (2001) Atmospheric nitrogen deposition to coastal estuaries and their watersheds. In: Valigura RA, Alexander RB, Castro MS, Meyers TP, Paerl HW, Stacey PE & Turner RE (Eds) Nitrogen Loading in Coastal Water Bodies: An Atmospheric Perspective (pp 53–76). American Geophysical Union, Washington, DC

Montgomery E, Coyne MS & Thomas GW (1997) Denitrification can cause variable NO_3- concentrations in shallow groundwater. Soil Science 162: 148–156

Mosier AR, Bleken MA, Chaiwanakupt P, Ellis EC, Freney JR, Howarth RB, Matson PA, Minami K, Naylor R, Weeks K & Zhao-liang Zhu (2001) Policy implications of human accelerated nitrogen cycling. Biogeochemistry 52: 281–320

Nusser SM & Goebel JJ (1997) The National Resources Inventory: a long-term monitoring programme. Environmental and Ecological Statistics 4: 181–204

Paustian K, Andren O, Janzen H, Lal R, Smith GT, Tiessen H, van Noordwijk M & Woomer P (1997a) Agricultural soil as a C sink to offset CO_2 emissions. Soil Use and Management 13: 230–244

Paustian K, Collins HP & Paul EA (1997b) Management controls on soil carbon. In: Paul EA, Paustian K, Elliott ET & Cole CV (Eds) Soil Organic Matter in Temperate Agroecosystems: Long-term Experiments in North America (pp 15–49). CRC Press, Boca Raton, FL, U.S.A.

Robertson GP & PM Vitousek (1981) Nitrification potentials in primary and secondary succession. Ecol. 62: 376–386

Scholefield D, Greenwood EA & Tichen NM (1988) The potential of management practices for reducing losses of nitrogen from grazed pastures. In: Jenkinson DS & Smith KA (Eds) Nitrogen Efficiency in Agricultural Soils (pp 62–72). Elsevier Applied Science, London and New York

Seely B, Lajtha K & Salvucci G (1998) The dynamics of N fluxes from canopy to ground water in a coastal forest ecosystem developed on sandy substrates. Biogeochemistry 42: 325–343

Seitzinger SP, Styles RV, Boyer EW, Alexander R, Billen G, Howarth RW, Mayer B & van Breemen N (2002) Nitrogen retention in rivers: model development and application to watersheds in the northeastern U.S.A. Biogeochemistry 57/58: 199–237

Shephard MA (1992) Effect of irrigation on nitrate leaching from sandy soil. Water and Irrigation Review 12: 19–22

Sollins P & McCorison FM (1981) Nitrogen and carbon solution chemistry of an old growth coniferous forest watershed before and after cutting. Water Resources Research 17: 1409–1418

U.S. Department of Commerce (1990) 1990 Census of Population: General population characteristics, United States. 1990-CP-1-1, Bureau of the Census. [online] URL: http://www.census.gov/main/www/cen1990.html

Valiela I, Collins G, Kremer J, Lajtha K, Geist M, Seely B, Brawley J & Sham JH (1997) Nitrogen loading from coastal watersheds to receiving estuaries: new method and application. Ecological Applications 7: 358–380

Velthof GL, Oenema O, Postma R & van Beusichem ML (1997) Effects of type and amount of applied fertilizer on nitrous oxide fluxes from intensively managed grassland, Nutrient Cycling Agroecosystems 46: 257–267

Vitousek PM, Aber JD, Howarth RW, Likens GE, Matson PA, Schindler DW, Schlesinger WH & Tilman DG (1997) Human alteration of the global nitrogen cycle: sources and consequences. Ecol. Appl. 7(3): 737–750

Watson CA, Fowler SM & Wilman D (1993) Soil inorganic-N and nitrate leaching on organic farms. The Journal of Agricultural Science 120: 361–369

Wyland LJ, Jackson LE, Chaney WE, Klonsky K, Koike ST & Kimple B (1996) Winter cover crops in a vegetable cropping system: Impacts on nitrate leaching, soil water, crop yield, pest and management costs. Agriculture, Ecosystems and Environment 59: 1–17

Zak DR, Grigal DF, Gleeson S & Tilman D (1990) Carbon and nitrogen cycling during old-field succession: constraints on plant and microbial biomass. Biogeochem 11: 111–129

Biogeochemistry **57/58**: 295–339, 2002.
© 2002 *Kluwer Academic Publishers. Printed in the Netherlands.*

A comparison of models for estimating the riverine export of nitrogen from large watersheds

RICHARD B. ALEXANDER[1,*], PENNY J. JOHNES[2], ELIZABETH W. BOYER[3] & RICHARD A. SMITH[1]
[1]*U.S. Geological Survey, 413 National Center, 12201 Sunrise Valley Drive, Reston, Virginia 20192, U.S.A.;* [2]*University of Reading, Aquatic Environments Research Centre, Department of Geography, Whiteknights Reading RG6 6AB, U.K.;* [3]*State University of New York, College of Environmental Science and Forestry, 1 Forestry Drive, Syracuse, NY 13210, U.S.A. (*author for correspondence, e-mail: ralex@usgs.gov)*

Key words: error analysis, model validation, nitrogen, watershed models

Abstract. We evaluated the accuracy of six watershed models of nitrogen export in streams ($kg\ km^2\ yr^{-1}$) developed for use in large watersheds and representing various empirical and quasi-empirical approaches described in the literature. These models differ in their methods of calibration and have varying levels of spatial resolution and process complexity, which potentially affect the accuracy (bias and precision) of the model predictions of nitrogen export and source contributions to export. Using stream monitoring data and detailed estimates of the natural and cultural sources of nitrogen for 16 watersheds in the northeastern United States (drainage sizes = 475 to 70,000 km^2), we assessed the accuracy of the model predictions of total nitrogen and nitrate-nitrogen export. The model validation included the use of an error modeling technique to identify biases caused by model deficiencies in quantifying nitrogen sources and biogeochemical processes affecting the transport of nitrogen in watersheds. Most models predicted stream nitrogen export to within 50% of the measured export in a majority of the watersheds. Prediction errors were negatively correlated with cultivated land area, indicating that the watershed models tended to over predict export in less agricultural and more forested watersheds and under predict in more agricultural basins. The magnitude of these biases differed appreciably among the models. Those models having more detailed descriptions of nitrogen sources, land and water attenuation of nitrogen, and water flow paths were found to have considerably lower bias and higher precision in their predictions of nitrogen export.

1. Introduction

Nitrogen inputs to terrestrial systems approximately doubled in the latter half of the 20th century, and riverine flux to coastal waters, where nitrogen is most limiting to primary production, increased by a similarly large factor (Vitousek et al. 1997). These increases have caused the eutrophication of many coastal and estuarine ecosystems worldwide, leading to chronic hypoxia, reductions in species abundance, and stressed fisheries resources. Although these problems are clearly related to cultural sources (Vitousek et al. 1997; Nixon et al. 1995), knowledge of the effects of specific nitrogen sources and watershed processes on riverine export is needed to better understand the likely consequences for coastal ecosystems of anticipated increases in N inputs related to population growth and economic development. The large increases that are expected to occur globally in multiple sources of nitrogen by early in the 21st century (i.e. a doubling of fertilizer and fossil-fuel N fixation; Galloway et al. 1994) underscore the need for improved understanding of nitrogen transport in large coastal watersheds. However, the increased complexity of sources and controlling processes in large watersheds potentially limits understanding at these scales. Nitrogen inputs to watersheds are removed at widely varying rates in streams and reservoirs and on the landscape through storage, denitrification, and interbasin transfers of agricultural products. Estimates of nitrogen removal in streams vary over two orders of magnitude (Seitzinger & Kroeze 1998; Alexander et al. 2000). The rates of nitrogen flux vary considerably in forested catchments (Johnson 1992), pastoral and agricultural catchments (Johnes 1996; Beaulac & Reckhow 1982), and in larger watersheds of mixed land use (Howarth et al. 1996; Caraco & Cole 1999; Seitzinger & Kroeze 1998). Knowledge of transport is also limited by the focus of most watershed studies on a relatively narrow range of environmental conditions in small watersheds.

Despite these difficulties, recent progress has been made in developing models of the mean-annual riverine yield or export (kg ha^{-1} yr^{-1}) from large watersheds. These models rely on relatively simple assumptions and descriptors of nitrogen sources and landscape characteristics. The models explain from 50 to 90 percent of the spatial variability in stream nitrogen export based on reported R^2 statistics. Nitrogen export varies by more than three orders of magnitude in major rivers of the world (Seitzinger & Kroeze 1998; Caraco & Cole 1999) and in individual countries such as the United States (Alexander et al. 2001). However, R^2, although frequently used as a measure of model performance, does not reliably describe the accuracy (bias and precision) of predictions of stream nitrogen export. R^2 is sensitive to various statistical properties of the explanatory and response variables (Montgomery & Peck 1982), making comparisons of model performance unreliable. The models also differ in their levels of spatial resolution and process complexity, which may affect the accuracy of model predictions of nitrogen export, processing rates, and source contributions to streams. Recent global (Seitzinger & Kroeze 1998; Caraco & Cole 1999) and regional (Howarth et al. 1996) models suggest that variations in export can largely be explained as a function of nutrient sources despite the large variability that occurs in nitrogen controlling processes in watersheds. By contrast, recent statistical (Smith et al. 1997; Preston & Brakebill 1999; Alexander et al. 2000, 2001) and export-coefficient models (Johnes 1996; Johnes & Heathwaite 1997) indicate that knowledge of spatial variations in watershed properties that influence nitrogen processing (e.g. surface water flow paths, soils, climate) can significantly improve the accuracy of estimates of stream export and source contributions over a broad range of watershed scales.

These export models rarely have been compared in systematic manner over a range of climatic conditions, sources, and watershed sizes. Such comparisons are needed to improve understanding of how model accuracy varies in different environmental settings and with the complexity of model descriptions of N sources and biogeochemical processes that affect nitrogen transport. This analysis provides an initial comparison of the performance of several prominent models for estimating riverine export of nitrogen. We begin by applying each of the selected models to 16 watersheds in the northeastern United States (see Figure 1). These watersheds have climatic conditions and nitrogen sources that lie within the range of watershed conditions used to calibrate the original models, and therefore, provide an appropriate collection of environmental settings for evaluating the models. The northeastern watersheds are the focus of SCOPE (Scientific Committee On Problems of the Environment) investigations of nitrogen cycling (e.g. see Boyer et al. 2002 and Van Breemen et al. 2002) for which detailed estimates of natural and cultural sources are available. This study is complementary to previous SCOPE investigations of N transport in watersheds (Howarth et al. 1996) and the SCOPE analyses in this volume that use certain of the models. The study also builds on previous studies that have evaluated a more limited number of models (Seitzinger & Kroeze 1998; Caraco & Cole 1999; Stacy et al. 2001; Alexander et al. 2001; National Research Council 2000).

The analysis is presented in eight sections. Following the introduction, section two summarizes the range of approaches that have been used to model nutrient export from large watersheds, noting their principle features and assumptions. Section three presents the methods and data used to evaluate the selected nutrient export models. We describe methods for quantifying the accuracy (bias, variability) of the model predictions, including the use of an error modeling technique to identify prediction biases caused by deficiencies in the export model descriptions of nitrogen sources and processes in watersheds. Section four presents the results of the error analysis. A discussion of the error analysis is given in section five. Model predictions of source contributions to stream export are compared in section six. Section seven

Figure 1. Location of 16 watersheds in the northeastern United States.

describes the model estimates of watershed attenuation of N, and the final section presents the summary and conclusions.

2. Background

A variety of deterministic, statistical, and hybrid methods have been used to model the transport of nitrogen in rivers basins. We provide a brief review of various models that have been applied to large watersheds, frequently several thousands of square kilometers in size or larger. We selected a subset of these models for use in this analysis (see section 3.1).

The simplest deterministic approaches (Howarth et al. 1996; Jaworski et al. 1992) are mass balance models that provide a static accounting of nitrogen inputs (e.g. fertilizer application, atmospheric deposition) and outputs (e.g. river export, crop N fixation and removal). Where major sources or sinks are difficult to measure independently (e.g. groundwater storage, denitrification), estimates are often obtained as a difference of measured terms. Recent refinements (Jordan & Weller 1996) have been made in budget methods to account for watershed imports and exports of nitrogen in food and feed. Additional sources and sinks have been quantified for the northeastern watersheds by several studies reported in this volume (e.g. Van Breemen et al. 2002; Boyer et al. 2002; Seitzinger et al. 2002) including N fixation and uptake in forests and N attenuation in streams and reservoirs. Selected results from these studies are used in this investigation. Where source inputs and sinks cannot be spatially referenced, budget terms are assumed to be uniformly distributed within watersheds with loss processes operating equally on all sources. In the absence of source-specific flux rates, estimates of the relative contributions of sources to surface waters must also be assumed to be proportional to the nitrogen inputs to watersheds. In extending these methods to large watersheds, uncertainties may exist over the rates of N supply and loss, which are based on the extension of literature estimates from experimental studies.

More complex deterministic models of nitrogen flux (e.g. HSPF, Bicknell et al. 1997; SWAT, Srinivasan et al. 1993; INCA, Whitehead et al. 1998; AGNPS, Young et al. 1995) describe transport and loss processes in detail by simulating nitrogen availability, transport, and attenuation processes according to mechanistic functions that include descriptions of the spatial and temporal variations in sources and sinks in watersheds. Simulation models of landscape processing of nitrogen, such as the biochemical dynamics of nitrogen in soils (e.g. CENTURY, Parton et al. 1988), have also been developed. The complexity of deterministic models often creates intensive data and calibration requirements, which generally limits their application in large watersheds; these models have been more commonly applied in small catchments. One deterministic surface-water nutrient model (SWAT, Srinivasan et al. 1993) has been recently applied in the watersheds of major regions of the United States (see Alexander et al. 2001), although this agricultural model does not account for all sources of nitrogen (e.g. atmosphere). In general, there are uncertainties involved in aggregating the components of fine-scale deterministic models in watershed applications (Rastetter et al. 1992) and in extrapolating the results of catchment models and field-scale measurements to larger spatial scales. Deterministic methods also often lack robust measures of uncertainty in model coefficients and predictions, which might otherwise be used to assess their accuracy in modeling N transport.

300

One common approach to estimating stream export (e.g. Delwiche & Haith 1983; Fisher & Oppenheimer 1991) has been to apply the reported yields (mass of N per unit drainage area) from small, homogeneous watersheds to the variety of land types contained within larger heterogeneous basins ('export coefficient' method). Because the nitrogen yields for given land types are highly variable (Beaulac & Reckhow 1982; Frink 1991; Johnson 1992; Johnes 1996), reflecting variations in climatic conditions, nutrient sources, and terrestrial and aquatic loss processes, these methods can produce imprecise and potentially biased estimates of export when extrapolated to other areas and larger scales (Beaulac & Reckhow 1982). Refinement of the export coefficient method in the United Kingdom has produced more robust models capable of accurately simulating the nutrient export response to temporal changes in nutrient source inputs and land and waste management practices at the watershed and regional scales (Johnes 1996; Johnes et al. 1996; Johnes & Heathwaite 1997). The model calculates the total N and P load delivered to a waterbody separately by source type as a function of the rates of nutrient input and the export potential of a watershed. The export potential is estimated according to the location of sources in watersheds and the landscape and climatic conditions. The export coefficients are expressed as a percentage of nutrient inputs (rather than mass per unit area as in conventional export methods), allowing the simulation of the effect of historical land-use changes and management in watersheds. One limitation of this model is that detailed information is required on the export potential of landscapes and the types and location of nutrient sources. Nevertheless, in the U.K., where source and monitoring information are available from the 1860s to date, the model has been extensively validated and applied nationally to all watersheds in England and Wales. The model accurately allows the scaling up of plot-scale experimental measurements to the watershed and regional scales, explaining up to 98% of the spatial variations in stream export.

Other watershed models have been developed that represent a mixture of deterministic and export-coefficient approaches. A recent model of watershed export combined the deterministic budget approach with literature-based export coefficients for different land types and source inputs (Castro et al. 2001). This model produced mixed results in applications to the drainages of 34 major estuaries of the United States (Castro et al. 2001; Stacy et al. 2001) with the method tending to overestimate the riverine export from agricultural watersheds. Other types of models described as 'loading functions' (GWLF; Haith & Shoemaker 1987) represent a compromise between the export coefficient method and the complexity offered by simulation models. These models have mechanistic water and sediment transport components (Howarth et al. 1991) with nutrient dynamics often described by simple empirical relations

(Haith & Shoemaker 1987). Model parameters may be obtained from the literature or statistically estimated if sufficient data are available. Applications have been made to eastern U.S. watersheds as large as several thousands of square kilometers (e.g. Lee et al. 2000; Howarth et al. 1991).

Statistical approaches to modeling nitrogen flux in coastal basins have their origins in simple correlations of stream nitrogen measurements with watershed sources and landscape properties. These methods assume limited *a priori* knowledge of biogeochemical processes, but provide empirical estimates of the aggregate supply and loss of nitrogen through the use of conventional stream monitoring data, which are often readily available in large watersheds with mixed land use. Anthropogenic N sources constitute the principle predictor variables in these models. Some of the models (e.g. Howarth et al. 1996) make use of literature rates of N processing (crop N fixation, N removal in crops) to estimate agricultural source inputs. Recent examples include regressions of nitrogen export from large watersheds on population density (Peierls et al. 1991), net anthropogenic sources (Howarth et al. 1996), atmospheric deposition (Jaworski et al. 1997; Howarth et al. 1996), and measures of per capita energy consumption by humans (Meybeck 1982). These models explain up to 80 percent or more of the spatial variations in nitrogen export. In contrast to complex deterministic models, these statistical methods have the advantage of being readily applied in large watersheds. Moreover, statistical approaches are capable of quantifying errors in model parameters and predictions. These simple correlative models are limited, however, in that they consider sources and sinks to be homogeneously distributed in space, do not separate terrestrial from in-stream loss processes, and rarely account for nonlinear interactions between sources and loss processes.

A recently developed hybrid approach (SPARROW; SPAtially Referenced Regression On Watershed attributes; Smith et al. 1997) expands on conventional regression methods by using a mechanistic model structure in correlating measured nitrogen flux in streams with spatial data on nitrogen sources, landscape characteristics (e.g. soil permeability, temperature), and stream properties (e.g. streamflow, water time of travel). The model, which separately estimates the quantities of nitrogen delivered to streams and the outlets of watersheds from point and diffuse sources, has been applied nationally in the United States (Smith et al. 1997) with separate studies of nitrogen flux in the Chesapeake Bay watershed (Preston & Brakebill 1999), the Mississippi River and its tributaries (Alexander et al. 2000), the watersheds of major U.S. estuaries (Alexander et al. 2001), and watersheds of New Zealand (McBride et al. 2000). By spatially referencing nitrogen sources and watershed attributes to surface water flow paths, defined according to a digital drainage network, and imposing a mass-balance constraints, the model has been shown to improve the accuracy of predictions of stream export and the interpretability of model coefficients (Smith et al. 1997; Alexander et al. 2000, 2001). Model estimates of in-stream nitrogen loss and stream nutrient export from watersheds of various land-use types are generally consistent with literature rates (Alexander et al. 2000). By comparison to the simple correlative approaches, this method requires more spatially detailed data on watershed characteristics, such as river drainage attributes (e.g. surface-water flow paths) and point and diffuse nitrogen sources.

Quasi-empirical models (Seitzinger & Kroeze 1998; Caraco & Cole 1999) of nitrate-nitrogen export from the largest rivers of the world were recently developed using empirical regression methods and literature-based rate coefficients. These models were developed for estimating N budgets at the continental scale and evaluating the effects of cultural sources on N export in some of the largest river basins of the world. These models indicate that the large variations in nitrate export among rivers worldwide can be largely explained by several major nitrogen sources and relatively simple descriptors of nitrogen removal on the landscape and in rivers. Nitrate export is modeled as a function of point sources (i.e. urban population and an assumed per capita discharge rate of 1.85 kg-N year^{-1}), the diffuse inputs of fertilizer and atmospheric deposition, statistically calibrated runoff (discharge per unit drainage area) coefficients, and a literature-based in-stream loss coefficient of 30%. Model predictions of export were highly correlated (r-squared > 80%) with measured export for 35 of the largest rivers of the world. Some of the principle outliers in the models were associated with watersheds containing reservoirs (Caraco & Cole 1999), which were not explicitly included in the models. The stronger correlations of nitrate with population density than observed for total dissolved nitrogen has been suggested as an indication that nitrate measurements more readily display the effects of anthropogenic activities on river export in these rivers (Caraco & Cole 1999). However, the accuracy of these global models is not well known for smaller watersheds and for rivers with higher ammonium and organic nitrogen loads.

3. Methods

3.1. *Export models*

We compared the performance of six empirical and quasi-empirical nitrogen watershed models (see Table 1) in 16 watersheds of the northeastern United States (see Figure 1). Watershed models classified as strictly deterministic (e.g. SWAT, HSPF) were not evaluated in this analysis (see comparisons in Alexander et al. 2001). Table 1 presents details of the model equations, the types of data required by each model, and characteristics of the calibration data. The data used in applying the SPARROW model are given in Alexander et al. (2000, 2001). Two of the models (SPARROW, HOWARTH) predict mean-annual total nitrogen export in streams. The remaining models predict mean-annual nitrate-N export; method performance is only evaluated for nitrate-N for these models. Two previously unpublished statistical models were estimated by applying ordinary least squares (LS) to data from the largest rivers of the world separately for 29 rivers (LS1-GLOBAL; Seitzinger & Kroeze 1998) and for 35 rivers (LS2-GLOBAL; Caraco & Cole 1999). These models offer several potential advantages. First, rather than assuming that the diffuse sources (i.e. fertilizer and atmospheric deposition) are supplied and transported at identical rates as in the quasi-empirical global models based on these data (Seitzinger & Kroeze 1998; Caraco & Cole 1999), the statistical models estimate separate rate coefficients for the diffuse sources, allowing any differences in the rates of supply and processing to be empirically determined. Second, the models allow the uncertainty of the model coefficients to be empirically determined and used to evaluate model fit. Finally, an intercept term can be evaluated in the models to determine whether any of the remaining variability in nitrogen export is potentially explained by a constant source. This source may potentially represent natural or background sources of nitrogen that are unexplained by the major cultural nitrogen sources explicitly specified by the models. Results for these two statistical models are described in section 4.1.

3.2. *Watershed characteristics*

Selected characteristics for the northeastern watersheds are presented in Table 2, including estimates of total nitrogen and nitrate-N stream export, drainage area, runoff (discharge per unit of drainage area), and land use. The data are compiled for selected years during the 1988 to 1993 time period (see Boyer et al. 2002, for details). A mixture of land use is represented in the drainages, but the watersheds are predominantly forested (range of 48 to 87%), with small to moderate amounts of cultivated land area (range from 2 to 40%). Developed land ranges from less than one percent to about 20% of the basin drainage area, but is commonly less than about 3%. Wetlands typically cover less than about 3% of the basin area. Population density ranges over nearly two orders of magnitude (Boyer et al. 2002). The drainage areas of the watersheds span more than two orders of magnitude from 475 to 70,000 km^2. Stream nitrogen export and runoff of the watersheds typically span about a factor of two to five. Nitrate nitrogen generally represents less than half of the total nitrogen in streams in most of the watersheds (median = 43%; interquartile range = 30 to 53%) although nitrate is the predominant fraction

Table 1. Stream nitrogen export models of total nitrogen (TN) and nitrate (NO_3)

Model Name	Reference	Export (kg km^2 yr^{-1})	Calibration Data Set	Export Equation	R^2
SPARROW	Alexander et al. (2000, 2001); Smith et al. (1997)	TN	374 sites in conterminous United States	$TN_i = \{\sum_{n=1}^{N} \sum_{j \in J(i)} S_{n,j} \beta_n \exp(-\alpha' Z_j) \exp(-k' T_{i,j})\} \varepsilon_i A_i^{-1}$ j is the reach belonging to the set of reaches (J(i)) located upstream of the downstream monitoring site in reach i, for which the export prediction is made $S_{n,j}$ = mass from source n (belonging to a total of N sources) input to the drainage of reach j β_n = source-specific coefficient (fertilizer use, livestock wastes, wet atmospheric deposition, nonagricultural land (diffuse runoff), industrial-municipal point sources)[a] $\exp(-\alpha' Z_j)$ = exponential function[b] giving the proportion of available nitrogen mass delivered to reach j as a function of land-to-water loss coefficients (defined by vector α) and associated landscepe characteristics (soil permeability-cm hr^{-1}; drainage density-km^{-1}; temperature-^{0}F.), Z_j, in the drainage to reach j $\exp(-k' T_{i,j})$ is the proportion of nitrogen mass in reach j transported to downstream reach i as a function of a first-order rate of nitrogen loss (k'), defined according to a vector of four discrete classes of channel size in units of reciprocal water travel time ($T_{i,j}$; days) ε is a multiplicative error term A_i is the drainage area of the basin (km^2)	0.88
HOWARTH	Howarth et al. (1996)	TN	9 regions (Canada, U.K., western Europe, eastern U.S.)	$TN = -120 + 0.79\ NO_y + 0.11\ NAI$ NO_y = atmospheric deposition (wet and dry oxidized forms – NO_3, HNO_3; kg km^{-2} yr^{-1}) NAI = net anthropogenic inputs (fertilizer + crop N fixation + food/feed imports – food exports; kg km^{-2} yr^{-1})	0.89

PEIERLS	Peierls et al. (1991)	NO_3	34 global large rivers	$NO_3 = 10^{1.51(0.64\log(PD))}$ PD = population density (people km^{-2})	0.51
GLOBAL	Seitzinger & Kroeze, (1998); Caraco & Cole, (1999)	NO_3	35 largest rivers in the world (only runoff coefficients calibrated)	$NO_3 = R_{export} (PI + WS_{export} (F + NO_y))$ $R_{export} = 0.7$; the average in-stream transport based on literature studies (fraction of N inputs to streams) PI = point source inputs (kg km^{-2} yr^{-1}); product of population density (people km^{-2}), fraction of population in urban areas, and a per capita N release of 1.85 kg N yr^{-1} WS_{export} = N loaded to streams (kg km^{-2} yr^{-1}); modeled as a function of runoff (R; discharge per unit drainage area; m yr^{-1}), where $WS_{export} = 0.4 R^{0.8}$ F = fertilizer use (kg km^{-2} yr^{-1})	0.84–0.89[c]
LS1-GLOBAL	Model fit with data from Caraco & Cole (1999)	NO_3	35 largest rivers in the world	$NO_3 = 2.81 PI' + \exp(-0.2785/R) (0.0816 F + 0.2969 NO_y)$ PI' = population density (people km^{-2}) in urban areas (fraction)	0.90
LS2-GLOBAL	Model fit with data from Seitzinger & Kroeze (1998)	NO_3	29 largest rivers in the world	$NO_3 = 0.34 PI' + \exp(-0.2960/R) (0.2316 F + 0.3762 NO_y)$	0.83

[a]All units are kg yr^{-1} except nonagricultural land area (ha). The land-to-water delivery function is equal to one for point-source inputs. Atmospheric deposition contributions to stream export are based on wet-fall deposition; land-to-water delivery fractions exceed unity indicating that additional atmospheric forms of nitrogen (e.g. ammonium, organic) are included (see Alexander et al. 2001).

[b]The product of the land-to-water delivery function (and its associated coefficients) and the nonpoint-source coefficients quantifies the fraction of the diffuse source inputs delivered to rivers.

[c]R^2 from correlation of predicted stream nitrogen export with measured stream export; only the runoff coefficients are statistically calibrated in the model.

in many of the more developed watersheds. The lowest nitrate contributions to total stream N export are found in the northern watersheds, where nitrate represents less than a third of total export. In the largest rivers of the world (Caraco & Cole 1999), the proportions of organic N and nitrate have been found to be roughly equivalent. Inputs of nitrogen sources, including fertilizer, atmospheric deposition (NO_y, the total wet and dry oxidized components of deposition – NO_3, HNO_3), and net anthropogenic sources, typically vary over a factor of two to three based on the interquartile range of the distribution, with the most extreme inputs of fertilizer and net anthropogenic nitrogen differing by a factor of about 10 (Boyer et al. 2002). NO_y-N is estimated to be about 65% of the total deposition in many of these watersheds (Boyer et al. 2002).

The cultural sources and climatic conditions of the northeastern watersheds lie within the range of conditions used to calibrate the original stream export models, and thus, the watersheds provide an appropriate set of locations for evaluating the models. Figure 2 compares conditions in the northeastern watersheds (i.e. NE US) with those in the calibration watersheds for selected explanatory variables of the models (i.e. runoff, population density, fertilizer use) and for stream nitrogen export (total nitrogen, nitrate-nitrogen). The range of the original data used to calibrate the various stream export models is inclusive of the full range of characteristics of the northeastern watersheds. Appreciable similarities exist in the watersheds given the considerable overlap in the interquartile ranges of many of the basin conditions (Figure 2). The drainage sizes of the northeastern watersheds are generally at the lower end of the range of watershed sizes used in the calibrations of most of the stream export models. Models that were applied to the largest rivers of the world (Peierls et al. 1991; Seitzinger & Kroeze 1998; Caraco & Cole 1999) used watersheds frequently larger than 0.2 million km^2; however, the calibration data include smaller river basins, such as the Susquehanna, Delaware, and Hudson, which are included in the set of 16 northeastern watersheds. The SPARROW model calibration used 374 U.S. watersheds ranging in size from about 10 to 2.9 million km^2 with an interquartile range of about 3,000 to 37,000 km^2 (approximately 10% of the watersheds are located in the northeastern United States, and include nine of the 16 northeastern watersheds). The HOWARTH model was calibrated for nine large regional watersheds in northern Europe and the eastern half of the United States and Canada, ranging in size from 0.3 to 3.2 million km^2.

3.3. *Error analysis*

We evaluated the performance of the models through an analysis of the errors in the predictions of stream nitrogen export. Prediction errors ($E_{i,k}$)

Table 2. Nitrogen export and watershed characteristics for the sites in the northeastern United States [from Boyer et al. 2002; TN = total nitrogen]

River Name	Drainage Area (km^2)	TN Export[a] (1988–1993) (kg km^2 yr^{-1})	NO$_3$ Export[a] (1988–1993) (kg km^2 yr^{-1})	Ratio NO$_3$/TN Export	Runoff (m yr^{-1})	Developed Land (%)	Cultivated Land (%)	Forested Land (%)	Other Land[b] (%)
Penobscot	20,109	317	66	0.21	0.59	0.4	1.5	83.8	14.4
Kennebec	13,994	333	87	0.26	0.57	0.9	5.9	79.6	13.6
Androscoggin	8,451	404	112	0.22	0.64	1.1	4.8	84.6	9.5
Saco	3,349	389	81	0.21	0.67	0.8	3.6	87.4	8.2
Merrimack	12,005	499	155	0.31	0.59	8.7	7.7	74.7	8.9
Charles	475	644	335	0.53	0.58	22.2	8.3	59.3	10.2
Blackstone	1,077	1,140	496	0.43	0.65	17.6	8.0	63.3	11.1
Connecticut	25,019	538	233	0.43	0.64	4.0	9.0	79.0	8.0
Hudson	11,942	428	222	0.53	0.62	2.7	10.4	80.8	6.0
Mohawk	8,935	826	427	0.53	0.55	4.7	28.0	63.1	4.2
Delaware	17,560	961	620	0.63	0.55	3.3	16.7	74.7	4.4
Schuylkill	4,903	1,755	1,419	0.83	0.49	10.2	38.4	48.1	3.3
Susquehanna	70,189	977	742	0.77	0.49	2.4	28.5	66.7	2.4
Potomac	29,940	897	392	0.43	0.33	2.6	34.6	60.8	2.0
Rappahannock	4,134	470	231	0.50	0.36	1.4	35.9	61.3	1.3
James	16,206	314	107	0.35	0.41	1.4	15.6	80.6	2.4
25th percentile	*4,711*	*400*	*111*	*0.30*	*0.49*	*1.3*	*7.3*	*61.2*	*3.0*
Median	*11,974*	*518*	*232*	*0.43*	*0.57*	*2.7*	*9.7*	*74.7*	*7.0*
75th percentile	*18,197*	*913*	*445*	*0.53*	*0.63*	*5.7*	*28.1*	*80.7*	*10.0*

[a] Mean-annual export.
[b] Includes open water, barren land, shrub land, and wetlands.

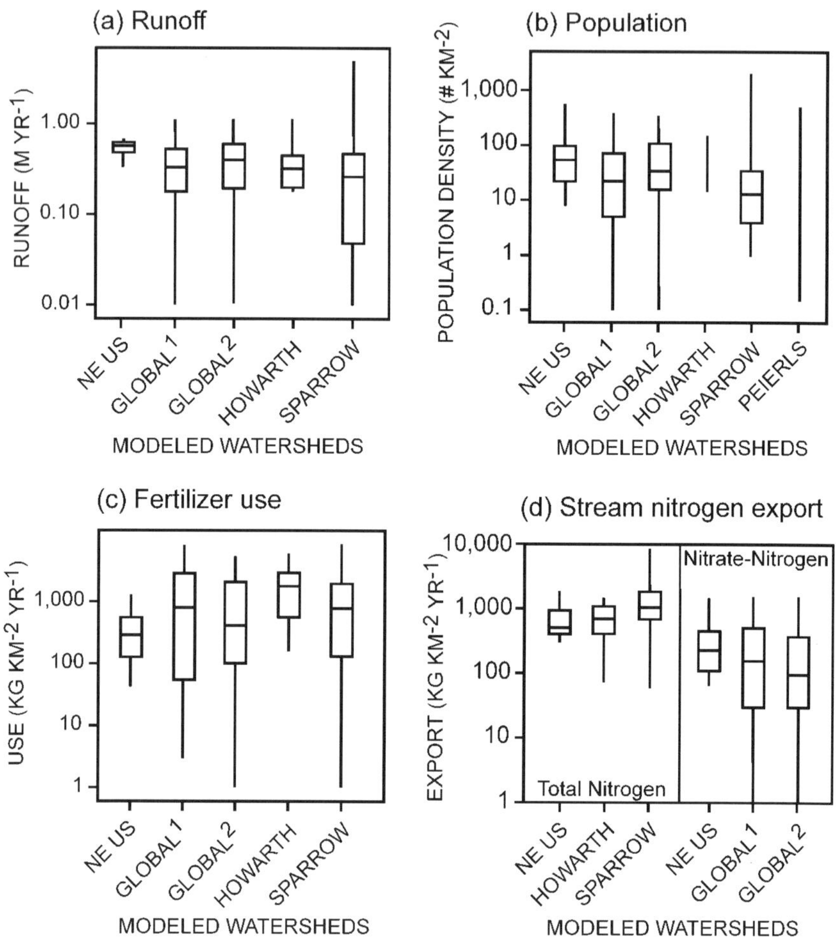

Figure 2. Box and whisker plots of the characteristics of the northeastern U.S. watersheds (NE US) in comparison to those of watersheds used to calibrate the original stream nitrogen export models: (a) runoff (discharge per unit drainage area), (b) population density, (c) fertilizer use, and (d) nitrogen export. Each box graphs the quartiles with the lower and upper edges representing the 25th and 75th percentiles, respectively. The midline plots the median. The upper and lower whiskers are drawn to the minimum and maximum values. The vertical lines in (b) give the range of the data. The models GLOBAL[1] (Caraco & Cole 1999) and GLOBAL[2] (Seitzinger & Kroeze 1998) were developed from data for the largest rivers of the world. The models are described in Table 1. Minimum values of zero are shown as 0.1 in (b) for the two global models and as 1.0 in (c) for the SPARROW and GLOBAL[2] models.

are computed for the ith watershed and kth model as the difference between the predicted ($P_{i,k}$) and 'measured' ($M_{i,k}$) stream nitrogen export, expressed as a percentage of the measured export, according to

$$E_{i,k} = \left[\frac{P_{i,k} - M_{i,k}}{M_{i,k}} \right] \cdot 100 \tag{1}$$

Errors with a positive sign indicate an over prediction of the measured export, whereas a negative sign indicates an under prediction of the measured stream export. The measured stream nitrogen export is based on the use of a rating curve technique (see Boyer et al. 2002; Cohn et al. 1989) to estimate the mean-annual nitrogen export for the period 1988 to 1993. The technique correlates instantaneous measurements of nitrogen flux (product of concentration and stream flow) with corresponding daily values of streamflow. By integrating the relation over the full record of daily flow, this method provides more precise estimates of stream export than can be obtained from a simple averaging of the instantaneous nitrogen export measurements. The standard error of the mean- annual export estimates for the 16 sites in this study was typically 5 to 10% of the mean. This magnitude of error is considered in forming conclusions about the comparative accuracy of the stream export models in Table 1.

The error analysis consists of two evaluations of the distribution of prediction errors for the 16 watersheds and six export models. First, we quantified the bias and variability of the model predictions of stream export according to robust statistical measures. The bias for each of the models was defined as the median of the prediction errors for the 16 watersheds from equation (1). The symmetry of the 25th and 75th percentile values of error about zero and the cumulative frequencies of the errors were also used as approximate indicators of the presence of bias in the prediction errors. The variability of the prediction errors was defined for each of the models as the interquartile range (IQR; i.e. difference between the 75th and 25th percentiles) of the prediction errors from equation (1). The model precision was defined as the reciprocal of the interquartile range. For selected models for which the original calibration data were available, we also quantified the bias and variability of the models for the original calibration set of watersheds, and compared these estimates with those observed for the 16 northeastern watersheds (see section 4.1).

Second, using multiple regression, we correlated errors in the predictions of stream nitrogen export with watershed characteristics that are associated with nitrogen sources and reflect terrestrial and aquatic processes affecting N mobilization and transport. The regression relation identifies 'factor-related' prediction biases that indicate potential deficiencies in the model descriptions of the supply and processing of nitrogen in the northeastern watersheds. Such deficiencies in model specification can be caused by watershed properties (sources or biogeochemical processes) that are not explicitly included in the export models or by model coefficients that poorly describe how N is supplied, mobilized, or transported in the northeastern watersheds. Either case is of interest and serves to highlight potential deficiencies in model performance. We fit separate regression models to the prediction errors associated with each of the six nitrogen export models by regressing the errors ($E_{i,k}$) from equation (1) on four characteristics of the watersheds that are measured with relatively high accuracy, according to the relation

$$E_{i,k} = \beta_{0,k} + \beta_{1,k} C_i + \beta_{2,k} D_i + \beta_{3,k} R_i + \beta_{4,k} BA_i + \varepsilon_{i,k} \tag{2}$$

where C_i is the cultivated land area (percent of total basin area) for the ith watershed, D_i is the developed (i.e. urban) land area (percent of total basin area), R_i is runoff (discharge per unit of drainage area; cm yr^{-1}), BA_i is the basin drainage area (km^2), $\beta_{o,k} \ldots, \beta_{4,k}$ are the regression parameters, and $\varepsilon_{i,k}$ describes the regression model error, assumed to be independent across watersheds. Cultivated and developed land area serve as indicators for a variety of nitrogen sources (e.g. fertilizer application, N fixation, urban washoff, and wastewater discharges) and landscape processes (e.g. N storage in forest biomass, crop harvesting, impervious cover) associated with these land uses. Evidence of a correlation between forestland and related processes in the error models would be expressed as a negative correlation with the cultivated and developed land area. Runoff and drainage basin area are potentially related to various biogeochemical processes that influence the mobilization and transport of nitrogen in terrestrial and aquatic environments. These may include biochemical reactions involving denitrification that are affected by water travel times through the subsurface and in channels. Runoff may reflect the effects on nitrogen export of such factors as climate (i.e. precipitation, evaporation), soil texture and moisture content, subsurface flow paths, stream properties (e.g. channel density and morphology, water velocity and flow), and water storage (e.g. lakes, reservoirs, groundwater). Climate-related effects on nitrogen flux may also reflect differences in vegetation that affect the natural supply of nitrogen to watersheds, although the land-use factors would be expected to reflect most major N sources. Drainage basin area provides a measure of the effects of spatial scale on nitrogen transport in watersheds, including scale-dependent properties of the rates of water flow, storage, and loss (e.g. length of surface and subsurface flow paths, water velocity, evaporation).

4. Results of the error analysis

We present the error analysis in three subsections. In section 4.1, we report the accuracy (bias, variability) of five of the nitrogen export models based exclusively on the original calibration data sets. Section 4.2 describes the bias and variability of the predictions of N export based on the application of the stream export models to the 16 northeastern watersheds and the calculation of prediction errors in equation (1). Section 4.3 gives the results of the regression-based models of the prediction errors (equation 2) associated with the application of each stream export model to the northeastern watersheds.

4.1. *Accuracy of the original model calibrations*

Table 3 gives the model parameters for the global models, LS1-GLOBAL, LS2-GLOBAL, based on a statistical fit to the two global data sets (Table 1). The overall fit of the models as measured by the R^2 values is similar or slightly improved in comparison to that of the original global models (Table 1). The model coefficients display weak ($p = 0.13$ to 0.19) to high ($p < 0.03$) levels of statistical significance indicating a range of uncertainty in the coefficient estimates. Point sources and atmospheric deposition are highly significant ($p < 0.03$) in the LS1-GLOBAL model, whereas fertilizer and atmospheric deposition sources are highly significant ($p < 0.02$) in the LS2-GLOBAL model. The mean coefficient values for runoff and atmospheric deposition are similar for the two models. By contrast, coefficient values for point sources and fertilizer differ greatly for the two models. Both models show consistently larger coefficient values for atmospheric deposition than for fertilizer suggesting a greater delivery of atmospheric N than fertilizer-related sources in the watersheds. According to the models, atmospheric inputs are delivered to watershed outlets from 1.5 (LS2-GLOBAL) to 3.5 (LS1-GLOBAL) times the rate of fertilizer related source inputs. A previous export model for North Atlantic watersheds (Howarth et al. 1996; Howarth 1998; NRC 2000) also found higher rates of delivery of atmospheric sources relative to other anthropogenic sources. Insignificant intercept terms ($p > 0.6$) were estimated for both models in preliminary regressions suggesting that virtually all of the nitrogen sources are accounted for by the three sources (i.e. point, fertilizer, and atmospheric) specified in the models. Additional sources of nitrogen that are accounted for by an intercept term represent apparently negligible quantities of nitrogen (e.g. <8% for the LS1-GLOBAL model). An intercept was not included in the final models because the standard errors of some of the source coefficients were inflated by its inclusion.

The prediction errors, based on the original calibration data sets, are presented separately for five of the export models in Table 4. The original

Table 3. Regression model parameters for the global data sets. The units of the regression coefficients are: fertilizer use (dimensionless), atmospheric deposition (dimensionless), runoff (m yr^{-1}), and point sources (kg yr^{-1})

| Model | R^2 | No. Sites | Model Coefficients | | | |
| | | | (p – *statistical significance*) | | | |
			Point Sources	Runoff	Fertilizer Use	Atmospheric Deposition[a]
LS1-GLOBAL[b]	0.90	35	2.805	0.2785	0.0816	0.2969
			(*<0.01*)	(*0.13*)	(*0.17*)	(*0.03*)
LS2-GLOBAL[c]	0.83	29	0.3442	0.2960	0.2316	0.3762
			(*<0.19*)	(*0.01*)	(*0.02*)	(*0.02*)

[a]NO_y total wet and dry oxidized components of atmospheric deposition (Caraco & Cole 1999; Seitzinger & Kroeze 1998).
[b]Model calibrated with data from Caraco & Cole 1999.
[c]Model calibrated with data from Seitzinger & Kroeze 1998.

data for large rivers of the world used to calibrate the PEIRLS model were not available. Several of the models have relatively unbiased predictions of export, and include the HOWARTH, SPARROW, and LS2-GLOBAL models. Median errors for these models are less than 5% and the interquartile range of each is relatively symmetrical, typically ranging from about –30% to 40%. Median errors in the other export models are larger by a factor of six or more, and indicate that the models tend to over predict export at most of their calibration sites. A tendency for over prediction is noted for the Seitzinger and Kroeze (1998; median = 46%; IQR = –1 to 66%) and the Caraco and Cole (1999; median = 18%; IQR = –25 to 82%) global models. Although the LS1-GLOBAL model has a very small median error (–0.6), the interquartile range is unbalanced (–34 to 61%) suggesting a tendency for over prediction at many of the sites. The lowest variability in prediction errors based on the calibration data is found for the HOWARTH (IQR = 27%), SPARROW (IQR = 65%), and GLOBAL (IQR = 66% based on the Seitzinger and Kroeze data) models. The variability of the global model prediction errors, based on the Caraco and Cole (1999) riverine data set of 35 sites, tends to be higher than that of the other global models, based on the Seitzinger and Kroeze (1998) data set.

4.2. *Accuracy of the nitrogen export models – northeastern US watersheds*

The errors in model predictions of stream nitrogen export are shown separately for each method in Figure 3 with summary statistics for the prediction errors listed in Table 5. Figure 3(a) shows the cumulative frequencies of the

Table 4. Errors in model predictions of stream nitrogen export for the original calibration data sets. Errors are computed as the difference between the predicted and measured values of stream nitrogen export expressed as a percentage of the measured export

Model	median	IQR[a]	Prediction Errors (%)			
			min.	25th percentile	75th percentile	max.
SPARROW	3.2	65.2	–95.2	–27.4	37.8	1210.6
HOWARTH	2.8	26.6	–29.3	–6.0	20.7	66.3
GLOBAL[b]	18.0	107.6	–77.3	–25.6	82.0	1204.8
LS1-GLOBAL	–0.6	94.8	–83.6	–34.0	60.8	1960.7
GLOBAL[c]	46.5	66.6	–83.6	–1.1	65.5	861.5
LS2-GLOBAL	–3.3	71.6	–80.6	–30.3	41.3	469.6

[a]Interquartile range (difference between the 25th and 75th percentiles of the distribution of errors).
[b]Caraco & Cole 1999; only runoff coefficients were calibrated in the original models.
[c]Seitzinger & Kroeze 1998; only runoff coefficients were calibrated in the original models.

prediction errors for each model. The prediction errors for the 16 watersheds were ranked in ascending order for each model and assigned a cumulative frequency value according to a Cunnane plotting position quantile (Cunnane 1978). Better performing models (i.e. those with smaller prediction errors) are associated with values of the percent error that plot closest to zero (i.e. the horizontal line in Figure 3(a)) and display a relatively shallow slope; the median error for a model corresponds to a cumulative frequency of 0.5. The box and whisker plots in Figure 3(b) display summary statistics (Table 5) for the prediction errors, including the median, interquartile range (difference between the 25th and 75th percentiles), and minimum and maximum values. The length of the box and the whiskers (i.e. interquartile range, and data range) provides information about the variability of the model predictions, whereas the location of the median and symmetry of the interquartile range relative to zero describe the magnitude of bias in model predictions for the watersheds.

Based on comparisons of the median and symmetry of the interquartile range of the model errors (Figure 3(b), Table 5) and their cumulative frequencies (Figure 3(a)), all of the methods show at least small amounts of bias. The median errors are within about 15% for five of the six models. The GLOBAL, SPARROW and LS1-GLOBAL models have the smallest median errors of less than about 5%. Prediction errors of this magnitude are within the standard error of the measured values of export (about 5 to 10%), and are therefore

314

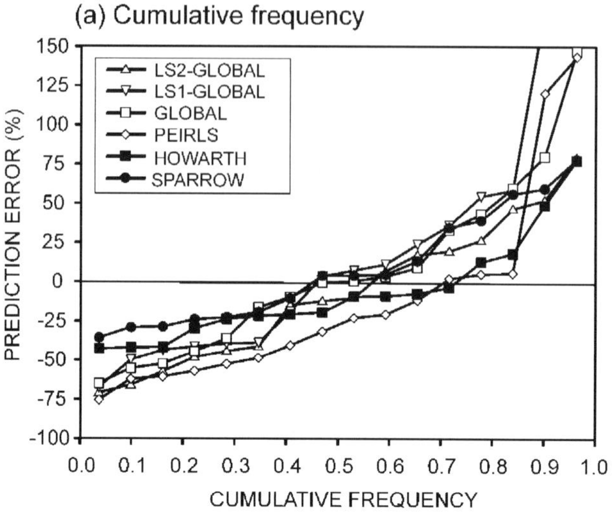

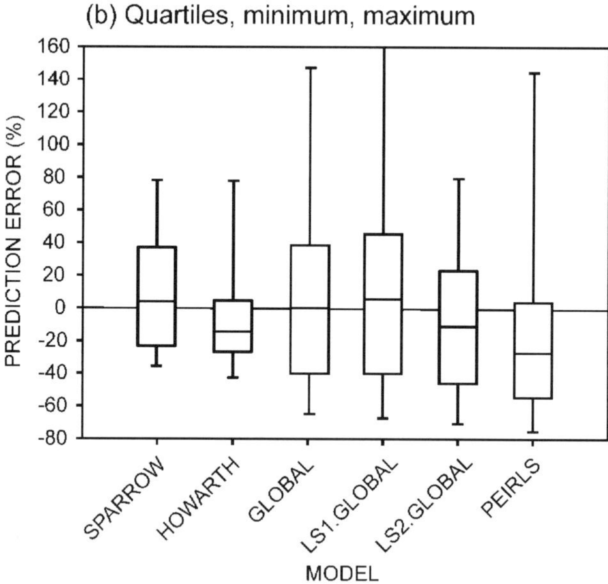

Figure 3. Errors in the predictions of stream nitrogen export from the application of the six watershed models to the 16 northeastern basins: (a) cumulative frequency plots (values of 174% and 373% for the LSI-GLOBAL model plot above the scale), and (b) box and whisker plots (the maximum of 373% for the LSI-GLOBAL model plots above the scale). Each box graphs the quartiles with the lower and upper edges representing the 25th and 75th percentiles, respectively. The midline plots the median. The upper and lower whiskers are drawn to the minimum and maximum values.

indistinguishable from zero. The LS2-GLOBAL and HOWARTH models have median errors of −11% and −14%, respectively. The greatest symmetry of the IQR about zero, an approximate measure of model bias, was observed for three of the models: GLOBAL (−43% to 35%), SPARROW (−23% to 36%), and LS1-GLOBAL (−43% to 48%). The HOWARTH, PEIERLS, and LS2-GLOBAL models all have a greater tendency to under predict stream export (the HOWARTH and PEIERLS models under predict export in 12 of the 16 watersheds; see Figure 3(a)). However, the HOWARTH model under predicts by a relatively small magnitude in comparison to other models. The large negative bias associated with the PEIERLS model predictions (median = −27%) is indicative of the tendency of this model to under predict stream export in the northeastern watersheds. The SPARROW model under predicts by the smallest magnitude of all of the models as evidenced by the proximity of the cumulative frequency curve to the zero line over the negative range of the prediction errors (i.e. probability <0.5) in Figure 3(a). The asymmetry in the 75th percentile (36%) as compared with the 25th percentile (−23%) indicates that the SPARROW model tends to over predict export by a somewhat larger magnitude than it under predicts export.

The smallest variability (i.e. highest precision) in the prediction errors as indicated by the magnitude of the interquartile range (IQR; Table 5) is found for the HOWARTH model (IQR = 26%), followed by the SPARROW (59%) and PEIERLS (60%) models. By comparison, the variability in prediction errors of the global models is about 30% higher than that for the SPARROW and PEIERLS models and about twice that of the HOWARTH model. The various global models differ only modestly in the magnitude of the variability in the prediction errors, with IQRs ranging from 66% to 80%. The LS2-GLOBAL model shows the smallest IQR, which is about 20% lower than that of the LS1-GLOBAL model.

Comparisons of the prediction errors for the northeastern watersheds (Table 5) with those for the original calibration data (Table 4) indicate that the greatest similarities among individual models are found in the magnitude of the variability of the prediction errors. For example, the variability (i.e. IQR) for the HOWARTH model predictions is 26.6% (Table 4) and 25.6% (Table 5), respectively, for the calibration and northeastern watersheds; variability in the prediction errors for the SPARROW model is 65.2% and 61.6%, respectively. Among the various global models, those based on the Seitzinger and Kroeze (1998) data display the most similar variability in prediction errors (GLOBAL IQR = 63.8 and 73.8; LS2-GLOBAL IQR = 71.6 and 66.3). Other comparisons of the model errors for the calibration and northeastern watersheds in Tables 4 and 5 are also of note. For example, distributions of prediction errors for the SPARROW model have a similar median (3.2% and

Table 5. Errors in model predictions of stream nitrogen export from applications in the northeastern watersheds. Errors are computed as the difference between the predicted and measured values of stream nitrogen export expressed as a percentage of the measured export (equation 1). The prediction errors are plotted in figure 3

Model	median	IQR[a]	Prediction Errors (%)			
			min.	25th percentile	75th percentile	max.
SPARROW	4.1	58.6	–35.6	–22.8	35.8	78.2
HOWARTH	–14.4	25.6	–42.7	–25.1	0.5	77.7
PEIERLS	–27.2	60.1	–75.3	–55.8	4.3	143.9
GLOBAL	0.1	73.8	–64.8	–42.8	35.3	147.2
LS1-GLOBAL	5.5	80.4	–67.3	–43.0	47.9	373.0
LS2-GLOBAL	–11.2	66.3	–70.7	–47.7	24.7	79.3

[a]Interquartile range (difference between the 25th and 75th percentiles of the distribution of errors).

4.1%) and quartile values. The various global models generally over predict stream export for the original calibration watersheds, whereas the GLOBAL and LS2-GLOBAL models tend to slightly under predict stream export in the northeastern watersheds. For the HOWARTH model, the prediction errors for the original calibration data display a slight positive bias (median = 3%; IQR from –6% to 21%) suggesting a tendency of the model to over predict export for the calibration watersheds. By contrast, this model has a somewhat greater tendency to under predict stream export in the northeastern watersheds.

4.3. *Factor-related model biases – northeastern US watersheds*

The results of the error analysis based on an application of the regression-based error models in equation 2 are shown in Table 6. For each model, we report the R^2, mean square error (MSE), regression coefficients for each of the explanatory variables, and the statistical significance (p-value) of the coefficients. In contrast to interpretations of conventional regression models, a high R^2 (also low MSE and small p values) for the error models identifies inaccuracies (i.e. 'factor-related' biases) in the specification of the original stream export models related to deficiencies in the model descriptions of nitrogen supply and processing. The coefficients of the error models quantify the magnitude of the change in prediction error that is expected to occur in response to a unit change in a watershed property, and may be directly compared among the six models. The interpretation of these coefficients in terms of the tendency for models to over or under predict stream nitrogen

Table 6. Results of the regression-based error models for the northeastern watersheds. The regression coefficients indicate the change in the percent prediction error per unit change in the land-area percentage (cultivated land or developed land), runoff (cm yr^{-1}), or drainage basin area ($\times$ 10^{-3} km^2). The intercept is expressed in units of percent prediction error

Model	R^2 (*p; statistical sig.*)	Mean Square Error (MSE)	Regression Coefficients (*p; statistical significance*)				
			Cultivated Land	Developed Land	Runoff	Basin Area	Intercept
SPARROW	0.36	1127.6	−2.49	0.42	−1.85	−0.29	150.9
	(0.260)		*(0.053)*	*(0.785)*	*(0.220)*	*(0.621)*	*(0.130)*
HOWARTH	0.43	854.4	−1.68	−0.066	−3.29	−0.437	206.4
	(0.157)		*(0.120)*	*(0.961)*	*(0.022)*	*(0.397)*	*(0.026)*
PEIERLS	0.72	1490.7	−4.30	6.45	−3.21	−0.22	200.5
	(0.005)		*(0.008)*	*(0.003)*	*(0.075)*	*(0.741)*	*(0.085)*
GLOBAL	0.72	1196.5	−4.84	3.89	−3.79	−0.65	280.4
	(0.004)		*(0.002)*	*(0.029)*	*(0.025)*	*(0.293)*	*(0.013)*
LS1-GLOBAL	0.79	3439.8	−7.10	13.01	−5.51	−0.30	378.5
	(0.001)		*(0.005)*	*(<0.001)*	*(0.048)*	*(0.770)*	*(0.038)*
LS2-GLOBAL	0.75	695.9	−3.93	−2.11	−2.64	−0.858	224.1
	(0.002)		*(0.001)*	*(0.102)*	*(0.038)*	*(0.082)*	*(0.010)*

export in response to changes in watershed characteristics will depend on the range of the prediction errors and their symmetry about zero. For models with less overall bias (as measured by the median error), a positive coefficient indicates that the prediction errors generally become more positive and larger in magnitude (tendency to over predict export) in response to increases in the magnitude of the factor. However, in the case of the PEIERLS and HOWARTH models, where the majority of the prediction errors are negative (i.e. under prediction), a positive coefficient, for example, would indicate that over much of the range of the explanatory factor the prediction errors become less negative and smaller (i.e. less under prediction) in response to increases in the watershed property. Statistical significance of the model coefficients is based on the application of a t-test. The residuals of the error models generally comply with the regression assumptions; they are approximately normal with relatively constant variance.

The R^2 of the error models span a large range, indicating that from 36% to 79% of the spatial variability in the prediction errors can be explained by the four watershed properties (Table 6). The MSE varies in response to how well the watershed characteristics explain the prediction error, but also as

318

a function of the magnitude of the variability in the prediction errors (i.e. the precision of the model predictions). The SPARROW and HOWARTH error models have the lowest R^2 values (0.36 and 0.43, respectively) among the various models with explanatory variables displaying moderate to very weak significance (p = 0.02 to 0.96). The small MSE for the HOWARTH model reflects the low variability (IQR) in the prediction errors. The other models show much higher R^2 values (0.72–0.79) with correspondingly larger coefficient values and higher levels of statistical significance. The signs of the explanatory coefficients are generally consistent among most of the models indicating that the direction of the biases are similar; however, the magnitudes of the coefficients vary indicating differences in the importance of these factors in potentially explaining prediction errors in the nitrogen export models.

Cultivated land area is found to be statistically significant in all models, but the magnitude of the effect based on the size of the coefficient value differs appreciably among the models. Cultivated land area is inversely correlated with the prediction errors, indicating that there is a tendency for most of the export models to under predict stream export in watersheds that are more highly agricultural and over predict in watersheds with less cropland and larger amounts of forested lands. The partial residual plots (Montgomery & Peck 1982) in Figure 4 display the relation between the prediction errors and cultivated land area for each of the models. The slope of the lines in Figure 4 corresponds to the slope coefficient of the cultivated land area term in the error models reported in Table 6. These plots show variations in prediction errors, adjusted for the variability that is explained by the other predictor variables in the model (i.e. developed land, runoff, and basin area); the partial error residuals (vertical axis) are also adjusted for the mean so that the residuals are centered about zero. The plots in Figure 4 illustrate the relative strength of these relations for the various export models. The strength varies in proportion to the magnitude of the slope of the fitted line relative to the variability of the partial residuals about the line. The fitted lines also illustrate for the various models how the prediction errors change in magnitude over the full range of cultivated land area in the watersheds. The PEIERLS and various global models show the strongest relations between the model prediction errors and cultivated land area as evidenced by the relatively steep slopes of the fitted lines and their relatively high statistical significance (Table 6). The coefficients for cultivated land in these models are typically one and a half to five times larger than that observed for the HOWARTH and SPARROW models, which display coefficients of –1.67% and –2.49% error per unit change in cultivated land percentage, respectively. Thus, the PEIERLS and global models under predict export by a much larger magnitude

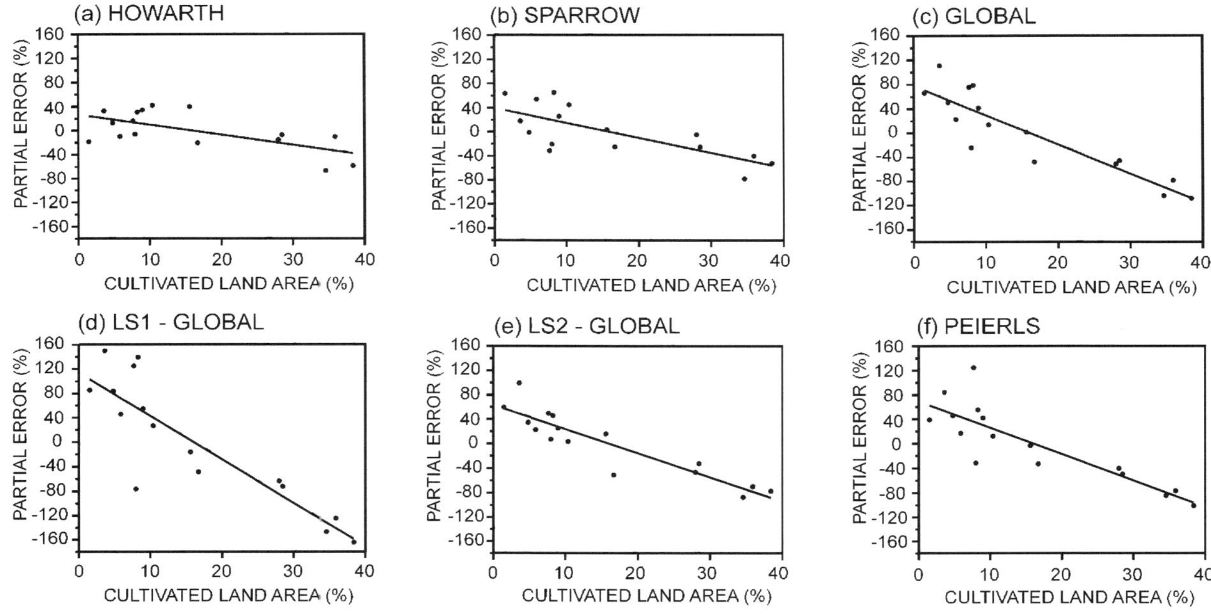

Figure 4. Partial residual plots of the relation between prediction errors and cultivated land area for the six error models: (a) HOWARTH, (b) SPARROW, (c) GLOBAL, (d) LSI-GLOBAL, (e) LS2-GLOBAL, and (f) PEIERLS. The slope of the lines corresponds to the slope coefficient of the cultivated land area variable in the error models in Table 6. The partial errors are adjusted for variability explained by the other model predictors (developed land, runoff, and basin area), and are centered about zero by subtracting the mean. The smaller slopes for the HOWARTH and SPARROW error models indicate that cultivated land area explains smaller amounts of the over and under prediction of nitrogen export in the watershed models.

319

in more highly cultivated watersheds than do the other models. In less cultivated basins, the global models also over predict export by a larger magnitude than the other models. The smaller slopes for the HOWARTH and SPARROW error models indicate that cultivated land area or related factors potentially explain smaller amounts of the over and under prediction of nitrogen export by the watershed models.

Developed land area is also found to be significant in explaining prediction errors for the PEIERLS and various global models. A positive relation is observed suggesting that there is a tendency for the stream export models to under predict in less developed basins (i.e. also basins with more area in forest and cultivated land). The magnitude of the coefficient values of the PEIERLS and LS1-GLOBAL error models is considerably larger than observed for cultivated land area in these models (i.e. 6.5 and 13% error per unit change in percentage land area). Developed land area is not statistically significant in the SPARROW ($p = 0.79$) and HOWARTH ($p = 0.96$) models.

Runoff is moderately to highly significant in all of the error models except SPARROW, which is statistically insignificant ($p = 0.22$). The negative correlation with the prediction errors suggests that there is a tendency for the stream export models to under predict in basins with high runoff and over predict in basins with low runoff. The coefficients for the SPARROW, LS2-GLOBAL, and HOWARTH error models are among the smallest (–1.85, –2.32, and –2.48% change in prediction error per unit change in runoff – cm yr^{-1}, respectively). Coefficients for the PEIERLS and global models range from –3.2 to –5.5%.

Drainage basin size is generally insignificant or only moderately significant in most of the error models. Four of the six models display the smallest change in prediction error per unit change in basin size (i.e. about –0.2% to –0.4% error per unit change in drainage area – 1000 km^2 – for the SPARROW, HOWARTH, PEIERLS, and LS1-GLOBAL models). None of these coefficients are statistically significant ($p > 0.30$). Two of the remaining models (GLOBAL, LS2-GLOBAL) display larger coefficients (–0.6% and –0.9% error per unit change in runoff – cm yr^{-1}, respectively), but have weak levels of statistical significance ($p > 0.08$).

5. Discussion of the error analysis

In evaluating the accuracy of the predictions of stream nitrogen export for the six watershed models, we presented two measures of model bias. First, we reported the overall bias (Table 5; Figure 3). This measure is defined as the median prediction error, and provides a measure of the typical bias in the model predictions of nitrogen export for the entire set of northeastern watersheds. The overall bias indicates whether there is evidence that the model predictions are systematically too high or low among the 16 basins (see Figure 3). Second, we reported the factor-related bias as quantified by the regression-based model coefficients (equation 2) in Table 6. These coefficients describe the magnitude of the change in the prediction error that occurs in response to unit changes in various watershed properties. Such changes are indicative of biases in the model predictions of nitrogen export caused by model mis-specification, including sources or delivery processes that are not explicitly included in the models or model coefficients that inaccurately describe the supply and transport of nitrogen. This information is useful for assessing model performance over a range of specific environmental conditions. Strong evidence of factor-related bias implies that improvements are feasible in the accuracy of the model predictions through improved calibrations or modifications of the model structure. As an additional note, a model with relatively little overall bias can display significant factor-related biases – a pattern that is evident for several of the models examined here. For example, although the GLOBAL and LS1-GLOBAL models display relatively small median prediction errors (<6%), large factor-related biases are detected that are related to specific watershed properties.

Based on the magnitude and statistical significance of the coefficients of the error models (Table 6), the prediction errors of the stream export models appear to be most strongly related to land use and runoff. The land-use factors indicate limitations in how well the stream export models account for nitrogen sources and/or attenuation processes associated with cultivated, developed or forested lands, which may be related to such factors as nitrogen fixation, N uptake by vegetation, or cultural inputs of N. Prediction errors are negatively correlated with cultivated land percentage in most of the error models indicating that the stream export models over predict nitrogen export in less agricultural basins and under predict in more agricultural basins. The magnitude of the prediction errors vary widely among the models over the range of cultivated land area in the watersheds (see Figure 4). The HOWARTH and SPARROW models display the smallest changes in prediction errors in response to changes in the cultivated land area; changes in the prediction errors of these models are at least 50% smaller than those of the other models. The lower cultivated land-related bias in the HOWARTH model may reflect the value of directly accounting for food/feed imports and exports and nitrogen fixation in crops, which are not explicitly included in the other models. The inclusion of N in livestock wastes in the SPARROW model accounts for feed consumption, which may help explain the model's relatively low cultivated land-related bias. The tendency of most of the models to over predict nitrogen export in less agricultural basins may also reflect limitations in the model estimates of nitrogen supply and attenuation on forested lands. Any effects of forestland would be expressed in the error models as a negative correlation with the percent cultivated land area. The prediction errors are also significantly correlated with developed land area for all models except for the SPARROW and HOWARTH models. The positive relation between the prediction errors and developed land area indicates a tendency of the models to over predict stream export in more developed basins and to under predict export in less developed basins. Less developed basins include watersheds that are more highly forested as well as those with cultivated lands. The measures of cultivated and developed land area in the 16 northeastern watersheds are uncorrelated ($r = -0.06$; variance inflation factors = 1), and therefore, provide independent land-use information in the error models. The use of additional land-use terms (e.g. forested land) in preliminary error models led to colinearity problems (variance inflation factors >9), and thus, were not used in the final models.

Runoff and drainage basin size are included in the error models to quantify the potential effects on stream N export of nitrogen mobilization processes that are independent of land use. Although runoff may reflect climate-related effects on stream export related to the supply of nitrogen in vegetation, the land-use terms in the error models should account for variations in many of the cultural and natural sources of N. In addition, natural sources of nitrogen, such as forest N fixation, are relatively minor contributors to the N budgets of the northeastern watersheds (Boyer et al. 2002; Table 8). Runoff and basin size are potentially related to many of the same hydrologic and time-dependent properties that influence nitrogen attenuation on the landscape and in streams (e.g. water velocity, flow, water time of travel); however, the two factors are relatively uncorrelated ($r = -0.27$; variance inflation factor = 1) for the northeastern watersheds, thereby providing independent explanatory measures in the error models.

The negative relation between the prediction errors and runoff indicates that the stream export models tend to under predict in watersheds where runoff is high and over predict in watersheds where runoff is low. These effects are least significant in the SPARROW model and most pronounced for the various global models. The global model predictions of stream export for major rivers of the world (Caraco & Cole 1999; Seitzinger & Kroeze 1998) also show a moderately significant negative relation between the prediction errors and runoff for the original calibration data. Runoff is sensitive to the effects of climate, geology, soils, and stream morphology on the rates of surface and subsurface flow. Runoff may influence the rates of nitrogen uptake and storage, and the permanent removal of nitrogen in terrestrial and aquatic environments by affecting water residence times and water contact with sites suitable for denitrification, such as anoxic soils, benthic stream sediments, channel hyporheic and riparian zones, and wetlands (Kelly et al. 1987; Hill 1996; Sauer et al. in press; Molot & Dillon 1993). The increased mobilization of N with runoff may explain the nearly proportional positive relation that has been observed between stream nitrogen export and runoff in developed (Sauer et al. in press; Behrendt 1996) and undeveloped (Lewis et al. 1999, in press) watersheds. The negative correlation ($r = -0.57$) between runoff and estimates of the total loss of nitrogen in the northeastern watersheds (computed as the difference between major N inputs and stream N exports; Boyer et al. 2002) is consistent with the effects of these nitrogen removal processes on stream nitrogen export; larger losses of nitrogen are observed in northeastern watersheds with relatively low runoff and smaller losses occur in watersheds with high runoff. Therefore, the negative relation between prediction errors and runoff suggests that the under prediction of stream export in watersheds with high runoff may be caused, in part, by the overestimation of the rates of nitrogen attenuation in the models. Conversely, the over prediction of stream export in watersheds with low runoff may reflect the underestimation of the rates of nitrogen attenuation rates in the export models.

Drainage basin size is found to be a relatively insignificant explanatory variable in nearly all of the error models. This suggests that the prediction errors are not strongly related to scale-dependent characteristics of the watersheds as measured by drainage size. It is possible that scale-dependent properties related to the rates of water flow and storage, such as water travel time, may be more clearly described by runoff for the northeastern watersheds, which may partially explain its importance in many of the error models. In contrast to the other models, drainage basin area is moderately significant in the LS2-GLOBAL error model, and is much larger in magnitude than observed for the other error models. Thus, this model displays direct evidence of a scale-dependency in the prediction errors related to basin size with a tendency of the model to under predict stream nitrogen export in large watersheds.

The error analysis indicates that the watershed models with more complex descriptions of nitrogen sources and attenuation processes have appreciably lower bias and higher precision in their predictions of nitrogen export. The two models with the most detailed descriptions of nitrogen sources, land and water N attenuation, and water flow paths (HOWARTH, SPARROW) show smaller factor-related biases in the predictions of stream nitrogen export. This is supported by both the smaller magnitude and statistical significance of the coefficients in the associated error models. These findings suggest that model complexity has a beneficial effect on prediction accuracy. The HOWARTH model gives a detailed accounting of agricultural sources, including fertilizer use, crop N fixation, and the import and export of nitrogen in food and feeds. The SPARROW model spatially references stream monitoring data, point and diffuse nitrogen sources, and landscape properties to surface water flow paths defined by a digital drainage network. Agricultural sources include fertilizers and livestock wastes. The model explicitly quantifies the rates of nitrogen removal on the landscape and in streams though the use of spatial referencing and mass-balance constraints. By contrast, the PEIERLS model lacks explicit point and diffuse source terms, and relies solely on population density as a predictor variable. This may contribute to the strong tendency of the model to under predict in more undeveloped (i.e. less populated) watersheds. In the global models, agricultural nitrogen sources are quantified exclusively as a function of fertilizer use and runoff; point-source contributions are a function of population density and urban land area. Neither the PEIERLS nor the global models account for the location of sources and water travel times in the watersheds.

6. Model predictions of source contributions

The predictions of source contributions to stream nitrogen export are reported for the stream export models in Table 7 and Figure 5. The classification of source contributions and the model assumptions and forms of nitrogen differ considerably among the various models, which affect the comparisons of model estimates. Unlike the previous evaluations of stream nitrogen export, there is no known measure of the magnitude of source contributions to streams, and thus, the predictions of nitrogen sources can only be compared among the models. Although the error models could be used to correct for model biases in the estimates of source shares, this would require assumptions about how the explanatory factors in the error models, especially developed and cultivated land, correspond to the source variables in the stream export models. The inverse relation between the prediction errors of most of the models and cultivated land area as previously discussed generally implies that the stream export models have a tendency to under predict in more agricultural watersheds and to over predict in more forested and less agricultural watersheds. This allows for the possibility that the effect of agricultural sources may be somewhat underestimated in more agricultural basins. It is also possible that prediction biases related to developed land area may affect the accuracy of the point source contributions estimated by some the models.

Estimates of nitrogen inputs from major watershed sources based on nitrogen budgets (see Boyer et al. 2002) are presented in Table 8 for comparison with the model predictions of source contributions at the watershed

Table 7. SPARROW predictions of source contributions to stream nitrogen export from the northeastern watersheds

River Name	Source Contributions to Export (% of export)[a]					Watershed Attenuation		
	Point	Fertilizer Use	Livestock Wastes	Atmosphere[b]	Nonagr. Nonpt.	In-stream Loss[c] (% of stream inputs)	In-stream Loss[d] (% of basin loss)	Landscape Loss[e] (% of basin loss)
Penobscot	1	5	2	31	61	47	54	46
Kennebec	2	3	5	36	55	46	37	63
Androscoggin	3	5	5	34	54	40	30	70
Saco	1	2	2	34	60	35	25	75
Merrimack	18	5	4	29	44	34	15	85
Charles	74	2	1	9	14	30	7	93
Blackstone	37	6	5	20	32	31	23	77
Connecticut	6	8	7	38	39	47	28	72
Hudson	4	9	8	39	41	45	23	77
Mohawk	13	12	15	36	25	40	21	79
Delaware	15	14	9	35	28	38	29	71
Schuylkill	37	17	15	17	13	28	17	83
Susquehanna	7	17	24	30	23	45	25	75

Table 7. Continued

River Name	Source Contributions to Export (% of export)[a]					Watershed Attenuation		
	Point	Fertilizer Use	Livestock Wastes	Atmosphere[b]	Nonagr. Nonpt.	In-stream Loss[c] (% of stream inputs)	In-stream Loss[d] (% of basin loss)	Landscape Loss[e] (% of basin loss)
Potomac	4	26	23	31	17	59	34	66
Rappahannock	1	26	23	27	23	59	18	82
James	1	11	14	39	35	58	18	82
25th Percentile	*2*	*5*	*4*	*28*	*23*	*35*	*18*	*71*
Median	*5*	*9*	*8*	*32*	*33*	*43*	*24*	*76*
75th Percentile	*16*	*15*	*15*	*35*	*47*	*47*	*30*	*82*

[a]Expressed as a percentage of the predicted stream nitrogen export from the watersheds.
[b]Atmospheric deposition contributions to stream export are based on wet-fall deposition. Land-to-water delivery fractions exceed unity indicating that additional atmospheric forms of nitrogen (e.g. ammonium, organic) are included (see Alexander et al. 2001).
[c]The in-stream loss of nitrogen in RF1 (river reach file 1) reaches.
[d]The mass of nitrogen removed in streams is estimated as (M/(1-S)) * S, where M is the measured stream nitrogen export in equation (1) and S is the in-stream nitrogen loss estimated by SPARROW and expressed as a fraction of the stream inputs. The mass of nitrogen removed in streams is expressed as a percentage of the total loss in the watershed (estimated as the differences between the watershed inputs – fertilizer, total atmospheric deposition, crop and forest N fixation, and net food/feed imports – and riverine N export; see Boyer et al. 2002).
[e]Computed as the complement of the in-stream loss percentage.

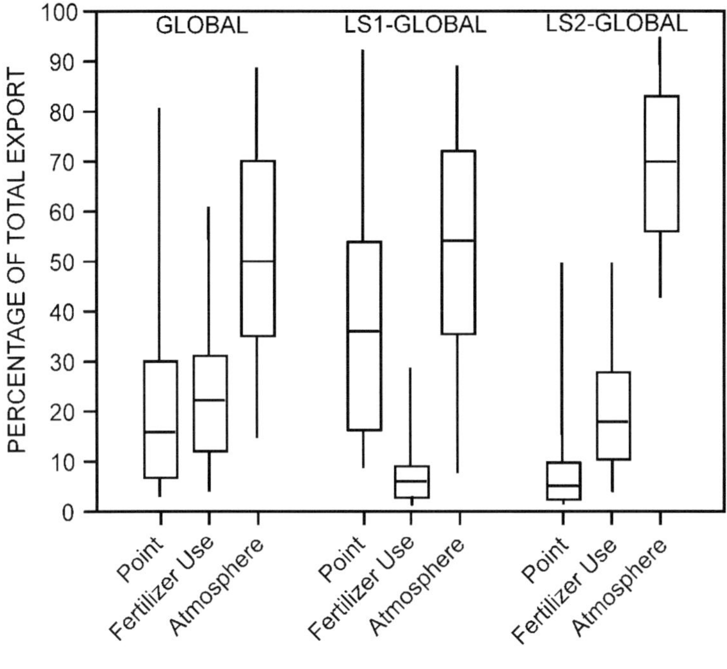

Figure 5. GLOBAL, LSI-GLOBAL, and LS2-GLOBAL model predictions of source contributions to stream nitrate-nitrogen export for the northeastern watersheds. Sources contributions are expressed as a percentage of the stream nitrate-nitrogen export. Each box graphs the quartiles with the lower and upper edges representing the 25th and 75th percentiles, respectively. The midline plots the median. The upper and lower whiskers are drawn to the minimum and maximum values.

outlets. The budget estimates would accurately describe nitrogen source contributions to stream export at the watershed outlets only if sources are uniformly distributed in the watersheds and loss processes are assumed to operate equally on all sources. Nevertheless, the budget estimates provide useful information on the major inputs of N for comparison with the model predictions. Atmospheric nitrogen typically represents from about 29% to 59% of the total inputs to the watersheds, based on the interquartile range (Table 8). With the exception of forest N fixation, which typically is less than 6% of the total inputs of nitrogen, much of the remaining portions of the nitrogen inputs originate from agricultural-related sources and products either applied directly to fields in fertilizers or consumed in food and feeds. Nitrogen inputs from food/feed imports are typically less than about 15% of the total inputs from major sources although N imports represent more than a third of the total nitrogen inputs in the more populated watersheds. Nitrogen in food imports make their way to streams via sewered and unsewered waste systems.

The source predictions of the various global models (GLOBAL, LS1-GLOBAL, LS2-GLOBAL) are shown in Figure 5. These models classify sources according to three types: point, atmospheric, and fertilizer (agricultural). Because these models predict nitrate rather than total nitrogen export as a function of the specified sources, use of the model to characterize sources in the northeastern watersheds assumes that the source shares for nitrate are equivalent for other nitrogen forms. In addition, other sources of nitrogen, such as natural fixation, are not explicitly described by the models, and would tend to be included by the three model sources. The GLOBAL model indicates that atmospheric deposition represents the predominant source in stream export (median = 50%; IQR = 35 to 70%), especially in the northern watersheds of the Penobscot, Kennebec, Androscoggin, and Saco where atmospheric contributions exceed 75%. Contributions from agricultural sources are second in importance in most of the watersheds (median = 22%; IQR = 12 to 31%). Agricultural sources are relatively large in all watersheds south of the Mohawk, but only rarely exceed 50%. Point sources (median = 16%; IQR = 7 to 30%) are only slightly smaller than agricultural sources, but contribute the largest shares (>50%) in the two smallest watersheds, the Charles and Blackstone, and in the Merrimack. The LS1-GLOBAL model suggests that atmospheric deposition has a similar, but slightly higher relative contribution than in the GLOBAL model (median = 54%; IQR = 35 to 72%). Larger differences are noted in the other sources, where point sources are typically higher (median = 36%; IQR = 16 to 54%) and agricultural sources are generally lower (median = 6%; IQR = 3 to 9%) than reported for the GLOBAL model. The LS2-GLOBAL model typically shows higher atmospheric contributions (median = 70%; IQR = 56 to 83%) than either of the other global models. The higher contributions are offset by lower contributions from both point and fertilizer-related sources; however, the LS2-GLOBAL model predictions of contributions from the fertilizer-related sources are generally similar to those predicted by the GLOBAL model.

The SPARROW model indicates that atmospheric sources and non-agricultural diffuse sources each contribute about one third of the nitrogen to stream export in most of the watersheds (Table 7). Some of the largest contributions from atmospheric nitrogen (>35%) are found in the Kennebec, Connecticut, Hudson, Mohawk, and James watersheds. Most watersheds have atmospheric contributions that are only slightly lower that this, and typically lie in the range from 28 to 35%. Agricultural sources (fertilizer plus livestock waste sources) are typically less than 30%, but contribute nearly 50% of the nitrogen in the Susquehanna, Potomac, and Rappahannock

Table 8. Estimates of nitrogen inputs from major sources and nitrogen losses in the northeastern watersheds

River Name	Nitrogen Inputs to Watershed[a] (% of total inputs)					In-Stream Loss[c] (% of stream inputs)	In-Stream Loss[d] (% of basin loss)	Landscape Loss[e] (% of basin loss)
	Atmosphere[b]	Fertilizer Use	Net Feed & Food Imports	Agric. N Fixation	Forest N Fixation			
Penobscot	70	11	2	9	7	68	130	–
Kennebec	64	5	11	15	5	63	74	26
Androscoggin	61	6	15	12	5	52	48	52
Saco	73	3	7	8	9	47	41	59
Merrimack	42	7	34	10	7	61	45	55
Charles	23	4	64	4	5	37	10	90
Blackstone	31	9	43	9	8	53	57	43
Connecticut	45	13	21	17	5	66	61	39
Hudson	55	11	9	20	5	58	38	62
Mohawk	34	13	11	40	2	60	48	52
Delaware	43	19	8	24	7	60	72	28
Schuylkill	22	23	28	23	4	52	48	52
Susquehanna	31	17	17	31	5	76	97	3

Table 8. Continued

River Name	Nitrogen Inputs to Watershed[a] (% of total inputs)					In-Stream Loss[c] (% of stream inputs)	In-Stream Loss[d] (% of basin loss)	Landscape Loss[e] (% of basin loss)
	Atmosphere[b]	Fertilizer Use	Net Feed & Food Imports	Agric. N Fixation	Forest N Fixation			
Potomac	17	23	27	27	6	69	53	47
Rappahannock	21	24	14	34	7	58	17	83
James	34	13	15	25	13	72	33	67
25th percentile	*29*	*7*	*10*	*10*	*5*	*53*	*40*	*37*
Median	*38*	*12*	*15*	*18*	*6*	*60*	*48*	*52*
75th percentile	*57*	*17*	*27*	*25*	*7*	*67*	*63*	*60*

[a]Expressed as a percentage of the total nitrogen inputs from major watershed sources (Boyer et al. 2002).
[b]NO_y total wet and dry oxidized components of atmospheric deposition (Boyer et al. 2002).
[c]The in-stream loss of nitrogen in RF1+RF3 scale reaches based on the RivR-N model (see Seitzinger et al. 2002).
[d]The mass of nitrogen removed in streams is estimated as (M/(1-S)) * S, where M is the measured stream nitrogen export in equation (1) and S is the in-stream nitrogen loss estimated by the RivR-N model (Seitzinger et al. 2002) and expressed as a fraction of the stream inputs. The mass of nitrogen removed in streams is expressed as a percentage of the total loss in the watershed (estimated as the differences between the watershed inputs – fertilizer, total atmospheric deposition, crop and forest N fixation, and net food/feed imports – and riverine N export; see Boyer et al. 2002).
[e]Computed as the complement of the in-stream loss percentage.

watersheds. Point sources (industrial and municipal) typically contribute less than 15%. Much larger point-source contributions (37 to 74%) are found in the more highly populated watersheds of the Charles, Blackstone, and Schuylkill. Non-agricultural diffuse sources are largest in the highly forested northern watersheds, contributing from 44 to 61%; in other watersheds, the contributions are less than about 30%. This source category is proportional to non-agricultural land area, and accounts for remaining sources of nitrogen that are not specified by the other source inputs in the model. These sources may include nitrogen in the surface and subsurface flows from urban, forested, wetlands, and barren lands. Nitrogen from forested lands may include biotic N fixation. Groundwater nitrogen is implicitly included in the agricultural and non-agricultural sources specified in the model, and may include older waters from a mixture of sources.

A comparison of the global and SPARROW models shows that the estimates of agricultural contributions are similar among the GLOBAL, LS2-GLOBAL, and SPARROW models. Atmospheric sources consistently contribute less to stream export according to the SPARROW model (about 30 to 40% less) than predicted by the various global models. Point source contributions are similar in the SPARROW and LS2-GLOBAL model, which are about one half of the magnitude of the point source shares predicted by the GLOBAL model. The SPARROW model classifies about a third of the nitrogen contributions as non-agricultural diffuse sources, which may include a mixture of sources in the groundwater and runoff from urban and rural lands. The point source shares from SPARROW and GLOBAL model are highly correlated with the percentage of land classified as developed ($r = 0.95$ for both models). The percentage of cultivated land is highly correlated with SPARROW estimates of agricultural contributions ($r = 0.92$); the GLOBAL model shows somewhat less correlation ($r = 0.53$). Municipal point sources in the SPARROW model expressed on a per capita basis have a median of 3.3 kg-N person^{-1} with an interquartile range from 1.8 to 5.8 kg-N person^{-1} (Alexander et al. 2001), and compare with a per capita rate of 1.85 kg-N person^{-1} for the GLOBAL model. Per capita rates for residential wastewater effluent in the United States have been previously estimated to range from 2.2 to 7 kg-N person^{-1} (Thomann 1972; US EPA 1980).

The estimation of source contributions to streams using the HOWARTH model is difficult because of uncertainty over how the intercept should be apportioned to each source and the lack of separate point and cultural diffuse inputs to the model. Although the model intercept of -120 provides a reasonably accurate adjustment to total stream export for additional N sources and watershed attenuation, adjustments for these factors cannot be reliably made to individual source terms. For example, the model does not ensure that the individual sources (atmospheric deposition, net antropogenic inputs) have positive nitrogen mass or that the mass contributions for sources are less than total stream export. This is not resolved by the use of any of several assumptions about how the intercept might be apportioned to the sources, including the assumption that the intercept is distributed to each source in proportion to the source's share of the net inputs of nitrogen to each watershed. Under this assumption, atmospheric contributions range from 67 to 115% and net anthropogenic sources range from –15 to 38%.

7. Model predictions of nitrogen attenuation in watersheds

We made separate estimates of the rates of nitrogen loss in streams and on the landscape (see Table 7) using empirically derived N loss coefficients in the SPARROW model (Alexander et al. 2000, 2001) and estimates of the total N loss in the northeastern watersheds, based on the difference between major N inputs and stream nitrogen export (see Boyer et al. 2002). The other nitrogen export models examined here lack explicit coefficients that quantify the rates of N removal in watersheds. We compared these estimates with those generated by an application of the RivR-N model to the northeastern watersheds (see Table 8; Seitzinger et al. 2002). RivR-N is a statistical in-stream loss model that was calibrated using literature observations from mass balance and denitrification studies for North American and European lakes and streams. The model was used to estimate the removal of nitrogen in streams and lakes of the northeastern watersheds as a function of the physical and hydraulic characteristics (i.e. depth, time of travel) of the water bodies.

SPARROW estimates of stream nitrogen losses, when expressed as a percentage of the total quantities of nitrogen removed in the watersheds (Boyer et al. 2002), range from 7% to 54% (median = 24%; IQR = 18 to 31%; Table 7). These estimates suggest that a majority – typically about 75% – of the nitrogen loss in watersheds can be explained by attenuation processes on the landscape (median = 76%; IQR = 71 to 82%). By comparison, RivR-N estimates of in-stream nitrogen loss are higher. When expressed as a percentage of the total quantities of nitrogen removed in the watersheds, are typically about 48% (IQR = 40 to 63%; Table 8). Thus, according to this model, landscape processes would typically account for 37 to 60% (IQR; median = 52%) of the total quantities of N removed in the watersheds.

The higher estimates of in-stream loss by the RivR-N model may primarily reflect differences in the spatial scale of the river networks used to derive the estimates. The RivR-N model includes additional nitrogen losses in streams smaller than RF1 streams (River Reach File; 1:500,000 scale; see Seitzinger et al. 2002) to which SPARROW is applied. The RivR-N model includes

1:100,000-scale reaches that are located upstream of RF1 streams. The RivR-N loss estimates, expressed as a percentage of external inputs to streams, are about 25 to 60% higher than the SPARROW estimates (median = 39%). By contrast, the RivR-N loss estimates for the 1:500,000-scale RF1 streams are only modestly larger than those for SPARROW (median ratio of RivR-N to SPARROW loss percentage = 1.11; IQR = 0.95 to 1.16; r = 0.75). Inclusion of the smaller 1:100,000-scale streams increases the loss percentages by about 10 to 20 percentage points (Seitzinger et al. 2002). As much as about 25% of the difference in stream loss between the two methods may relate to non-uniformities in the geographic distribution of sources in the watersheds. When SPARROW estimates of in-stream loss are derived under an assumed uniform spatial distribution of sources (identical to that assumed by the RivR-N model), the estimated losses are typically larger (median = 11%; IQR = 5 to 25%) than those in Table 7, which are based on the actual reach locations of sources. This suggests that a larger proportion of the point and diffuse sources are probably located in the lower portions of the watersheds (e.g. urban sources) and undergo less decay during the shorter travel times to watershed outlets.

8. Summary and conclusions

We evaluated the accuracy (bias and precision) of six nitrogen export models having varying levels of spatial resolution and process complexity and representing various empirical and quasi-empirical models that have been applied to large watersheds. Four of the models were previously described in the literature; two models were statistically calibrated in this study using published data sets for the largest rivers of the world. Many of the models were previously shown to explain large portions of the spatial variability in nitrogen export from rivers in major continents of the world according to reported R^2 statistics. However, the accuracy (bias and precision) of model predictions of stream export, which R^2 alone does not reliably measure, has not been previously reported and compared among the models. This study illustrates the value of using more reliable methods than R^2 to evaluate model performance. We validated the models using detailed data on stream nitrogen export, land-use, and natural and cultural inputs of nitrogen for 16 northeastern watersheds in the United States. The watersheds cover a sufficient portion of the range of the conditions present in the original calibration watersheds so as to provide an appropriate set of locations for evaluating the models. The analysis improves understanding of how the models perform over a range of environmental settings and how model complexity affects prediction accuracy.

Most of the models predicted stream nitrogen export to within 50% of the measured export in a majority of the watersheds; however, all models showed at least small amounts of bias in the model predictions. The three models with the smallest bias (SPARROW, LS1-GLOBAL, and GLOBAL) have a median prediction error of less than 5%. The PEIERLS model had the largest bias (median error = –27%) followed by the HOWARTH model (median error = –14%); both of these models under predicted nitrogen export in 12 of the 16 watersheds. The lowest variability in the prediction errors (i.e. most precise estimates of stream export) was observed for the HOWARTH model, followed by the SPARROW and PEIERLS models.

We developed regression-based models of the prediction errors to determine whether biases in model predictions are potentially caused by mis-specification of the models in relation to various watershed characteristics (i.e. 'factor-related' bias). Such biases may be caused by sources or delivery processes that are not explicitly included in the models or model coefficients that inaccurately describe the supply and transport of nitrogen. This measure of bias provides information about the performance of the nitrogen export models in specific environmental settings. Evidence of factor-related bias implies that improvements are feasible in the accuracy of the model predictions through improved calibrations or modifications of the model structure.

The two nitrogen export models with the smallest factor-related biases (SPARROW, HOWARTH), as evidenced by small coefficient values for each of the four watershed properties (cultivated land, developed land, runoff, drainage size) evaluated in the error models, had prediction biases that were at least 50% and smaller. The prediction biases were also less statistically significant than those detected for the other models. Because these models have more detailed descriptions of nitrogen sources, land and water attenuation, and water flow paths than the other models, the results suggest that model complexity has a beneficial effect on the accuracy of the predictions of stream export. The HOWARTH model gives a detailed accounting of agricultural sources, including crop N fixation and the import and export of foods and feeds. SPARROW spatially references stream monitoring data, point and diffuse nitrogen sources, and landscape properties to surface water flow paths and imposes mass-balance constraints to empirically estimate the rates of nitrogen transport on the landscape and in streams.

The evaluations of factor-related biases indicated that the prediction errors of all of the export models are inversely correlated with cultivated land area. Thus, there is a tendency for the models to under predict stream export in watersheds that are more highly agricultural and over predict in watersheds with less cropland and larger amounts of forested lands. The lower cultivated land-related bias for the HOWARTH model may reflect the value of a more detailed accounting of the supply and transport of nitrogen in agricultural watersheds, including nitrogen fixation in crops, feed imports, and crop exports. These are not explicitly accounted for in the various global models. The inclusion of the livestock waste source in the SPARROW model provides additional specification of agricultural sources that may account for its relatively low cultivated land-related bias.

All of the export models except for SPARROW showed a statistically significant negative correlation between prediction errors and runoff. In view of the effects of runoff on stream export and nitrogen attenuation in watersheds, this finding suggests that the models may need to account more effectively for nitrogen loss processes (e.g. denitrification, storage) at the watershed scale related to the rates of water transport through surface and subsurface pathways. The rates of nitrogen removal on the landscape and in streams may be mediated by various hydrogeologic factors related to runoff (e.g. channel density, stream morphology, water velocity, soil texture, groundwater storage). More explicit descriptions of these factors in the models may improve prediction accuracy.

Comparisons of the model predictions of source contributions to stream export displayed the greatest consistency in the results for agricultural sources; notable differences were found in the estimates of point sources and atmospheric contributions. Although there are uncertainties as to the specific effects of the prediction biases on the estimates of source contributions, the error analysis suggests that many of the models may underestimate the contributions of agricultural sources in more highly agricultural watersheds. Some of the models may overestimate N contributions from point sources in more highly developed watersheds.

The study represents an initial effort to validate the reliability of several prominent stream nitrogen export models, and provides information for guiding future applications and enhancements to the models. The regression based error analysis illustrated here can be readily applied in future evaluations of stream export models. It provides a reasonable approach for validating and possibly correcting watershed models. The method also identifies factor-related biases that can potentially be eliminated through improved model calibrations. Future assessments of model errors would benefit from evaluations of additional stream export models, such as GWLF (Haith & Shoemaker 1987) and export-coefficient models (Johnes 1996), as well as the inclusion of larger numbers of watersheds representing a more diverse range of climate, land uses, and nitrogen sources. Because of the importance of landscape attenuation and nitrogen processing related to specific land uses, future error assessments should make use of deterministic landscape models in evaluating stream export models. Improvements in the modeling of landscape sources and sinks may yield important gains in prediction accuracy of regional export models, and provide insight into ways to scale up catchment fluxes more reliably. Research in this area may also lead to improvements in the ability to combine mechanistic descriptions of processes in deterministic models with statistical methods of empirically estimating flux rates at the watershed scale.

Acknowledgements

This work was initiated as part of the International SCOPE Nitrogen Project, which received support from both the Mellon Foundation and from the National Center for Ecological Analysis and Synthesis. Support was provided by the U.S. Geological Survey National Water Quality Assessment Program (NAWQA). We thank LJ Puckett and M Focazio of the USGS and two anonymous reviewers for comments on the manuscript.

References

Alexander RB, Smith RA & Schwarz GE (2000) Effect of stream channel size on the delivery of nitrogen to the Gulf of Mexico. Nature 403: 758–761

Alexander RB, Smith RA, Schwarz GE, Preston SD, Brakebill JW, Srinivasan R & Pacheco PA (2001) Atmospheric nitrogen flux from the watersheds of major estuaries of the United States: An application of the SPARROW watershed model. In: Valigura R, Alexander R, Castro M, Meyers T, Paerl H, Stacey P & Turner RE (Eds) Nitrogen Loading in Coastal Water Bodies: An Atmospheric Perspective, American Geophysical Union Monograph 57, pp 119–170

Beaulac MN & Reckhow KH (1982) An examination of land use – nutrient export relationships. Wat. Res. Bull. 18: 1013–1024

Behrendt H (1996) Inventories of point and diffuse sources and estimated nutrient loads – a comparison for different river basins in central Europe. Water Sci. Tech. 33: 99–107

Bicknell BR, Imhoff JC, Kittle JL Jr., Donigian AS Jr & Johanson RC (1997) Hydrological simulation program – fortran user's manual for release 11. U.S. Environmental Protection Agency, Environmental Research Laboratory, Athens, Georgia, U.S.A. EPA/600/R–97/080

Boyer EW, Goodale CL, Jaworski NA & Howarth RW (2002) Anthropogenic nitrogen sources and relationships to riverine nitrogen export in the northeastern U.S.A. Biogeochemistry 57/58: 137–169

Caraco NF & Cole JJ (1999) Human impact on nitrate export: An analysis using major world rivers. Ambio. 28: 167–170

Castro MS, Driscoll C, Jordan TE, Reay W, Seitzinger S, Stiles R & Cable J (2001) Assessment of the contribution made by atmospheric nitrogen deposition to the total nitrogen load to thirty-four estuaries on the Atlantic and Gulf coasts of the United States, In: Valigura R, Alexander R, Castro M, Meyers T, Paerl H, Stacey P & Turner RE (Eds)

Nitrogen Loading in Coastal Water Bodies: An Atmospheric Perspective, American Geophysical Union Monograph 57, pp 77–106

Cohn TA, DeLong LL, Gilroy EJ, Hirsch RM & Wells DK (1989) Estimating constituent loads. Wat. Resour. Res. 25: 937–942

Cunnane C (1978) Unbiased plotting positions – a review. J. Hydrology 37: 205–222

Delwiche LL & Haith DA (1983) Loading functions for predicting nutrient losses from complex watersheds. Water Resour. Bul. 19: 951–959

Fisher DC & Oppenheimer M (1991) Atmospheric nitrogen deposition and the Chesapeake Bay estuary. Ambio. 20: 102–108

Frink CR (1991) Estimating nutrient exports to estuaries. J. Environ. Qual. 20: 717–724

Galloway JN, Levy II H & Kasibhatla PS (1994) Year 2020: consequences of population growth and development on the deposition of oxidized nitrogen. Ambio. 23: 120–123

Haith DA & Shoemaker LL (1987) Generalized watershed loading functions for stream flow nutrients. Water Resources Bulletin 23: 471–478

Hill AR (1996) Nitrate removal in stream riparian zones. J. Envir. Qual. 25: 743–755

Howarth RW (1998) An assessment of human influences on fluxes of nitrogen from the terrestrial landscape to the estuaries and continental shelves of the North Atlantic Ocean. Nutrient Cycling in Agroecosystems 00: 1–11

Howarth RW, Billen G, Swaney D, Townsend A, Jaworski N, Lajtha K, Downing JA, Elmgren R, Caraco N, Jordan T, Berendse F, Freney J, Kudeyarov V, Murdoch P & Zhu Zhao-liang (1996) Regional nitrogen budgets and riverine N & P fluxes for the drainages to the North Atlantic Ocean: natural and human influences. Biogeochem. 35: 75–139

Howarth RW, Fruci JR & Sherman D (1991) Inputs of sediment and carbon to an estuarine ecosystem: Influence of land use. Ecological Applications 1: 27–39

Jaworski NA, Groffman PM, Keller AA & Prager JC (1992) A watershed nitrogen and phosphorus balance: the Upper Potomac River basin. Estuaries 15: 83–95

Jaworski NA, Howarth RW & Hetling LJ (1997) Atmospheric deposition of nitrogen oxides onto the landscape contributes to coastal eutrophication in the northeast United States. Environ. Sci. Techno. 31: 1995–2004

Johnes PJ (1996) Evaluation and management of the impact of land use change on the nitrogen and phosphorus load delivered to surface waters: the export coefficient modelling approach. J. of Hydrology 183: 323–349

Johnes PJ & Heathwaite AL (1997) Modelling the impact of land use change on water quality in agricultural catchments. Hydrological Processes 11: 269–286

Johnes PJ, Moss B & Phillips GL (1996) The determination of water quality by land use, livestock numbers and population data – testing of a model for use in conservation and water quality management. Freshwater Biology 36: 951–473

Johnson DW (1992) Nitrogen retention in forest soils. J. Envir. Qual. 21: 1–12

Jordan TE & Weller DE (1996) Human contributions to terrestrial nitrogen flux: assessing the sources and fates of anthropogenic fixed nitrogen. Bioscience 46: 655–664

Kelly CA, Rudd JWM, Hesslein RH, Schindler DW, Dillon PJ, Driscoll CT, Gherini SA & Hecky RE (1987) Prediction of biological acid neutralization in acid-sensitive lakes. Biogeochem. 3: 129–141

Lee KY, Fisher TR, Jordan TE, Correll DL & Weller DE (2000) Modeling the hydrochemistry of the Choptank River Basin using GWLF and Arc/Info: 1. Model calibration and validation. Biogeochem. 49: 143–173

Lewis WM Jr, Melack JM, McDowell WH, McClain M & Richey JE (1999) Nitrogen yields from undisturbed watersheds in the Americas. Biogeochemistry 46: 149–162

Lewis WM Jr (2002) Yield of nitrogen from minimally disturbed watersheds of the United States. Biogeochemistry 57/58: 375–385

McBride GB, Alexander RB, Elliot AH & Shankar U (2000) Regional scale modelling of water quality. Water and Atmosphere, Vol. 8 (pp 29–31). National Institute of Water and Atmospheric Research (NIWA), Auckland, New Zealand

Meybeck M (1982) Carbon, nitrogen and phosphorus transport by world rivers. Amer. J. of Sci. 282: 401–450

Molot LA & Dillon PJ (1993) Nitrogen mass balances and denitrification rates in central Ontario Lakes. Biogeochem. 20: 195–212

Montgomery DC & Peck EA (1982) Introduction to linear regression analysis. John Wiley & Sons, New York, NY

National Research Council (2000) Clean coastal waters: Understanding and reducing the effects of nutrient pollution, Ocean Studies Board and Water Science and Technology board, National Academy of Sciences, Washington, DC

Nixon SW (1995) Coastal marine eutrophication: a definition, social causes, and future concerns. Ophelia 41: 199–219

Parton WJ, Stewart JWB & Cole CV (1988) Dynamics of C, N, P, and S in grassland soils: a model. Biogeochem. 5: 109–131

Peierls BL, Caraco NF, Pace ML & Cole JJ (1991) Human influence on river nitrogen. Nature 350: 386–387

Preston SD & Brakebill JW (1999) Application of spatially referenced regression modeling for the evaluation of total nitrogen loading in the Chesapeake Bay watershed. U.S. Geological Survey Water Resources Investigations Report 99–4054

Rastetter EB, King AW, Cosby BJ, Hornberger GM, O'Neill RV & Hobbie JE (1992) Aggregating fine-scale ecological knowledge to model coarser-scale attributes of ecosystems. Ecological Apps. 2: 55–70

Sauer TJ, Alexander RB, Brahana JV & Smith RA (in press) The importance and role of watersheds in the transport of nitrogen. In: Follett, RF & JL Hatfield (Eds) Nitrogen in the Environment: Sources, Problems, and Management. Elsevier Science Publishers, the Netherlands

Seitzinger SP & C Kroeze (1998) Global distribution of nitrous oxide production and N inputs in freshwater and coastal marine ecosystems. Global Biogeochem. Cycles 12: 93–113

Seitzinger SP, Renee VS, Boyer EA, Alexander RB, Billen G, Howarth RW, Bernhard MB & van Breemen N (2002) Nitrogen retention in rivers: Model development and application to watersheds in the northeastern U.S.A. Biogeochemistry 57/58: 199–237

Smith RA, GE Schwarz & RB Alexander (1997) Regional interpretation of water-quality monitoring data. Wat. Resour. Res. 33: 2781–2798

Srinivasan R, Arnold JG, Muttiah RS, Walker D & Dyke PT (1993) Hydrologic unit modeling of the United States (HUMUS). In: Yang S (Ed) Advances in Hydro-Science and Engineering, Vol. I, Part A (pp 451–456). Washington, D.C., U.S.A.

Stacy PE, Greening HS, Kremer JN, Peterson D & Tomasko DA (2001) Contributions of atmospheric nitrogen deposition to U.S. estuaries: Summary and conclusions. In: Valigura R, Alexander R, Castro M, Meyers T, Paerl H, Stacey P & Turner RE (Eds) Nitrogen Loading in Coastal Water Bodies: An Atmospheric Perspective, American Geophysical Union Monograph 57, pp 187–226

Thomann (1972) Systems analysis and water quality management. Environmental Research and Applications, Inc., New York, NY

United States Environmental Protection Agency (1980) Design manual: Onsite wastewater treatment and disposal systems. Office of Water Programs Operations, Washington, DC, EPA 625/1-80-012

Van Breemen N, Boyer EW, Goodale CL, Jaworski NA, Paustian K, Seitzinger SP, Lajtha K, Mayer B, van Dam D, Howarth RW, Nadelhoffer KJ, Eve M, & Billen G (2002) Where did all the nitrogen go? Fate of nitrogen inputs to large watersheds in the northeastern U.S.A. Biogeochemistry 57/58: 267–293

Vitousek PM, Aber JD, Howarth RW, Likens GE, Matson PA, Schindler DW, Schlesinger WH & Tilman DG (1997) Human alteration of the global nitrogen cycle: sources and consequences. Ecological Applications 7: 737–750

Whitehead PG, Wilson EJ & Butterfield D (1998) A semi-distributed Integrated Nitrogen model for multiple source assessment in Catchments (INCA): Part I – model structure and process equations. Sci. of the Total Environ. 210/211: 547–558

Young RA, Onstad CA & Bosch DD (1995) AGNPS: An agricultural nonpoint source model. In: Singh VP (Ed) Computer Models of Watershed Hydrology. Water Resource Publications, Highlands Ranch, Colorado

Biogeochemistry **57/58**: 341–374, 2002.
© 2002 *Kluwer Academic Publishers. Printed in the Netherlands.*

Regional analysis of inorganic nitrogen yield and retention in high-elevation ecosystems of the Sierra Nevada and Rocky Mountains

JAMES O. SICKMAN[1]*, JOHN M. MELACK[1] & JOHN L. STODDARD[2]
[1]*Department of Ecology, Evolution and Marine Biology, University of California, Santa Barbara, and Donald Bren School of Environmental Science and Management, University of California, Santa Barbara 93106;* [2]*United States Environmental Protection Agency, Western Ecology Division 200 SW 35th Street, Corvallis, Oregon*
(*author for correspondence, e-mail: sickman@icess.ucsb.edu.)*

Key words: catchment features, nitrogen, nitrogen retention, nitrogen yield, Rocky Mountains, Sierra Nevada

Abstract. Yields and retention of dissolved inorganic nitrogen (DIN: $NO_3^- + NH_4^+$) and nitrate concentrations in surface runoff are summarized for 28 high elevation watersheds in the Sierra Nevada of California and Rocky Mountains of Wyoming and Colorado. Catchments ranged in elevation from 2475 to 3603 m and from 15 to 1908 ha in area. Soil cover varied from 5% to nearly 97% of total catchment area. Runoff from these snow-dominated catchments ranged from 315 to 1265 mm per year. In the Sierra Nevada, annual volume-weighted mean (AVWM) nitrate concentrations ranged from 0.5 to 13 μM (overall average 5.4 μM), and peak concentrations measured during snowmelt ranged from 1.0 to 38 μM. Nitrate levels in the Rocky Mountain watersheds were about twice those in the Sierra Nevada; average AVWM NO_3^- was 9.4 μM and snowmelt peaks ranged from 15 to 50 μM. Mean DIN loading to Rocky Mountain watersheds, 3.6 kg ha^{-1} yr^{-1}, was double the average measured for Sierra Nevada watersheds, 1.8 kg ha^{-1} yr^{-1}. DIN yield in the Sierra Nevada, 0.69 kg ha^{-1} yr^{-1}, was about 60% that measured in the Rocky Mountains, 1.1 kg ha^{-1} yr^{-1}. Net inorganic N retention in Sierra Nevada catchments was 1.2 kg ha^{-1} yr^{-1} and represented about 55% of annual DIN loading. DIN retention in the Rocky Mountain catchments was greater in absolute terms, 2.5 kg ha^{-1} yr^{-1}, and as a percentage of DIN loading, 72%.

A correlation analysis using DIN yield, DIN retention and surface water nitrate concentrations as dependent variables and eight environmental features (catchment elevation, slope, aspect, roughness, area, runoff, soil cover and DIN loading) as independent variables was conducted. For the Sierra Nevada, elevation and soil cover had significant ($p < 0.1$) Pearson product moment correlations with catchment DIN yield, AVWM and peak snowmelt nitrate concentrations and DIN retention rates. Log-linear regression models using soil cover as the independent variable explained 82% of the variation in catchment DIN retention, 92% of the variability in AVWM nitrate and 85% of snowmelt peak NO_3^-. In the Rocky Mountains, soil cover was significantly ($p < 0.05$) correlated with DIN yield, AVWM NO_3^- and DIN retention expressed as a percentage of DIN loading (%DIN retention). Catchment mean slope and terrain roughness were positively correlated with steam nitrate concentrations and negatively related to %DIN retention. About 91% of the variation in DIN yield and 79% of the variability in AVWM NO_3^- were explained by log-linear models based on soil cover. A log-linear regression based on soil cover explained 90% of the variation of % DIN retention in the Rocky Mountains.

Introduction

Regional analyses of N budgets have shown that watershed characteristics such as runoff, elevation and atmospheric N deposition are related to watershed N export and provide a means of estimating N yields over broad areas (Howarth et al. 1996; Lewis in press). Lewis et al. (1999) found that yields of all N fractions were strongly related to runoff and runoff explained nearly 85% of the variance in yield of N in undisturbed forested and savanna catchments ranging in area from 0.1 to 5×10^6 km. Six alpine and subalpine watersheds from the Sierra Nevada were included in these analyses, but were found to form a distinctive cluster owing to their exceedingly low rate of N yield (i.e., less than 10% of the mean yield measured for the other study catchments).

In another analysis, using data from undisturbed temperate-zone watersheds of the United States (United States Geological Survey Hydrologic Benchmark Network), Lewis (in press) found an equally strong relationship between runoff and N yield. Based on this study, the rate of dissolved inorganic nitrogen (DIN: $NO_3^- + NH_4^+$) yield from Sierra Nevada watersheds is less than 40% of the rate found in temperate zone watersheds with similar runoff and atmospheric N deposition. The finding that alpine and subalpine catchments export less N than temperate or tropical watersheds with similar characteristics is surprising since ecosystem retention of N may be constrained in high-elevation regions. Short growing seasons, extensive and deep snowcover and sparse vegetation result in low N retention capacity and a large temporal disconnection between N availability (spring snowmelt) and vegetative N demand (summer). Thus, alpine and subalpine watersheds seem to comprise a distinct population for studying the relationships between catchment characteristics and N yield and for evaluating the impact of increased N deposition.

Episodic declines in acid neutralizing capacity (ANC) have been observed in alpine and subalpine catchments and results largely from ionic dilution following an initial pulse of nitrate and base cations (Melack & Stoddard 1991; Stoddard 1995); episodic acidification (ANC values < 0) may occur when nitrate pulses are sufficiently large (Stoddard 1995; Leydecker et al. 1999). Increasing N deposition to alpine and subalpine ecosystems in the

Colorado Front Range has resulted in increases of DIN in surface waters and current modeling studies suggest that alpine tundra and subalpine forests may experience nitrogen saturation at N deposition greater than 4–6 kg N ha^{-1} yr^{-1} (Baron et al. 1994; Williams et al. 1996a; Heuer et al. 1999; Williams & Tonnessen 2000). To date, ecological changes from N deposition appear to be restricted to the Front Range, but as urbanization increases in and near the Rocky Mountains the extent of N-affected ecosystems may increase. In the Sierra Nevada, recent shifts in limitation of algal growth at Lake Tahoe and Emerald Lake have been associated with alterations in both N and P supply (Jassby et al. 1994; Sickman & Melack 1998; Sickman et al. unpublished data). Given the current status of high elevation ecosystems in the western United States and the likelihood that N deposition will increase (Galloway et al. 1994), it would be valuable to predict, on a regional basis, the N retention capacity of these ecosystems. If critical and target loads for nitrogen are to be determined, data from a regionally extensive set of catchments are required (Williams 1997; Williams & Tonnessen 2000). To date, however, there have been few process-level studies on N cycling in alpine and subalpine watersheds (e.g., Brooks et al. 1996, 1998; Meixner et al. 1998, 1999) of sufficient detail to model accurately the impact of increased N loading. Furthermore, it will be difficult to extrapolate results from plot-scale modeling studies to larger regions despite recent improvements in biogeochemical modeling (Baron et al. 1994; Kiefer & Fenn 1997; Magill et al. 1997) given the potentially large temporal and spatial variability of N sources, sinks and transformations at the landscape scale.

In contrast, there is a wealth of catchment-scale data on the input and loss of nitrogen from alpine and subalpine watersheds in the Sierra Nevada and Rocky Mountains. We propose that this information can provide a basis for predicting the N retention capacity of high elevation ecosystems over large areas. Nitrogen budgets for alpine and subalpine watersheds in the western United States have been accumulating since the early 1980s and the dataset is now of a size that allows for a statistical analysis of environmental and catchment features influencing the N retention capacity of high elevation ecosystems. Similar analyses, using variables such as runoff, catchment area and elevation, have been successful in predicting elemental fluxes and chemical concentrations in surface runoff across large regions and over a broad range of conditions (Meybeck 1982; Hedin et al. 1995; Howarth et al. 1996; Lewis et al. 1999).

Using previously published and unpublished data from high elevation watersheds in the western United States, we investigate the relationships between catchment N export and retention and seven watershed variables: elevation, watershed area, runoff, % soil cover, inorganic nitrogen loading,

344

and catchment aspect, slope and roughness. Our goal is to test the hypothesis that nitrogen yields, retention capacity and surface water chemistry (NO_3^-) can be predicted on the basis of general environmental and terrain variables in high elevation ecosystems. If successful, these variables will provide a basis for assessing the sensitivity of high elevation ecosystems to increased N deposition and may prove useful in regional-scale modeling of N biogeochemistry and setting of critical nitrogen loads.

Methods

Our statistical analyses are restricted to alpine and subalpine catchments of the Sierra Nevada and Rocky Mountains and to inorganic nitrogen budgets, i.e., inputs and losses of nitrate and ammonium. Little data are available on the fluxes of organic nitrogen in high elevation catchments, although there is growing evidence that organic N is an important component in atmospheric deposition and ecosystem nitrogen losses (Church 1999; Neff et al. in press). Current studies show that forested watersheds at low to middle elevations have high N retention rates and little DIN yield and, for that reason, are not included in our analyses.

Chemical data used were drawn primarily from previously published studies (Tables 1 and 2). For some Sierra Nevada catchments, fluxes were computed based on unpublished records of stream chemistry, stream discharge and loading using methods from Melack et al. (1998) (Table 1). In all cases the raw data were evaluated for completeness and quality. All catchments had comprehensive estimates of annual inorganic N loading in wet deposition and in some instances dry deposition (Table 3). In cases where no dry deposition estimates were available we conservatively assumed that dry N loading was 25% of wet inorganic N deposition; we based this percentage on dry deposition measurements made at Niwot Ridge and Emerald Lake (Sievering et al. 1996; Williams et al. 1995; Sickman et al. in press). Outflow DIN losses were based on at least biweekly chemistry during snowmelt runoff (the period of greatest N yield) and periodic sampling during the remainder of the year; for the majority of the Sierra Nevada catchments, automated samplers were used to collect samples every 1–2 days during snowmelt runoff. Data had to span at least one annual cycle to be included and in most cases several years were available (Table 1). Data from sub-regions ($\geq$ 10 ha) of larger catchments were included in the analysis (e.g., Andrews Creek and Icy Brook) if measurements of N fluxes and surface water chemistry were available.

Table 1. Landscape characteristics of high elevation watersheds in the Sierra Nevada. Soil cover is expressed as a percentage of total catchment area. Mean slope and mode aspect are in degrees. Mean roughness is dimensionless

Catchment	Elevation m	Area ha	Runoff mm yr^{-1}	Soil Cover	Mean Slope	Mode Aspect	Mean Roughness	Years of Record	Sources[1]
Crystal	2951	135	424	53%	21	105	42	1990–93	A
Emerald	2800	120	1120	22%	29	278	38	1985–98	A
Lost	2475	25	1210	36%	14	214	34	1990–93	A
Marble Fork-Kaweah	2621	1908	1245	40%	18	278	34	1993–94	A
Pear	2904	136	703	22%	24	281	37	1990–93	A
Ruby	3390	441	507	18%	27	108	42	1990–94	A
Spuller	3131	97	789	33%	22	60	37	1990–94	A
Topaz	3218	178	696	41%	10	108	32	1990–98	A
High	3603	15	811	5%	17	93	45	1993–94	B
Low	3444	225	926	8%	26	103	42	1993–94	B
M1	3078	106	1265	20%	18	318	36	1993–94	B
M2	3188	90	995	18%	11	315	39	1993–94	B
M3	3249	67	986	10%	11	360	39	1993–94	B
Mills	3554	177	912	6%	26	82	43	1993–94	B
Treasure	3420	175	636	10%	29	101	42	1993–94	B
Sierran Mean =	3135	260	882	23%	20	187	39		

[1] Sources: A: Melack et al. 1998; B: Stoddard 1995, Sickman and Stoddard unpublished data

346

Independent variables

Watershed features used as independent variables in the statistical analyses, i.e., elevation, area, runoff and soil cover, were chosen because they were obtainable and are surrogates for complex environmental processes that are known to control N cycling in catchments. These processes include both the size of and fluxes between the major watershed nitrogen pools, the transit time and pathways for water movement and the degree of soil and ground-water flushing. Elevation (at catchment outlet) captures several catchment features including, vegetation biomass and type, length of growing season and vegetative N demand (Fisk et al. 1998). Area is a proxy for time and distance of N transport in a watershed (Lovett et al. 2000) and may provide a surrogate for hydrologic flowpaths and variable source-area dynamics; all of which exert control on nitrogen cycling in watersheds (Creed & Band 1998). Runoff reflects the amount of flushing experienced by catchment soil, the amount of water available to vegetation and soil moisture properties that may affect N processes such as denitrification; runoff is also highly correlated with precipitation. The rationale behind including soil cover in the analysis is based on several recent studies suggesting that soil microbial processes control N cycling in high elevation ecosystems (Brooks et al. 1999; Brooks & Williams 1999; Heuer et al. 1999). Soil cover was computed as a percentage of total catchment area. Soil depths and development are most likely positively related to soil area, thus soil area may approximate soil volume, soil N content and the magnitude of soil microbial N processes. Inorganic nitrogen loading (expressed in units of kg ha^{-1} yr^{-1}) was included, because it provides a basis for testing whether current N loads are affecting surface water chemistry and N yields, and sets the baseline against which potential future increases in N loading may be gauged.

Three additional terrain indices, mean slope, mode aspect and mean roughness, were computed from the U.S. Geological Survey National Elevation Dataset (NED), and used as independent variables in the correlation analysis. The NED is a seamless, 30 m-resolution, gridded elevation dataset that has been filtered to minimize artifacts. Slope was calculated by fitting a plane to the elevation values of a 3 × 3 neighborhood of cells around each NED cell; the direction the fitted plan faces is the aspect for the cell. Terrain roughness (Andrew et al. 1999) reflects variation in slope and aspect at each cell of the NED and was computed as follows:

$$R_{ij} = ((V_s/V_m)^*100) + ((A_n/8)^*100)$$

Where R_{ij} is the roughness at cell row i, column j; V_s is the standard deviation of slope in a 3 × 3 cell neighborhood around cell ij; V_m is the maximum standard deviation in slope for any 3×3 cell neighborhood for all of the 28 study watersheds; and A_n is the number of different aspect classes (binned into eight, 45 degree sectors) found within each 3×3 cell neighborhood. Any NED cell with a high variation in slope and many different aspect classes within the 3×3 cell neighborhood would have a high roughness value. The mean roughness value for each of the 28 watersheds was used in the correlation analysis.

Slope was included in the correlation analyses as a measure of the steepness of the catchment, which may influence hydrologic residence time or flow-routing in mountainous terrain (Clow & Sueker 2000). Aspect controls the input and distribution of solar radiation in a catchment (Dozier & Frew 1990) and may capture variations in the relative timing of snowmelt (Cline et al. 1998) and patterns of soil moisture which could effect N cycling (Sickman et al. in press). Mean roughness is a measure of the relative terrain complexity among the study sites and may provide an index for time and distance of N transport in a watershed, hydrologic flowpaths and residence time, and variable source-area dynamics.

Dependent variables

Five dependent variables were used in the statistical analyses: dissolved inorganic nitrogen yield (DIN: $NO_3^- + NH_4^+$), annual volume-weighted mean (AVWM) nitrate concentration, peak snowmelt nitrate concentration, and DIN retention (both net and % change). In cases where there was more than one year of data, we averaged the annual estimates to obtained a single value for each variable. Averaging was necessary in order to balance the influence of catchments with many years of data (i.e., Emerald and Loch Vale) with catchments with few years of data.

DIN yield is the amount of dissolved inorganic nitrogen exported via catchment outflow and was expressed in kg N ha^{-1} yr^{-1} (i.e., nitrogen fluxes are expressed in terms of the mass of elemental N and not compound mass). With the exception of Green Lakes #4, DIN yield was computed by the authors of the original study. DIN yield at Green Lakes #4 was computed from raw data (discharge and chemical concentrations) obtained from the Niwot Ridge LTER database. DIN yield estimates from the Hourglass catchments include only nitrate losses and were included because ammonium concentrations in high elevation watersheds are typically at or near the detection limit (Landers et al. 1987). Annual volume-weighted mean nitrate concentrations are discharge-weighted averages of outflow nitrate concentrations. In the case of Snake River and Deer Creek, AVWM nitrate was computed from nitrate yields and catchment runoff. For Green Lakes #4 we computed AVWM nitrate from raw data (discharge and chemical concentrations) obtained from the Niwot Ridge LTER database. Peak nitrate concentrations were determined from time-series data during snowmelt runoff when available; the average of all available years was used for each catchment. The intensity of chemical sampling allowed us to make accurate estimates of peak concentrations at all catchments since peak concentrations occurred only slightly before peak runoff (i.e., 1 to 3 weeks). Nitrate concentrations were included in the analyses because they provide a means for judging the N saturation status of catchments and the degree of strong acid-anion acidification during snowmelt.

Inorganic nitrogen retention was computed by subtracting DIN yield from DIN loading. For the analyses we expressed retention both in absolute terms (net DIN retention: kg N ha^{-1} yr^{-1}) and as a fraction of loading (% DIN retention: % of DIN loading). Expressing retention as a fraction of loading allowed us to compare the N retention efficiency of catchments with different N loading.

Correlation and regression procedures

Pearson product moment correlations were used to measure the strength of association between the dependent and independent variables within the Sierra Nevada and Rocky Mountain datasets. The Pearson correlations were tested with Bonferroni's method to evaluate the statistical significance of the associations. Due to the conservative nature of the test we assigned a threshold of $p < 0.1$ to determine whether variables were significantly correlated. Once significant correlations were identified, linear and log-linear models were developed between the dependent and independent variables using standard regression and multiple regression procedures. In the multiple regression analysis, multi-colinearity between independent variables was assessed by computing a variance inflation factor (VIF) to ensure that independent variables were not significantly correlated to one another.

We also performed a regression tree analysis (least squares fitting method: Systat version 7.01) on the pooled dataset (Rocky Mountain plus Sierra Nevada, n = 26 to 28 depending on dependant variable; see Table 3) to determine whether the watershed and terrain variables could explain differences in dependant variables at larger spatial scales. Owing to our relatively small sample size, tree growth was severely constrained. Regression trees were limited to 5 end-nodes with a minimum of 4 catchments per node. The minimum proportional reduction in error allowed at any branch in the tree was 0.1.

General site descriptions

The catchments used in the analysis are located in the alpine and subalpine zones of the Sierra Nevada of California and Rocky Mountains of Colorado and Wyoming. They capture a wide range of the geographic, geologic and hydrochemical variation among high elevation watersheds in the western United states (Tables 1 and 2). For the Sierra Nevada watersheds, elevations ranged from 2,475 m to 3,603 m and the mean elevation was 3,135 m (Table 1). The Rocky Mountain catchments were of similar elevation with an overall average outlet elevation of 3,186 m (Table 2). Soil coverage in the Sierra Nevada watersheds tended to be lower than in the Rocky Mountains; in all of the Sierra catchments, including those with higher soil coverage such as Crystal, most of the watershed was above treeline. The overall average soil percentage in Sierra Nevada catchments was 23% and ranged from 5 to 53% (Table 1). In the Rocky Mountains, average soil cover was 59% with a range from 5 to 97% (Table 2). In catchments with low soil coverage, talus and bedrock comprise the majority of the watershed area. Mean slope of the study catchments ranged from 10° to 29° in the Sierra Nevada and from 6° to 35° in the Rocky Mountains; the overall mean slope in each data set was 20° (Tables 1 and 2). Catchments in both mountain ranges had a wide variety of aspects (Tables 1 and 2). On average the Sierra Nevada catchments had higher terrain roughness (mean = 39) than the Rocky Mountain watersheds (mean = 34), although the most topographically complex watershed, Andrews Creek (R = 47), is located in the Rocky Mountains (Tables 1 and 2).

At all sites, precipitation fell predominately as snow during the winter and the accumulated snowpack underwent little melt or evaporative losses until spring snowmelt (Williams & Melack 1991; Leydecker & Melack 1999; Baron 1992). Rainfall was sparse, comprising on average about 10–25% of annual precipitation. The snowmelt period accounted for nearly all stream discharge and solute export; winter snowmelt in the Sierra Nevada accounted for less than 5% of annual runoff (Melack et al. 1998); we assume a similar relationship is true for the Rocky Mountains owing to comparable environmental conditions. Average catchment runoff was slightly higher in the Sierra Nevada (mean 882 mm) than in the Rocky Mountains (755 mm). The Emerald, Pear, Topaz and M-site watersheds are all located along the western slope of the southern Sierra Nevada within the Tokopah Valley of Sequoia National Park. This valley comprises the headwaters of the Marble Fork of the Kaweah River. Crystal and Spuller watersheds lie along the eastern slope of the central Sierra. Lost watershed is situated near the crest of the Sierra Nevada near Lake Tahoe. The remainder of the Sierra Nevada watersheds are located along the eastern slope within Rock Creek canyon. Mills and Low are nested subcatchments within the Ruby watershed.

Table 2. Landscape characteristics of high elevation watersheds in the Rocky Mountains. Soil cover is expressed as a percentage of total catchment area. ND = no data available. Mean slope and mode aspect are in degrees. Mean roughness is dimensionless

Catchment	Elevation m	Area ha	Runoff mm yr^{-1}	Soil Cover	Years of Record	Mean Slope	Mode Aspect	Mean Roughness	Sources[1]
Loch Vale	3050	660	750	18%	1984–93	33	5	44	C
Icy Brook	3225	290	815	15%	1992	34	311	44	C, D
Andrews Creek	3300	160	1082	5%	1992	35	310	47	C, D
East Glacier	3282	29	670	81%	1988–90	10	171	35	E
West Glacier	3276	61	1591	39%	1988–90	17	120	38	E
Rabbit Ears Pass	2910	200	609	95%	1991–92	6	198	32	F
Hourglass-Alpine	3192	99	1150	ND	1986–87	14	9	26	G
Hourglass-Subalpine	2871	924	720	ND	1986–87	16	9	24	G
Green Lakes #4	3550	200	857	50%	1985–93	27	341	39	H, I
East St. Louis	2878	803	315	95%	1987–88	18	310	29	J
Fool Creek Alpine	3180	67	400	97%	1987–88	13	27	25	J
Snake River	3350	1040	430	65%	1996	22	279	30	K
Deer Creek	3350	1170	420	85%	1996	18	327	30	K
Rocky Mt. Mean =	3186	439	755	59%		20	186	34	

[1]Sources: C: Baron & Campbell 1997; D: Campbell et al. 1995; E: Reuss et al. 1995; F: Peters & Leavesley 1995, N.E. Peters personal communication; G: Stednick 1989; H: Williams et al. 1996a; I: Niwot Ridge Long-term Ecological Database (BIR 9115097); J: Stottlemyer & Troendle 1992, R. Stottlemyer personal communication; K: Heuer et al. 1999.

Loch Vale watershed and its two subcatchments, Icy Brook and Andrew Creek, are located in Colorado Front Range of Rocky Mountain National Park. East and West Glacier watersheds are in the Glacier Lakes Ecosystem Experiment Site (GLEES) area of southeastern Wyoming. Rabbit Ears Pass watershed is situated in the North Fork Walton Creek basin southeast of Steamboat Springs, Colorado. The two Hourglass catchments are tributaries of the Cache la Poudre River and lie outside the northern boundary of Rocky Mountain National Park. Green Lake #4 is one of a series of lakes located near Niwot Ridge in the Colorado Front Range near Denver, Colorado. East St. Louis and Fool Creek are study areas in the Fraser Experimental Forest (FEF), 137 km west of Denver. The Snake and Deer Creek catchments are located west of the continental divide near FEF.

Results

Nitrate chemistry, DIN yields and DIN retention

On the whole both AVWM and peak nitrate concentrations were higher in the Rocky Mountains than in the Sierra Nevada. Average AVWM nitrate for the Sierra Nevada watersheds was 5.4 μM and for the Rocky Mountain catchments it was 9.4 μM (Table 3). Peak snowmelt concentrations averaged 14 μM in the Sierra and 27 μM in the Rocky Mountains. There was, however, a large overlap in these concentrations. Several of the highest elevation sites in the Sierra Nevada, High Lake, Low Lake and the M-sites, had nitrate concentrations greater than Rocky Mountain catchments located in Wyoming and west of the continental divide i.e., the GLEES watersheds, the Snake River and Dear Creek watershed. For the entire dataset, Loch Vale watershed and its subcatchments had the highest AVWM nitrate levels. Peak concentrations were greatest at Rabbit Ears Pass in the Rocky Mountains, 50 μM, and at High Lake watershed in the Sierra Nevada, 38 μM.

Atmospheric deposition of nitrogen in the Rocky Mountain dataset, 3.6 kg ha^{-1} yr^{-1}, was double the rate measured for the Sierra Nevada catchments, 1.8 kg ha^{-1} yr^{-1} (Table 3). Atmospheric N deposition to catchments along the Front Range of the Rocky Mountains has increased over the past decade, and at Niwot Ridge, N loading as high as 7 kg ha^{-1} yr^{-1} has been measured in recent years (Fenn et al. 1998).

DIN export from the Rocky Mountain catchments, 1.1 kg ha^{-1} yr^{-1}, was greater than the rate of 0.69 kg ha^{-1} yr^{-1} measured for the Sierra Nevada watersheds. The Loch Vale watersheds and subcatchments stand out with yields in the range of 1.7 to 3.1 kg ha^{-1} yr^{-1}. In the Sierra Nevada, relatively high DIN yields, 1.2 to 1.5 kg ha^{-1} yr^{-1}, were measured at High Lake, Low

352

Table 3. Nitrogen chemistry and fluxes in high elevation watersheds in the Sierra Nevada and Rocky Mountains. Units for nitrate concentration are μM. Units for inorganic N (DIN) and dissolved inorganic N (DIN) are kg N ha^{-1} yr^{-1}. Data for outflow mean nitrate are annual volume-weighted means. Outflow peak nitrate is the highest nitrate concentration measured during the annual snowmelt nitrate pulse. ND = no data available

Catchment	Outflow Mean NO_3^-	Outflow Peak NO_3^-	DIN Load	DIN Yield	Net DIN Retention	% DIN Retention
Sierra Nevada:						
Crystal	0.5	1.0	2.0	0.03	2.0	98%
Emerald	4.9	7.0	2.6	0.80	1.8	69%
Lost	0.6	1.8	2.1	0.13	2.0	94%
Marble Fork-Kaweah	2.4	6.0	2.0	0.43	1.5	78%
Pear	4.0	9.0	2.5	0.40	2.1	84%
Ruby	4.1	11	1.5	0.32	1.2	79%
Spuller	4.1	13	1.8	0.44	1.4	76%
Topaz	1.8	1.5	2.4	0.18	2.3	93%
High	13	38	1.2	1.5	–0.3	–24%
Low	9.6	24	1.2	1.3	–0.1	–7%
M1	4.6	17	2.1	0.98	1.1	53%
M2	6.5	16	1.9	0.83	1.1	57%
M3	7.1	22	1.9	0.95	1.0	51%
Mills	9.3	22	1.2	1.2	0.0	0%
Treasure	8.9	17	1.1	0.82	0.3	27%
Sierran Mean =	5.4	14	1.8	0.69	1.2	55%
Rocky Mountains:						
Loch Vale	16	27	[c] 3.9	1.7	2.2	56%
Icy Brook	22	32	[c] 3.9	2.2	1.7	43%
Andrews Creek	24	38	[c] 3.9	3.1	0.8	21%
East Glacier	0.6	15	[ab] 2.6	0.08	2.5	97%
West Glacier	4.9	30	[ab] 4.9	1.25	3.6	74%
Rabbit Ears Pass	9.9	50	[ab] 2.8	0.69	2.1	75%
Hourglass-Alpine	11	ND	ND	1.8	ND	ND
Hourglass-Subalpine	5.2	ND	ND	0.55	ND	ND
Green Lakes #4	13	30	[ab] 5.9	1.6	4.3	73%
East St. Louis	2.1	ND	[a] 3.2	0.14	3.1	96%
Fool Creek Alpine	1.0	ND	[a] 3.9	0.14	3.7	96%
Snake River	5.7	5.7	[b] 2.3	0.54	1.8	77%
Deer Creek	7.1	16	[b] 1.9	0.39	1.5	79%
Rocky Mt. Mean =	9.6	27	3.6	1.1	2.5	72%

[a]Dry deposition was not measured directly but assumed to equal 25% of wet deposition.
[b]DIN loading was estimated by a combination of snow surveys and NADP data.
[c]DIN loading was estimated from NADP data.

Lake and Mills; these catchments are adjacent to one another and located along the eastern slope of the Sierra Nevada in the Rock Creek drainage. Other Rock Creek catchments such as Ruby and Treasure, had yields similar to watersheds along the western slope of the Sierra: < 1.0 kg ha^{-1} yr^{-1}.

Despite higher rates of N loading, the Rocky Mountain catchments were more efficient at retaining DIN than the Sierra Nevada watersheds. Overall net DIN retention for the Rocky Mountain dataset was 2.5 kg ha^{-1} yr^{-1}, which represents 72% of loading. In the Sierra Nevada, overall DIN retention was 1.2 kg ha^{-1} yr^{-1} or 55% of DIN loading. At several locations, including the GLEES watersheds, catchments in the Fraser Experimental Forest (East St. Louis and Fool Creek), and the Crystal, Lost and Topaz watersheds, DIN retention was greater than 90%. At the other extreme, three Sierra Nevada watersheds, High, Low and Mills, had no retention or had a net export of DIN, i.e., losses of DIN exceeded inputs. The negative retentions at Low Lake watershed are within the expected errors for the N budgets; however, the net DIN export at High Lake is well outside these errors (errors for fluxes were estimated by combining error in analytical chemistry, waters fluxes and sampling frequency using standard error propagation techniques, see Sickman et al. in press and Melack et al. 1998). For the Rocky Mountain sites, the Loch Vale catchments retained the lowest percentage of DIN loading, i.e., 21 to 56%.

Correlations and regression analysis

Prior to using the independent variables in the correlation and regression analyses, we tested for significant correlations among these variables (Tables 4 & 5). For the Sierra Nevada, elevation was found to be negatively correlated with DIN loading (Pearson r = –0.755, Bonferroni p = 0.032) and positively correlated with catchment roughness (Pearson r = 0.71, Bonferroni p = 0.079). The relationship between elevation and roughness is intuitive and demonstrates that topographic complexity generally increases with elevation in the Sierra Nevada. The negative correlation between elevation and DIN loading probably results because of the cluster of watersheds in the Rock Creek basin (i.e., Ruby, Low, Mills, Treasure, High) which are at high elevation but receive lower rates of DIN loading. The correlation between elevation and soil cover in the Sierra Nevada (Pearson r = –0.693, p = 0.118) was nearly significant and suggests that soil cover generally decreases with elevation.

In the Rocky Mountains soil cover was negatively correlated with both mean slope (Pearson r = –0.885, Bonferroni p = 0.008) and mean roughness (Pearson r = –0.933, Bonferroni p = 0.001), suggesting that steeper, more topographically complex watersheds contain less soil (Table 5). Mean

Table 4. Summary of Pearson Product Moment correlations and Bonferroni probabilities among catchment landscape features for high elevation watersheds of the Sierra Nevada. Significant correlations ($p < 0.1$) are underlined

	Elevation	Area	Runoff	Soil Cover	DIN Loading	Mean Slope	Mode Aspect
Pearson Correlation:							
Area	−0.348						
Runoff	−0.448	0.272					
Soil Cover	−0.693	0.299	−0.103				
DIN Loading	*−0.755*	−0.015	0.138	0.654			
Mean Slope	0.185	−0.004	−0.306	−0.262	−0.208		
Mode Aspect	0.132	−0.032	0.250	−0.283	0.010	−0.238	
Mean Roughness	*0.713*	−0.293	−0.457	−0.598	*−0.716*	0.482	−0.035
Bonferroni Probability:							
Area	1.000						
Runoff	1.000	1.000					
Soil Cover	0.118	1.000	1.000				
DIN Loading	*0.032*	1.000	1.000	0.227			
Mean Slope	1.000	1.000	1.000	1.000	1.000		
Mode Aspect	1.000	1.000	1.000	1.000	1.000	1.000	
Mean Roughness	*0.079*	1.000	1.000	0.520	*0.075*	1.000	1.000

slope was also positively correlated with mean roughness (Pearson r = 0.778, Bonferroni p = 0.048).

The correlation analysis showed that soil cover was strongly related to stream nitrate concentrations, DIN yield and DIN retention for watersheds in both the Sierra Nevada and Rocky Mountains (Tables 6 and 7). In addition, elevation showed strong correlations with nitrate concentrations and DIN retention for Sierra Nevada catchments. No significant relationships were found between elevation and any dependent variables in the Rocky Mountains. As was the case with the correlation between DIN loading and elevation, the cluster of sites in the Rock Creek basin is probably responsible for the negative correlation between DIN loading and nitrate concentrations observed within the Sierra dataset (Table 6).

In the Rocky Mountains, mean slope was positively correlated with the DIN yield and AVWM nitrate, and mean roughness was positively related to AVWM nitrate; both of these topographic indices were negatively correlated with % DIN retention (Table 7). In contrast, there were no statistically significant correlations between the topographic indices and dependant variables in the Sierra Nevada (Table 6).

DIN yield was positively related to elevation in the Sierra Nevada, although the linear model did not explain a majority of the variation in DIN yield (Figure 1(a)). Soil cover was negatively correlated with DIN yield. Log-linear models using soil cover were good predictors of DIN yield for both the Sierra Nevada and Rocky Mountain watersheds; 82–91% of the variation in yield was explained by these log-linear equations (Figure 1(b)). The slopes of the regression equations between soil cover and DIN yield were significantly different (p < 0.05) and show that DIN yield in the Rocky Mountains increased more rapidly as soil cover declined than it did in the Sierra Nevada Net DIN retention was inversely related to elevation and positively related to soil cover in the Sierra dataset (Figures 2(a) & (b)). No significant relationship was found between net DIN retention and catchment features in the Rocky Mountains. For the Sierra Nevada catchments, asymptotes of DIN retention ($\sim$2.0 kg ha^{-1} yr^{-1}) occurred in catchments below ca. 3000 m elevation and > 25% soil cover. Zero or negative retentions were found in high elevation catchments with sparse soils.

In the Sierra Nevada, %DIN retention generally decreased with elevation (Figure 3(a)). The effect of soil cover on DIN retention was similar in the Sierra Nevada and Rocky Mountains when DIN retention was expressed as a percentage of DIN loading (Figure 3(b)). Percent DIN retention declined with decreasing soil cover in a logarithmic fashion with a high degree of overlap between the two mountain ranges. Natural logarithmic models using soil cover explained about 87% and 90% of the variation in %DIN reten-

Table 5. Summary of Pearson Product Moment correlations and Bonferroni probabilities among catchment landscape features for high elevation watersheds of the Rocky Mountains. Significant correlations (p < 0.1) are underlined

	Elevation	Area	Runoff	Soil Cover	DIN Loading	Mean Slope	Mode Aspect
Pearson Correlation:							
Area	−0.202						
Runoff	0.257	−0.552					
Soil Cover	−0.303	0.179	−0.655				
DIN Loading	0.313	−0.566	0.619	−0.480			
Mean Slope	0.391	0.019	0.221	*−0.885*	0.433		
Mode Aspect	0.006	0.398	−0.140	−0.295	0.263	0.504	
Mean Roughness	0.427	−0.389	0.428	*−0.933*	0.518	*0.778*	−0.065
Bonferroni Probability:							
Area	1.000						
Runoff	1.000	1.000					
Soil Cover	1.000	1.000	0.800				
DIN Loading	1.000	1.000	1.000	1.000			
Mean Slope	1.000	1.000	1.000	*0.008*	1.000		
Mode Aspect	1.000	1.000	1.000	1.000	1.000	1.000	
Mean Roughness	1.000	1.000	1.000	*0.001*	1.000	*0.048*	1.000

Table 6. Summary of Pearson Product Moment correlations and Bonferroni probabilities between N fluxes, N retention and nitrate concentrations and catchment landscape features for high elevation watersheds of the Sierra Nevada. Significant correlations (p < 0.1) are underlined. No correlations are shown between DIN loading and DIN retention because loading is used in the computation of retention

	DIN Yield	AVWM NO_3^-	Peak NO_3^-	Net DIN Retention	% DIN Retention
Pearson Correlation:					
Elevation	0.644	*0.787*	*0.750*	*–0.769*	*–0.740*
Area	–0.193	–0.242	–0.240	0.104	0.175
Runoff	0.284	–0.008	0.061	–0.084	–0.088
Soil Cover	*–0.867*	*–0.901*	*–0.848*	*0.836*	*0.829*
DIN Loading	–0.665	*–0.781*	–0.747	—	—
Mean Slope	0.134	0.223	0.051	–0.189	–0.201
Mode Aspect	0.027	0.231	0.319	–0.175	–0.097
Mean Roughness	0.570	0.716	0.671	–0.700	–0.686
Bonferroni Probability:					
Elevation	0.385	*0.020*	*0.052*	*0.032*	*0.065*
Area	1.000	1.000	1.000	1.000	1.000
Runoff	1.000	1.000	1.000	1.000	1.000
Soil Cover	*0.001*	*0.000*	*0.003*	*0.004*	*0.005*
DIN Loading	0.273	*0.024*	*0.056*	—	—
Mean Slope	1.000	1.000	1.000	1.000	1.000
Mode Aspect	1.000	1.000	1.000	1.000	1.000
Mean Roughness	1.000	0.107	0.248	0.147	0.189

tion for the Sierra Nevada and Rocky Mountains, respectively. The slopes of the equations were significantly different (p < 0.05) and show that retention increased more rapidly in the Sierra Nevada with expanded soil cover than in the Rocky Mountains. Based on the log-linear models, 80% retention was reached in the Sierra Nevada with catchment soil cover of 30%, whereas this threshold was reached in the Rocky Mountains when soils covered 60% of catchment area.

Annual VWM nitrate concentrations were predictable on the basis of elevation and soil cover in the Sierra Nevada and on the basis of soil cover in the Rocky Mountains (Figures 4(a) and (b)). In the Sierra Nevada, AVWM nitrate increased with elevation ($R^2 = 0.62$). In both mountain ranges, AVWM nitrate decreased in a logarithmic fashion as soil cover increased; these models

Table 7. Summary of Pearson Product Moment correlations and Bonferroni probabilities between N fluxes, N retention and nitrate concentrations and catchment landscape features for high elevation watersheds of the Rocky Mountains. Significant correlations (p < 0.1) are underlined. No correlations are shown between DIN loading and DIN retention because loading is used in the computation of retention

	DIN Yield	AVWM NO_3^-	Peak NO_3^-	DIN Retention	% DIN Retention
Pearson Correlation:					
Elevation	0.292	0.124	–0.514	0.104	–0.172
Area	–0.341	–0.096	–0.598	–0.367	0.110
Runoff	0.624	0.361	0.391	0.166	–0.481
Soil Cover	*–0.924*	*–0.840*	–0.213	0.326	*0.881*
DIN Loading	0.510	0.294	0.396	—	—
Mean Slope	*0.765*	*0.823*	0.036	–0.301	*–0.811*
Mode Aspect	0.294	0.360	–0.130	–0.000	–0.280
Mean Roughness	0.741	*0.842*	0.419	–0.279	*–0.847*
Bonferroni Probability:					
Elevation	1.000	1.000	1.000	1.000	1.000
Area	1.000	1.000	1.000	1.000	1.000
Runoff	0.911	1.000	1.000	1.000	1.000
Soil Cover	*0.002*	*0.049*	1.000	1.000	*0.014*
DIN Loading	1.000	1.000	1.000	—	—
Mean Slope	*0.092*	*0.074*	1.000	1.000	*0.098*
Mode Aspect	1.000	1.000	1.000	1.000	1.000
Mean Roughness	0.149	*0.046*	1.000	1.000	*0.040*

explained about 80–90% of the variation in AVWM. The increase in AVWM nitrate with declining soil cover was more rapid in the Rocky Mountains relative to the Sierra Nevada. The inverse relationship observed between DIN loading and AVWM nitrate concentrations is counter-intuitive and is likely an artifact of the cluster of watersheds in the Rock Creek basin which exhibit high nitrate concentrations while receiving lower rates of DIN deposition.

The regression-tree results are summarized in Table 8. In the case of DIN yield, peak nitrate and %DIN retention, DIN loading and soil cover were first and second branching-variables, respectively, in tree growth; these models explained from 72 to 87% of the variation in the dependant variables. A five node tree using DIN loading, elevation and soil cover explained 92% of the variation in DIN retention. For AVWM nitrate, mean roughness was the primary branch-variable in the regression tree.

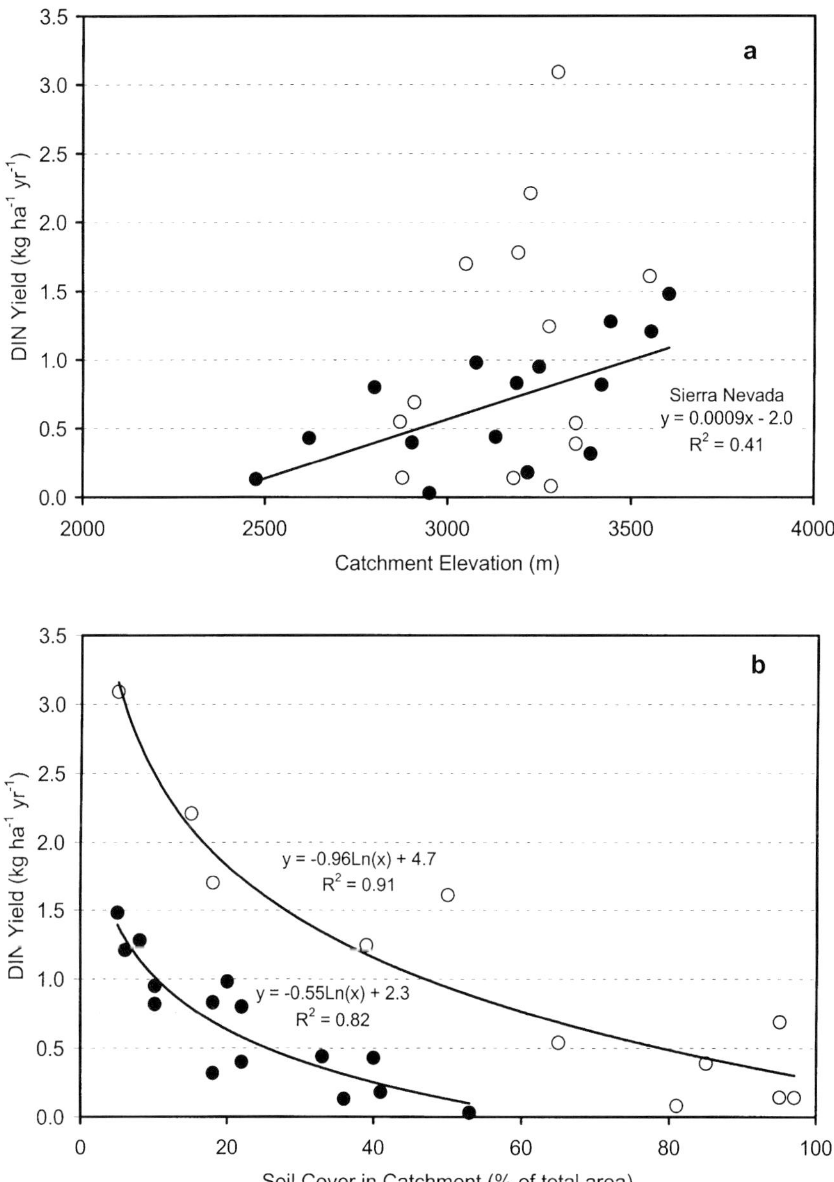

Figure 1. Relationship between catchment DIN yield and elevation and soil cover for high elevation watersheds of the Sierra Nevada (●) and Rocky Mountains (○).

360

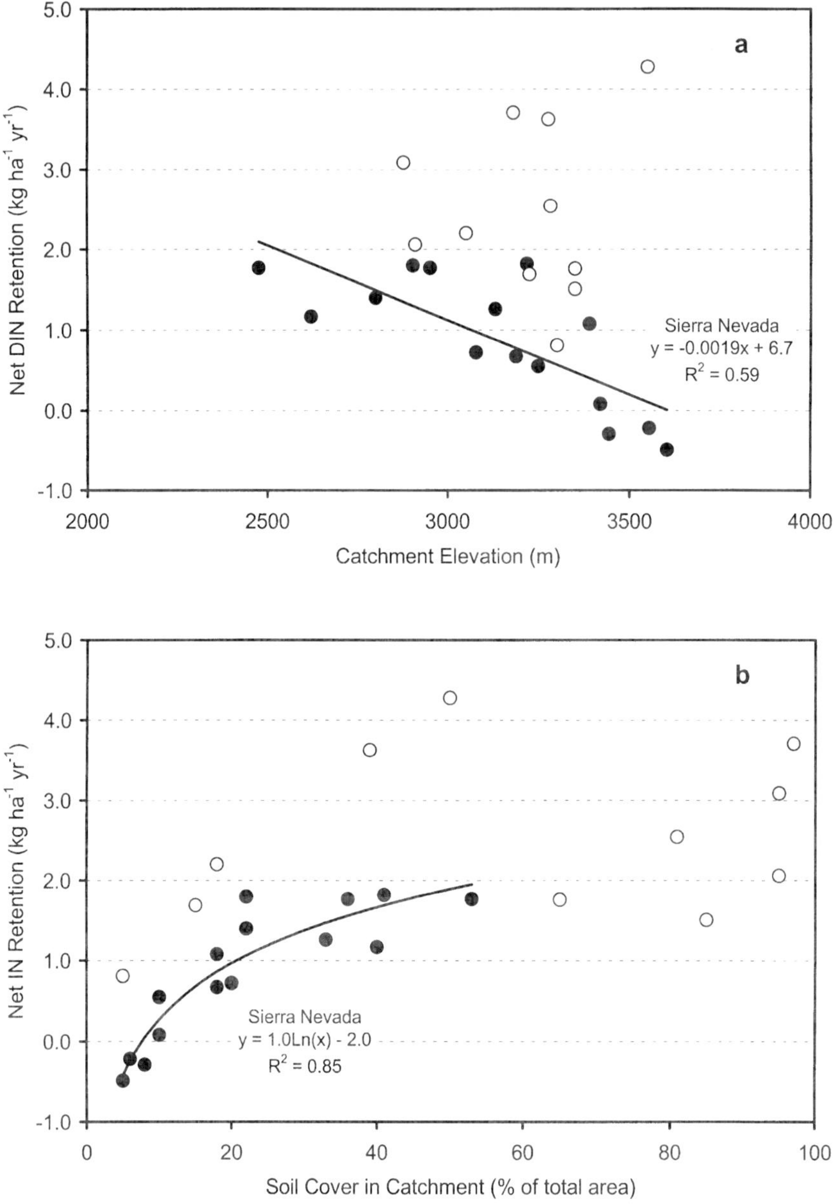

Figure 2. Relationship between net catchment IN retention (i.e., IN loading – DIN yield) and elevation and soil cover for high elevation watersheds of the Sierra Nevada (•) and Rocky Mountains (○).

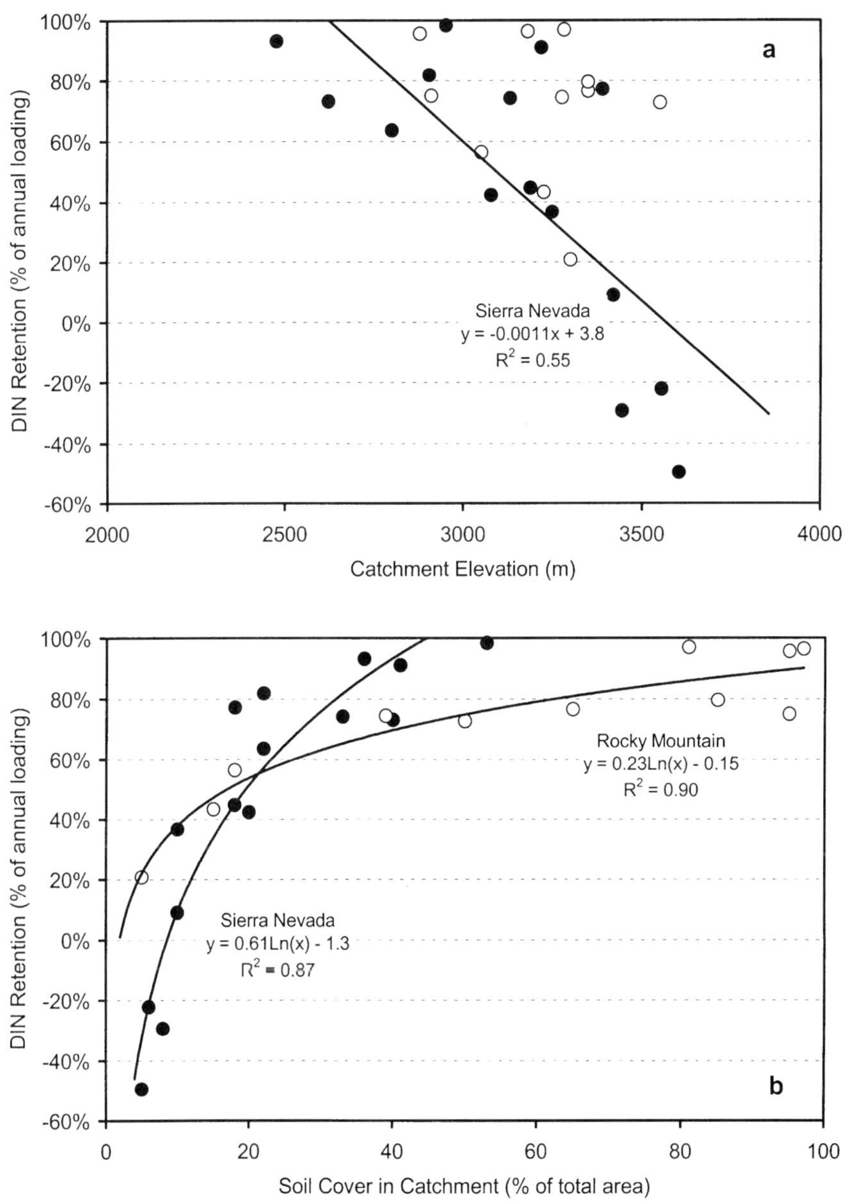

Figure 3. Relationship between percent catchment IN retention (i.e., net I N retention ÷ IN loading) and elevation and soil cover for high elevation watersheds of the Sierra Nevada (•) and Rocky Mountains (○).

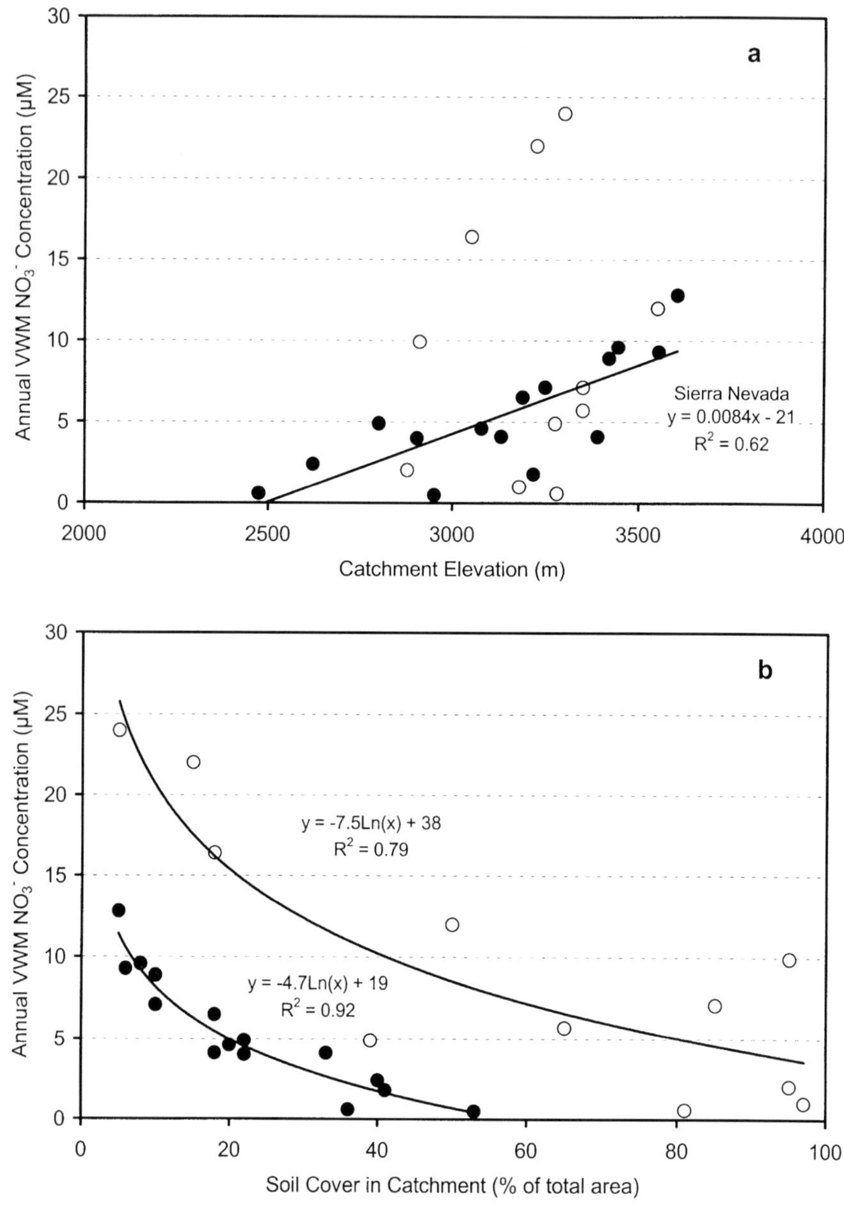

Figure 4. Relationship between the annual volume-weighted mean nitrate concentration in catchment outflow and elevation and soil cover for high elevation watersheds of the Sierra Nevada (•) and Rocky Mountains (○).

Table 8. Summary of regression-tree analysis of pooled Rocky Mountain and Sierra Nevada data sets (n = 26 to 28). Independent variables used in the analysis were: DIN loading (L), elevation (E), %soil cover (S), terrain roughness (R), area, slope, and runoff. Branching variables are shown in order. Tree growth was limited to 5 end-nodes and a minimum of 4 catchments per end-node. The minimum proportional reduction in error allowed at any tree-branch was 0.1

Dependant Variable	Branching Variables	# of End Nodes	Model Fit
DIN Yield	L-S	3	0.73
AVWM NO_3^-	R-E	3	0.79
Peak NO_3^-	L-S	3	0.72
DIN Retention	L-E-S-L	5	0.92
% DIN Retention	L-S	3	0.87

Discussion

Landscape controls on N cycling in alpine and subalpine ecosystems

At the catchment scale, soil cover and elevation had substantial predictive value for stream chemistry and N fluxes in alpine and subalpine ecosystems of the Rocky Mountains and Sierra Nevada. High nitrate concentrations and low inorganic nitrogen retention rates were measured in watersheds with little soil and at high altitudes. Neither catchment runoff or area, which were hypothesized to act as surrogates for hydrologic controls on N cycling, had statistically significant relationships to the watershed-scale N parameters used in our analysis. More sophisticated indices of catchment topography (i.e., slope, aspect and roughness) were only useful in predicting nitrate concentration and DIN retention in the Rocky Mountains; however, multiple regression analysis showed that most of these relationships were due to covariance of slope and topographic roughness with soil cover (see Table 5).

Our findings are consistent with general ecological theories of environmental controls on biological sequestration and release of nitrogen in alpine soils (Stanton et al. 1994; Fisk et al. 1998; Bieber et al. 1998; Brooks & Williams 1999). Elevation influences the extent and timing of snowcover (snow regime) in high elevation systems. Snow regime in turn, through its effect on moisture and temperature patterns in soils, exerts control on plot-to-catchment scale rates of microbial N transformations in soils and N sequestration by plants (Schimel et al. 1996; Brooks et al. 1999; Sickman et al. in press). Lack of soil-cover constrains N uptake in both higher plants and soil microbial populations by limiting the absolute size of these N pools in the Rocky Mountains and Sierra Nevada. In the Sierra Nevada, increasing elevation results in shorter growing seasons for plants through longer snow-lie and colder and perhaps drier soil conditions, thereby reducing plant N uptake. Short-term N storage (in labile N pools) is enhanced during years with high snowfall, because N mineralization and nitrification in snow-covered soils continue later into the spring as a result of delayed snowmelt. The combination of lower N uptake by plants and greater labile N in soil results in higher stream nitrate concentrations and lower DIN retention during years with deep, late-melting snowpacks (Sickman et al. in press). For the Sierra Nevada watersheds, we hypothesize that low soil cover and high altitude worked synergistically in curtailing DIN retention by reducing the size of catchment N reservoirs and by decreasing the total flux between these reservoirs and atmospheric N deposition.

Soil cover exerted a quantitatively similar effect on net DIN retention and AVWM nitrate concentrations in both the Sierra Nevada and Rocky Mountains (Figures 1(b) and 4(b)). The similarity of these relationships in Sierra Nevada and Rocky Mountains indicates a consistent effect of soil N processes across the alpine and subalpine regions of the western United States and over a 5 to 6 fold variation in DIN loading rates. The intercepts of the Rocky Mountain equations were about double the Sierra Nevada intercepts, which may reflect the overall 2x higher rate of DIN loading to alpine systems in the Rocky Mountains.

Current DIN yields and AVWM nitrate levels in the Rocky Mountain watersheds may be a forecast of conditions in the Sierra Nevada if atmospheric DIN loading were to double. No simple relationship likely exists between DIN deposition and stream water nitrate at a single catchment or on a year-to-year basis because there are so many factors governing the susceptibility of alpine watersheds to N saturation. However, our regional analysis suggests there may be a relationship between loading and N dynamics at a large spatio-temporal scale and that site specific changes in concentration are lost when examining regional variations. A similar argument is made by Williams & Tonnessen (in press) to justify their estimates of critical N loads in the Rocky Mountains. Annual variation in nitrate concentrations is driven by hydrological and biological factors at the catchment-scale (e.g., Creed & Band 1998, and the present study), but the influence of deposition may emerge when looking at N dynamics at the regional or continental scale over a number of years.

Recent studies of functionally-similar catchments have demonstrated that inter-site differences in nitrate export behavior can exist without variations in DIN loading rates (Creed & Band 1998; Lovett et al. 2000; Clow & Sueker 2000). The regions examined in these analyses ranged in area from 10 to 2000 km^2. Similarly, in our analysis, DIN loading was not positively correlated with nitrate concentrations of DIN yield in either the Sierra Nevada or Rocky Mountains; regions on the order of 50,000 km^2 in area. However, when we examined these relationships at a larger spatial scale with the regression-tree analysis (> 1,000,000 km^2), small-scale variability was eliminated and a large-scale pattern emerged. DIN loading explained more of the differences in N dynamics for the combined data sets than any of the other terrain or topographic variables we considered. In an analysis of undisturbed watersheds in North America, Lewis (in press) found a positive relationship between catchment DIN loading and DIN yield; this study examined watersheds in a region > 5,000,000 km^2. These findings suggest that the concept of representative elementary area (REA), proposed by Wood et al. (1988) may apply when examining the regional variability of N dynamics. The REA can be considered the scale at which a statistical treatment of spatial variability can replace a deterministic description. For empirical modeling of the relationship between DIN loading and yield or stream nitrate concentrations, we suggest that studies should examine regions greater than 100,000 km^2 to form valid conclusions.

Topographic and terrain modeling of N biogeochemistry

Current concerns over the impact of nitrogen deposition on natural ecosystems has led to the need for evaluating global N biogeochemical cycles and for predicting the sensitivity of ecosystems over large regions (e.g., Fenn et al. 1998; Williams & Tonnessen 2000). In particular, there has been considerable effort to: 1) relate simple catchment features such as area, elevation and runoff to N yield from river basins in the context of global biogeochemical cycles (Meybeck 1982; Howarth et al. 1996; Lewis et al. 1999; Lewis in press) and (2) use more complex terrain parameters (e.g., slope, aspect, bedrock geology, vegetation, soil area, DIN deposition and variable source-area dynamics) to predict N yield, retention and surface water nitrate concentrations in smaller watersheds (e.g., Creed & Band 1998; Clow & Sueker 2000). The goal of both types of analyses is to develop empirical models to describe complex biogeochemical processes that can currently only be modeled at small scales.

Empirical models based on catchment features have had mixed success in predicting stream nitrogen concentration in small catchments. Clow & Sueker (2000) were able to explain 97% of the variation in nitrate chemistry of nine subalpine catchment in Rocky Mountain National Park on the basis of regression equations based on catchment slope and surficial geology (i.e., extent of talus). However, when these equations were tested with existing synoptic stream-survey data from the Rocky Mountains the model could only explain 19% of the variation in nitrate concentrations. The authors attribute the model's poor performance to the fact that the synoptic-survey data contain a high proportion of small, high-elevation catchments with limited areas of subalpine soils compared to the calibration data.

Catchment land-cover was used by Cooper et al. (2000) in modeling long-term stream chemistry in the Tywi catchment of South Wales, United Kingdom. In this study, the authors developed empirical relationships between stream chemistry and landscape types (i.e., based on catchment soil and vegetation) and used these relationships along with the spatial distribution of landscape types and a stream-mixing algorithm to model stream chemistry over a 2000 km^2 region. The coefficient of determination in a regression between measured and modeled nitrate concentrations was 0.65.

Artificial neural networks (ANN) were used by Lek et al. (1999) to predict stream DIN and TN concentration at 927 sites throughout the United States that were impacted by non-point source pollution. Independent variables used as inputs to the ANNs included catchment area, precipitation, runoff, livestock density and various landscape descriptors (forest, wetland, urban, agricultural). The ANNs were validated using data not used in the training procedure and were shown to explain about 70% of the variation in stream N concentrations.

Lovett et al. (2000) found that variations in stream nitrate concentrations among 39 streams in the Catskill Mountains of New York could not be explained by differences in catchment DIN loading, watershed topography or groundwater inputs. Instead, differences in forest composition which were induced by past land-use practices were believed to have produced the observed variation in nitrate concentrations. However, the variety of topography and DIN loading in these watersheds was much lower than in our analyses and in the previously mentioned modeling studies; the region examined may be below the REA for modeling stream nitrate concentrations from DIN loading or topography. Thus, care must be taken in scaling the findings of Lovett et al. (2000) to larger montane regions of the United States (cf. Stoddard et al. 1998, 1999).

Current N saturation status in Rocky Mountains and Sierra Nevada

Overall, catchment DIN retention is higher in the Rocky Mountain watersheds than in the Sierra Nevada. We suggest that this difference is due primarily to greater soil cover in the Rocky Mountains and not due to greater rates of DIN retention per unit soil area. We base this conclusion on the relationship between DIN retention and soil cover which demonstrates that Sierra Nevada catchments with 20 to 40% soil cover are retaining equal amounts and percentages of DIN to catchments in the Rocky Mountains with > 60% soil cover (Figures 2(b) and 3(b)).

While it is possible that variations in climate and soil properties explain these differences, the data may imply that soils in the Rocky Mountains are less N limited because of higher rates of DIN loading. Alternatively, environmental conditions in the Rocky Mountains may be more severe than in the Sierra Nevada (e.g., greater extent of frozen soils), therefore terrestrial ecosystems in the Rocky Mountains may be less able to prevent N losses. Current ecological theory suggests that terrestrial communities are N limited because of N losses that are not under control of biota [Vitousek & Field 1999]; these losses include leaching of dissolved organic N and denitrification [Vitousek et al. 1998]. The persistence of N limitation in high elevation ecosystems and the inability of biotic communities to prevent episodic nitrate losses may be related to microbial and hydrologic processes which conspire to induce temporal and spatial disconnections between inorganic N availability and demand.

Stoddard (1994) provided a framework to assess the degree to which ecosystems are affected by N deposition that is based on seasonal patterns in surface water nitrate concentrations. Our analyses suggest that rates of catchment-scale DIN retention are also indicative of N-saturation status and correspond well with this framework. Four stages were used in Stoddard's framework to describe the N saturation status of watersheds. At Stage 0, maximum spring episode concentrations are less than precipitation concentrations and growing season concentrations are near the detection limit. Watersheds that meet this criteria include the Crystal, Topaz, Lost, and Marble Fork basins in the Sierra Nevada and East Glacier, Dear Creek, East St. Louis and Fool Creek Alpine basins in the Rocky Mountains. Inorganic nitrogen retention for these stage 0 catchments ranged from 80–100%.

At the next step in the sequence towards N-saturation, Stage 1, nitrate concentrations in spring episodes exceed concentrations in precipitation and there is a delay in the decline of nitrate levels to later in the growing season. In the Sierra Nevada, catchments at Stage 1 of N saturation would include Spuller, Ruby, Pear, and Emerald. Examples in the Rocky Mountains would include West Glacier. These catchments have DIN retention rates in the range of ca. 70–80%.

Stage 2 of N-saturation includes higher episodic concentrations and elevated nitrate concentrations well into and through the growing season. In the Sierra Nevada, the M-sites and Treasure watersheds can be classified at this stage. These catchments retained from ca. 20 to 60% of DIN loading. Stage 2 watersheds in the Rocky Mountains include Green Lakes #4, Rabbit Ears Pass, Snake River and the Loch Vale watersheds. These Rocky Mountain basins had variable rates of %DIN retention; the overall range was from ca. 20–75%.

Stage 3 of N-saturation differs from stage 2 in that the watershed becomes a net source of N rather than a sink. Two watersheds in the Sierra meet this criteria, High and Low, and one catchment, Mills, is on the verge of stage 3. In all three of these catchments DIN export equals or exceeds DIN inputs from atmospheric deposition. In the case of Low, negative DIN retention is within the expected errors of the N budgets, hence it is possible that the catchment is also still on the verge of stage 3. In the case of High the amount of net DIN export from the basin, is beyond expected errors in flux estimates. Some of the net export can explained by organic nitrogen in precipitation, but this input is more than balanced by organic and particulate nitrogen losses from the basin (Sickman et al. in press).

The conceptual model of Stoddard (1994) is based on data from forested temperate watersheds, primarily in the Northeastern U.S. and Europe. At first exposure, it may seem dubious to apply Stoddard's N saturation stages to alpine watersheds, where the basins are above timberline, soils are thin (when present at all) and the annual hydrologic cycle is dominated by snow accumulation and rapid melt. Yet much of the recent data from alpine watersheds suggests strongly that the same processes that Stoddard used to explain the progression from Stage 0 to Stage 3 in forested watersheds are controlling N export from the alpine zone. In forested watersheds, N is largely immobilized by biotic uptake in soils (Tietema et al. 1998; Nadelhoffer et al. 1995), especially the organic layer of soils (Gundersen et al. 1998). In alpine watersheds, organic soils seem to play a role similar to the one they play in forested watersheds (as partially indicated by the relationships between N retention and soil cover reported in this paper and by earlier studies; Williams et al. 1996b), as do talus fields (Williams et al. 1997; Williams et al. 1995), although they are largely unrecognizable to most scientists as soils. Studies indicate that the NO_3^- leaching from watersheds during snowmelt has an isotopic signature largely attributable to soil transformation (e.g., dominated by nitrification, rather than by atmospheric isotope ratios), in both forested and alpine watersheds (Kendall et al. 1995). It seems likely that similarities in N behavior between forested and alpine watersheds outweigh the dissimilarities. The types of pools and processes governing N retention and N leaching are nearly identical; it is only the size of the pools that differ. Smaller N pools in the limited soils of alpine watersheds create the potential for nitrogen saturation to occur at deposition rates that seem trivial when compared to those in the eastern U.S. and Europe.

Nitrogen deposition along the eastern slope of the Sierra Nevada is less than 1.5 kg ha^{-1} yr^{-1}. This rate of N loading is low compared to current inputs to other North American catchments experiencing adverse effects of N deposition (Fenn et al. 1998). At the High watershed, episodic acidification occurred during snowmelt (ANC < 0) and net export of ANC was exceeded by hydrogen ion export (Stoddard 1995, Sickman and Stoddard unpublished data). The Ruby watershed is adjacent to the High catchment and receives similar levels of N deposition, yet it did not experience acidic episodes and was a sink for N loading (Sickman & Melack 1998; Melack et al. 1998).

Differences in N cycling between the High and Ruby catchments are probably explained by greater soil cover in the Ruby watershed and a proportionally higher percentage of talus and boulders in the High watershed. Substantial pools of DIN nitrogen have been measured in talus deposits in the Rocky Mountains (Williams et al. 1997; Bieber et al. 1998). In addition, leaching from these pools may represent a large component of the nitrate exported from alpine watersheds such as Andrews Creek and Icy Brook (Campbell et al. 1995; Kendall et al. 1995). The fact that High watershed is exporting DIN in excess of atmospheric loading might be explained by release of N that has been held in long-term storage within the talus. Nitrogen inputs from dry deposition and organic N substrates supplied by small mammals (i.e., waste products and nesting materials) have the potential to build up and persist within talus since there is little or no N utilization by plants and denitrification is unlikely. However, more research, possibly employing detailed analyses of stable isotopes of C and N, will be needed to more fully understand N dynamics within talus fields.

Summary

The correlation analysis confirms that watershed features such as elevation and soil cover are good surrogates for complex N processes controlling catchment-scale N retention. Soil cover was an especially good predictor for catchment DIN yield, stream nitrate concentrations and DIN retention in alpine and subalpine ecosystems in both the Sierra Nevada and Rocky Mountains. The regression models provide a basis for predicting the status of high elevation ecosystems over large regions and under varying inputs of atmospheric N loading. Because the equations quantify the effect of DIN loading on surface water chemistry and nitrogen retention, they may also be useful for evaluating critical N loads in the western United States.

Acknowledgements

This synthesis was possible due to the efforts of many people working under arduous field conditions. We wish to thank the field and laboratory personnel whose labors went into collecting the data required for the nitrogen budgets in the Sierra Nevada and Rocky Mountains. In particular we wish to acknowledge Pete Kirchner, Kevin Skeen and Sage Root for their many years of field work in the Sierra Nevada. We thank Delores Lucero and Frank Setaro for laboratory measurements of Sierra Nevada samples and Randy Comeleo for computing the terrain indices. We are grateful to Jake Peters, Bob Stottlemyer and Bob Musselman for additional information on their study sites. Al Leydecker provided help with all aspects of our work and we thank him for his efforts and ideas over many years. Helpful comments on the manuscript were provided by Josh Schimel, Mark Williams, Tom Meixner, Bill Lewis, Jill Baron and an anonymous reviewer. Financial support was provided by NASA and the California Air Resources Board. Data for Green Lake 4 was provided by the Niwot Ridge Long-term Ecological Database (BIR 9115097). This manuscript was initiated as part of the International SCOPE Nitrogen Project, which received support from both the Mellon Foundation and from the National Center for Ecological Analysis and Synthesis.

References

Andrew N, Bleich V & August P (1999) Habitat selection by mountain sheep in the Sonoran desert: Implications for conservation in the United States and Mexico. California Wildland Conservation Bull. 12: 1–33

Baron JS (Ed) (1992) Biogeochemistry of Subalpine Ecosystem: Loch Vale Watershed. Springer-Verlag, New York

Baron JS & Campbell DH (1997) Nitrogen fluxes in a high elevation Colorado Rocky Mountain Basin. Hydrological Processes 11: 783–799

Baron JS, Ojima DS, Holland EA & Parton WJ (1994) Analysis of nitrogen saturation potential in Rocky Mountain tundra and forest: Implications for aquatic systems. Biog. 27: 61–82

Bieber AJ, Williams MW, Johnson MJ & Davinroy TC (1998) Nitrogen transformations in alpine talus fields, Green Lakes Valley, Front Range, Colorado, USA. Arctic and Alpine Research 30: 266–271

Brooks PD, Williams MW & Schmidt SK (1996) Microbial activity under alpine snowpacks, Niwot Ridge, Colorado. Biogeochemistry 32: 93–113

Brooks PD, Williams MW & Schmidt SK (1998) Soil inorganic nitrogen and microbial biomass dynamics before and during spring snowmelt. Biog. 43: 1–15

Brooks PD, Campbell DH, Tonnessen KA & Heuer K (1999) Natural variability in N export from headwater catchments: snow cover controls on ecosystem N retention. Hydrological Processes 13: 2191–2201

Brooks PD & Williams MW (1999) Snowpack controls on nitrogen cycling and export in seasonally snow-covered catchments. Hydrological Processes 13: 2177–2190

Campbell DH, Clow DW, Ingersoll GP, Mast MA, Spahr NE & Turk JT (1995) Processes controlling the chemistry of two snowmelt-dominated streams in the Rocky Mountains. Water Resources Research 31: 2811–2821

Church TM (1999) Atmospheric organic nitrogen deposition explored at workshop. Eos, Transactions of the American Geophysical Union 80: 355

Cline DW, Bales RC & Dozier J (1998) Estimating the spatial distribution of snow in mountain basins using remote sensing and energy balance modeling. Water Resources Research 34: 1275–1285.

Clow DW & Sueker JK (2000) Relations between basin characteristics and stream water chemistry in alpine/subalpine basins in Rocky Mountain National Park, Colorado. Water Resources Research 36: 49–61

Cooper DM, Jenkins A, Skeffington R & Gannon B (2000) Catchment-scale simulation of stream water chemistry by spatial mixing: Theory and application. J. of Hydrology 233: 121–137

Creed IF & Band LE (1998) Export of nitrogen from catchments within a temperate forest: Evidence for a unifying mechanism regulated by variable source area dynamics. Water Resources Research 34: 3105–3120

Dozier J & Frew J (1990) Rapid calculation of terrain parameters for radiation modeling from digital elevation data. IEEE Transactions on Geoscience and Remote Sensing 28: 963–969

Fenn ME, Poth MA, Aber JD, Baron JS, Bormann BT, Johnson DW, Lemly AD, McNulty G, Ryan DF & Stottlemyer R (1998) Nitrogen excess in North American Ecosystems: predisposing factors, ecosystem responses, and management strategies. Ecological Applications 8: 706–733

Fisk MC, Schmidt SK & Seastedt TR (1998) Topographic patterns of above – and below-ground production and nitrogen cycling in alpine tundra. Ecology 79: 2253–2266

Galloway JN, Levy H III & Kasibhatla PS (1994) Year 2020: consequences of population growth and development on deposition of oxidized nitrogen. Ambio 23: 12–123

Gundersen P, Emmett BA, Kjonaas OJ, Koopmans CJ & Tietema A (1998) Impact of nitrogen deposition on nitrogen cycling in forests: A synthesis of NITREX data. Forest Ecology and Management 101: 37–55

Hedin LO, Armesto JJ & Johnson AH (1995) Patterns of nutrient loss from unpolluted, old-growth temperate forests: Evaluation of biogeochemical theory. Ecology 76: 493–509

Heuer K, Brooks PD & Tonnessen KA (1999) Nitrogen dynamics in two high elevation catchments during spring snowmelt 1996, Rocky Mountains, Colorado. Hydrological Processes 13: 2203–2214

Howarth RW, Billen G, Swaney D, Townsend A, Jaworski N, Lajtha K, Downing JA, Elmgren R, Caraco N, Jordan T, Berendse F, Freney J, Kudeyarov V, Murdoch P & Ahao-Liang S (1996) Regional nitrogen budgets and riverine N and P fluxes for the drainages to the North Atlantic Ocean: Natural and human influences. Biog. 38: 1–96

Jassby AD, Reuter JE Axler RP Goldman CR & Hackley SH (1994) Atmospheric deposition of nitrogen and phosphorus in the annual nutrient load of Lake Tahoe (California, Nevada). Water Resources Research 30: 2207–2216

Kendall C, Campbell DH, Burns DA, Shanely JB, Silva SR & Chang CCY (1995) Tracing sources of nitrate in snowmelt runoff using the oxygen and nitrogen isotopic compositions of nitrate. In: Tonnessen KA, Williams MW & Tranter M (Eds) Biogeochemistry of Seasonally Snow-Covered Catchments (pp 339–347). IAHS Publication no. 228

Kiefer JW & ME Fenn (1997) Using vector analysis to assess nitrogen status of Ponderosa and Jeffrey pine along deposition gradients in forests of southern California. Forest Ecology and Management 94: 47–59

Landers DH, Eilers J, Brakke W, Overton W, Kellar M, Silverstein R, Schonbrod R, Crowe R, Linthurst J, Omernik S, Teague S & Meier E (1987) Western Lake Survey, Phase I: Characteristics of Lakes in the Western United States. Volume I. Population Descriptions and Physico-chemical Relationships. U.S. Environmental Protection Agency. Washington, DC

Lek S, Guiresse M & Giraudel JL (1999) Predicting stream nitrogen concentration from watershed features using neural networks. Water Research 33: 3469–3478

Lewis WM Jr, Melack JM, McDowell WH, McClain M & Richey JE (1999) Nitrogen yields from undisturbed watersheds in the Americas. Biog. 46: 149–162

Lewis WM Jr (in press) Yield of nitrogen from undisturbed watersheds of the United States and its relationship to nitrogen deposition. Biog.

Leydecker A, Sickman JO & Melack JM (1999) Episodic lake acidification in the Sierra Nevada, California. Water Resources Research 35: 2793–2804

Lovett GM, Weathers KC & Sobczak WV (2000) Nitrogen saturation and retention in forested watersheds of the Catskill Mountains, New York. Ecological Applications 10: 73–84

Magill AH, Downs MR, Nadelhoffer KJ, Hallet RA & Aber JD (1997) Biogeochemical response of forest ecosystem to simulated chronic nitrogen deposition. Ecological Applications 7: 402–415

Meixner T, Brown A & Bales RC (1998) Importance of biogeochemical processes in modeling stream chemistry in two watersheds in the Sierra Nevada, California. Water Resources Research 34: 3121–3133

Meixner T, Gupta HV, Bastidas LA & Bales RC (1999) Sensitivity analysis using mass flux and concentration. Hydrological Processes 13: 2233–2244

Melack JM & Stoddard JL (1991) Sierra Nevada, California. In: Charles DF (Ed) Acidic Deposition and Aquatic Ecosystems (pp 503–530). Springer-Verlag, New York, NY

Melack JM, Sickman JO, Leydecker A & Marrett D (1998) Comparative analysis of high-altitude lakes and catchments in the Sierra Nevada: Susceptibility to acidification. Final report, contract A032-188, California Air Resources. Board, Sacramento, California

Meybeck M (1982) Carbon, nitrogen, and phosphorus transport by world rivers. American J of Sci. 282: 401–450

Nadelhoffer, KJ, Downs MR, Fry B, Aber JD, Magill AH & Melillo JM (1995) The fate of ^{15}N-labelled nitrate additions to a northern hardwood forest in eastern Maine, USA. Oecologia 103: 292–301

Neff JC, Holland EA, Dentener FJ, McDowell WH & Russel KM (in press) Atmospheric organic nitrogen: Implications for the Global N cycle. Biog.

Peters NE & Leavesley GH (1995) Biotic and abiotic processes controlling water chemistry during snowmelt at Rabbit Ears Pass, Rocky Mountains, Colorado, USA. Water, Air and Soil Poll. 79: 171–190

Reuss JO, Vertucci FA, Musselman & Sommerfeld (1995) Chemical fluxes and sensitivity to acidification of two high-elevation catchments in southern Wyoming. J. of Hydrology 173: 165–189

Schimel JP, Kielland K & Chapin FS (1996) Nutrient availability and uptake by tundra plants. In: Reynolds JF & Tenhunen JD (Eds) Landscape function: Implications for Ecosystem Response to Disturbance: A Case Study in Arctic Tundra (pp 201–221). Springer-Verlag New York, NY

Sickman JO & Melack JM (1998) Nitrogen and sulfate export from high elevation catchments of the Sierra Nevada, California. Water, Air and Soil Poll. 105: 217–226

Sickman JO, Leydecker A & Melack JM (in press) Nitrogen mass balances and abiotic controls on N retention and yield in high-elevation catchments of the Sierra Nevada, California, U.S.A. Water Resources Research

Sievering H, Rusch D & Caine N (1996) Nitric acid, particulate nitrogen and ammonium in the continental free troposphere: nitrogen deposition to an alpine tundra ecosystem. Atmospheric Environ. 30: 2527–2537

Stanton ML, Rejmanek M & Galen C (1994) Changes in vegetation and soil fertility along a predictable snowmelt gradient in the Mosquito Range, Colorado, U.S.A. Arctic and Alpine Research 26: 364–374

Stednick JD (1989) Hydrochemical characterization of alpine and alpine-subalpine stream waters, Colorado Rocky Mountains, U.S.A. Arctic and Alpine Research 21: 276–282

Stoddard JL (1994) Long-term changes in watershed retention of nitrogen. In: Baker LA (Ed) Environmental Chemistry of Lake and Reservoirs (pp 223–284). Adv. Chem. Ser. No. 237. American Chemical Society, Washington, DC

Stoddard JL (1995) Episodic acidification during snowmelt of high elevation lakes in the Sierra Nevada Mountains of California. Water, Air and Soil Poll. 85: 353–358

Stoddard JL, Driscoll CT, Kahl JS & Kellogg JP (1998) Can site-specific trends be extrapolated to a region? An acidification example for the northeast. Ecological Applications 8: 288–299

Stoddard JL, Jeffries DS, Lukewille A, Clair TA, Dillon PJ, Driscoll CT, Forsius M, Johannessen M, Kahl JS, Kellogg JH, Kemp A, Mannio J, Monteith DT, Murdoch PS, Patrick S, Rebsdorf A, Skjelkvale BL, Stainton MP, Traaen T, van Dam H, Webster KE, Wieting J & Wilander A (1999) Regional trends in aquatic recovery from acidification in North America and Europe. Nature 401: 575–578

Stottlemyer R & Troendle CA (1992) Nutrient concentration patterns in streams draining alpine and subalpine catchments, Fraser Experimental Forest, Colorado. J. of Hydrology 140: 179–208

Tietema A, Emmet BA, Gunderen P, Kjonaas OJ & Koopmans C (1998) The fate of ^{15}N-labelled nitrogen deposition in coniferous forest ecosystems. Forest Ecology and Management 101: 19–27

Williams MW, Bales RC, Brown AD & Melack JM (1995) Fluxes and transformations of nitrogen in a high-elevation catchment, Sierra Nevada. Biog. 28: 1–31

Williams MW, Baron JS, Caine N, Sommerfeld R & and Sanford R Jr. (1996a) Nitrogen saturation in the Rocky Mountains. Environ. Science and Technol. 30: 640–646

Williams MW, Brooks PD, Mosier A & Tonnessen KA (1996b) Mineral nitrogen transformations in and under seasonal snow in a high-elevation catchment in the Rocky Mountains, United States. Water Resources Research V32: 3161–3171

Williams MW (1997) Nitrogen cycling and critical loads in high-elevation catchments of the Colorado Front Range. Eos, Transactions of the American Geophysical Union, S168.

Williams MW, Davinroy T & Brooks PD (1997) Organic and inorganic nitrogen pools in talus fields and sub-talus water, Green Lakes Valley, Colorado Front Range Hydrological Processes 11: 1747–1760

Williams MW & Melack JM (1991) Precipitation chemistry in and ionic loading to an alpine basin, Sierra Nevada. Water Resources Research 27: 1563–1574

Williams MW & Tonnessen KA (2000) Critical loads for inorganic nitrogen deposition in the Colorado Front Range, USA. Ecological Applications 10(6): 1648–1665

Wood EF, Sivapalan M, Beven K & Band L (1988) Effects of spatial variability and scale with implications to hydrological modeling. J. of Hydrology 102: 29–47

Vitousek PM, Hedin LO, Matson PA, Fownes JH & Neff JC (1998) Within-system element cycles, input-output budgets and nutrient limitation. In: Groffman PM & Pace ML (Eds) Successes, Limitation and Frontiers in Ecosystem Science (pp 432–451). Springer, New York, NY
Vitousek PM & Field CB (1999) Ecosystem constraints to symbiotic nitrogen fixers: a simple model and its implications. Biogeochemistry 46: 179–202

Biogeochemistry **57/58**: 375–385, 2002.
© 2002 *Kluwer Academic Publishers. Printed in the Netherlands.*

Yield of nitrogen from minimally disturbed watersheds of the United States

WILLIAM M. LEWIS, JR.
*Center for Limnology, Cooperative Institute for Research in Environmental Sciences,
University of Colorado, Boulder, Colorado 80309-0216, U.S.A.*

Key words: dissolved organic nitrogen, nitrate, nitrogen cycle, nitrogen yields

Abstract. Watersheds of the US Geological Survey's Hydrologic Benchmark Network program were used in estimating annual yield of total nitrogen and nitrogen fractions (ammonium, nitrate, dissolved organic N, particulate N) in relation to amount of runoff, elevation, and watershed area. Only watersheds minimally disturbed with respect to the nitrogen cycle were used in the analysis (mostly natural vegetation cover, no point sources of N, atmospheric deposition of inorganic N < 10 kg ha^{-1} y^{-1}). Statistical analysis of the yields of total nitrogen and nitrogen fractions showed that elevation and watershed area bear no significant relationship to nitrogen yield for these watersheds. The yields of total nitrogen and nitrogen fractions are, however, strongly related to runoff ($r^2 = 0.91$ for total N). Annual yield increases as runoff increases, but at a rate lower than runoff; annual discharge-weighted mean concentrations decline as annual runoff increases. Yields of total nitrogen and most nitrogen fractions bear a relationship to runoff that is nearly indistinguishable from a relationship that was documented previously for minimally disturbed watersheds of the American tropics. Overall, the results suggest strong interlatitudinal convergence of yields and percent fractionation for nitrogen in relation to runoff.

Introduction

Transport of nitrogen in rivers on a continental or global scale can be estimated from monitoring data for large rivers (Meybeck 1982). Because transport rates have been strongly perturbed anthropogenically in many parts of the world, however, it is difficult to reconstruct global or continental transport under natural conditions. While information is available for numerous watersheds that are unaffected by anthropogenic sources of nitrogen on the ground, the widespread distribution of anthropogenically enhanced nitrogen deposition from the atmosphere has changed the nitrogen supply to many watersheds that are otherwise unperturbed (Aber et al. 1989; Howarth et al. 1996).

A recent review of nitrogen yield data for the Americas (Lewis et al. 1999) showed that exclusion of data sets potentially affected by mobilization of nitrogen either on the ground or through the atmosphere restricts the consideration of data representing background conditions primarily to the American tropics. An analysis of these data for annual nitrogen yield in relation to watershed area, elevation, and runoff showed a very strong relationship between annual runoff and annual yield of total nitrogen ($r^2 = 0.85$) and of individual nitrogen fractions. The analysis also showed secondary but significant relationships between elevation and annual yield of dissolved inorganic N, and between watershed area and ratios of nitrogen fractions. It was not clear whether background nitrogen yields from watersheds at temperate latitudes would follow trends similar to those that are characteristic of the American tropics.

Recent studies in Europe as part of the NITREX network have shown that naturally vegetated watersheds show virtually no perturbation of inorganic nitrogen yield in runoff up to a threshold of approximately 10 kg ha^{-1} y^{-1} of nitrogen deposition (Dise & Wright 1995). At deposition rates between 10 and 25 kg ha^{-1} y^{-1}, many of these watersheds show increased output of inorganic N; above 25 kg ha^{-1} y^{-1}, all watersheds show higher output of inorganic N (organic N was not included in the analysis). Thus the European NITREX studies indicate that temperate watersheds to be used in statistical analysis of yield per unit area for undisturbed conditions could in fact include those that receive some nitrogen enrichment from the atmosphere, provided that this enrichment does not exceed a threshold in the vicinity of 10 kg ha^{-1} y^{-1}. If this is the case, then the number of watersheds in North America from which data could be taken would be great enough to support a statistical analysis of nitrogen yields under background conditions.

The purpose of this paper is to use watersheds of North America having primarily natural vegetative cover, lacking point sources of nitrogen, and having nitrogen deposition below 10 kg ha^{-1} y^{-1} as a means of estimating background yield of total nitrogen and nitrogen fractions. Data for these watersheds also are analyzed for relationships between the yield of total nitrogen and nitrogen fractions in relation to variables that were used in the analysis of similar data for the American tropics (elevation, runoff, watershed area).

The present analysis is based on watersheds that are part of the United States Geological Survey (USGS) Hydrologic Benchmark Network (HBN) program. Other sites could be considered as well (e.g. Williams & Melack 1997; Sickman & Melack 2002), but there are several advantages to dealing exclusively with the benchmark watersheds. First, they are sufficiently numerous and well distributed (Figure 1) to make an appropriate basis for

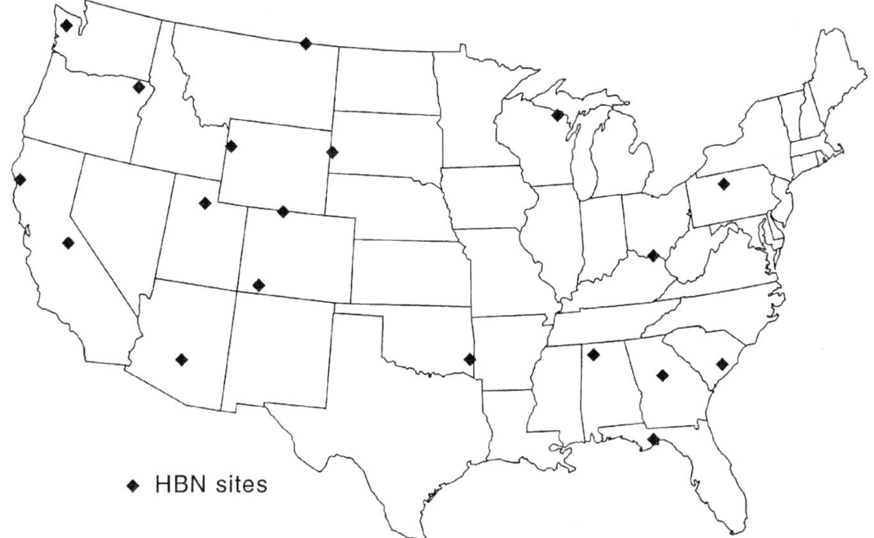

Figure 1. Location of the US Geological Hydrologic Benchmark Network watersheds that were used in the study.

statistical analysis. In addition, all of the sampling and analyses occurred through the use of the same protocols. Finally, the hydrologic documentation for these watersheds is excellent, as required for computation of annual yield of total nitrogen and nitrogen fractions.

Methods

Characteristics of the HBN data were obtained from USGS sources (Alexander et al. 1996), and then screened for combinations of sites and years for which information was available on all nitrogen fractions to be considered in this analysis: ammonium, nitrate, dissolved inorganic N (DIN), dissolved organic N (DON), total dissolved N (TDN), particulate N (PN), and total N (TN). Although the benchmark program includes 63 sites, 40 of these lack complete coverage for all nitrogen fractions over any extended interval or with sufficient frequency, and four additional ones have significant anthropogenic N sources (including atmospheric deposition, as described below). Of the remaining 19, all nitrogen fractions were measured consistently over the two years between October of 1980 and September of 1982. These 19 sites were used in the analysis (Figure 1).

Estimates were made of the nitrogen deposition for each one of the benchmark watersheds as part of the screening process for stations. The National

378

Atmospheric Deposition Program (NADP) monitoring site nearest to each of the benchmark watersheds was located, and the deposition at this site was considered to be representative of deposition to the benchmark watershed. Deposition records for ammonium and nitrate during the years 1981–1982 (spanning most of the months used in calculating yields) were tabulated. The NADP data do not include dry deposition or wet deposition of DON or PN.

Methods of field collection and laboratory analysis are as described by Alexander et al. (1996, 1998). Inspection of the raw data raised only a few problematic issues. Particulate nitrogen, which was estimated as the difference between paired Kjeldahl N, nitrate, and nitrite analyses on filtered and unfiltered subsamples, sometimes showed negative concentrations. This is an expected result of random error in analysis of samples that contain low concentrations of particulate nitrogen. Because elimination of the negative data points would bias the weighted means, the negative numbers were left in the data set. Two apparently erroneous, extreme concentrations of Kjeldahl nitrogen were excluded. For ammonium, mean values below the detection limit (10 μg/L) were set to half the detection limit.

Daily discharge records for the 19 watersheds under analysis were retrieved from USGS sources (Alexander et al. 1996). Individual chemical analyses then were matched with discharge data. A single chemical analysis was considered representative of a span of days midway between the most recent previous analysis and the subsequent analysis. Discharge from this interval of time was applied to the concentration data for each nitrogen constituent in the computation of discharge-weighted means for each constituent. In this manner, a discharge-weighted mean concentration for each nitrogen fraction was obtained over the entire two years for each watershed. The sampling frequency for nitrogen analysis typically was monthly (one site bi-monthly; occasional missing data for some sites).

The USGS data on ammonium and organic nitrogen are biased. Over the period relevant to the present study, the USGS was using mercuric chloride tablets to preserve samples intended for ammonium and Kjeldahl analyses. In response to queries about the possibility of contamination caused by these tablets, the USGS arranged for two of its laboratories to conduct tests for bias caused by contamination. The results, which are reported along with the documentation for the HBN data set (Alexander et al. 1996) indicated that one of the laboratories found a small amount of contamination (3 to 15 μg/L) and another found negligible contamination. A detailed statistical analysis of field data for waters in Texas, however, provided strong circumstantial evidence for more significant bias caused by contamination (Schertz et al. 1994); bias was higher for Kjeldahl nitrogen analyses than for ammonium analyses.

Table 1. Characteristics of sites used in the analysis. Runoff is the mean for the two years considered in the analysis (see text)

Site	Latitude (N)	Longitude (W)	Elev m asl	Area km^2	Runoff mm/yr
Young Womans Creek, PA	41	77	238	119	541
Scape Ore Swamp, S.C.	34	80	50	249	266
Falling Creek, GA	33	83	367	186	219
Sopchoppy, FL	30	84	3	264	580
Sipsey Fork, AL	34	87	540	238	470
Upper Twin Creek, OH	38	83	538	31	245
Popple River, WI	45	88	429	360	299
Rock Creek, MT	48	106	771	850	25
Castle Creek, SD	44	103	5920	205	41
Encampment River, WY	41	106	2521	189	548
Kiamichi River, OK	34	94	270	104	775
Vallecito Creek, CO	37	107	2410	186	682
Wet Bottom Creek, AZ	34	111	707	93	94
Red Butte Creek, UT	40	111	5400	18	198
Merced River, CA	37	119	1224	469	886
Elder Creek, CA	39	123	1391	18	1333
Andrews Creek, WA	48	120	4300	57	536
Cache Creek, WY	43	110	2057	28	449
Minam River, OR	45	117	774	622	799
Mean	–	–	1574	226	473
Standard Error	–	–	412	50	76

The data for the present study were tested statistically for evidence of bias caused by contamination associated with the mercuric chloride preservation method. Mean values for ammonium concentration for each station were taken over the interval October 1980 through September 1986, when mercuric chloride preservation was used. A comparison then was made with means for each station over the interval October 1986 to September 1994, when the same protocols were used for analysis, but without use of mercuric chloride tablets for preservation. The comparison showed evidence of bias. Ammonium concentrations averaged 37 μg/L higher across all stations over the interval when mercuric chloride tablets were used (standard error, 7 μg/L). A similar analysis was conducted for the Kjeldahl nitrogen measurements leading to estimates of dissolved organic nitrogen; the mean bias was 150 μg/L (standard error, 25 μg/L). These biases are within the range reported by Schertz et al. (1994) for stations in Texas. Because handling and storage of the tablets were the probable causes of contamination, bias is expected to vary from one station to another. Therefore, mean concentrations of total nitrogen and nitrogen fractions for any given station were corrected for the bias associated with that station prior to statistical analysis of the data.

Results

As a first step in the statistical analysis, the three independent variables (elevation, area, runoff) were compared with each other. A correlation matrix showed that the independent variables are not significantly associated statistically ($p > 0.05$). As a second step, all three of the independent variables (logarithmically transformed here and in all other analyses) were entered into a stepwise multiple regression analyses of nitrogen yield. There were seven such multiple regressions: one for each of the nitrogen components listed in Table 2. In no case is watershed area or elevation significantly related to yield of total nitrogen or nitrogen fractions in these multiple regressions. For this reason, the remaining analyses focus only on runoff.

Table 3 summarizes the relationships between runoff and yield of total nitrogen and nitrogen fractions; all relationships are highly significant ($p < 0.001$) and account for high amounts of variance (Figure 2). Yield of total nitrogen and all nitrogen fractions increases with runoff. In all cases, the rate of increase in yield is less than the rate of increase in runoff (slope <1.0). The result is a decline in concentrations with increasing runoff, even though yield per unit area is increasing with increasing runoff (Table 3).

Discussion

Runoff explains a very high percentage of variance in the yield of total nitrogen and nitrogen fractions among minimally disturbed HBN watersheds. Lewis et al. (1999) reached the same conclusion for undistubed tropical watersheds. The equations that were developed for tropical watersheds are very similar to those developed for the benchmark watersheds of the U.S. (Figures 2). Although rigorous statistical comparison of the relationships for tropical and temperate watersheds is problematic because the tropical watersheds cover a much greater span of physical conditions, the indication of the comparisons shown in Figure 2 is that tropical-temperate differences in relationships between runoff and yield are small and possibly insignificant.

Table 2. Summary of yield data (kg ha^{-1} y^{-1}) for the benchmark watersheds and estimates of atmospheric deposition for the same watersheds (kg ha^{-1} y^{-1}, inorganic N)

Site	NH$_4^+$-N	NO$_3^-$-N	DIN	DON	TDN	PN	TN	Deposition
Young Womans Creek, PA	0.05	1.72	1.77	0.87	2.63	0.65	3.28	7.5
Scape Ore Swamp, S.C.	0.06	0.48	0.54	0.86	1.40	0.14	1.54	4.5
Falling Creek, GA	0.16	0.25	0.41	0.49	0.90	0.21	1.11	3.5
Sopchoppy, FL	0.26	0.44	0.70	2.34	3.04	1.45	4.49	3.5
Sipsey Fork, AL	0.15	0.29	0.43	0.37	0.80	1.59	2.40	5.5
Upper Twin Creek, OH	0.08	1.07	1.15	0.50	1.65	0.19	1.84	5.5
Popple River, WI	0.15	0.35	0.50	1.43	1.93	0.02	1.95	4.0
Rock Creek, MT	0.02	0.11	0.13	0.11	0.24	0.12	0.37	1.5
Castle Creek, SD	0.01	0.05	0.07	0.10	0.17	0.02	0.19	1.5
Encampment River, WY	0.33	0.45	0.78	2.09	2.87	0.88	3.75	1.5
Kiamichi River, OK	0.31	0.62	0.93	2.47	3.40	0.41	3.81	4.5
Vallecito Creek, CO	0.29	1.09	1.37	2.91	4.28	1.07	5.35	1.5
Wet Bottom Creek, AZ	0.05	0.08	0.13	0.16	0.29	0.13	0.42	1.0
Red Butte Creek, UT	0.12	0.21	0.33	0.73	1.06	0.12	1.18	1.5
Merced River, CA	0.44	0.74	1.17	1.27	2.45	2.11	4.55	1.5
Elder Creek, CA	0.36	1.03	1.39	1.83	3.21	0.90	4.11	1.5
Andrews Creek, WA	0.11	0.41	0.52	2.47	2.99	0.58	3.57	1.5
Cache Creek, WY	0.32	0.47	0.79	0.99	1.78	0.42	2.20	1.5
Minam River, OR	0.22	0.59	0.80	1.53	2.34	1.42	3.75	1.0
Mean	0.18	0.55	0.73	1.24	1.97	0.65	2.62	2.8
Standard Error	0.03	0.10	0.11	0.21	0.27	0.14	0.36	0.4

Similarity in the relationships between runoff and nitrogen yield for temperate and tropical watersheds is somewhat surprising. The nitrogen cycles of tropical latitudes are, except at the highest elevations, not interrupted by prolonged periods of low temperature. The interlatitudinal comparisons of Figure 2 indicates that the temperate winter, while highly significant seasonally to the rates of many processes affecting the nitrogen cycle, is of low significance to annual mass balance of N fractions and TN.

The study of tropical watersheds by Lewis et al. (1999) indicated that ratios of DIN to DON and PN to TN were influenced by watershed size. No such trend appears in the data for benchmark watersheds, but the test for such relationships by use of the data for benchmark watersheds is weak because the range of watershed sizes is small (2 orders of magnitude) by comparison with the range of data available for the tropics (7 orders of magnitude). Unfor-

Table 3. Relationship (log-log) of yield (kg ha^{-1} y^{-1}) to runoff (mm/yr) for total N and N fractions. Predicted yields and concentrations of total N and N fractions are shown for a range of runoff values. All relationships are significant at $p < 0.001$

Function	Constant	Slope	r^2	Yield as N, kg ha^{-1} y^{-1}				Concentration as N, μg/L			
				100	250	500	1000 mm	100	250	500	1000 mm
Ammonium N	−2.93	0.81	0.74	0.05	0.10	0.18	0.32	49	41	36	32
Nitrate N	−2.19	0.71	0.66	0.17	0.33	0.53	0.87	171	131	107	87
Diss. Inorg. N	−2.12	0.74	0.77	0.23	0.45	0.75	1.25	228	179	149	125
Diss. Org. N	−2.32	0.89	0.78	0.29	0.66	1.23	2.27	292	264	245	227
Tot. Diss. N	−1.94	0.84	0.84	0.54	1.15	2.05	3.66	536	460	410	366
Part. N	−2.90	0.97	0.54	0.11	0.27	0.53	1.03	112	108	106	103
Tot. N	−1.90	0.87	0.91	0.69	1.53	2.79	5.09	692	612	558	509

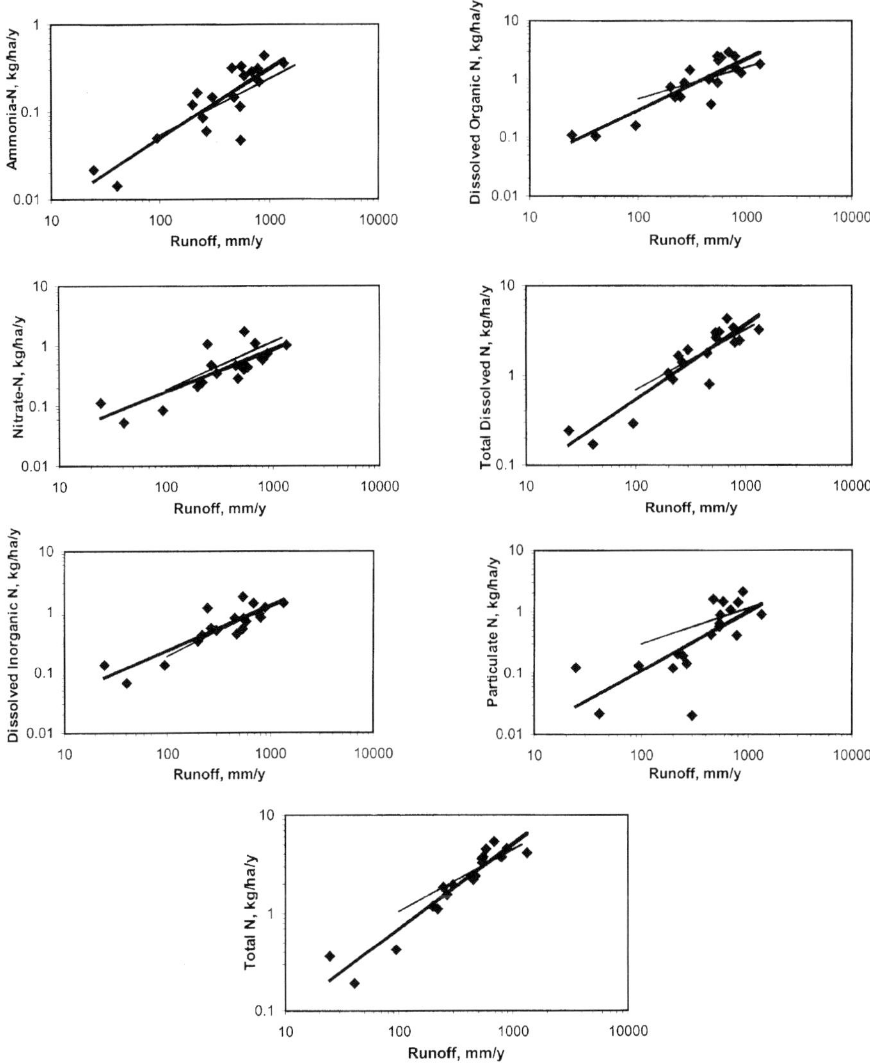

Figure 2. Relationships of runoff to total N and N fractions expressed as annual watershed yield for the 19 benchmark watersheds. The trend lines for tropical watersheds (narrow lines) from Lewis et al. (1999) are shown for comparison.

tunately, background nitrogen yields cannot be estimated empirically for very large watersheds at temperate latitudes.

Variables not included in this analysis (e.g. vegetation type, soil type, land-use history) may explain additional variance in nitrogen yields beyond what is explained by the three variables that were used here. Unfortunately, there are several impediments to the analysis of additional variables. Most important is that a large amount of variance (91% for total N) is already accounted for, indicating that an accounting of the small residual variance would require a much larger data set. Also, some of the additional variables most likely to relate to nitrogen yields (e.g. vegetation type) are categorical in nature and thus are more difficult to quantify.

The data for these watersheds confirm, along with data for the American tropics, that runoff is a master variable explaining interwatershed variance of yield for N and N fractions. The role of runoff in explaining N transport and N fractionation is remarkably insensitive to latitude.

Acknowledgements

This work was initiated as part of the International SCOPE N Project, which received support from both the Mellon Foundation and from the National Center for Ecological Analysis and Synthesis. I thank James Saunders for assistance with the statistical analysis, Sujay Kaushal for assistance with NADP data, Richard Alexander for providing me with access to and comments on the HBN data, Bob Howarth for encouragement of this project, and three reviewers for helpful comments.

References

Aber JD, Nadelhoffer KJ, Streudler P & Melillo J (1989) Nitrogen saturation in northern forest ecosystems. Bioscience 39: 378–386

Alexander RB, Slack JR, Ludtke AS, Fitzgerald KK & Schertz TL (1996) Data from selected U.S. Geological Survey National Stream Water Quality Monitoring Networks (WQN). US Geological Survey Digital Data Series, DDS-37

Alexander RB, Slack JR, Ludtke AS, Fitzgerald KK & Schertz TL (1998) Data from selected US Geological Survey National Stream Water Quality Monitoring Networks. Water Resources Research 34: 2401–2405

Dise NB & Wright RF (1995) Nitrogen leaching from European forests in relation to nitrogen deposition. Forest Ecology Management 71: 153–161

Howarth RW, Billen G, Swaney D, Townsend A, Jaworski M, Lajtha K, Downing JA, Elmgren R, Caraco N, Jordan T, Berendse F, Freney J, Kudeyarov V, Murdoch P & Ahao Liang S (1996) Regional nitrogen budgets and riverine N & P fluxes for the drainages to the North Atlantic Ocean: Natural and human influences. Biogeochemistry 38: 1–96

Lewis WM Jr, Melack JM, McDowell WH, McClain M & Richey JE (1999) Nitrogen yields from undisturbed watersheds in the Americas. Biogeochemistry 46: 149–162

Meybeck M (1982) Carbon, nitrogen, and phosphorus transport by world rivers. American Journal of Science 282: 401–450

Schertz TL, Wells FC & Ohe DJ (1994) Sources of trends in water quality data for selected streams in Texas, 1975–1989 water years. U.S. Geological Survey Water Resources Investigation Report 94-4213

Sickman JO & Melack JM (2002) Regional analysis of inorganic nitrogen yield and retention in high-elevation ecosystems of the Sierra Nevada and Rocky Mountains. Biogeochemistry 57/58: 341–374

Williams MR & Melack JM (1997) Atmospheric deposition, mass balances and processes regulating streamwater solute concentrations in mixed conifer catchments of the Sierra Nevada, California. Biogeochemistry 37: 111–144

Biogeochemistry **57/58**: 387–403, 2002.
© 2002 *Kluwer Academic Publishers. Printed in the Netherlands.*

Nitrogen budgets for the Republic of Korea and the Yellow Sea region

V.N. BASHKIN[1,2,*], S.U. PARK[1], M.S. CHOI[3] & C.B. LEE[3]
[1]*Department of Atmospheric Sciences, Seoul National University, Seoul, Republic of Korea;* [2]*JGSEE, School of Energy and Materials, KMUTT, 91 Pracha-uthit Rd, Rasburana, Tungkru Bangkok 10140, Thailand;* [3]*Department of Oceanography, Seoul National University, Seoul, Republic of Korea (*author for correspondence, corresponding address: Geography Faculty, Moscow State University, Vorobyovy Gory, Moscow 119899, Russia, e-mail: bashkin@rc.ru)*

Key words: anthropogenic loading, biogeochemical cycling, Korea, nitrogen deposition, Northeast Asia, regional budget

Abstract. Growing populations in northeast Asia have greatly altered the nitrogen cycle, with increases in agricultural production to feed the population, and with increases in N emissions and transboundary air pollution. For example, during the 1900's over 50% of the N deposition over Republic of Korea was imported from abroad. In this paper, we present biogeochemical budgets of N for the South Korean peninsula (the Republic of Korea) and for the Yellow Sea region. We quantify N inputs from atmospheric deposition, fertilizers, biological fixation, and imports of food, feed, and products. We quantify outputs in riverine export, crop uptake, denitrification, volatilization, runoff, sedimentation and sea water exchange. Calculations were conducted using mean values from 1994–1997. All of the nitrogen budgets were positive, with N inputs exceeding outputs. The excess N inputs gave rise to increases in N storage in landfills and in groundwater. Annual accumulation of N in the Yellow sea, including inputs from South Korea and other drainage areas, was 1229 kt yr^{-1} with a residence time for N of approximately 1.5 years, thus doubling N content in marine waters every 3 years during 1994–1997. The human derived N inputs leads to excessive eutrophication and pollution of the Yellow Sea.

Introduction

The estimation of regional fluxes of pollutants provides a powerful tool for understanding changes associated with human activities. Several examples of regional-scale N budgets include studies of N cycling in the North Atlantic Ocean and its watershed (Howarth et al. 1996) and in the Mediterranean watershed (Bashkin et al. 1997). The modification of the nitrogen cycle by humans has been well documented (e.g. Howarth et al. 1996). Nitrogen (N) is a key element of many biogeochemical processes and can be both a nutrient limiting the productivity of terrestrial and aquatic ecosystems and a pollutant where excessive accumulation in biogeochemical food chains leads to many environmental problems. The main anthropogenic sources of N pollution are related to fertilizer application, waste production and emission of gaseous species.

Modern projections suggest that large increases in emissions may occur during the next 25–50 years in East Asia associated with planned development patterns. Estimates assuming the current growth rate of energy consumption predict that N emissions in East Asia will surpass the emissions of North America and European combined by the year 2020. The primary man-made source of acidifying and greenhouse compounds in East Asia is fossil fuel of low quality, with high content of sulfur (up to 7% in Thai lignite, Chinese brown coal etc.) and heavily oil. The multiple effects of acidification and increased N deposition may cause decreases in base cations, leading to nutrient imbalances in forest vegetation which in turn increases forest vulnerability to diseases, attacks of insects and parasites. Nitrogen deposition changes natural vegetation, bacterial and mycorrhizal composition into more nitrophilic communities, with less of diversity of species, especially endemic ones. The NO_x compounds can transport to the great distance in East Asia. The spatial scale of atmospheric NH_x pollution and its effects depend on its form: gaseous NH_3 is important near sources up to 100 km; aerosol NH_4 is transported over longer distances leading to effects up to 1000 km (Carmichael et al. 1997). At present, N_2O warrants also a priority on the policy and research institutes. The increasing N_2O concentration in the atmosphere mainly results from nitrification and denitrification processes due to increasing application of mineral fertilizers in various regions of the Earth and especially in East Asia. This N species contributes to the greenhouse effect (about 6%) and to the destruction of stratospheric ozone (Erisman et al. 1999). There is agreement both nationally and internationally that long-range transboundary air pollution is not limited to the geographical limits of individual East Asian countries. It is known that during the winter the major weather patterns in East Asia facilitate the transboundary transport of air pollutants from west to east, from land to sea and the reverse in summer. Pollutants can thus be transported from country to country in the whole region of East Asia. For instance, it has been calculated that during 1990's about 50% of oxidized and about 60% of reduced N deposition over Republic of Korea were imported from abroad. It is therefore impossible for individual countries to solve the problem of air pollution and acid rain alone. There is need for regional intergovernmental cooperation. Currently, regional/sub-regional agreements on the issue of pollutant emission abatement strategy do not exist at all or are in the initial stages (Bashkin & Park 1998).

The abatement strategy for reduction of N emissions and deposition and for decreasing N losses from agroecosystems due to excessive application of fertilizers could be based on regional biogeochemical budget calculations which quantify the dominant N fluxes in terrestrial and aquatic ecosystems. The goal of this paper is to establish biogeochemical budgets for N for both the South Korean peninsula and for the Yellow Sea region.

Site description

The Republic of Korea

The Republic of Korea (or South Korea) is in the southern part of mountain Korean peninsula between 126° E–130° E and 34° N–38° N. Geographically, this is the northern temperate zone of the Eastern Hemisphere. The overall area of Republic of Korea (ROK) comprises 99,022 km^2. The dominant vegetation in ROK consists of forest trees with varying undergrowth of shrubs and small plants. The forest vegetation is divided according climate into temperate and subtropical zone forests. The secondary vegetation, which occurs in the extensive areas in the western and southern regions, consists mostly of shrubs, grasses and conifers, associated with deciduous trees. Grassland with shrubs also appears commonly as a kind of climax vegetation in the higher elevation and plateau remnants. Agricultural land uses occupy about 20% of total South Korean area.

Approximately two-thirds of the total South Korean topography is mountainous. The mountain range of Taebaeg on the Gangweon-Do runs southward along the east coast with lateral branches and spurs expending in a south westerly direction. The slopes to the east are steep while those to the west are gentle. The mountain range slopes towards the south, thus making the southern part of the country fairly level. In the contrary, the northern part is mountainous and hilly land. About 80% of the ROK lands are in regions where the altitude of the summit ranges from 300 to more than 1,000 meters. The land bears a strongly dissected relief reworked by numerous erosion cycles. The low lands include both coastal plains clayey materials and the continental alluvial plains and valley flood plans of the interior.

According to USDA Soil Taxonomy, the soils in ROK are classified as 6 orders and 14 suborders (Um 1985). The dominant soils are inceptisols with 4 suborders: Andepts, Aquepts, Ochrepts and Umbrepts (5,840,441 ha). The Entisol order consists of 4 suborder (Psamments, Aquents, Fluvents and Orthents) and occupied 2,849,102 ha. The Altisol order is presented by

Aqualfs and Udalts suborder on the area of 309,677 ha and Histosol order is devided to Saprist and Hemist suborder with total area of 384 ha.

The Yellow-Bohai Seas region

The Yellow Sea is a typical epicontinental sea surrounded by the continent of China and the Korean peninsula and connected with the East China Sea to the south and with Bohai Sea in the north (Figure 1). The mean water depth is about 44 m and the surface area is 420,000 km^2. The rivers (the Huanghe, the Aprock, the Han, the Keum, the Haihe, the Luanhe) drain freshwater of about >160 km^3 and suspended materials of 1.1×10^9 ton annually into Bohai and Yellow Sea system. Many industrial complexes and large cities are along the coastlines, from which great quantities of pollutants are discharged into rivers or directly into coastal waters. Further, the westerly and northwesterly winds, which prevail in winter and spring, deliver great amount of mineral dust and anthropogenic material. As this coastal system receives large amounts of terrestrial material produced by natural weathering and human activities, it is a suitable study site for the investigation of the biogeochemical cycle of N in the regional scale.

The Yellow Sea is a semi-enclosed basin. Wide coastal areas (<40 m water depth) are located along shorelines nearby both continents with a channel (>60 m water depth), which is developed in NW-SE direction. The southward flows in both coastal areas and northward flows of warm water in the channel are the general circulation pattern in winter, but northward flow may be disappeared in summer, which results in the formation of the Yellow sea cold water in the channel. The tidal fronts near the boundary of shallow coastal area and channel are developed during summer when thermal stratification is established in the channel and vertically homogenous water mass by strong tidal currents is sustained in shallow coastal area (Seung & Park 1990). These fronts may affect to the transport of terrestrial materials (including N) to offshore locations, and hence the biogeochemical activity.

Billions of tons of terrestrial materials are discharged annually through rivers (including the Huaghe, Aprock, Han, Keum, and others) and tens of million tons of mineral dusts (otherwise known as 'yellow sand') are deposited annually into surface seawaters from the atmosphere. Most of the particles derived from the Huanghe River are deposited in the Bohai Sea. It has been suggested that the atmospheric dust flux to the Yellow Sea may be comparable to the river input (Gao et al. 1992; Zhang et al. 1992; Choi et al. 1998).

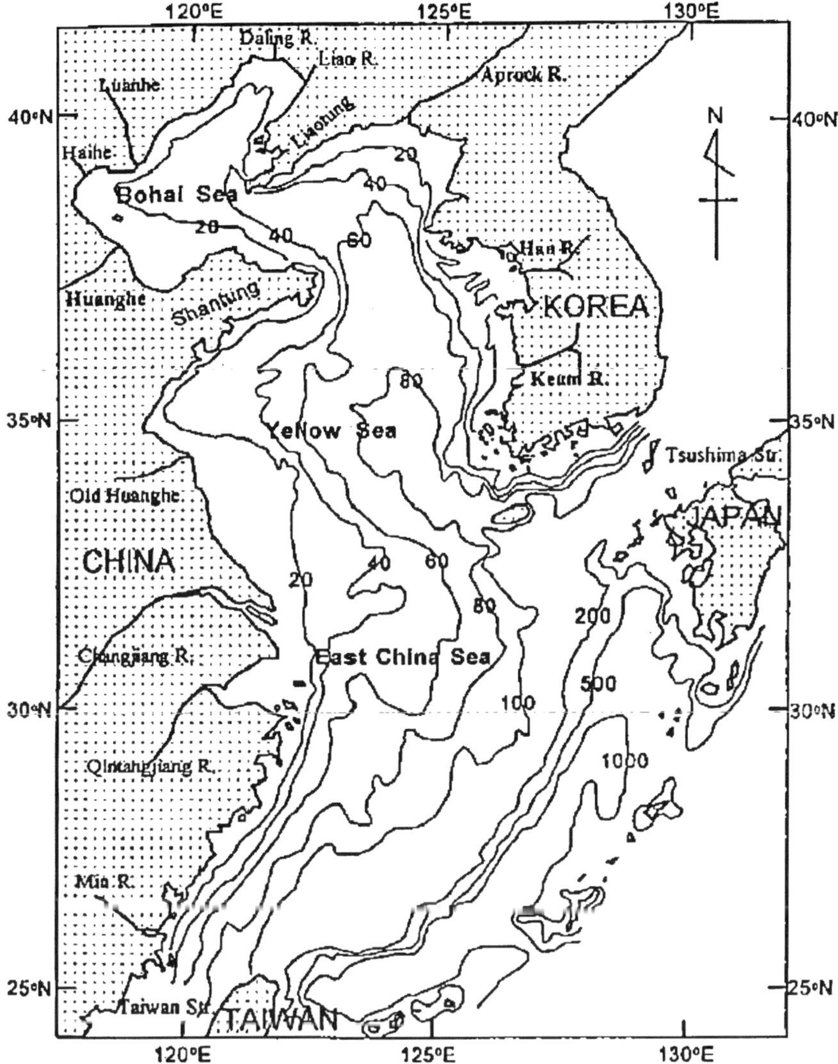

Figure 1. Bathymetry and geographic setting including the East China Sea, the Yellow Sea and Bohai Sea.

Results and discussion

Application of biogeochemical mass balance approaches to regional studies

Any type of budget for biogeochemical turnover of pollutants depends on the availability of data. The statistical characteristics for regions in East Asia were extracted from both national and international sources (e.g. Environ-

Table 1. Components of the nitrogen cycle quantified for each of the budgets presented in this paper. (+ estimated; − not estimated)

Nitrogen Budget Term	Republic of Korea, agricultural lands	Republic of Korea, entire country	Yellow & Bohai Seas region
Deposition	+	+	+
Fertilizers	+	+	−
N fixation	+	+	+
Import	−	+	−
Riverine N	−	−	+
Crops	+	−	−
Denitrification	+	+	+
Volatilization	+	+	+
Discharge	+	+	−
Sedimentation	−	−	+
Sea exchange	−	−	+

mental Statistics Yearbook 1998; UNESCO 1978; ESCAP 1998). Data on content of N species in river waters were selected both from literature studies. The specific data sources used are referenced along with the presentation of results from the individual studies below.

As is typical, more precise calculations can be made for small watersheds with homogenous (Moldan & Cerny 1994) or variable (Bashkin 1984; Gunderson & Bashkin 1994) land uses. With currently available data, it is unable to fully account for the fate of both natural and anthropogenic N added to the Yellow Sea basin. We have more complete data at smaller scales allowing us to make more detailed estimates for the agricultural ecosystems in ROK and for the total land area or the ROK. Table 1 summarizes budget terms estimated in this study, for each of the N budgets that we establish, including one for the agricultural lands in South Korea, one for the entire land area of the South Korean Peninsula within the Republic of Korea, and one for the region of the Yellow-Bohai Seas.

Quantifying N fluxes

We established a biogeochemical budget for regions in the ROK. Nitrogen input terms included atmospheric deposition, mineral and organic fertilizers, and biological fixation (Table 1). Output items considered included crop uptake, river discharge, denitrification and volatilization. All calculations were conducted for 1994–1997.

Input of N from atmospheric deposition was calculated from two models, HEMISPHERE (Sofiev 1998) and MOGUNTIA (Dentener & Crutzen 1994; Zimmerman 1998), and showed similar values (10.7–11.0 kg ha^{-1} yr^{-1}). Inputs from the application of mineral fertilizers averaged 226 kg ha^{-1} yr^{-1} in South Korean agriculture, which was the maximum input source term in the N budget for the region. Nitrogen inputs from biological fixation (nonsymbiotic only, since the area under symbiotically fixed crops was very small in ROK) was carried out using data from the Environmental Statistics Yearbook 1998; Cleveland et al. 1999; Zhu et al. 1997, as follows. In agricultural lands, the rate of fixation in the land area in rice plantations (1009560 ha) was taken to be 45 kg ha^{-1} yr^{-1}, yielding an annual flux of 45430 tons. Fixation rates in other cropland areas (966280 ha) were assumed to be 15 kg ha^{-1} yr^{-1}, yielding an annual flux of 14494 tons. In forested lands (5072600 ha) the assumed fixation rate (1 kg ha^{-1} yr^{-1}) yielded an annual flux of 5072 tons. Total N inputs to agricultural and forest lands from fixation, therefore, equaled 64996 tons yr^{-1}.

N losses due to denitrification were calculated using data from Environmental Statistics Yearbook 1998; Freney 1996; Lin et al. 1996; Mosier et al. 1998; Zhu et al. 1997, as follows. Denitrification loss in the agricultural areas of rice plantation (1009560 ha) was calculated as 32% of the fertilizer use rate, yielding an annual flux of 73011 tons. Denitrification loss in upland crop areas (966280 ha) was calculated as 15% of the fertilizer use rate, yielding an annual flux of 31887 tons. Denitrification loss from manure was calculated as 13% of the manure N application rate, yielding an annual flux of 20528 tons. Denitrification losses from soils were assumed to be 3 kg ha^{-1} yr^{-1}, yielding an annual flux of 5916 tons. Agricultural recycled N was considered for regional biogeochemical budget in South Korea agroecosystems as organic fertilizer N. The values of organic fertilizer N were assessed using the statistical data on human and animal/poultry population and rates of N in excreta (Table 2). Losses from anthropogenic NH_3 emissions were estimated in an earlier study (Park 1998). The modified European calculation factors (IPCC 1997) were applied. The average total value was 142123 ton and NH_3 emission from fertilizers was predominant (35% from the total value).

In addition to the input/output items for agroecosystems, we estimated the N fluxes with river runoff for calculating the N budget for the whole South Korean area. The mean annual water discharge was 61.6×10^{12} L. In accordance with statistical data, about half of wastewater was untreated in ROK in 1994–1997. As a consequence, the content of reduced N in surface waters was almost the same as the content of oxidized N. Nitrite-N was also monitored in South Korea rivers and its mean content was 0.045 mg/L.

Table 2. Annual accumulation of nitrogen in human and animal excreta in Republic of Korea (using mean values for 1994–1997)

Items	Rate, kg per capita yr^{-1}	Population, thousand	Tons $year^{-1}$
HUMAN			
Population (adults)			
– Urban	0.44	33745.0	14848
– Rural	0.69	5494.0	3791
Subtotal			18639
LIVESTOCK			
Cattle	11.35	3267.0	37085
Horse	9.79	6.70	6552
Pig	3.22	6691.0	21525
Sheep	0.70	1.6	1
Goat	0.70	653.0	458
Poultry	1.00	85623.0	88483
Subtotal			154113
TOTAL			172752

The dissolved organic N (DON) content in the most monitored rivers and water reservoirs was negligible (<1 mg L^{-1}) due both to intensive mineralization and to algae uptake as well as the low content of organic matter in South Korean soils. The fluxes of suspended matter were significant, totaling 1.1×10^9 ton yr^{-1}, especially during summer monsoon period, with discharge-weighted mean N content of 0.085%. The total fluxes of dissolved inorganic N (DIN) and solid and particulate N (N-SPM) were 193142 tons per year in 1994–1997 (Table 3).

The quantitative parameterization of different input and output components of N biogeochemical cycle in South Korean peninsula allows us to make up the calculation of two mass balance estimates: one for South Korean agroecosystems ($\sim2.0 \times 10^6$ ha) and one for the whole area of ROK, including input and output estimates for the components as indicated in Table 1.

Nitrogen budget for South Korean agroecosystems

The overall N budget for agroecosystems in South Korea is illustrated in Table 4. Since South Korean agriculture is characteristic of both developed countries (applying great amounts of synthetic fertilizers) and developing ones (N recycling using organic fertilizers), it was of interest to compare the

Table 3. Annual riverine fluxes of nitrogen from the area of Republic of Korea (using mean values for 1994–1997)

N species	N content, mg N L^{-1}	N fluxes, tons year^{-1}
N-NO$_3^-$	1.470	90552
N-NO$_2^-$	0.045	2784
N-NH$_4^+$	1.540	94864
N-SPM	0.085	5220
TOTAL		193142

Table 4. Nitrogen budget in agroecosystems of Republic of Korea (using mean values for 1994–1997)

N budget term	Tons year^{-1}	% of input
INPUT		
Fertilizers	446081	65.0
Manure	157904	23.0
Biological fixation	59924	8.7
Deposition	21692	3.3
Subtotal	685601	100.0
OUTPUT		
Crop production	259779	37.8
Denitrification	132211	19.3
NH$_3$ volatilization	142123	20.7
Subtotal	524113	77.8
BUDGET*	+151488	22.2

*Excess N accumulated in the landscape is distributed between surface runoff, leakage to groundwater, and increasing N content in crops (mainly vegetables).

corresponding values of N mass balance in some European countries and China with local figures (Table 5). The comparison of mass balance values between South Korean agroecosystems and those for developed European countries reveals similarities in the N budgets such as a great surplus of N (typically more than 100 kg ha^{-1} yr^{-1}). N crop uptake efficiency in Asian countries is less than in European ones and ROK is the lowest with only 38% efficiency. The large amount of surplus N in the budgets supports the idea that that an increasing non-sustainability within the agriculture, human nutrition

Table 5. Nitrogen mass balance (kg ha^{-1} yr^{-1}) and N crop uptake efficiency (% from total input) in different developed and developing countries*

Country	Agricultura 1 area, 10^6 ha	Input	Output	Surplus	Crop uptake efficiency
Denmark	2.9	217	30	187	59
Germany	12.0	215	51	164	73
U.K.	18.1	127	17	110	—
Netherlands	2.3	463	96	365	63
Norway	1.0	147	80	67	71
Sweden	3.7	121	21	100	63
S. Korea	2.0	347	51	296	38
China	94.9	294	95	199	51

*Data for European countries from Isermann (1991) and for China from Xing & Zhu (2002).

Table 6. Distribution accumulated nitrogen between various waste treatment types in the Republic of Korea (using mean values for 1994–1997)

Items	Tons year^{-1}	% of total
Landfill	207740	75
Incineration	13849	5
Agricultural use	5540	2
Recycling	27699	10
Damping at sea	22159	8

and waste management complex has occurred both in European and Asian countries. It has been leading to a disturbance of N biogeochemical cycle.

In order to estimate the fate of N accumulated in agroecosystems, we assessed the annual N accumulation in municipal waste for the urban area of the country (Table 6) using data from the Environmental Statistics Yearbook (1998). The annual N accumulation in sewage and wastes of 6 kg per capita was calculated from the population (46,164 thousand people) yielding an annual accumulation of 276987 tons yr^{-1}. The majority of this N was deposited in landfills with subsequent transformation, denitrification, and leaching to surface and ground water. Thus human waste leads to further pollution of drinking water, eutrophication of surface waters and increasing input of N_2O to atmosphere. The DIN contents in many South Korean water reservoirs are

Table 7. Biogeochemical budget of nitrogen for the Republic of Korea (using mean values for 1994–1997)

N budget term	Tons year^{-1}	% of input
INPUT		
Deposition	108160	13.3
Fertilizers	446081	54.8
N fixation	64996	8.0
Import		
Foods	184110	22.6
Goods	10377	1.3
Subtotal	813724	100.0
OUTPUT		
River discharge	189124	23.2
Denitrification	132211	16.2
NH$_3$ volatilization	142123	17.5
Sea waste damping	22159	2.7
Subtotal	485617	59.7
BUDGET	+328107	+40.3

4–10 mg N L^{-1} in summer season and most of these reservoirs are eutrophic (Environmental Statistics Yearbook 1998).

Nitrogen budget for the Republic of Korea

Taking into account the values of various input/out items of biogeochemical N cycle as well as literature data (Environmental Statistics Yearbook 1998; Park 1998; Freney 1996; Mosier et al. 1998; Zhu et al. 1997), the regional mass budget was calculated for the whole South Korean territory (Table 7). The dominant N inputs were related to the application of mineral fertilizers and import of food and goods (about 80% of total input). Deposition (including about 55% from abroad from transboundary air pollution) and non-symbiotic N fixation were responsible for the other 20% of input. N outputs were associated with N volatilization via direct NH$_3$ volatilization and biological denitrification (33.7% of total input) and river discharge (23.2% of total input). Total outputs were only about 60% of the inputs, indicating other storage or loss of N in the landscape. The fate of this excess N (40.3% of total N inputs) in the landscape is shown in Table 8. As it has been shown already during the analysis of N balance in agroecosystems, the main part

398

Table 8. Distribution of accumulated nitrogen in the Republic of Korea (using mean values for 1994–1997)

Accumulation	Ton year^{-1}	% of total
Landfill	207740	63.2
Forest uptake	16455	5.0
Groundwater leakage*	103912	31.8

*calculated by difference.

of excessive N is stored at landfills with corresponding prolonged problems of environment pollution. The values of N forest uptake were calculated on a basis of data on net primary productivity and N content in tree stems and branches. Groundwater leakage was calculated as the difference between total N accumulation and the sum of annual landfill storage and plant uptake in forest ecosystems. In general, the decomposition of waste residues in landfills lasts longer than 1 year and thus this approach to estimating groundwater leakage can be used.

Nitrogen Budget for the Yellow-Bohai Sea

In accordance with approaches shown in Table 1, the total riverine N fluxes were assessed for the Yellow-Bohai sea system (UNESCO 1978; Zhu 1997; ESCAP 1998; Choi 1998; Cha et al. 1998). These data are shown in Table 9. Previous regional analyses of the SCOPE-N Project suggest that without the influence of humans on the landscape, the flux of N from land to coastal waters would be on the order of 130 kg N km^{-2} yr^{-1} when expressed per area of watershed (Howarth et al. 1996; Lewis et al. 1999). In comparison, actual fluxes from Yellow-Bohai Seas drainage basin (areas of China and Koreas) were some 8-fold larger than this for drainages from China and 13-fold larger for the drainages from South Korean area in 1994–1997. Due to the huge amount of SPM transport in Yellow river, the N discharge from China area was dominated by solid matter (87%). The opposite pattern was true for South Korea, where only 6% of total N flux was discharged with as solid and particulates and 94% was discharged as DIN. Comparing these N budget estimates with data reported in the literature (Cha et al. 1998, Choi 1998, Park 1998, Nixon et al. 1996), we calculated the N fluxes in the entire Yellow Sea region (Table 10).

The majority of both the soluble N (53%) and particulate N (>99%) inputs to the Yellow Sea originated in China, despite the fact that only 30% of the riverine discharge was contributed from China. Total N transported to the

Table 9. Assessment of riverine fluxes of nitrogen to the Yellow-Bohai seas system

Watershed	Discharge, km^3/yr	N species	N content, mg L^{-1}	N fluxes, tons yr^{-1}
Han	25.0	NO_3^-	1.330	33188
		NH_4^-	1.310	32800
		NO_2^-	0.037	930
		N-SPM	0.064	1600
		Total		68518
Keum	6.4	NO_3^-	1.610	10317
		NH_4^+	1.540	9859
		NO_2^-	0.045	298
		N-SPM	0.056	1040
		Total		21514
Aprock	33.6	NO_3^-	0.200	6920
		NH_4^+	0.410	13840
		N-SPM	0.098	3460
		Total		24220
Liaohe	14.8	NO_3^-	0.200	2960
		NH_4^+	0.390	5920
		N-SPM	0.160	2664
		Total		11544
Haihe	22.8	NO_3^-	0.500	11400
		NH_4^+	0.650	14820
		N-SPM	0.180	4104
		Total		30324
Yellow	59.2	NO_3^-	0.870	51495
		NH_4^+	1.390	76947
		N-SPM	18.300	1083177
		Total		1211619
Yellow & Bohai seas	161.8	DIN		271614
		N-SPM		1096045
		Total		1367959

Yellow & Bohai Seas in riverine export was 52% of the total N inputs (2581 kt yr^{-1}). Both wet and dry atmospheric N deposition provided significant inputs (20 and 22%) to the region, while inputs from fixation in marine waters were small (6%). Denitrification is the major output from the Yellow & Bohai Seas systems, (37%), with similar values of sedimentation and water exchange

Table 10. Biogeochemical budget of nitrogen for the Yellow-Bohai seas system

N budget term	Tons year^{-1}	% of input
INPUT		
Wet deposition	504000	20
Dry deposition	557000	22
N fixation	152880	6
Riverine soluble N	271614	10
Riverine SPM-N	1096045	42
Subtotal	2581539	100
OUTPUT		
Denitrification	946680	37
Losses as N_2O	23520	1
Sedimentation	170000	7
Water exchange	212000	8
Subtotal	1352200	53
ACCUMULATION	+1229339	+47
N pool in marine water	1818000	
Residence time, yr	1.47	

with East China Sea (7% and 8% of total input) and with negligible values of losses as N_2O (<1%). The annual accumulation of N in Yellow sea was 1229 kt yr^{-1} (+47% of total N inputs) and the residence time of N was 1.5 years. This means that the N content in marine water was doubling every 3 years during 1994–1997.

Conclusions

We quantified N budgets for agroecosystems in South Korea; the entire landscape of the South Korea peninsula (in the borders of the Republic of Korea), and for the Yellow Sea region. The nitrogen budget for South Korean agroecosystems was positive, with N inputs exceeding outputs by about 22%. Most of the N input was from the use of fertilizers (65%). Of the N inputs to the agricultural landscape, 38% was accounted for in crop production, 19% was lost to denitrification, 21% was lost to ammonia volatilization, and the remaining 22% accumulated in the landscape.

Similarly, the N budget for the whole of the South Korea (ROK) was positive, with inputs exceeding outputs by 47%. The accumulation of N in the landscape is due to rates of fertilization, atmospheric deposition, and N imports that were not matched by output rates in river discharge, gaseous losses, and sea water damping. The excess N inputs increased N storage in landfills and in groundwater and gradually increased riverine N discharge to Yellow sea.

The majority of both soluble (53%) and particulate (>99%) N inputs to the Yellow-Sea region were contributed from Northern China, while this area contributes only 30% of the streamflow discharge to the Yellow Sea Basin. Delivery of N to the Yellow Sea region in river loads accounted for 52% of the total N inputs. Both wet and dry deposition inputs were substantial, accounting for 20% and 22% of the total inputs, respectively. Fixation in marine waters accounted for only 6% of the inputs. A substantial fraction (37%) of the N inputs to the Yellow Sea region is lost to denitrification. Sedimentation and water exchange with East China Sea were smaller, yet significant loss terms (accounting for 7% and 8% from total input, respectively). Losses of N inputs as N_2O were negligible (<1%). Therefore, the annual N accumulation of in the Yellow-Bohai seas was 1229 kt yr^{-1} (+47% of total inputs) and the residence time of N was 1.5 years, doubling the N content in marine water every 3 years during 1994–1997. The human-derived N inputs have led to excessive eutrophication and pollution of Yellow sea.

Acknowledgements

This work was initiated as part of the International SCOPE Nitrogen Project, which received support from both the Mellon Foundation and from the National Center for Ecological Analysis and Synthesis. We thank the Korean Science and Technology Federation for granting Prof. Vladimir Bashkin sabbatical leave in Seoul National University in 1998–1999, Mr. Y. Steklov (UN ESCAP, Thailand) for help in information sources and fruitful discussion of results, and Ministry of Environment of Republic of Korea for financial support of this project (G-7 grant). We thank G. Xing and an anonymous reviewer for helpful comments that improved the manuscript.

References

Bashkin VN (1984) Study of landscape-agrogeochemical balance of nutrients in agricultural regions, II: Potassium. Water, Air and Pollution 21: 97–103
Bashkin VN, Erdman LK, Abramychev AY, Sofiev MA, Priputina IV & Gusev A (1997) The input of anthropogenic airborne nitrogen to the Mediterranean Sea through its watershed. MAP Technical Reports Series No. 118, UNEP, Athens, 95 pp

Bashkin V & Park SU (Eds) (1998) Acid deposition and ecosystem sensitivity in East Asia, NovaScience Publisher, 427 pp

Carmichael GR, Hong MS, Ueda H, Chen LL, Murano K, Park JK, Lee H, Kim Y, Kang C & Shim S (1997) Aerosol composition at Cheju Island, Korea. Journal of Geophysics Research 102: 6047–6061

Cha HJ, Kim JY, Koh CH & Lee CB (1998) Temporal and spatial variation of nutrient elements in surface seawater of the west coast of Korea. The Journal of the Korean Society of Oceanography 3: 25–33

Choi MS (1998) Distribution of trace metals in the riverine, atmospheric and marine environments of the western coast of Korea. PhD thesis, Department of Oceanography, Seoul National University, 338 pp

Cleveland CC, Townsend AR, Schimel DS, Fisher H, Howarth RW, Hedin LO, Perakis SS, Latty EF, von Fischer JC, Eleseroad A & Wasson MF (1999) Global patterns of terrestrial biological nitrogen fixation in natural ecosystems. Global Biogeochemical Cycles 13: 623–646

Dentener FJ & Crutzen PJ (1994) A three dimensional model of the global ammonia cycle. Journal of Atmospheric Chemistry 19: 331–369

Environmental Statistics Yearbook (1998) Ministry of Environment, Republic of Korea, 581 pp

Erisman JW, Brydges T, Bull K, Cowling E, Grennfelt P, Nordberg L, Satake K, Scheider T, Smeulders S, van der Hoek K, Wisniewski J & Wisniewski J (1999) Summary statement. In: International Nitrogen conference, Elsevier Science

ESCAP (1998) Sources and Nature of Water Quality Problems in Asia and the Pacific, New York, United Nations, 164 pp

ESCAP (1997) Sustainable Development of Water Resources in Asia and the Pacific: an overview, New York, United Nations, 162 pp

Gao Y, Arimoto R, Duce A, Lee DS & Zhou MY (1992) Input of atmospheric trace elements and mineral matter to the Yellow Sea during the spring of a low-dust year. Journal of Geophysics Research 97: 3767–3777

Gundersen P & Gashkin VN (1994) Nitrogen cycling. In: Moldan & Cherny J (Eds) Biogeochemistry of Small Catchments. John Wiley and Sons, pp 253–277

Freney JR (1996) Control of nitrogen emission from agriculture. In: Lin HC et al. (Eds) Proceedings of SCOPE/ICSU Nitrogen Workshop: The Effect of Human Disturbance on the Nitrogen Cycle in Asia, pp 85–99

Howarth RW (Ed) (1996) Nitrogen Cycling in the North Atlantic Ocean and its Watersheds. Kluwer Academic Publishers, 304 pp

IPCC (1997) Guidelines for National Greenhouse Gas Inventories. OECD/ICDE, Paris

Iserman K (1991) Share of agriculture in nitrogen and phosphorus emission into the surface waters of Western European against the background of their eutrophication. Fertilizer Research 26: 253–269

Lewis WM, Melack JM, McDowell WH, McClain ME & Richey JE (1999) Nitrogen yields from undisturbed watershed in the Americas. Biogeochemistry 46: 149–162

Lin H-C, Yang S-S, Hung T-C & Chou C-H (Eds) (1996) The Effect of Human Disturbance on the Nitrogen Cycle in Asia, Proceedings of SCOPE/ICSU Nitrogen Workshops

Moldan B & Cherny J (Eds) (1994) Biogeochemistry of Small Catchments. John Wiley and Sons, 424 pp

Mosier A, Abrahamsen G, Bouwman L, Bockman O, Drange H, Forlking S, Howarth R, Kroeze C, Oenema O, Smith K & Bleken M (Eds) (1998) International workshop on dissipation of N from the human N cycle and its role in present and future N_2O emissions to the atmosphere. Nutrient Cycling in Agroecosystems. Kluwer Academic Publishers, 313 pp

Nixon SW, Ammerman JW, Atkinson LP, Berounsky VM, Billen G, Boicourt WC, Boynton WR, Church TM, Ditoro DM, Elmgren R, Garber JM, Giblin AE, Jahnke RA, Owens NJP, Pilson JH & Seitzinger SP (1996) The fate of nitrogen and phosphorus at the land-sea margin of the North Atlantic Ocean. In: Howarth RW (Ed) Nitrogen Cycling in the North Atlantic Ocean and its Watersheds. Kluwer Academic Publishers, pp 141–180

Park SU (Ed) (1998) Research and Development on Basic Technology for Atmospheric Environment in Global Scale: Development of Technology for Monitoring and Prediction of Acid Rain. Ministry of Environment, Republic of Korea, 602 pp

Seung YH & Park YC (1990) Physical and environmental character of the Yellow Sea. In: Park CH, Kim DH & Lee SH (Eds) The Regime of the Yellow Sea. Institute of East and · West Staudies, Seoul, pp 9–38

Sofiev MA (1998) Numerical modeling for Acid deposition on Eurasian continent. In: Bashkin VN & Park SU (Eds) Acid Deposition and Ecosystem Sensitivity in East Asia. NovaScience Publishers, U.S.A., pp 5–48

Um KT (1985) Soils of Korea. Soil Survey Materials No. 11. Agricultural Science Institute, Rural Administration, Korea, 66 pp

UNESCO (1978) World Water Balance and Water Resources of the Earth. Gidrometeoizdat Publishers, Leningrad, 666 pp

Zhang J, Huang WW, Liu SM & Wang JH (1992) Transport of particulate heavy metals towards the China Sea: a preliminary study and comparison. Marine Chemistry 40: 161–178

Zhu ZL (Ed) (1997) Nitrogen Balance and Cycling in Agroecosystems of China. Kluwer Academic Publisher, 355 pp

Zimmermann PH (1998) Moguntia: A handy global tracer model. In: van Dop H (Ed) Air Pollution Modeling and its Applications VI. Plenum, New York

Xing GX & Zhu ZL (2002) Regional nitrogen budgets for China and its major watersheds. Biogeochemistry 57/58: 405–427

Biogeochemistry **57/58**: 405–427, 2002.

Regional nitrogen budgets for China and its major watersheds

G.X. XING & Z.L. ZHU
Institute of Soil Science, Chinese Academy of Sciences, Nanjing 210008, P.R. China

Key words: analysis and estimation, China watershed, major river valley, nitrogen budgeting

Abstract. Since the Changjiang River, Huanghe River and Zhujiang River are the three major rivers in China that are flowing into the Pacific Ocean, this paper addresses nitrogen budgeting, source (input) and sink (output and storage), in these three river valleys, and the China watershed as well. In the China watershed, the anthropogenic reactive N has far exceeded the terrestrial bio-fixed N in nature, and human activities have significantly altered the N cycling in this region. In 1995, the total amount of anthropogenic reactive N in China reached 31.2 Tg with 22.2 Tg coming from synthetic fertilizers and 4.18 Tg from NO_x emission from fossil fuel combustion, and the input of recycling N amounted to 30.5 Tg, consisting mainly of human and animal excrement N, reflecting the intensity of the human activity. The sink of N includes N in the harvested crop, denitrification and storage in agricultural soils, transportation into waterbodies and volatilization of NH_3. N output and storage in soil reached up to 48–53 Tg. Of this amount, 14 Tg was in the harvested crops, 12 Tg stored in agricultural soils, 11 Tg transported into waterbodies, 5 – 10 Tg denitrified in the soils and a limited amount exported through food/feed.

In this paper – besides the N budget in the China watershed – the N budgets and especially N transports into waterbodies in the Changjiang, Huanghe and Zhujiang river valleys are estimated.

Introduction

Human activity has markedly altered N cycling in nature, with the anthropogenic reactive N far exceeding the bio-fixed N in amount in the natural terrestrial ecosystem (Galloway et al. 1995). As a result of the increase in anthropogenic N, emission of N_2O, NO_y and NH_x intensifies during the biogeochemical N cycling. For instance, N_2O, a long-lived greenhouse gas and a participant in the atmospheric photochemical reaction that leads to depletion of stratospheric ozone, has been increasing drastically since the pre-industrial era. N_2O emission, due to increasing application of synthetic N fertilizers and other agricultural activities, contributes 6.30 Tg N yr^{-1} (IPCC 1996; Mosier et al. 1998). The amount of NO_3^- transported from terrene to water has also increased rendering a negative impact on water quality and the aquatic ecosystem process. As a result, eutrophication of waterbodies and red tides in lakes and estuaries occur frequently in China and in other parts of the world as well.

NO_x emitted from fossil fuel combustion is not only a second source of N_2O, but also a major cause of acid rain, which affects adversely forest and aquatic ecosystem process, corrodes architecture and degrades soil quality. The increase in anthropogenic N has significantly disturbed the natural biogeochemical N cycling on a global scale, thus attracting more and more attention from scientists and governments the world over.

China is the second biggest country on the Eurasian Continent, with a terrestrial area of 9.6×10^6 km^2, accounting for 19% of the continent's and the biggest watershed on the west Pacific Ocean. Besides, China has the largest population in the world. To meet the food and fiber demands of the 1.2 billion people, application of synthetic N fertilizers has been increasing drastically, reaching 22.20 Tg N in 1995 (China Agricultural Yearbook 1980 – 1996), about one fourth of the world's total. Moreover, as China is still a developing country, coal remains the chief energy source in use. In 1995, China consumed approximately 1.4×10^9 tons of coal, accounting for 74.6% of the total energy materials used (China Statistical Yearbook 1996). Combustion of coal generates and emits large amounts of NO_x into the atmosphere, some of which will later return to the earth through dry and wet deposition and enter into N cycling in the terrestrial and hydrologic ecosystems.

Although N_2O emission, NH_3 volatilization and NO_3^- transport from cropland (Xing & Zhu 1997; Xing 1998; Xing & Yan 1999; Xing & Zhu 2000), NO_x emission from fossil fuel combustion (Wang et al. 1996), anthropogenic NH_3 volatilization (Wang et al. 1997), production and emission of anthropogenic reactive N (Galloway et al. 1996) and N input to estuaries from the Changjiang, Huanghe and Zhujiang Rivers (Duan et al. 2000) have been estimated, the integrated effects of human activities on N cycling in the China watershed are less investigated (Xing & Zhu 2000). We regard the terrestrial part of China as a watershed on the west Pacific Ocean. By using the statistics and conversion factors available, we present here an estimated nitrogen budget of the China watershed.

The Changjiang, Huanghe and Zhujiang Rivers are the three major outflowing rivers located in different climatic zones in China. The regions in the middle and lower reaches of the three rivers and on the west coast of the Pacific Ocean are the most developed areas in China, and bear the strongest impact of human activities. However, little work has been done to estimate nitrogen budgets on a valley scale in China. Besides nitrogen budgeting on the scale of the China watershed, we also analyzed and estimated nitrogen

Table 1. Hydrological data in the China watershed and the Changjiang, Huanghe and Zhujiang River valleys (Ren et al. 1980; Wu 1998; Cui 1999)

Country and river	Drainage area	River length	Annual precipitation	Average flow	Total runoff
	10^6 km^2	10^3 km	10^3 mm yr^{-1}	10^3 m^3/s	10^{11} m^3 yr^{-1}
China	9.6		0.63		27.15
Changjing	1.81	6.38	1.05	31.06	9.79
Huanghe	0.75	5.46	0.49	1.82	0.57
Zhujiang	0.45	2.20	1.44	11.07	3.49

Table 2. Cultivated land area, consumption of chemical N fertilizers in the China watershed and the Changjiang, Huanghe and Zhujiang River valleys*

Country and river	Cultivated area (10^7 ha)			Chemical fertilizer-N consumption (Tg N)		
	Upland	Paddy field	Total area	Upland	Paddy field	Total amount
China	7.01	2.49	9.50	17.2	5	22.2
Changjing	1.14	1.27	2.41	3.18	4.12	7.30
Huanghe	0.84	0.029	0.86	1.71	0.07	1.78
Zhujiang	0.18	0.31	0.49	0.51	0.96	1.47

*Calculated on the basis of the data cited from China Agricultural Yearbook (1996).

input, output and storage in the Changjiang, Huanghe and Zhujiang River valleys in this paper.

Basic data and valley characters

Some basic data of the China watershed and the Changjiang, Huanghe and Zhujiang River valleys are listed in Tables 1, 2, 3 and 4 including area of the river valleys, length and mean flow of the rivers, total amount of runoffs, population and density, land area, area of cultivated land, consumption of chemical N fertilizers; cultivation area of crops; and size of livestock.

The average total runoff in China is 27.15×10^{11} m^3 yr^{-1} (Wu 1998). China has a number of out-flowing rivers that cover an area of 6.12×10^6 km^2, accounting for 63.8% of the total terrestrial part of China, whereas the drainage areas of the rivers flowing out into the Pacific amount to 5.64×10^6 km^2, making up 56.7% of the total. Among them, the Changjiang, Huanghe

408

Table 3. Population and Population density in the China watershed and the Changjiang, Huanghe and Zhujiang River valleys*

Country and river	Population (10^8)			Population density
	Rural area	City	Total	Individual per km^2
China	9.02	2.82	11.84	123
Changjing	3.23	0.86	4.09	226
Huanghe	0.57	0.19	0.76	101
Zhujiang	0.76	0.21	0.97	215

*Calculated on the basis of the data cited from the China Agricultural Yearbook (1996).

Table 4. Size of livestock in the China watershed and the Changjiang, Huanghe and Zhujiang River valleys*

Country and river	The number of the domesticated animals (10^4 head)					
	Cattle	Milk cows	Horses & others animals	Sheep	Pigs	Poultry
China	12789	417	7656	27695	44169	410858
Changjing	3868	37	302	3924	19464	99447
Huanghe	1110	34	291	3640	2364	37156
Zhujiang	1237	4.5	66	221	4479	32025

*Calculated on the basis of the data cited from the China Agricultural Yearbook (1996).

and Zhujiang Rivers are the three major ones (Figure 1), covering in total an area of 3.01×10^6 km^2, about 31.1% of the total (Ren et al. 1980), with a total runoff amount of 13.8×10^{11} m^{3}yr^{-1}, about 51.1% of the total.

The Changjiang River, zigzagging eastwards between 24°–30° Lat. N, is the largest out-flowing river in China, covering an area of 1.81×10^6 km^2. With its main stream being 6.38×10^3 km long, the river runs through 18 provinces, metropolises and autonomous regions, that is, Tibet, Qinghai, Yunnan, Sichuan, Guizhou, Hubei, Hunan, Jiangxi, Anhui, Jiangsu, Shanghai, Shaanxi, Henan, Gansu, Zhejiang, Fujian, Guangxi and Guangdong. Its middle and lower reaches stretch from east to west through the subtropics in the central and northern parts, where the annual precipitation is averaged 1.05×10^3mm. The average annual runoff in the Changjiang River valley reaches 9.79×10^{11} m^3, more than one-third of the total in China (Table 1). About one-fourth of the country's cultivated land is within the valley. Of the 2.41×10^7 ha, 1.27×10^7 ha is paddy and 1.14×10^7 ha upland (Table 2).

Figure 1. Changjiang, Huanghe and Zhujiang river in China.

The Huanghe River, the second largest out-flowing river, winds between $32.5°–41.7°$ Lat. N, running 5.46×10^3 km into the Pacific. It passes through 8 provinces and autonomous regions, i.e. Qinghai, Gansu, Ningxia, Inner Mongolia, Shaanxi, Shanxi, Henan and Shandong, covering an area of 0.75×10^6 km^2, mostly arid and semi-arid regions, where the annual precipitation is 0.49×10^3 mm. So the average annul runoff in the valley is only 0.57×10^{11} m^3, the lowest among the three river valleys (Table 1). The Huanghe River valley has a total of 0.86×10^7 ha of cultivated land, of this amount, 0.029×10^7 ha are paddy fields and 0.84×10^7 ha uplands (Table 2), constituting 9.1% of the total in China.

The Zhujiang River, the fourth largest out-flowing river with 2.2×10^3 km length, flows through 6 provinces and autonomous regions, namely, Guangxi, Guangdong, Yunnan, Guizhou, Hunan and Jiangxi, stretching over an area of 0.45×10^6 km^2 in the southern subtropics between $20°–26°$ Lat. N. The river valley enjoys a humid climate with an annual rainfall of 1.44×10^3 mm. Its annual runoff reaches 3.49×10^{11} m^3, almost half of that in the Changjiang River valley and over 6 times as much as that in the Huanghe River valley (Table 1). The cultivated land in the valley consists of 0.31×10^7 ha of paddy fields and 0.18×10^7 ha of uplands, adding up to 0.49×10^7 ha (Table 2), about 5.1% of the total of the country. The high proportion of paddy fields in the Zhujiang River valley is incomparable in the Changjiang River and the Huanghe River valleys.

Methods

Computation of the basic data of the valleys

All the basic data used in computing N inputs and outputs in the Changjiang, Huanghe and Zhujiang Rivers valleys were gathered from the 1996 Statistical Yearbook and 1996 Agricultural Yearbook of the related provinces, metropolises and autonomous regions, which, however, were usually compiled on the basis of an administrative region, a province, a metropolis or an autonomous region, rather than a river valley. In order to calculate N budgets of the valleys, the area of the part that the river flows through and its percentage of the total land area of each region should first be calculated. Based on the percentage, basic data, like cultivated land area, sown area, population, size of livestock, consumption of chemical N fertilizers, consumption of fossil fuel, and area of major crops in the region were calculated. Here in the paper, the percentage of the area of River valley within each province to the area of that province where the corresponding river passes was cited from Zhang (1986), Yi (1957), Xi et al. (1994), the Geological Department of Zhengzhou Normal College (1959), the Research Committee of China Natural Resources (1992) and Sun (1959).

Sources of N input

The N inputs of the China watershed and the Changjiang, Huanghe and Zhujiang River valleys can be sorted into two groups, anthropogenic reactive N and recycling N. Sources of the former include application of synthetic fertilizer N, NO_x emission from combusting fossil fuel, symbiotic N fixation by leguminous crops, N fixation by azotobacteria in farmlands and N imported with food/feed. Nevertheless, in calculating N input to the Changjiang, Huanghe and Zhujiang River valleys, the last portion was not taken into account. And sources of the recycling N include excrements from human and animals, atmospheric wet deposition, crop residues left in the farmland as manure and NO_x emission from burning of crop residues in the fields or in kitchens. Atmospheric dry deposition, however, was not considered in the group.

Output and storage of N

In calculating N output, the following factors were counted, i.e. N in the harvested crops and N storage in the farmlands, denitrification of soil N in the fields, NH_3 volatilization, NO_3^- transport into waterbodies and N exported with food/feed. In calculating N fluxes to the waterbodies, according to Howath et al. (1996), the anthropogenic reactive N and that of human wastes were reckoned separately. The computation of the N flux from the Changjiang, Huanghe and Zhujiang River valleys did not cover the N exported with food/feed.

NO_x emission from combusting fossil fuel

Because of the huge variety of fossil fuels and the sharp difference in conversion factor for calculating NO_x emission of the same kind of fuels used in different industries (Wang et al. 1996), the consumption of the fuels by different industries in China and the Changjiang, Huanghe and Zhujiang River valleys were worked out separately on a kind-by-kind basis. And then the NO_x fluxes from combustion of coal, coke, crude petroleum, petrol, diesel oil, kerosene, residual oil, liquefied gas and natural gas consumed in different industries were figured out and summed up separately according to their attribute under three categories, coal, petroleum and natural gas. Eventually, the NO_x flux in China in 1995 from the consumption of coal, petroleum and natural gas was obtained, individually.

Storage and denitrification

As chemical N differs from organic N in storage rate and denitrification rate in soil and their rates also vary from paddy fields to uplands, N storage from chemical fertilizers and organic manure in paddy fields and uplands was calculated, separately. However, N loss through denitrification of organic manure N was calculated regardless of whether paddy fields or uplands.

Atmospheric deposition

N input through atmospheric wet deposition to the Changjiang River and Zhujiang River valleys in South China was reckoned separately from that to the Huanghe River valley in North China, because NO_x and NH_3 concentrations in the atmospheric wet deposition show significant zonal difference between north and south.

412

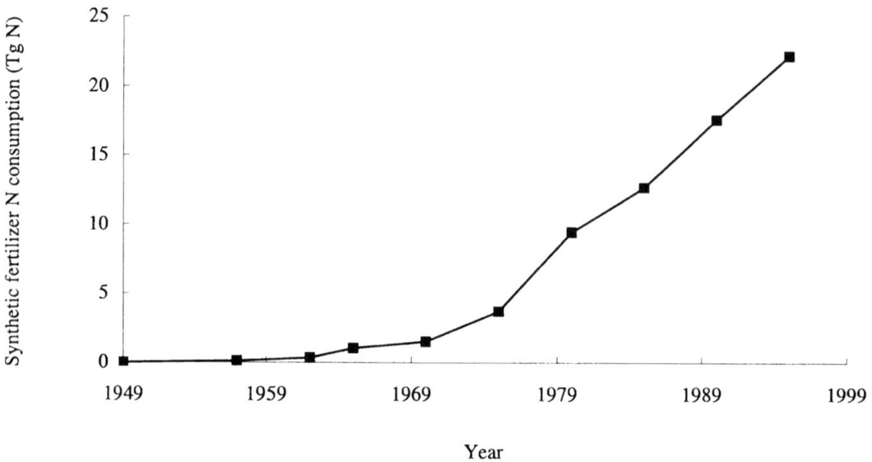

Figure 2. Consumption of synthetic N fertilizers in China from 1949 to 1995.

Conversion coefficients

The calculations of the N budgets in the China watershed and the Changjiang, Huanghe and Zhujiang River valleys were based on the same conversion coefficients except that for N in atmospheric wet deposition in the three valleys.

Results and discussion

Input N

N input in the China watershed
Synthetic N fertilizers: Application of synthetic N fertilizers is the largest N source in the China watershed, contributing 22.2 Tg in 1995. However, before 1950, the consumption of synthetic N fertilizers was rather limited and only 6×10^{-3} Tg in 1949, about 0.03% of that in 1995. It multiplied by 3.7×10^3 times during the 46 years from 1949 to 1995 (Figure 2).

NO_x *emission from combusting fossil fuel*: With the development of industry and agriculture in China, the consumption of fossil fuel has been increasing drastically. Based on the conversion factors Wang et al. (1996) suggested for calculating NO_x emitted from combustion of different fossil fuels, the NO_x flux in China was worked out on a year-by-year basis (Figure 3). In 1995, 4.18 Tg N was emitted into the atmosphere in the form of NO_x from combusting fossil fuel in China (Table 5).

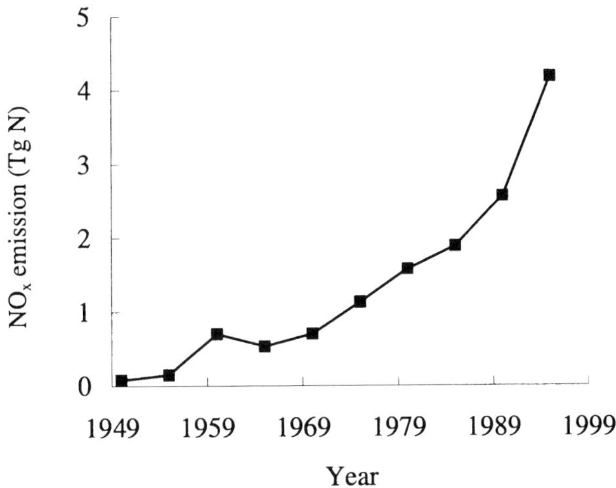

Figure 3. Increasing trend of NO_x emission from combustion of fossil fuels in China.

Table 5. NO_x emission from fossil fuel combustion in 1995 in China.

Type of fossil fuel	NO_x emission (Tg N)	Percentage to total
Coal (including coking)	3.47	83.0
Petrol (including refining)	0.70	16.8
Natural gas	0.01	0.2
Total	4.18	100

Table 5 list NO_x fluxes from consumption of coal, petroleum and natural gas in China in 1995, showing clearly that coal is the biggest contributor, responsible for 83% of the total.

Symbiotic N fixation by leguminous crops: In reckoning symbiotic N fixation by leguminous crops in China, only soybean, peanut, pulses and also leguminous green manure crops were taken into account. Based on the data cited from the China Agricultural Yearbooks of 1980, 1981, 1986, 1991 and 1996, and the conversion factors suggested by Zhu (1997), the symbiotic N fixation by leguminous crops during the period from 1949 to 1995 was calculated and is shown in Figure 4. There is an increasing trend from 1949 to 1976 and then a decline from 1976 to 1980 due to population pressure leading to displacement of green manure with grain crops. The gap in N supply was filled by chemical N fertilizers, which rapidly increased in application rate. Figure 4 also shows that the symbiotic N fixation by leguminous crops began

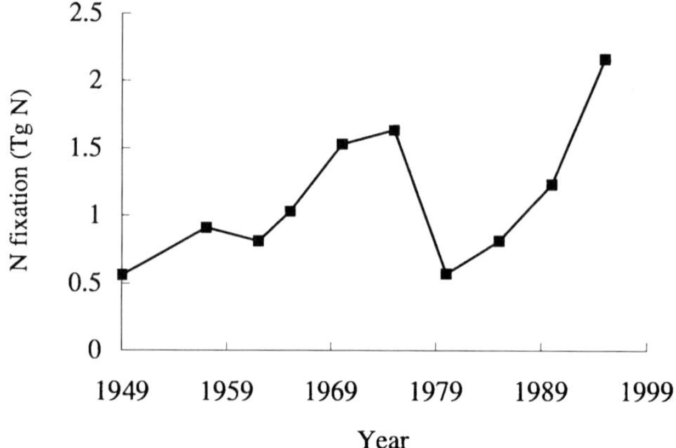

Figure 4. Change in the N fixation by legumes.

to turn upward after 1980 due to expansion of the production of soybean, peanut and pulses. In 1995, the symbiotic N fixation by leguminous crops reached 2.16 Tg N in China (Table 6).

Non-symbiotic N fixation: In China, paddy fields account for 26% of its total cultivated land and one-fifth of the world's total paddy field area. Moreover, N fixation in the paddy field is higher than in the upland. So in calculating N input, non-symbiotic N fixation was also reckoned as a source. As is suggested by Zhu (1997), the conversion coefficient used for non-symbiotic N fixation in the paddy field and upland was 45 kg $N \cdot ha^{-1} \cdot yr^{-1}$ and 15 kg $N \cdot ha^{-1} \cdot yr^{-1}$, respectively. Based on the conversion coefficient, the basic data of the acreage of paddy fields and uplands in 1995 were converted into non-symbiotic N fixation in the farmland, 2.17 Tg N, among which 1.12 Tg in the paddy field and 1.05 Tg in the upland (Table 6).

N import in food/feed: Based on the quantities of agricultural and animal products imported from other countries (China Statistical Yearbook 1996) and N contents in different products (China Agricultural Technical and Economic Manual 1983), the 1995 N inputted from food/feed import was estimated to be only 0.52 Tg N (Table 6).

Recycling N: Human and animal wastes in the rural areas in China are commonly used as manure. Based on the conversion coefficient suggested by Xing and Yan (1999), the total amount of the wastes in 1995 was estimated at 18 Tg N, among which 13.7 Tg was from livestock, 3.07 Tg from the rural population and 1.14 Tg the urban residents (Table 6). About 40% of the

Table 6. Estimated N input in China (1995)

Items	Amount (Tg N)
Anthropogenic reactive N	
– Synthetic fertilizer N	22.2
– NO_x-N formed during fossil fuel combustion	4.18
– N fixation by legumes in agricultural field	2.16
– Non-symbiotic N fixation in agricultural field	2.17
– Food/feed import	0.52
Subtotal	31
Recycling N	
– N from the excrements of domesticated animal and rural human	
N from the excrements of the domesticated animals	13.7
N from the excrements of rural people	3.07
N from excrements of urban people	1.14
– N from crop residue used as fertilizer	1.43
– NO_x-N formed during the combustion of crop residue	0.12
– N from atmospheric wet deposition	
N from NO_x	2.68
N from NH_x	8.28
Subtotal	31
Total	62

wastes were used as manure applied into crop fields. And also about 38% of the crop residues were utilized for the same purpose. Based on the work of Xing and Yan (1999), the amount of the residues used in this field was estimated to be 1.43 Tg (Table 6). The rest of the crop residues, about 62%, were burnt partly in farm fields and partly in kitchens as cooking fuel. On the basis of the conversion coefficient, 0.35g $N \cdot kg^{-1}$ dm (Delmas et al. 1995) for estimating NO_x flux from burning crop residues, the total amount of NO_x generated from burning of crop residues was reckoned roughly at 0.12 Tg N in 1995 (Table 6). Reports from 27 year-round monitoring posts scattered all over the country (Wang 1994), from the 2 similar posts in Shandong Province (Zhang & Liu 1994) and from the one in Qinghai Province, West China (Yang et al. 1991) indicated that the mean value of the concentrations of NO_3 and NH_4 in the precipitation monitored was 4.36×10^2 mg N m^{-3} and 1.35×10^3 mg N m^{-3} in China, respectively. And based on the data of the mean annual precipitations from 1951 to 1990 (Table 1), we estimated the atmospheric wet deposition N in the early 1990s to be 11 Tg, of which

2.68 Tg was NO_3 and 8.28 Tg NH_4 (Table 5), showing a NO_3:NH_4 ratio of 1:3. Obviously the proportion of NH_4 in the ratio is much higher in China than in other parts of the world (Galloway 1985; Loye-pilot & Morelli 1988; Weijer & Vugts 1990), which might be attributed to the high proportion of ammonium dicarbonate, about 50%, in the makeup of the chemical fertilizers in China. This type of fertilizer suffers higher NH_3 loss through volatilization than urea and any other types of N fertilizers. Moreover, inadequate handling facilities for human and animal wastes also result in higher volatilization of NH_3 from the wastes in China.

The above-described fluxes of anthropogenic reactive N and recycling N to the China watershed are summarized in Table 6, which shows that the total amount of N fluxes to the China watershed is estimated at 62 Tg, of which 31 Tg, about 50%, is anthropogenic reactive N and 31 Tg, about 50%, recycling N.

N input to the Changjiang, Huanghe and Zhujiang River valleys

The input of anthropogenic reactive N to the Changjiang, Huanghe and Zhujiang River valleys was 9.5 Tg, 2.3 Tg and 1.9 Tg, respectively, accounting for 31%, 8% and 6%, respectively, of the total to the China watershed in 1995. Synthetic chemical N fertilizers were also the main source of N input to the valleys and estimated at 7.30 Tg, 1.77 Tg and 1.47 Tg (Table 7), respectively, accounting for 77%, 76% and 79% of their corresponding total of anthropogenic reactive N input. The N fluxes from the other sources of anthropogenic reactive N to the valleys are all shown in Table 7.

The flux of recycling N to the Changjiang, Huanghe and Zhujiang Rivers was 9.5 Tg, 2.2 Tg and 2.8 Tg, respectively (Table 7). Being the predominant source of recycling N, human and animal wastes in the three valleys contained 5.32 Tg N, 1.45 Tg N and 1.36 Tg N, respectively, accounting for 56%, 66% and 49% of their corresponding total of recycling N input.

The atmospheric wet deposition N to the Changjiang, Huanghe and Zhujiang River valleys was estimated on the basis of two different groups of values, because the concentrations of NO_3 and NH_4 in the precipitation differ from North China to South China (Wang 1994). By integrating the monitoring results of different researchers (Wang 1994; Zhang & Liu 1994; and Yang et al. 1991), yielded were the mean concentrations of NO_3 and NH_4 in the precipitation, 3.35×10^2 mg N·m^{-3} and 1.42×10^3 mg N·m^{-3}, respectively, in North China and 5.11×10^2 mg N·m^{-3} and 1.47×10^3 mg N·m^{-3}, respectively, in South China. As the Huanghe River valley is located in North China, the atmospheric wet deposition of NO_3 and NH_4 there was estimated to be 0.12 Tg N and 0.52 Tg N, respectively (Table 7). Whereas the Changjiang River valley and the Zhujiang River valley are

Table 7. Estimated N input in the Changjiang, Huanghe and Zhujiang River valleys (1995)

Items	Amount (Tg N)		
	Changjiang	Huanghe	Zhujiang
Anthropogenic and naturally reactive N			
– Synthetic fertilizer N	7.30	1.77	1.47
– NO_x formed during fossil fuel combustion	0.91	0.28	0.16
– N fixation by legumes in agricultural field	0.56	0.15	0.10
– Non-symbiotic N fixation in agricultural field	0.74	0.14	0.16
Subtotal	9.5	2.3	1.9
Recycling N			
– Excretia N of humans and raised animals			
N from the excrements of raised animals	3.93	1.24	1.03
N from the excrements of the rural population	1.10	0.14	0.26
N from the excrements of the urban population	0.29	0.07	0.07
Subtotal	5.32	1.45	1.36
– N from crop residue used as fertilizer	0.43	0.14	0.14
– NO_x-N formed from crop residue combustion	0.04	0.011	0.01
– Atmospheric wet deposition N			
NO_x-N	0.96	0.12	0.33
NH_x-N	2.79	0.52	0.96
Subtotal	9.5	2.2	2.8
Total	19.0	4.6	4.7

in South China, and their atmospheric wet deposition of NO_3 and NH_4 was determined to be 0.96 Tg N and 2.79 Tg N, and 0.33 Tg N and 0.96 Tg N, respectively (Table 7). The NO_3:NH_4 ratio in the atmospheric wet deposition in the Changjiang, Zhujiang and Huanghe River valleys was 1:3, 1:3 and 1:4, respectively.

N fluxes from other sources to the Changjiang, Huanghe and Zhujiang River valleys are also listed in Table 7.

N output and storage

N output and storage in the China watershed
N in the harvested crops: Based on the data from the China Agricultural Yearbook (1996) about yields of 30 major agricultural crops (including 7 species of vegetables) and the data about N contents in the crops and N distribution ratio between straw (leaves and stems) and grains (edible parts) listed in a number of agricultural handbooks and fertilizers handbooks (Lu &

Table 8. Conversion factors for N loss through denitrification in agricultural fields of China

Fertilizer type	Conversion factor	Reference
Synthetic fertilizer N		
– Rice paddy fields	33–41%	Zhu (1997)
– Uplands	13–29%	Cai et al. (1998)
Organic fertilizer N	10–30%	Wen et al. 1988; Shi et al. 1991; Chen et al. 1994; He et al. 1994

Shi 1982; China Agricultural Technology and Economy Manual 1983; Pang 1994; Huang et al. 1996), the N in the harvested crops was worked out to be 14 Tg, of which 10.5 Tg in grains and edible parts and 3.5 Tg in straws (Table 11).

Denitrification in farmland: N loss through denitrification differs in rate between paddy fields and uplands, being higher in the former than in the latter. And it also differs between from chemical fertilizers and from organic manure (Table 8). The data in Table 8 are all based on micro-plot field experiments carried out with chemical fertilizers, green manure, rice straw, ping dung and sheep droppings, all labeled with [15]N. In 1995, China applied 5 Tg N to the paddy fields and 17 Tg N to the uplands in chemical fertilizer, and 10 Tg N in organic manure to the farmlands. By using the data in Table 8 as conversion coefficients, the N losses through denitrification in 1995 from chemical fertilizers in paddy fields and in uplands organic manure in cropland were figured out. The N loss was estimated at 1.65–2.05 Tg N from chemical fertilizers in paddy fields, 2.23–4.99 Tg N from chemical fertilizers in uplands, and 1.03–3.10 Tg N from organic manure in cropland. So the total N loss through denitrification from the farmland reached 5–10 Tg in China in 1995 (Table 11).

N storage in farmland: The N detention rate of chemical N fertilizers varies from paddy fields to uplands. On average, it is higher in the former than in the latter. It also differs from that of organic manure, which has a higher residue rate. Zhu* summarized the results of 142 field and greenhouse experiments with ^{15}N-labeled chemical N fertilizers and ^{15}N-labeled organic manure in different soils (Table 9). Based on the data in Table 9 as conversion factors, the N retention was worked out as 12 Tg N in China in 1995, indicating this was the most important access to sink.

Table 9. Estimated N storage in agricultural soils in China (1995)

N type	Application amount (Tg N)	Storage ratio* (%)	N storage (Tg N)
Synthetic N			
– Paddy fields	5.00	20 (12–30)	1.00
– Uplands	17.20	27 (11–68)	4.64
Organic N	12.04	49 (27–78)	5.9
Total	34.97		12

*The data in unpublished paper of Zhu ZL.

N transport to waterbodies: In calculating N transport to waterbodies in the China watershed, it can be divided into two portions, anthropogenic reactive N and human wastes discharged direct into waterbodies, to be calculated separately. The former includes synthetic chemical N fertilizers, symbiotic N fixation by leguminous crops and green manure crops, N imported with food and feed, N in atmospheric wet deposition and also non-symbiotic N fixation in the cropland, because paddy fields account for 26% of the country's total farmland and, what is more, non-symbiotic N fixation in the paddy field is much higher than in the upland. And the latter encompasses wastes from urban residents and rural population. In calculation, only 60% of the wastes from rural population were counted because about 40% of the wastes in the rural area are used as manure.

The anthropogenic reactive N transported into waterbodies through leaching and runoff was estimated at 8.20 Tg by using IPCC (1996) default value, 30% (ranging between 10 and 80%), as the conversion factor, whereas the N in human wastes discharged direct into waterbodies was figured out as 2.7 Tg in China in 1995 on the basis that the contribution rate of human wastes to N load in waterbodies was 3.3 kg $N \cdot yr^{-1}$ per person (Meybeck 1989). So the total N transport into waterbodies was 11 Tg. In the calculation, only N in atmospheric wet deposition was covered, but not that in atmospheric dry deposition. Besides, the rural area still had 60% of the animal wastes left unused. It is very hard to reckon how much N was transported into waterbodies from it. Consequently, the N transport into waterbodies in China might be underestimated. Nevertheless, China differs from other countries in climate and farming system. Under the significant influence of monsoon, in most agricultural regions, rainfalls concentrate in summer and occur grudgingly in winter and spring. Moreover the major agricultural regions with humid and warm climate can grow crops all year round. In addition, China has 26% of its farmland growing rice. Thus, using 30% of the N applied as conversion factor for calculation of N transport into waterbodies

Table 10. NH$_3$ volatilization conversion coefficient

Sources	Emission factors (kg N kg^{-1}N)		References
Animal and human excrement N		0.20	IPCC, 1996
Synthetic fertilizer N	Upland:	0.08 for urea and 0.10 for NH$_4$HCO$_3$	Xing & Zhu, 2000
	Paddy field:	0.22 for urea and 0.28 for NH$_4$HCO$_3$	

in the China watershed might lead to overestimation. It seems that there is much uncertainty in calculating N transport into waterbodies in the China watershed.

NH$_3$ volatilization: China does not have many varieties of chemical N fertilizers. Urea and ammonium bicarbonate are the two dominant ones, sharing half and half in the makeup of chemical N fertilizers in 1995. Things have been changing since 1995, with the latter declining in proportion and being replaced by the former. Studies reveal that N loss rate through NH$_3$ volatilization varies, with ammonium bicarbonate higher than urea and with upland higher than paddy field. Human and animal wastes are another source of NH$_3$ volatilization. In calculating its N loss through NH$_3$ volatilization, IPCC-recommended default value, 0.2kg N (NH$_3$ + NO$_x$) per kg of N in human and animal wastes, was used as conversion factor. However, the NO$_x$ content in the wastes is rather limited. Schimel et al. (1986) reported that NH$_3$ volatilization from the animal wastes applied into the farmland accounted for about 20% of the N in the wastes, whereas it was 25% in the report by Van der Hoek (1994). In Table 10 listed are the IPCC-recommended default value and conversion factors used in calculating NH$_3$ volatilization from animal and human excrements N. Based on the conversion factors listed in Table 10, the NH$_3$ volatilization in the agriculture of China was estimated at 6.1 Tg N, of which 3.35 Tg originated from human and animal wastes (Table 11).

Food/feed exports: Although China is the country that turns out the most agricultural products, the demand of such a big population greatly limits its export of agricultural products. According to the statistical data about the export of agricultural products in the China Agricultural Yearbook and the data about N contents in these products, the total estimated N exported with food/feed was limited to 0.11 Tg (Table 11).

From Table 11, it can be inferred that N in the harvested crops and N storage in soils are the largest sink of input N in the China watershed,

Table 11. Estimation of N output and storage in China (1995)

Items	Amount (Tg N)
N in the harvested crops	
– N in the grains (edible parts)	10.5
– N in the straws	3.5
Subtotal	14
Denitrification	
– Synthetic fertilizer N	
Rice fields	1.65–2.05
Uplands	2.23–4.99
– Organic fertilizer N	1.03–3.1
Subtotal	5–10
Storage in agricultural land	
– Synthetic fertilizer N	
Rice fields	1.0
Uplands	4.6
– Organic fertilizer N	5.9
Subtotal	12
N transported into waterbodies	
– Anthropogenic reactive N	8.2
– Excretia N directly from human in urban and rural area	2.7
Subtotal	11
NH_3 volatilization	
– From chemical fertilizer N	2.71
– From excretia of raised animal and human being	3.35
Subtotal	6.1
Food/feed exports	0.11
Total	48–53

followed by N transport into waterbodies, denitrification in agricultural soils and NH_3 volatilization.

N output and storage in the Changjiang, Huanghe and Zhujiang River valleys

The N output and storage in the Changjiang, Huanghe and Zhujiang River valleys was estimated at 16–18 Tg, 3.6–4.1 Tg, and 4.1–4.4 Tg, respectively (Table 12).

Based on the statistical data about the yields of the 30 major agricultural crops on a province-by-province basis in the China Agricultural Yearbook (1996) and the method described in the previous paragraphs, the N in the harvested crops in the Changjiang, Huanghe and Zhujiang River valleys were worked out. By using the data in the agricultural handbooks and fertilizers handbooks about N contents and distribution ratio of grains and straws, N in the harvested crops from the Changjiang, Huanghe and Zhujiang River valleys were figured out to be 4.34 Tg N, 1.04 Tg N and 1.07 Tg N, respectively.

Of the output and storage N from the Changjiang, Huanghe and Zhujiang River valleys, 3.51 Tg, 0.95 Tg and 1.16 Tg was stored in soil, respectively (Table 12). The calculation was performed based on the conversion factors listed in Table 9.

The N transport to waterbodies in the Changjiang, Huanghe and Zhujiang River valleys was calculated in the same way as that in the China watershed. It was divided into two portions for calculation, anthropogenic reactive N and human wastes discharged direct into waterbodies. The same conversion factors were used, the IPCC (1996)-recommended 30% default value for the former and Meybeck-suggested (1989) 3.3 kg N yr^{-1} per person for the latter. The total N transport into waterbodies in the Changjiang, Huanghe and Zhujiang River valleys was estimated at 3.79 Tg, 0.83 Tg and 0.84 Tg, respectively (Table 12). Though the acreage of the Zhujiang River valley is only 60% that of the Huanghe River valley, the N transport into waterbodies differs little between them. This is because the Zhujiang River valley is in the tropical and subtropical South China with much higher precipitation and runoff than in the Huanghe River valley located in the semi-arid temperate North China. Due to the humid and warm climate, the former can have 2–3 croppings a year whereas the latter can only have one cropping and a vast area of grassland. As a result, the per-unit application rate of chemical N fertilizers in the former is much higher than that in the latter. Moreover, the population density of the former is almost twice that of the latter (Table 3).

Based on the information about DIN concentrations in the three rivers and related hydrological data gathered in 1980–1989 from three observation posts located at each of the lower reaches of the rivers, Duan et al (2000) figured out the DIN transported into the estuaries as 0.78 Tg, 0.06 Tg and 0.15 Tg, respectively. Though the results did not cover organic N, they are much lower than our results (Table 12), indicating that there is much uncertainty in calculating N transport from the valleys into estuaries, which calls for further studies.

Denitrification of soil N in farmland is another access to sink for input N in the Changjiang, Huanghe and Zhujiang River valleys. By using the

Table 12. Estimated N output and N storage in the Changjiang, Huanghe and Zhujiang River valleys (1995)

Items	Amount (Tg N)		
	Changjiang	Huanghe	Zhujiang
N in the harvested crops	4.34	1.04	1.07
Denitrification in agricultural soils			
– Synthetic fertilizer N			
Rice fields	1.36–1.69	0.02–0.03	0.32–0.39
Uplands	0.41–0.92	0.22–0.50	0.07–0.10
– Organic fertilizer N	0.37–1.12	0.10–0.29	0.09–0.28
Subtotal	2.14–3.73	0.34–0.82	0.48–0.77
Storage in agricultural land			
– Synthetic fertilizer N			
Rice fields	0.82	0.012	0.19
Uplands	0.86	0.46	0.14
– Organic fertilizer N	1.83	0.48	0.83
Subtotal	3.51	0.95	1.16
N transported into the waterbodies			
– Anthropogenic and natural reactive N from input sources	2.87	0.65	0.62
– N from the excretia of people in cities and rural areas	0.92	0.18	0.22
Subtotal	3.79	0.83	0.84
NH_3 volatilization			
– From chemical fertilizer N	1.32	0.17	0.29
– From excretia N of raised animals	1.0	0.28	0.27
Subtotal	2.32	0.45	0.56
Total	16–18	3.6–4.1	4.1–4.4

conversion coefficients listed in Table 6, the N loss through denitrification of soil N in farmland in the Changjiang, Huanghe and Zhujiang River valleys was estimated at 2.14–3.73 Tg, 0.34–0.82 Tg and 0.48–0.77 Tg (Table 12), accounting for 18–30%, 13–27% and 16%–25%, respectively, of the N output and N storage in the valleys. The percentages of the N loss through denitrification against N output in the Changjiang River valley and the Zhujiang River valley were quite close to each other, but higher than in the Huanghe River valley. This is because the former two valleys are the major rice growing regions of China, while the latter has only a limited acreage of paddy fields

(Table 2), and what is more, denitrification of soil N is much higher in paddy fields than in uplands.

The calculation of the NH_3 volatilization of synthetic chemical N fertilizers and human and animal wastes in the three valleys was carried out based on the conversion factors listed in Table 10. The N loss through NH_3 volatilization in the Changjiang, Huanghe and Zhujiang Rivers was estimated at 2.32 Tg, 0.45 Tg and 0.56 Tg, respectively.

The N output and N storage in the Changjiang, Huanghe and Zhujiang River valleys are summarized and listed in Table 12.

Conclusion

In China – with the biggest population in the world – anthropogenic activities have been significantly altering the N biogeochemical cycling in the region with the rapid development of industry and agriculture. The total anthropogenic reactive N was estimated at 31 Tg in 1995, of which 22.2 Tg came from application of synthetic chemical fertilizers, 4.18 Tg from NO_x emission from combusting fossil fuel, 2.16 Tg from N fixation by leguminous crops and 0.52 Tg from food/feed import. Non-symbiotic N fixation in farmlands was estimated to be 2.17 Tg. Although N fixation in forests, grasslands, natural wetland and lightning N are not counted in, it is indubitable that human-created N far exceeds terrestrial biological N fixation. Synthetic chemical N fertilizer is the dominant source of anthropogenic reactive N. NO_x emission from combusting fossil fuel is another major source since China is one of the biggest coal consumers in the world.

The amount of recycling N reflects acceleration of the N cycling as influenced by human activities. The analysis and evaluation of the recycling N in the China watershed shows that this part of N could not be negligible, and totaled 31 Tg in 1995, with 17.9 Tg coming from human and animal wastes.

The total N sink in the China watershed amounted to 48–53 Tg in 1995, with N in the harvested crops and N storage in agricultural soils being the major sinks and amounting to 14 Tg and 12 Tg, respectively. The total N loss through denitrification of soil N was estimated at 5–10 Tg in 1995, and the total N transport into waterbodies at 11 Tg, which was an estimate with much uncertainty.

The N budgeting on the basis of valley show that the N input and N output in the Changjiang, and Zhujiang River valleys, which are the well-developed regions of China, were much higher than that in the Huanghe River valley located in the arid and semi-arid warm temperate zone.

This estimation was performed mainly based on agricultural N recycling and NO_x emission from combusting fossil fuel. The N sources and sinks in forests, grasslands and natural wetlands are not yet integrated into our N budgeting in China. However, from the viewpoint of evaluating the influence of human activities on N cycling, it is reasonable to estimate N input and sinks on the basis of agricultural N cycling.

Acknowledgements

This work was funded by the National Natural Science foundation of China (39790110). It was also initiated as part of the International SCOPE N project, which received support from both the Mellon Foundation and from the National Center for Ecological Analysis and Synthesis.

References

1996 Agricultural Yearbook of Tibet, Qinghai, Gansu, Ningxia, Inner Mongolia, Shaanxi, Shanxi, Henan, Shandong, Yunnan, Guizhou, Sichuan, Hubai, Hunan, Jiangxi, Anhui, Jiangsu, Shanghai, Zhejiang, Fujian, Guangdong and Guangxi provinces, municipalities and autonomous regions (1996), China Agricultural Press. Beijing

1996 Statistical Yearbook of Tibet, Qinghai, Gansu, Ningxia, Inner Mongolia, Shaanxi, Shanxi, Henan, Shandong, Yunnan, Guizhou, Sichuan, Hubai, Hunan, Jiangxi, Anhui, Jiangsu, Shanghai, Zhejiang, Fujian, Guangdong and Guangxi provinces, municipalities and autonomous regions (1996), China Statistic Publishing House. Beijing

Cai GX, Fan XH & Zhu ZL (1998) Gaseous loss of nitrogen from fertilizers applied to wheat on a calcareous soil in North China Plain. Pedosphere 8: 45–32

Chen BC, Wen BQ, Tang JY & Liu ZZ (1994) Effect of *Azolla* on N balance in rice paddy fields. Acta Agriculturae Nucleatae Sinica 8: 97–102

China Agricultural Technology and Economy Manual (1983) Agriculture Press, Beijing, pp 105–291

China Agricultural Yearbook (1980–1996) Agricultural Press, Beijing

China Statistical Yearbook (1996) China Statistic Publishing House, Beijing

Cui DC (1999) China Agroclimatology. Zhejiang Science and Technology Press, Hangzhoou, pp 63–78

Delmas R, Lacaus JP & Brocard D (1995) Determination of biomass burning emission factors: Methods and results. Envrion. Monit. and Assess 38(2–3): 181–204

Duan SW, Huang HY & Zhang S (2000) Transport of soluble inorganic nitrogen of the main rivers to estuaries in China. Nutrient Cycling in Agroecosystems (in press)

Galloway JN (1985) The deposition of sulfur and nitrogen from the remote atmosphere. In: Galloway JN, Carlson RJ, Andreae MO & Rodhe H (Eds) The Biogeochemical Cycling of Sulfur and Nitrogen in the Remote Atomsphere (pp 143–173). Reidel, Dordrecht

Galloway JN, Howarh RW, Michaels AE, Nixon SW Prospero JM & Dentener FJ (1996) Nitrogen and phosphorus budgets of the North Atlantic Ocean and its watershed. Biogeochemistry 35: 3–25

Galloway JN, Schlesinger WH, Levy IH, Michaels A & Schnoor JL (1995) Nitrogen fixation: anthropogenic enhancement-environmental response. Global Biogeochem. Cycles 9: 235–252

Galloway JN, Zhao DW, Thomson VE & Chang LH (1996) Nitrogen mobilization in the United States of America and the People's Republic of China. Atmospheric Environment 30: 1551–1561

He DY (1994) Cycling of nitrogen in the paddy soil-crops-domestic animals system. Acta Ecologica Sinica 14: 113–120

Howarth RW, Billen G, Swaney D, Townsend A, Jaworski N, Lajtha K, Dourning JA, Elmgren R, Caraco N, Jordan T, Berende F, Freney J, Kudeyarov V, Murdoch P & Zhu ZL (1996) Regional nitrogen budgets and riverine N and P fluxes for the drainages to the North Atlantic Ocean: Natural and human influences. Biogeochemistry 35: 75–139

Huang JL, Luo JX & Gong CY (1996) Handbook of Applied Fertilizers. Science and Technology Press of Hunan Province, Changsha, pp 262–269

IPCC (1996) Guidelines for National Greenhouse Gas Inventories. OECD/OCDE, Paris

Loye-pilot MD & Morelli J (1988) Fluctuations of ionic composition of precipitations collected in Corsica related to changes in the origins of incoming aerosols. J.Aerosol Science 19: 577–585

Lu RK & Shi TJ (1982) Agricultural Chemistry Handbook. Science Press, Beijing, pp 29–38

Meybeck M, Chapman DV & Helmer R (1989) Global Freshwater Quality: A First Assessment. World Health Organization/United Nations Environment Programme Basil Blackwell, Inc., Cambridge, MA

Mo SX & Qian JF (1983) Study of alfalfa transformation and its availability to rice plants. Acta Pedologica Sinica 20: 12–22

Mosier AR, Kroeze C, Oenema O, Seitzinger S & van Cleemput O (1998) Closing the global atmospheric N_2O budget: Nitrous oxide emissions through the agricultural nitrogen cycle. Nutrient Cycling in Agroecosystems 52: 225–248

Pang CZ (1994) Handbook of Applied Fertilizers. Science and Technology Press of Guangxi Province, Nanning, pp 338–358

Ren ME, Yang RZ & Pao HS (1980) Essentials of Natural Geography in China. Commercial Publish House, Beijing, pp 68–103

Schimel DS, Parton WJ, Adamsen FJ, Woodmansee RG, Senft RL & Stillwell MA (1986) The role of cattle in the volatile loss of nitrogen from a shortgrass steppe. Biogeochem. 2: 39–52

Shi SL, Liao HQ, Wen QX, Xu XQ & Pan ZP (1991) Fate of N from green manures and ammonium sulfate. Pedosphere 1: 219–227

Sun JZ (1959) Economic Geography in South China. Science Press, Beijing, pp 9–64

The Geological Department of Zhengzhou Normal College (1959) The Geography in Henan Province. Commercial Publish House, Beijing, pp 54–55

The Research Committee of China Natural Resources (1992) A Study on Development Strategy in Exploiting the Natural Resources in West China. China Science and Technology Press, Beijing, pp 187–189

Van der Hoek KW (1994) Berekeningsmethodiek ammoniakemissie in Nederland voor de jaren 1990, 1991 en 1992. RIVM report No. 773004003

Wang WX (1994) Study on the cause of acid rainfall formation in China. China Environmental Science 11: 321–329

Wang WX, Lu XF, Pang YB, Tang DG & Zhang WH (1997) Geographical distribution of NH_3 emission intensities in China. Acta Scientiaf Circumstentiae 17: 3–7

Wang WX, Wang W, Zhang WH & Hong SX (1996) Geographical distribution of SO_2 and NO_x emission intensities and trends in China. China Environmental Sciences 16: 161–167

Weijers E & Vugts HF (1990) An observation of study on precipitation chemistry data as a function of surface wind direction. Water Air Soil Pollut. 52: 115–132

Wen QX, Cheng LL & Shi QL (1988) Decompostion of *Azolla* in the field and availability of *Azolla* nitrogen to plants. Azolla utilization (pp 241–254). IRRI

Wu CJ (1998) China Economic Geography. Science Press, Beijing, pp 6–91

Xing GX & Yan XY (1999) Direct nitrous oxide emission from agricultural field in China, as estimated by the Phase II IPCC methodology. Environmental Science and Policy 2(3): 355–361

Xing GX & Zhu ZL (1997) Preliminary studies on N_2O emission fluxes from upland soils and paddy soils in China. Nutrient Cycling in Agroecosystems 49: 17–32

Xing GX & Zhu ZL (2000) An assessment of N loss from agricultural fields to the environment in China. Nutrient Cycling in Agroecosystems 57: 67–73

Xing GX (1998) N_2O emission from cropland in China. Nutrient Cycling in Agroecosystems 52: 249–254

Yang LY, Ren YX & Jia L (1991) Preliminary study of chemical compositions of precipitation of Wudaoliang, Qinghai province. Plateau Meteorol. 10: 209–216

Yi YQ (1957) The summary of Geography in Shaanxi province. Shaanxi Press, Taiyuan, p 15

Zhang J & Liu MG (1994) Observations on nutrient elements and sulphate in atmospheric wet depositions over the northwest Pacific coastal oceans-yellow sea. Marine Chemistry 47: 173–189

Zhang JM (1986) The mountainous regions and hilly land in Shandong province. Shandong Science and Technology Press, Jinan, pp 20–21

Zhu ZL (1997) Nitrogen Balance and Cycling in Agroecosystems of China. In: Zhu ZL, Wen QX & Freney JR (Eds) Nitrogen in Soil of China (pp 323–330). Kluwer Academic Publishers, Dordrecht/Boston/London

Biogeochemistry **57/58**: 429–476, 2002.
© 2002 *Kluwer Academic Publishers. Printed in the Netherlands.*

Landscape, regional and global estimates of nitrogen flux from land to sea: Errors and uncertainties

PENNY J. JOHNES & DAN BUTTERFIELD
University of Reading, Aquatic Environments Research Centre, Department of Geography, Whiteknights, Reading RG6 6AB, U.K.

Key words: error propagation, global N cycling, modelling uncertainty, models, nitrogen

Abstract. Regional to global scale modelling of N flux from land to ocean has progressed to date through the development of simple empirical models representing bulk N flux rates from large watersheds, regions, or continents on the basis of a limited selection of model parameters. Watershed scale N flux modelling has developed a range of physically-based approaches ranging from models where N flux rates are predicted through a physical representation of the processes involved, through to catchment scale models which provide a simplified representation of true systems behaviour. Generally, these watershed scale models describe within their structure the dominant process controls on N flux at the catchment or watershed scale, and take into account variations in the extent to which these processes control N flux rates as a function of landscape sensitivity to N cycling and export. This paper addresses the nature of the errors and uncertainties inherent in existing regional to global scale models, and the nature of error propagation associated with upscaling from small catchment to regional scale through a suite of spatial aggregation and conceptual lumping experiments conducted on a validated watershed scale model, the export coefficient model. Results from the analysis support the findings of other researchers developing macroscale models in allied research fields. Conclusions from the study confirm that reliable and accurate regional scale N flux modelling needs to take account of the heterogeneity of landscapes and the impact that this has on N cycling processes within homogenous landscape units.

Introduction

In developing any model to simulate environmental behaviour the primary step is to determine the questions it needs to answer and the scale, both spatial and temporal, at which the answers are required. In the case of global N cycling the impetus to construct models is both to aid our understanding of global N cycling rates and processes, and to provide a means of assessing the origins of terrestrial, freshwater and oceanic N enrichment resulting from anthropogenic disruption of the global N balance. Oceanic N enrichment has both direct ecological consequences in the oceans and wider implications for global climatic function through the coupling of ocean-climate systems and the disruption of the global N balance (see for example, Larsson et al. 1985; Jaworski et al. 1989; Turner & Rabelais 1991; Law et al. 1992; Nixon 1995; Vitousek et al. 1997). There is, therefore, a need to develop models with the ability to distinguish accurately both the total rates of N flux from land to ocean, and the specific spatial origins and delivery zones of the N load. Associated with this requirement for spatial discrimination of N flux sources is a requirement for some estimate of the relative accuracy of the gross N flux estimates and the attribution of this flux to specific terrestrial or atmospheric sources.

A range of regional to global scale N flux models have been developed to date (see for example Meybeck 1982 as adapted in Seitzinger & Kroeze 1998; Peirls et al. 1991; Howarth et al. 1996; Seitzinger & Kroeze 1998; Caraco & Cole 1999). Each has been designed to generate an estimate of annual N flux from land to ocean from the major contributing areas on a country by country or continent by continent basis, and each provides different estimates of N flux to oceans from these contributing areas. In each of these models a subtly different selection of parameters is employed to describe the rate of N flux from land to ocean as it varies over 3–4 orders of magnitude across the globe (see discussion in Alexander et al. 2002, for details). The goodness-of-fit of each model is then reported as a function of the r^2 value of a regression relationship between observed and predicted N flux rates for a selection of major world rivers. Given the magnitude of variation in N flux rates globally it is not, perhaps, surprising that a good fit with a high r^2 value is reported for most of the models. Thus it could be argued that these models provide a reasonably good indication of the general rates of N flux from land to ocean on a global scale. What is not clear, however, is which of these models is providing the most robust and reliable estimates of N flux rates from land to ocean, and whether the regional or watershed scale estimates generated by these models are providing a reliable indication of the rates of N enrichment experienced locally by the biota in adjacent coastal waters.

In plot to catchment scale modelling the accuracy and precision of model predictions is normally ascertained through calibration and validation of model estimates against a known measure of environmental behaviour, typically field monitoring data. In physically based models ranges of values represented by the probability density function (PDF) can be generated for each model parameter based on assumptions about the likely statistical distribution of values for that parameter, under local and current environmental conditions (see, for example, Entekhabi & Eagleson 1989; Bergström & Graham 1998). Approaches such as Monte Carlo simulation within the range of uncertainty for the model parameters or the sectioning method of Addiscott and Wagenet (1985) can then be used to generate an estimate of the mathematical uncertainty associated with model predictions and the sensitivity of the model to its parameters. However, even with the statistical checks and balances possible in plot to catchment scale model development there remains the problem that the relationships between processes and controlling environmental conditions are usually observed in few spatial locations and over short periods of time. As Blöschl and Sivapalan (1995) argue these relationships are often used to form the basis of model equations describing system behaviour over a wider spatial scale and/or longer periods of time. In both cases the original observations may be insufficiently representative of the true range of variation in environmental behaviour of those parameters. Thus there is an element of uncertainty associated with the application of these parameter values to other watersheds or regions if the conditions in those systems do not directly mimic those in the original modelled catchments. As a result, upscaling of parameter ranges from plot or watershed scale to regional or global scale cannot be justified without some estimate of the relative accuracy of model output owing to the unknown and variable nature of the uncertainty associated with the model parameter ranges. Addiscott and Tuck (1996) also argue that the validity of model parameters depends on the range of values for each parameter remaining constant in time. In the case of process equations developed for physically-based models this may be a valid assumption to make given the temporal and spatial scales over which the model is expected to operate. However, for models used to generate hindcast or forecast estimates of environmental behaviour and for regional to global scale models where N flux rates are described by simple indices of human population density and extent of economic development, such assumptions would be unlikely to hold true over the spatial and temporal scales of operation for these models.

So what is the solution? How can we develop an index of robustness or a measure of the uncertainty associated with N flux estimates generated by the various regional to global scale models? The very nature of the spatial scale at which existing global N cycling models operate means that there is no robust means of validating model output against field monitoring data. Even where measurements of total N concentrations have been collected for major world rivers or estuaries, the physical scale of such water bodies means that fully representative water sample collection both in space and time is prohibited within the confines of present technology. There remains, therefore, a high level of uncertainty associated with the N flux estimates generated by these regional to global scale N cycling models.

One means of assessing model performance is to run a cross comparison of the N flux estimates generated by the regional to global scale models against

432

predictions of N flux produced by validated watershed-scale models for data-rich areas of the world. On the basis of relative model performance, guidance can then be given on the likely errors associated with the regional and global scale models, and the limits to their robust application although as Addiscott (1998) argues, validation at one spatial or temporal scale does not confirm model validity at other scales. Alexander et al. (2002) attempt to address this question by conducting a multi-way inter-model comparison for a range of the regional to global scale N flux models compared with model estimates generated by a validated watershed scale model (SPARROW: see Smith et al. 1997) and observed N flux rates for 16 of watersheds draining the NE United States (see also Boyer et al. and Van Breemen et al. both 2002). Most of the models compared in this exercise are less than reliable in predicting watershed scale variations in N flux rates, but in terms of its representation of within-region variations in N flux rates the model developed by Howarth and colleagues (1996) for prediction of N flux to the North Atlantic Basin from its major watersheds is undoubtedly the most robust and reliable (see Alexander et al. 2002, for details). It is not perhaps surprising that this is also the most complex of the regional to global scale models, with the greatest range of model parameters. In a multi-way inter-model comparison of N flux models applied to the watersheds of nine shallow estuaries on Cape Cod, MA, Valiela and colleagues concluded that the more complex models performed better than the simpler models (Valiela et al. submitted).

For the simpler empirical models of N flux developed for global scale application and often based on an upscaling of concepts developed to explain watershed scale response, a poorer performance is also evident when model estimates are compared with observed N loading data for individual water-sheds. Thus, for example, Seitzinger and Kroeze (1998) took the watershed scale model developed by Caraco and Cole (1999) and scaled this up for application on a $1° \times 1°$ grid, using global databases. This upscaled model predicts an average DIN flux rate from England to coastal waters of 0.1 kg N ha^{-1} (reported as 10 kg N km^{-2} watershed yr^{-1}), whereas the rate calculated from a watershed scale model for the U.K. (this paper) validated against historical N flux observations for 38 U.K. watersheds, estimates an average total N flux rate of 26.4 kg N ha^{-1} from England to its coastal waters in 1991. Seitzinger and Kroeze (1998) then applied their modelled estimate of 0.1 kg N ha^{-1} as DIN flux for England to the watershed area of the U.K. Tamar Estuary (a watershed too small to be explicitly defined the $1° \times 1°$ grid of their model adaptation) to allow subsequent calculation of N_2O emission rates from the estuary. Whilst Seitzinger and Kroeze (1998) report the resultant N_2O flux rate of 5×10^3 kg N yr^{-1} as being within a factor of 5 of emission rates measured for the Tamar Estuary by Law et al. (1992), the

N loading rate is a factor of 280 lower than the observed nitrate loading rate of 28 kg N ha^{-1} for the Tamar at its upper tidal limit (Uncles et al. in press). The total N flux rate to the upper tidal limit of the Tamar calculated by the validated watershed scale model is 39 kg N ha^{-1} (Uncles et al. submitted). Conversely, in the application of their model at watershed scale to 35 major world rivers, Caraco and Cole (1999) report a predicted 11.2 kg N ha^{-1} of nitrate export from the U.K. Thames watershed per year. Given detailed observations of the proportion of nitrate contributing to the total N load in the River Thames and its tributaries (nitrate = 60% total N; Johnes 1996; Johnes & Burt 1991), the Caraco and Cole model estimate converts to a total N export rate of 18.7 kg N ha^{-1} yr^{-1}. This compares reasonably well with the total N loading predicted by the validated watershed model (Johnes 1996; this paper) for the Thames watershed, which estimates total N flux from the Thames watershed as 22.1 kg N ha^{-1} in 1991. Thus there are marked differences in the level of agreement between the Caraco and Cole (1999) model compared to the watershed-scale model application to the Thames watershed, and the global scale Seitzinger and Kroeze (1998) model adaptation predictions for the Tamar watershed compared to both the watershed-scale model application to the Tamar Watershed and observed N flux data for the River Tamar.

The cause of these differences lies, probably, in the fact that the 1° × 1° resolution adopted by Seitzinger and Kroeze (1998) to allow global scale application led to a loss of valuable spatial resolution and precision in both the input data and model parameter values, whereas the original Caraco and Cole (1999) model was designed to be applied at major watershed scale, taking account of sub-grid scale spatial variations in the model parameter values. The difference in estimates of N flux to the Tamar estuary generated by the two models derive from the spatial scale of the modelling unit. In the Seitzinger and Kroeze (1998) model, input data and model parameter values are averaged for each 1° × 1° grid cell and the model is not physically based and cannot take account of sub-grid scale variations in the dominant process controls on N cycling and flux. In their application of this model to the Tamar estuary, the average estimate of N flux for the 1° × 1° grid cell within which England lay was assumed to provide a reliable estimate of N flux rates for the 924 km^2 Tamar catchment. In the validated watershed scale model, the basic modelling unit is the parish, an administrative unit averaging 14 km^2 in area and the model is physically based. Model estimates of N flux from each parish unit within any watershed are then lumped together to generate a watershed scale N flux estimate, allowing local variations in the dominant process controls on N flux rates to be taken into account. However, the precise nature of the poor fit between the N flux estimates generated for the Tamar estuary cannot be conclusively assigned from this simple comparison.

As Troutman (1983) argues there are two basic forms of modelling error. First there are errors where, although the input data are correct and provide a robust representation of the true environmental behaviour and variability of each model parameter, the model itself as a simplified representation of true systems behaviour is insufficiently detailed or fails to describe within its structure the dominant process controls at a particular spatial scale. These might be termed *process errors*. Second there are errors generated where the model is appropriately parameterised for the questions it has been constructed to answer at the selected spatial and temporal scale of application. In this case the errors then arise from incorrect input data describing the environmental behaviour of model parameters at the selected spatial or temporal scale, either through inaccurate field measurement or observation, or incomplete representation of the statistical properties of the distribution for each parameter over the range of environments to which it has been applied. These might be termed *input errors*. In the case of the simple empirical models developed to date to describe regional to global scale patterns of N flux from land to ocean both forms of error are likely to be present to some extent. This does not mean that these models are necessarily inappropriate if they are used solely to answer some of the macroscale questions relating to the global N cycling imbalance, particularly where they are used to assess the general origins of oceanic N enrichment and the spatial patterns in N flux as they vary from continent to continent. Indeed, as Arnell (1999) argues, the increasing interest in answering macroscale questions relating to environmental function and response over wide geographic domains requires the development of models capable of being applied at the macroscale without watershed scale calibration. However, Arnell also argues that if we wish these models to have a reliable predictive capability, then they need to be based on a physical representation of the processes involved. In this case, some of the simpler descriptive models of N flux produced to date, lacking any physically based description of the environmental controls on N cycling and flux processes within landscapes, would be unsuitable for use as tools to evaluate the specific spatial origins of oceanic N enrichment or the likely impact of management strategies to ameliorate the rate of N flux to the oceans.

Thus the primary source of error and uncertainty associated with the existing regional to global scale N flux models appears to be generated by the spatial scale or grid at which the models are applied, and the fact that none of these models is physically based. Because they are not physically based, they cannot take account of subgrid scale or landscape scale variations in local environmental controls on the dominant processes controlling N cycling and flux within particular terrestrial or freshwater environments.

They cannot, therefore, deliver spatially sensitive or robust estimates of N flux at the subgrid or landscape scale.

This problem is not restricted to N cycling models, but has been exercising scientists developing global climate and water balance models and, in particular, the hydrological components of these models for some time (see, for example, Entekhabi & Eagleson 1989; Federer et al. 1996; Bergström & Graham 1998; Vörösmarty et al. 1998; Boulet et al. 1999). As Boulet and colleagues (1999) argue regional land surfaces are not necessarily homogenous in terms of the processes controlling regional water balance components, and in their analysis the simple 1-dimensional SISPAT model described by Braud et al. (1995) showed a non-linear response to the spatial variability of particular parameters when upscaled to regional scale. They suggest that this heterogeneity and the problems it generates when upscaling for macroscale application can be resolved by adopting a mosaic approach in which the land surface is divided into homogenous patches, with fluxes calculated for individual patches and then lumped together to give an aggregate flux rate for the region. Becker and Braun (1999) present a similar argument, based on case studies on scaling, disaggregation and aggregation in predicting hydrological response characteristics in watersheds in Northern Germany and Bavaria. They suggest, from their analysis, that the subdivision or disaggregation of the land surface into smaller units displaying homogenous or quasi-homogenous hydrological behaviour is critical to the development of appropriate models for macroscale hydrological modelling. This argument is equally pertinent when applied to the development of solute flux models. They also argue that the behaviour of these units, which they term hydrotopes, needs to be modelled separately using unit-specific parameter values, with the calculated unit fluxes then aggregated to regional scale. They demonstrate the veracity of their contention by comparison of the errors generated where models or their parameters are extrapolated to run across large heterogenous landscape units. Similar arguments have also been presented at much finer modelling scales, as exemplified by the work of Famigletti and Wood (1995) who explored the effect of explicit patterns of environmental characteristics on areally averaged evapotranspiration at scales ranging from local to watershed scale. They concluded from their analysis that a threshold spatial scale could be defined for evapotranspiration modelling, termed the Representative Elementary Area (REA), below which fine scale spatial variations in the environmental factors controlling evapotranspiration rates would be important in describing watershed scale evapotranspiration fluxes, but above which the natural variability of environmental factors could be represented by PDFs describing the statistical distribution for each parameter. If this argument is applied to modelling in general, then the size of the REA, hydrotope, or landscape unit which can be defined as homogenous for the purposes of model application will be dependent on the dominant process controls operating on the modelled variable and their variance at the particular spatial or temporal scale at which the model is being developed.

Increased N loading results in deleterious impacts on coastal systems (such as the increased incidence and duration of anoxic/hypoxic events, and the decline of eel grass coverage). In order to effectively manage N enrichment of coastal waters, a tool is needed which can predict reliably the spatial variation in N delivery to the oceans, probably from individual watersheds as they drain to major estuaries, constricted coastal waters such as the Baltic Sea, the North Sea or the Gulf of Mexico, or directly to the oceans. For this, simple descriptive models based on regression of measured TN loading against a limited range of parameters in a limited range of environments will be insufficient for the purpose. Thus a model which provides an estimate of N flux to coastal waters as a function of the range of economic activity and agricultural intensity across a selection of major world rivers is unlikely to be able to distinguish between the rates of N flux to coastal waters generated directly as a result of high intensity agricultural production, as opposed to similar N flux rates generated where there is lower intensity of agricultural production but a lower intrinsic N retention capacity within the landscape (as a function, for example, of wetter winters, steeper slopes, impeded drainage). Instead, models which include a wider range of parameters reflecting environmental sensitivity to N flux are required. If a landscape unit can be defined as a spatial unit representing similar functional behaviour then representation of N flux at regional to global scale may be improved by defining a series of landscape units, modelling these separately with unit-specific parameter values and then lumping the model estimates together to provide regional to continental scale predictions of N flux. If this approach were to be adopted for regional to global scale N flux modelling the critical principal is that the landscape units need to be defined as homogenous in terms of the key controls on N cycling if they are to represent a valid unit for modelling. The units should never be defined by traditional political, cultural or socio-economic divisions (nations, for example), as these will rarely represent homogenous landscape units as they respond to N cycling and flux.

Another issue which arises is how the REA can be determined for regional to global scale N flux modelling, or at what level of spatial resolution the landscape units should be defined. Many of the data sets used by the existing regional to global scale N flux models have constrained these models to run at national or major watershed scale by virtue of the fact that the databases on which these models are reliant are widely available only for individual and well-researched major watersheds, usually within the developed world, or for individual nations within FAO and other statistical compilations. To avoid this constraint a landscape unitary approach would need to construct databases from the primary data sources from which these national statistics were derived. However, the original scale at which the data were collected may well, in many instances, be too fine for sensible application at regional to global scale. In the U.K. , for example, detailed statistics relating to agricultural land use and livestock production have been collected annually from 1866 to date and are available for the entire agricultural area of the U.K. in units termed parishes, each parish representing an area of land averaging 14 km^2 in area. Clearly this is too fine a scale for application across the entire global land surface. As a result, some upscaling is required if these data are to be utilised at regional to global scale. This in itself introduces a further range of issues relating to the order in which model parameters are averaged, interpolated or aggregated and the error and uncertainties that this scaling introduces to model estimates of N flux. As Addiscott and Tuck (1996) argue averaging or interpolating a parameter before running a non-linear model does not give the same result as running the model and then averaging or interpolating the results. This discrepancy between interpolating the output and interpolating the parameter is important because it raises an uncertainty in simulations. Stein et al. (1992) investigate this further. Sivapalan and Kalma (1995) raise a similar issue, arguing that the difference between lumping (aggregating) the entire mosaic of units across a landscape as opposed to representing the land surface as a combination of units acting in parallel may generate different modelling outcomes and further uncertainty in model estimates. Thus as we scale up from parish scale to major watershed or landscape unit scale, in U.K. terms, the sequence of modelling steps will be critical in determining the level of error and uncertainty associated with our ultimate regional scale N flux estimates.

In this paper we investigate the nature of the errors and uncertainties generated by upscaling in N flux models, utilising a watershed-scale model which utilises the major geoclimatic regions of the U.K. as the basis for assigning region-specific values to the model parameters to generate estimates of N (and P) flux from land to water at parish to watershed scale. The model used was the National Export Coefficient Model developed by Johnes and colleagues at the Aquatic Environments Research Centre, U.K. (for details see Johnes 1996; Johnes et al. 1996; Heathwaite & Johnes 1996; Johnes & Heathwaite 1997; Johnes & Hodgkinson 1998; Johnes et al. 1998a, 1998b; Johnes 1999; Johnes et al. 2000; Johnes 2000; Uncles et al. in press). This has the benefits that it has been (a) rigorously calibrated and validated in numerous applications at the watershed scale including multi-way inter-model comparisons and (b) developed and then applied to entire landscape units in data-rich areas of the world, including a recent multi-way inter-model comparison conducted for the Cape Cod watersheds, U.S.A. (Valiela et al. submitted).

The export coefficient model

Export coefficient modelling is a watershed or catchment scale, semi-distributed approach which calculates mean annual total N (and total P) loading delivered to a water body (freshwater or marine) as the sum of the nutrient loads exported from each nutrient source in the catchment. The model equations and modelling procedures are detailed in full in Johnes (1996) and can be summarised thus:

$$L = \sum_{i=1}^{n} E_i (A_i (I_i)) + p$$

Where   L = Loss of nutrients
E = Export coefficient for nutrient source i
A = Area of catchment occupied by land use type i, or Number of livestock type i, or of people
I = Input of nutrients to source i
p = Input of nutrients from precipitation

The export coefficient (E_i) expresses the rate at which nitrogen or phosphorus is exported from each land use type in the catchment. For animals, the export coefficient expresses the proportion of the wastes voided by the animal which will subsequently be exported from stock houses and grazing land in the catchment to the drainage network, taking into account the amount of time each livestock type will spend in stock housing, the proportion of the wastes voided which are subsequently collected and applied to the land in the catchment, and the loss of nitrogen through ammonia volatilisation during storage of manures. For human wastes, the export coefficient reflects the use of phosphate rich detergents and dietary factors in the local population, and is adjusted to take account of any treatment of the wastes prior to discharge to a water body using the following equation:

$$E_h = D_{ca} * H * 365 * M * B * R_s * C$$

Where E_h = Annual export or N or P from human population (kg a^{-1})
D_{ca} = Daily output of nutrients per person (kg d^{-1})
H   = Number of people in the catchment

365 = Days per year

M   = Coefficient for mechanical removal of nutrients during treatment (range 0.85–0.9, reflecting removal of 10–15% of the nutrient load)

B   = Coefficient for biological removal of nutrients during treatment (range 0.8–0.9, reflecting removal of 10–20% of the nutrient load)

R_s = Retention coefficient of the filter bed (range 0.1–0.8, reflecting retention of 20–90% of the nutrient load)

C   = Coefficient for removal of P if phosphorus stripping takes place (range 0.1–0.2, reflecting removal of 80–90% of the P load)

Initially based on models developed in eutrophication research in the 1980s, this approach has been developed, refined and tested on 38 U.K. catchments over the past 10 years in research programmes funded by the U.K. Natural Environment Research Council, the National Rivers Authority (NRA), the Environment Agency (EA) and the Ministry of Agriculture, Fisheries and Food (MAFF; now DEFRA, the Department for Environment, Food and Rural Affairs). The nutrient source categories taken into account are:

(a) the area of land cultivated for cereal crops, other arable crops, bare fallow land, permanent grassland, temporary (ley or rotational) grassland, and fertiliser N applications to this land, the area of rough grazing land (unfertilised), and the area of woodland, and the rates of N fixation to all crops, grass and non-agricultural land;

(b) the total number of cattle, pigs, sheep and poultry, including young animals, the average amount of N produced per animal annually, and the nature of animal waste handling;

(c) the total number of people, the average amount of N produced per person annually and the nature and extent of sewage and wastewater treatment facilities;

(d) N input to the catchment from atmospheric deposition.

In the U.K. detailed information is available for each of these nutrient source categories from the Annual Agricultural Census Returns (1866 to date), the Decadal Population Census (1851 to date), the Surveys of Fertiliser Practice commissioned by MAFF, roughly on a quinquennial basis (1969 to date), and detailed models of atmospheric N deposition developed as part of national research programmes on Surface Waters Acidification (SWAP) and Global Atmospheric N Enrichment (GANE) (see Whitehead et al. 1998; Johnes 1999 for further details). In addition to these data sources detailing the nature and extent of N sources within the U.K. landscape there are also routine monthly observations of nitrate and DIN flux in all 1400 major surface water catchments in England and Wales from 1973 to date, and sometimes for earlier periods, available from the Environment Agency Monitoring Programme archive, and continuous records of flow in these catchments from the Surface Water Archive extending over the same time period and earlier. In this sense, the U.K. can be described as a data-rich region for which it could not be argued that regional N flux models were limited in their reliability through inadequate description of nutrient sources. Development of the export coefficient modelling approach has benefited from this rich data archive on which the model has been constructed and tested.

This technique has been developed at a number of spatial scales to suit different management objectives. At its finest scale it has been applied to individual watersheds from 5 to 1200 km^2 in area. At its coarsest scale (this paper) it has been applied to the 3 major drainage basins of England and Wales, representing the watersheds draining to the North Sea (62318 km^2), the North Atlantic (77937 km^2) and the English Channel (10892 km^2). At its finest scale it uses the field as the spatial modelling unit, providing output on an annual basis, and has predicted within $\pm$ 5% of observed N (and P) loadings for all sites modelled at this scale. Recently, a simpler version has been developed for the U.K. Environment Agency for application at a national scale. In this, the model was adapted to allow estimates of N and P flux to be calculated for all watersheds within England and Wales without the need for basin-specific calibration in all 1400 watersheds. The model structure was simplified to run for a limited number of landscape units types sharing similar functional behaviour in terms of process controls on N cycling and flux. These were defined based on the major classes of geoclimatic region identified in the 1st Land Utilisation Survey of Britain (1931–1940). The landscape units or geoclimatic regions defined for England and Wales are shown in Figure 1. These represent areas with broadly similar climate, geology, soil types, topography and natural vegetation cover which have, therefore, similar ranges of nutrient export potential (and nutrient retention capacity) as a function of flow volume, timing and routing from land to stream. Generic sets of export coefficients (unit-specific parameter values) were derived for each of these geoclimatic regions, which could then be applied to parish scale census data for any parish lying within each region type. The coefficients were selected to reflect the intrinsic nutrient retention capacity of each region. The coefficients selected for each nutrient source in each region were validated in a rigorous multiple validation procedure (see Johnes et al. 1998b). The validated model is very robust, producing a close fit with observed data from water quality archives for a wide variety of landscape types and production systems (r^2 = 0.98 for both N and P for 38 catchments and >90 pairs of data). The 38

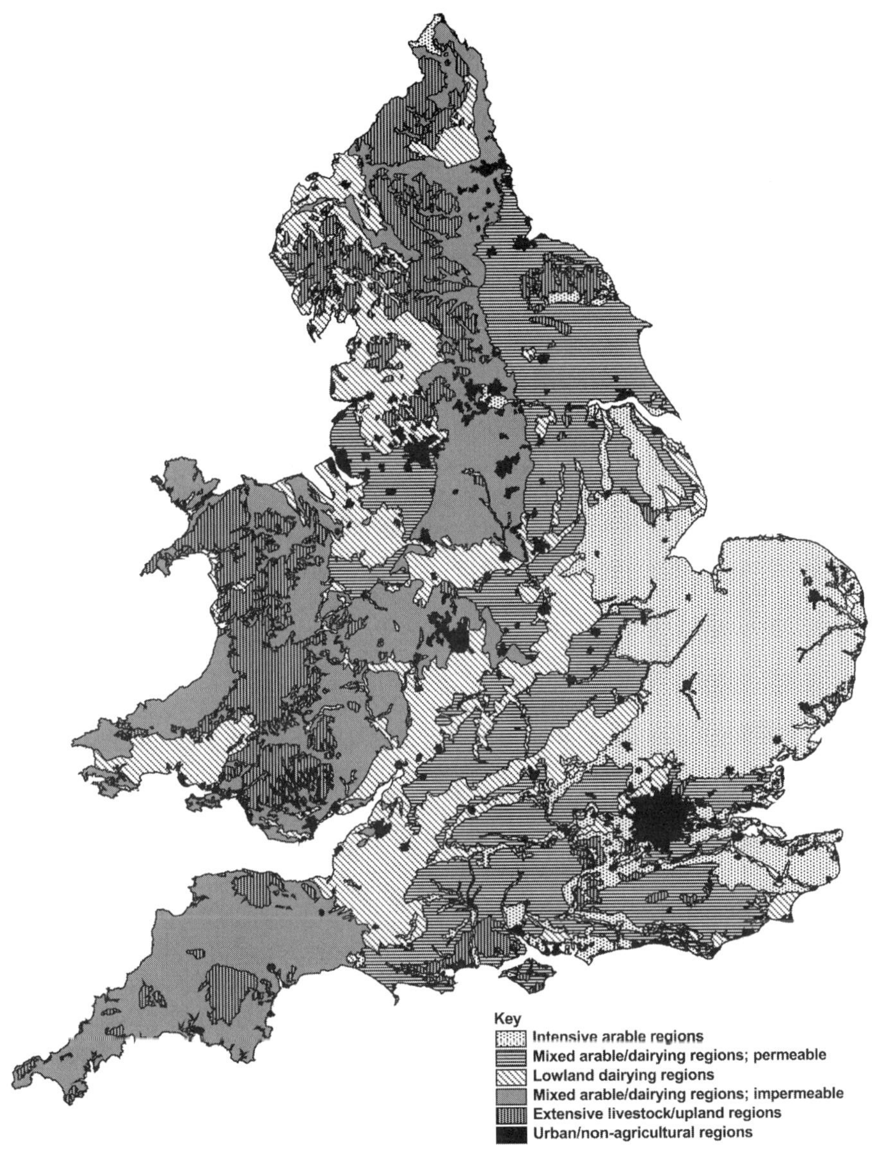

Figure 1. Characteristic Geoclimatic Region Types in England and Wales (after Johnes et al. 1998a).

catchments in which the model was validated against observed N flux data are shown in Figure 2, coded to indicate the dominant geoclimatic region type for each catchment.

In the Environment Agency project the model was run based on 1931 parish scale Agricultural Census data to provide a baseline estimate of N

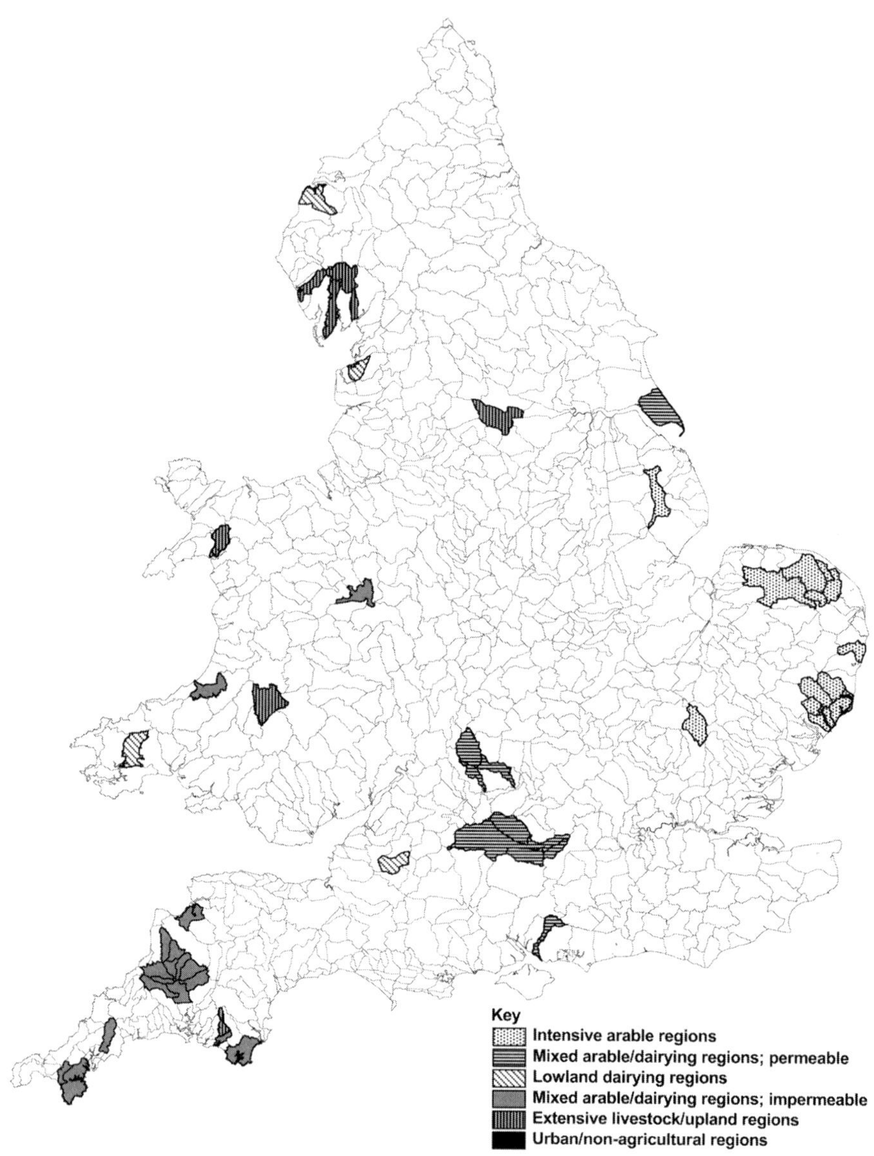

Figure 2. Export coefficient modelling applications to surface water catchments in England and Wales.

(and P) flux to U.K. lakes against which to gauge the present extent of N (and P) enrichment of these waters (see Johnes et al. 1994, 1998a, 1998b for details). In work funded by MAFF the model was updated (see Johnes et al. 2000, for details). Parish scale Agricultural Census data for 1991 were run through the model to provide a direct comparison with the earlier 1931 model predictions, allowing estimation of the rates and sources of changes in N and P loss from agriculture to water over the past 60 years. Model output was separated into its contributing layers to indicate the relative contribution of different nutrient sources to the total N and P load exported from land to water. Overall, the export coefficient model predicted a 136% increase in N flux from England and Wales to coastal waters, from an average rate of 11.2 kg N ha^{-1} in 1931 to an average of 26.4 kg N ha^{-1} in 1991.

The parish scale output from the 1991 model run is presented in Figure 3, showing the spatial variations and patterning in N flux estimates generated by the model at parish scale and the total N flux rate for adjacent coastal waters. Clear patterns are apparent, with the lowest rates of N flux predicted for the upland areas of England and Wales which support low density sheep grazing on moorland and cattle grazing on the shallower slopes at the foot of the moors. This reflects the fact that despite the abundance of runoff (averaging 1200–2000 mm annually) and the relatively high proportion of overland flow and near-surface lateral quickflow generated across this region as a function of thin soils overlying impermeable bedrock with moderate to steep slopes, the landscape is used relatively un-intensively (in the U.K. context). Thus the high N export potential of these landscapes is not translated into high N flux rates. Low to moderate rates of N flux are predicted from the flat dry counties of East Anglia, despite the use of this region for intensive arable production with associated high rates of fertiliser N applications to crops and grass. This reflects the fact that despite a high rate of N input to this landscape, the flatness of the landscape and the low rates of runoff (averaging <200 mm per year) generate a low N export potential. The highest rates of N flux are estimated for the wetter hill country of the west of England and Wales with typical annual runoff rates of 500–1000 mm per year, where higher fertiliser application rates (averaging 200 kg N ha^{-1} applied annually to grassland), and high stocking densities for dairy and beef cattle production lead to a combination of high N input rates and high N export potential. The potential distinctions made by this form of modelling, based on the landscape unitary approach, are even greater than they appear in Figure 3 because this also takes account of the spatial distribution of people in England and Wales. As a coarse generalisation, human population density is highest in the flatter, drier lands of the south and east of the U.K. and lowest in the north and west. Thus very low N flux rates from non-point agricultural sources to the

444

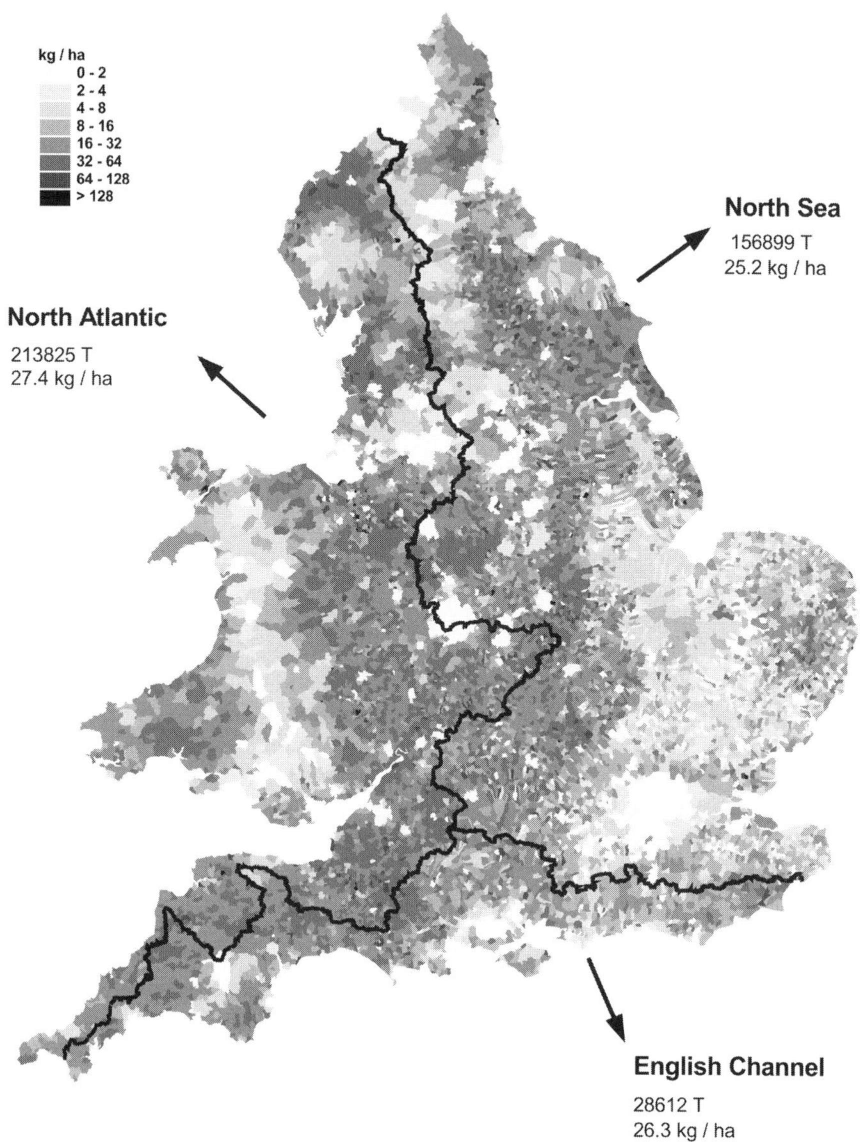

Figure 3. Export coefficient model prediction of TN export (1991) using 6 spatially distributed sets of export coefficients and parish scale input data. TN export to coastal waters calculated from aggregated parish scale TN export estimates.

North Sea and, to a lesser extent, the English Channel are compensated for in the average N flux rates to coastal waters shown in Figure 3 by higher N flux rates from the human population in each of these major drainage units. Lower population densities (generally) in the north and west mean that the total N flux estimate for the North Atlantic drainage unit is largely attributed to N flux from non-point agricultural and atmospheric sources.

The model makes a non-linear discrimination between the rates of N input to the system and the rates of N flux from land to stream by taking account of the intrinsic N retention or N export potential of the landscape of England and Wales. Within each of the regional sub-models the model is linear, but the overall national model is non-linear. The spatial distinctions that the model makes are important and real in terms of the observed rates of N flux from land to water within England and Wales, and provide valuable guidance for informing environmental management and government policy in relation to N (and P) flux from non-point and point sources to U.K. waters. The question then arises of how much of this spatial resolution would be lost, and what scale of error would be associated with N flux estimates generated by the model at watershed to regional scale if this landscape unitary approach, the spatial scale of the modelling units (currently parish scale) or the parameterisation of the model itself were to be simplified in line with the forms of model structure associated with existing regional to global scale models. In the existing model structure there are 7 classes of land use, for which fertiliser N input and N fixation rates are separately applied, atmospheric N deposition, and 4 classes of livestock, with people separately accounted for on a per capita basis. In addition there are further modifiers incorporated on a geoclimatic region basis, relating to the natural environmental characteristics of each region, particular land management practices, manure handling and management, and sewage treatment facilities. This degree of detail in accounting for N inputs to land and the intrinsic N retention capacity of landscape units is unparalleled in the existing regional to global scale models. How much of this detail would need to be retained in order to generate accurate, robust and spatially discrete estimates of N flux from land to ocean at regional to global scale is the subject of the scaling analysis conducted in this paper.

Quantifying the errors associated with upscaling from parish to regional scale

The export coefficient model was used, therefore, to estimate the errors and uncertainties inherent in the regional and global scale models as a function of scale. This provided an insight into the inherent errors built into the regional and global scale N models generated by scaling up from catchment scale studies to global applications. Two forms of scaling error have been investigated relative to their impacts on the accuracy of predictions of N flux to U.K. coastal waters (North Sea, North Atlantic, English Channel): (1) Spatial aggregation (lumping) of input data and model output and (2) Conceptual lumping of model parameters.

To achieve this the parish scale National Export Coefficient Model predictions of N export in England and Wales were systematically scaled up, coarsening either the scale of the input data, or the range of the export coefficients built into the model, or both. Parish scale units were aggregated to catchment scale, based on the 1400 watershed units routinely monitored by the Environment Agency (see Figure 2 for catchment boundaries). The spatial scale at which modelling took place was therefore scaled up from parish units averaging 1355 ha in area to catchments of an average 155 km^2 in area. These were then further aggregated, running the model using input/output units representing the major drainage basins of the England and Wales (e.g. the Thames basin, the Severn, the Trent, the Great Ouse and so on, averaging 18893 km^2 in area). The complexity of the model framework itself was also reduced, sequentially aggregating the source type categories and the degree of landscape sensitivity reflected in the number of different export coefficient groups, until the final model run used two categories of nutrient source (agricultural and non-agricultural land), one set of export coefficients for the entire land mass, irrespective of landscape sensitivity to N export and three spatial units: the major drainages to the North Sea, the North Atlantic and the English Channel. At each stage model output was then lumped together, based on the major drainage unit boundaries, to predict overall N flux from the land mass of England and Wales to the North Sea (62318 km^2), the North Atlantic (77937 km^2) and the English Channel (10892 km^2). In total, a matrix of 8 scales of spatial aggregation from the original parish scale model estimates with 5 sequentially aggregated sets of model parameters gave 40 different estimates of N flux from land to coastal waters for each of the 3 major drainage units. The forms of aggregation were as follows:

Spatial aggregation categories

1. *Lumped output (parish scale input data)*:
   TN export to coastal waters calculated from aggregated parish scale TN export estimates.

2. *Lumped output (catchments from parish output)*:
   Catchment scale export rates calculated from aggregated parish scale TN export estimates, TN export to coastal waters calculated from aggregated catchment scale TN export estimates.

3. *Lumped output (major catchments from parish output)*:
Major catchment scale export rates calculated from aggregated parish scale TN export estimates, TN export to coastal waters calculated from aggregated major catchment scale TN export estimates.

4. *Lumped output (major catchments from catchments output)*:
Major catchment scale export rates aggregated from catchment scale TN export estimates calculated from catchment scale input data (aggregated from parish scale input data), TN export to coastal waters calculated from aggregated major catchment scale TN export estimates.

5. *Lumped output (major catchments from catchment output from parish output)*:
Major catchment scale export rates aggregated from catchment scale TN export estimates aggregated from parish scale TN export estimates, TN export to coastal waters calculated from aggregated major catchment scale TN export estimates.

6. *Lumped parish input to catchments, catchment output lumped to major drainages*:
Catchment scale export rates calculated from catchment scale input data aggregated from parish scale input data, TN export to coastal waters calculated from aggregated catchment scale TN export estimates.

7. *Lumped parish input to major catchments, major catchment output lumped to major drainages*:
Major catchment scale export rates calculated from major catchment scale input data aggregated from parish scale input data, TN export to coastal waters calculated from aggregated major catchment scale TN export estimates.

8. *Lumped parish input to major drainages*:
TN export to coastal waters calculated from major drainage scale TN export estimates, based on aggregated parish scale input data.

Conceptual lumping categories

1. *Original method (6 spatially distributed sets of export coefficients) applied to*:
land use units (cereals, other arable crops, bare fallow, permanent grass, temporary grass, rough grazing, orchards and woodland);
livestock units (cattle, pigs, sheep, poultry); people; atmospheric N deposition

2. *Original coefficients (6 spatially distributed sets) with amalgamated input data units*:
land use units (crops, grass, moorland, woodland);

livestock units (cattle, pigs, sheep, poultry); people; atmospheric N deposition

3. *Aggregated coefficients (1 set) with amalgamated input data units*:
land use units (crops, grass, moorland, woodland);
livestock units (cattle, pigs, sheep, poultry); people; atmospheric N deposition

4. *Original coefficients (6 spatially distributed sets) with coarsely amalgamated input data units*:
land use units (agricultural land, semi-natural vegetation);
livestock units (cattle, pigs, sheep, poultry); people; atmospheric N deposition

5. *Aggregated coefficients (1 set) with coarsely amalgamated input data units*:
land use units (agricultural land, semi-natural vegetation);
livestock units (cattle, pigs, sheep, poultry); people; atmospheric N deposition

Results of spatial aggregation:

The model estimates of TN flux from each of the 3 drainage units to the North Sea, the North Atlantic and the English Channel, generated by this matrix of model forcing are presented in Table 1 (units are kg ha^{-1} and tonnes per annum for 1991). By comparing the estimates of landscape scale N flux to each of these coastal waters with the initial estimates produced by the parish scale National Export Coefficient Model, it was then possible to derive an estimate of the relative loss of model accuracy (compared to estimates generated with the original model parameterisation and input scale) in each step of scaling in the prediction of N flux from land based sources to the adjacent oceans for England and Wales. The results of this analysis are presented in Table 2. A final comparison was made to assess the impact of scaling and coarsening of model parameters on the spatial discrimination of the model estimates for the 3 major drainage units draining to coastal waters. One of the strengths of the original model is that it allows accurate representation of the spatial variations in N flux estimates as a function of both the rates of N input to the system, and the intrinsic nutrient retention capacity or nutrient export potential of landscape units as defined by those environmental variables controlling N cycling and hydrological transport efficiency within each landscape unit. By comparing the range of variation in model estimates generated at each stage of spatial aggregation or conceptual lumping it was possible to determine the relative loss of spatial discrimination associated with the scaling process. The results of this analysis are presented in Tables

Table 1. Model estimates of TN from land to coastal waters generated through scaling from parish to major drainage scale and from coarsening model structure

Drainage to...	Area (km^2)	Original Method (tonnes)	Original coefficients, amalgamated input data (tonnes)	Aggregated coefficients, amalgamated input data (tonnes)	Original coefficients, coarsely amalgamated input data (tonnes)	Aggregated coefficients, coarsely amalgamated input data (tonnes)	Original Method (kg ha^{-1})	Original coefficients, amalgamated input data (kg ha^{-1})	Aggregated coefficients, amalgamated input data (kg ha^{-1})	Original coefficients, coarsely amalgamated input data (kg ha^{-1})	Aggregated coefficients, coarsely amalgamated input data (kg ha^{-1})
Lumped output (Parishes):											
North Sea	62318	156899	149774	171746	141503	145317	25.2	24.0	27.6	22.7	23.3
North Atlantic	77937	213825	204463	194170	219854	187535	27.4	26.2	24.9	28.2	24.1
English Channel	10892	28612	27776	29753	26487	26834	26.3	25.5	27.3	24.3	24.6
Lumped output (Catchments from Parish output):											
North Sea	62318	153662	146866	168054	138834	142485	24.7	23.6	27.0	22.3	22.9
North Atlantic	77937	206776	197966	187335	212731	180845	26.5	25.4	24.0	27.3	23.2
English Channel	10892	27465	26732	28542	25540	25824	26.2	25.5	27.2	24.3	24.6
Lumped output (Major Catchments from Parish output):											
North Sea	62318	156866	149741	171723	141473	145300	25.2	24.0	27.6	22.7	23.3
North Atlantic	77937	213802	204417	194118	219800	187483	27.4	26.2	24.9	28.2	24.1
English Channel	10892	28612	27775	29753	26487	26834	26.3	25.5	27.3	24.3	24.6
Lumped output (Major Catchments from Catchment output):											
North Sea	62318	150098	144345	166409	136168	141241	24.1	23.2	26.7	21.9	22.7
North Atlantic	77937	196661	190467	183696	203996	177273	25.2	24.4	23.6	26.2	22.7
English Channel	10892	26913	26189	28283	24989	25619	25.6	25.0	27.0	23.8	24.4

Table 1. Continued

Drainage to…	Area (km^2)	Original Method (tonnes)	Original coefficients, amalgamated input data (tonnes)	Aggregated coefficients, amalgamated input data (tonnes)	Original coefficients, coarsely amalgamated input data (tonnes)	Aggregated coefficients, coarsely amalgamated input data (tonnes)	Original Method (kg ha^{-1})	Original coefficients, amalgamated input data (kg ha^{-1})	Aggregated coefficients, amalgamated input data (kg ha^{-1})	Original coefficients, coarsely amalgamated input data (kg ha^{-1})	Aggregated coefficients, coarsely amalgamated input data (kg ha^{-1})
Lumped output (Major Catchments from Catchment output from Parish output):											
North Sea	62318	153123	146358	167581	138382	142085	24.6	23.5	26.9	22.2	22.8
North Atlantic	77937	205307	196645	186284	211237	179819	26.3	25.2	23.9	27.1	23.1
English Channel	10892	27465	26731	28542	25540	25824	26.2	25.5	27.2	24.3	24.6
Lumped Parish input to Catchments, Catchment output lumped to Major Drainages:											
North Sea	62318	150643	144860	166881	136629	141641	24.2	23.2	26.8	21.9	22.7
North Atlantic	77937	197901	191663	184658	205348	178210	25.4	24.6	23.7	26.3	22.9
English Channel	10892	26913	26189	28283	24989	25619	25.6	25.0	27.0	23.8	24.4
Lumped Parish input to Major Catchments, Major Catchment output lumped to Major Drainages:											
North Sea	62318	153356	145562	170348	136583	144264	24.6	23.4	27.3	21.9	23.1
North Atlantic	77937	203551	192131	191027	205119	184437	26.1	24.7	24.5	26.3	23.7
English Channel	10892	28433	27134	29356	25652	26505	26.1	24.9	27.0	23.6	24.3
Lumped Parish input to Major Drainages:											
North Sea	62318	153818	147543	170445	137524	144310	24.7	23.7	27.4	22.1	23.2
North Atlantic	77937	198218	191727	190996	204567	184484	25.4	24.6	24.5	26.2	23.7
English Channel	10892	28086	27135	29356	25652	26505	25.8	24.9	27.0	23.6	24.3

Table 2. Error propagation resulting from scaling from parish to major drainage scale, and from coarsening model structure. Percentage change is calculated relative to the sum of parish scale estimates of TN export to coastal waters generated by the original model

Drainage to...	Original Method (% change)	Original coefficients, amalgamated input data (% change)	Aggregated coefficients, amalgamated input data (% change)	Original coefficients, coarsely amalgamated input data (% change)	Aggregated coefficients, coarsely amalgamated input data (% change)
Lumped output (Parishes):					
North Sea	0.00	−4.54	9.46	−9.81	−7.38
North Atlantic	0.00	−4.38	−9.19	2.82	−12.30
English Channel	0.00	−2.92	3.99	−7.43	−6.22
Lumped output (Catchments from Parish output):					
North Sea	−2.06	−6.40	7.11	−11.50	−9.19
North Atlantic	−3.28	−7.40	−12.40	−0.50	−15.40
English Channel	−0.37	−3.03	3.53	−7.35	−6.33
Lumped output (Major Catchments from Parish output):					
North Sea	−0.02	−4.56	9.45	−9.83	−7.39
North Atlantic	0.01	−4.38	−9.20	2.81	−12.30
English Channel	0.00	−2.93	3.99	−7.43	−6.22
Lumped output (Major Catchments from Catchment output):					
North Sea	−4.34	−8.00	6.06	−13.2	−9.98
North Atlantic	−8.01	−10.90	−14.10	−4.58	−17.10
English Channel	−2.38	−5.00	2.60	−9.35	−7.07
Lumped output (Major Catchments from Catchment output from Parish output):					
North Sea	−2.41	−6.72	6.81	−11.80	−9.44
North Atlantic	−3.97	−8.02	−12.90	−1.19	−15.90
English Channel	−0.37	−3.03	3.53	−7.35	−6.33
Lumped Parish input to Catchments, Catchment output lumped to Major Drainages:					
North Sea	−3.99	7.67	6.36	−12.90	−9.73
North Atlantic	−7.45	−10.40	−13.60	−3.96	−16.70
English Channel	−2.38	−5.00	2.60	−9.35	−7.07
Lumped Parish input to Major Catchments, Major Catchment output lumped to Major Drainages:					
North Sea	−2.26	−7.23	8.57	−12.90	−8.05
North Atlantic	−4.79	−10.10	−10.60	−4.06	−13.70
English Channel	−0.63	−5.17	2.60	−10.30	−7.37
Lumped Parish input to Major Drainages:					
North Sea	−1.96	−5.96	8.63	−12.30	−8.02
North Atlantic	−7.30	−10.30	−10.70	−4.33	−13.70
English Channel	−1.84	−5.17	2.60	−10.30	−7.37

Table 3. Statistical properties of TN export estimates generated through scaling from parish to major drainage scale, and from coarsening model structure: interquartile range

Drainage to…	Number of modelling units (n)	Lower Quartile (Q1)					Upper Quartile (Q3)				
		Original Method (kg ha^{-1})	Original coefficients, amalgamated input data (kg ha^{-1})	Aggregated coefficients, amalgamated input data (kg ha^{-1})	Original coefficients, coarsely amalgamated input data (kg ha^{-1})	Aggregated coefficients, coarsely amalgamated input data (kg ha^{-1})	Original Method (kg ha^{-1})	Original coefficients, amalgamated input data (kg ha^{-1})	Aggregated coefficients, amalgamated input data (kg ha^{-1})	Original coefficients, coarsely amalgamated input data (kg ha^{-1})	Aggregated coefficients, coarsely amalgamated input data (kg ha^{-1})
Lumped output (Parishes):											
North Sea	6500	10.50	10.30	11.50	10.20	10.90	60.10	57.80	55.30	56.30	47.00
North Atlantic	3615	8.14	8.00	9.01	8.43	8.66	64.0	61.80	50.20	68.80	49.90
English Channel	1390	13.90	13.70	15.60	13.70	14.50	59.50	56.90	53.70	60.70	50.80
Lumped output (Catchments from Parish output):											
North Sea	465	11.50	11.40	12.70	11.50	11.90	42.50	41.00	36.90	41.40	33.90
North Atlantic	313	8.24	8.08	8.58	8.25	8.45	41.60	39.90	33.20	45.60	33.70
English Channel	147	15.80	15.50	16.90	15.60	16.70	44.00	43.20	41.90	47.50	42.20
Lumped output (Major Catchments from Parish output):											
North Sea	5	13.50	12.30	13.10	11.90	11.40	33.20	32.10	30.00	33.60	28.40
North Atlantic	5	19.70	15.80	14.40	16.80	13.80	30.40	29.00	26.90	30.80	26.30
English Channel	3	13.90	14.10	14.20	14.00	13.80	37.10	35.70	4.30	37.70	33.20
Lumped output (Major Catchments from Catchment output):											
North Sea	5	22.30	21.40	22.90	20.00	20.20	32.30	31.30	29.50	32.30	27.60
North Atlantic	5	20.30	19.80	20.50	20.10	20.20	28.50	27.50	25.50	28.80	24.20
English Channel	3	25.40	24.70	26.50	23.70	24.00	35.10	34.20	32.90	36.10	31.90

Table 3. Continued

Drainage to…	Number of modelling units (n)	Lower Quartile (Q1)					Upper Quartile (Q3)				
		Original Method (kg ha^{-1})	Original coefficients, amalgamated input data (kg ha^{-1})	Aggregated coefficients, amalgamated input data (kg ha^{-1})	Original coefficients, coarsely amalgamated input data (kg ha^{-1})	Aggregated coefficients, coarsely amalgamated input data (kg ha^{-1})	Original Method (kg ha^{-1})	Original coefficients, amalgamated input data (kg ha^{-1})	Aggregated coefficients, amalgamated input data (kg ha^{-1})	Original coefficients, coarsely amalgamated input data (kg ha^{-1})	Aggregated coefficients, coarsely amalgamated input data (kg ha^{-1})
Lumped output (Major Catchments from Catchment output from Parish output):											
North Sea	5	22.60	21.60	23.00	20.30	20.20	33.10	32.00	29.60	33.10	27.70
North Atlantic	5	21.60	20.60	21.40	20.90	20.50	29.20	28.00	25.80	29.60	24.50
English Channel	3	25.90	25.20	26.70	24.20	24.20	36.20	34.80	33.30	36.90	32.30
Lumped Parish input to Catchments, Catchment output lumped to Major Drainages:											
North Sea	465	12.20	12.10	13.20	11.90	12.30	42.00	40.90	36.90	41.40	33.70
North Atlantic	313	7.69	7.62	8.70	7.63	8.58	41.00	40.20	33.20	45.30	33.50
English Channel	147	15.90	15.70	16.80	15.20	16.50	44.10	43.30	41.90	47.50	42.20
Lumped Parish input to Major Catchments, Major Catchment output lumped to Major Drainages:											
North Sea	5	15.60	15.50	15.00	14.90	12.80	32.50	30.90	29.90	32.10	28.10
North Atlantic	5	9.78	9.46	10.00	9.58	9.63	29.60	27.70	26.50	29.40	25.20
English Channel	3	13.70	13.60	13.90	13.40	13.60	36.20	34.40	33.80	36.20	32.70
Lumped Parish input to Major Drainages:											
North Sea	1	24.70	23.70	27.40	22.10	23.20	24.70	23.70	27.40	22.10	23.20
North Atlantic	1	25.40	24.60	24.50	26.20	23.70	25.40	24.60	24.50	26.20	23.70
English Channel	1	25.80	24.90	27.00	23.60	24.30	25.80	24.90	27.00	23.60	24.30

454

Table 4. Statistical properties of TN export estimates generated through scaling from parish to major drainage scale, and from coarsening model structure. % deviation of quartiles from quartiles of data distribution generated by original method, operating at parish scale

Drainage to…	Number of modelling units (n)	Lower Quartile (Q1)					Upper Quartile (Q3)				
		Original Method (%)	Original coefficients, amalgamated input data (%)	Aggregated coefficients, amalgamated input data (%)	Original coefficients, coarsely amalgamated input data (%)	Aggregated coefficients, coarsely amalgamated input data (%)	Original Method (%)	Original coefficients, amalgamated input data (%)	Aggregated coefficients, amalgamated input data (%)	Original coefficients, coarsely amalgamated input data (%)	Aggregated coefficients, coarsely amalgamated input data (%)
Lumped output (Parishes):											
North Sea	6500	0.00	−1.90	9.52	−2.86	3.81	0.00	−3.83	−7.99	−6.32	−21.80
North Atlantic	3615	0.00	−1.72	10.70	3.56	6.39	0.00	−3.44	−21.60	7.50	−22.00
English Channel	1390	0.00	−1.44	12.20	−1.44	4.32	0.00	−4.37	−9.75	2.02	−14.60
Lumped output (Catchments from Parish output):											
North Sea	465	9.52	8.57	21.00	9.52	13.30	−29.30	−31.80	−38.60	−31.10	−43.60
North Atlantic	313	1.23	−0.74	5.41	1.35	3.81	−35.00	−37.70	−48.10	−28.80	−47.30
English Channel	147	13.70	11.50	21.60	12.20	20.10	−26.10	−27.40	−29.60	−20.20	−29.10
Lumped output (Major Catchments from Parish output):											
North Sea	5	65.80	51.10	60.90	46.20	40.00	−48.10	−49.80	−53.10	−47.50	−55.60
North Atlantic	5	41.70	13.70	3.60	20.90	−0.72	−48.90	−51.30	−54.80	−48.20	−55.80
English Channel	3	32.40	34.30	35.20	33.30	31.40	−38.30	−40.60	−92.80	−37.30	−44.80
Lumped output (Major Catchments from Catchment output):											
North Sea	5	60.40	54.00	64.70	43.90	45.30	−45.70	−47.40	−50.40	−45.70	−53.60
North Atlantic	5	93.30	88.60	95.20	91.40	92.40	−52.60	−54.20	−57.60	−52.10	−59.70
English Channel	3	212.00	203.00	226.00	191.00	195.00	−45.20	−46.60	−48.60	−43.60	−50.20

Table 4. Continued

Drainage to…	Number of modelling units (n)	Lower Quartile (Q1)					Upper Quartile (Q3)				
		Original Method (%)	Original coefficients, amalgamated input cata (%)	Aggregated coefficients, amalgamated input data (%)	Original coefficients, coarsely amalgamated input data (%)	Aggregated coefficients, coarsely amalgamated input data (%)	Original Method (%)	Original coefficients, amalgamated input data (%)	Aggregated coefficients, amalgamated input data (%)	Original coefficients, coarsely amalgamated input data (%)	Aggregated coefficients, coarsely amalgamated input data (%)
Lumped output (Major Catchments from Catchment output from Parish output):											
North Sea	5	115.00	106.00	119.00	93.30	92.40	−44.90	−46.80	−50.70	−44.90	−53.90
North Atlantic	5	165.00	153.00	163.00	157.00	152.00	−54.40	−56.30	−59.70	−53.80	−61.70
English Channel	3	86.30	81.30	92.10	74.10	74.10	−39.20	−41.50	−44.00	−38.00	−45.70
Lumped Parish input to Catchments, Catchment output lumped to Major Drainages:											
North Sea	465	49.90	48.60	62.20	46.20	51.10	−34.40	−36.10	−42.30	−35.30	−47.30
North Atlantic	313	−44.70	−45.20	−37.40	−45.10	−38.30	−31.10	−32.40	−44.20	−23.90	−43.70
English Channel	147	51.40	49.50	60.00	44.80	57.10	−26.60	−28.00	−30.30	−21.00	−29.80
Lumped Parish input to Major Catchments, Major Catchment output lumped to Major Drainages:											
North Sea	5	12.20	11.50	7.91	7.19	−7.91	−45.40	−48.10	−49.70	−46.10	−52.80
North Atlantic	5	−6.86	−9.90	−4.76	−8.76	−8.29	−50.70	−53.90	−55.90	−51.10	−58.10
English Channel	3	68.30	67.10	70.80	64.60	67.10	−43.40	−46.30	−47.20	−43.40	−48.90
Lumped Parish input to Major Drainages:											
North Sea	1	135.00	126.00	161.00	110.00	121.00	−58.90	−60.60	−54.40	−63.20	−61.40
North Atlantic	1	212.00	202.00	201.00	222.00	191.00	−60.30	−61.60	−61.70	−59.10	−63.00
English Channel	1	85.60	79.10	94.20	69.80	74.80	−56.60	−58.20	−54.60	−60.30	−59.20

3 and 4 in the form of the interquartile range (Q1 and Q3) for the range of estimates of N flux to coastal waters mapped at each stage. In interpreting these results it is important to bear in mind that land draining to the North Atlantic Ocean generally has a lower intrinsic nutrient retention capacity than land draining to the English Channel and the North Sea. This reflects the fact that land in the North and West of the U.K. tends to be wet, cold and steep (in the U.K. context), with drier, warmer and flatter landscapes towards the South and East. These characteristics are, of course, represented in the geoclimatic region units defined for England and Wales (see Figure 1), but will nevertheless have a bearing on how the model performs through the sequence of spatial aggregation and conceptual lumping steps, particularly where the regional sub-model structure is modified.

Figures 4 to 10 illustrate column one of each of Tables 1 to 4 (Original Method), showing the effects of spatial aggregation on modelled estimates of N flux from the 3 major drainage units to coastal waters. These estimates are based on the original model parameterisation with 7 categories of land use unit, 4 categories of livestock unit, people, atmospheric N deposition, and the 6 sets of spatially distributed export coefficients reflecting the intrinsic N export potential of the 6 landscape unit categories for England and Wales. What is apparent from these figures is that by lumping either the parish scale input data, or the parish, catchment or major catchment scale N flux estimates, the estimate of total N flux to coastal waters from each of the three major drainage units decreases, even where the original model parameterisation is maintained (see column 1, Tables 1 and 2). This probably reflects the fact that by upscaling the model input data the coincidences where areas with a low intrinsic nutrient retention capacity (perhaps as a function of high rainfall and/or steep slopes) are combined with high intensity agriculture are lost in the accounting process. The greatest errors associated with spatial aggregation occur in the estimates of TN export from the North Atlantic drainage unit, with a maximum error of 8% (Table 2, column 1, row 11) associated with running the original model where the parish scale input data were lumped into the catchment units, and the resultant catchment scale TN estimates aggregated into the major catchments and then into the major drainage units. This is probably a function of the greater range of variation in both the parish scale N flux estimates and the comparatively low intrinsic nutrient export potential of the landscape units in the North Atlantic Drainage unit, compared to the drainages for the English Channel and North Sea. Even so this form of spatial aggregation also produced the maximum errors for the estimates of TN export from the English Channel (2.38%; Table 2, column 1, row 12) and North Sea (4.34%; Table 2, column 1, row 10) drainage units. The analysis, conducted solely on the original model parameterisation, suggests that the minimum

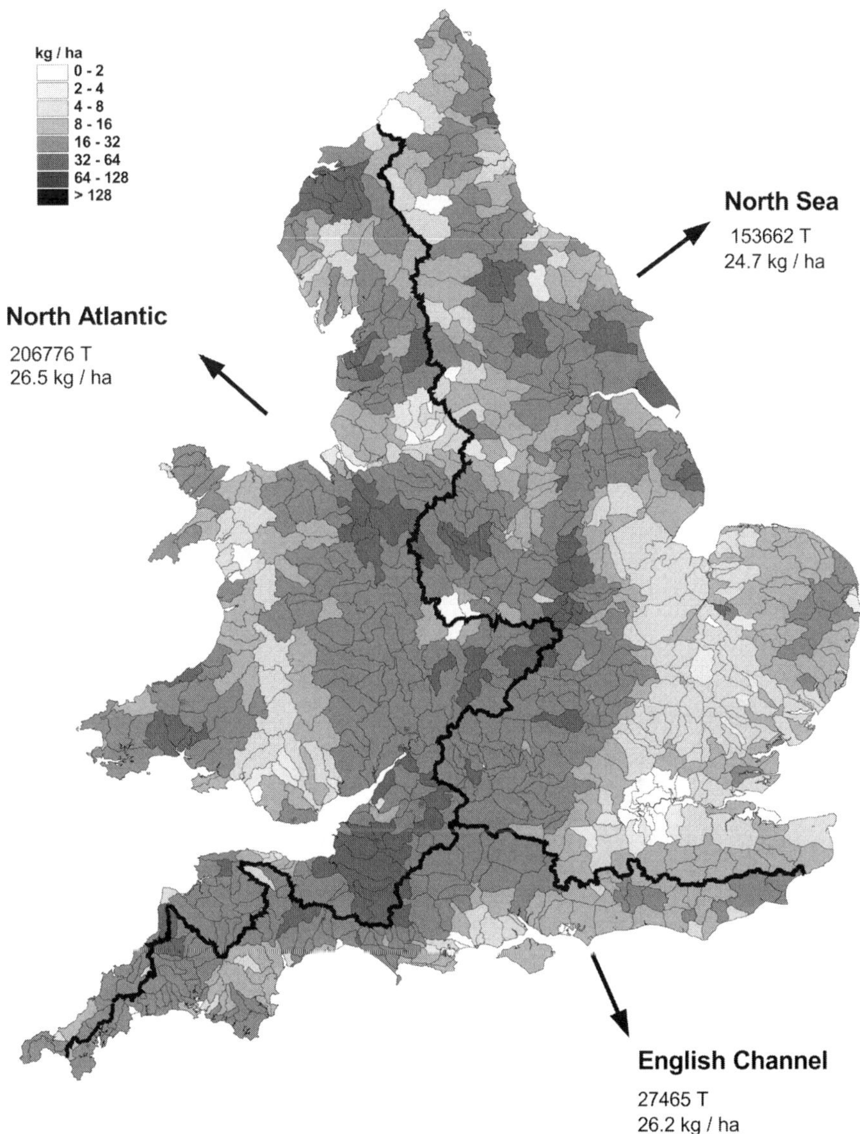

Figure 4. Export coefficient model prediction of TN export (1991) using 6 spatially distributed sets of export coefficients. Catchment scale export rates aggregated from distributed parish scale TN export estimates. TN export to coastal waters calculated from aggregated catchment scale TN export estimates.

458

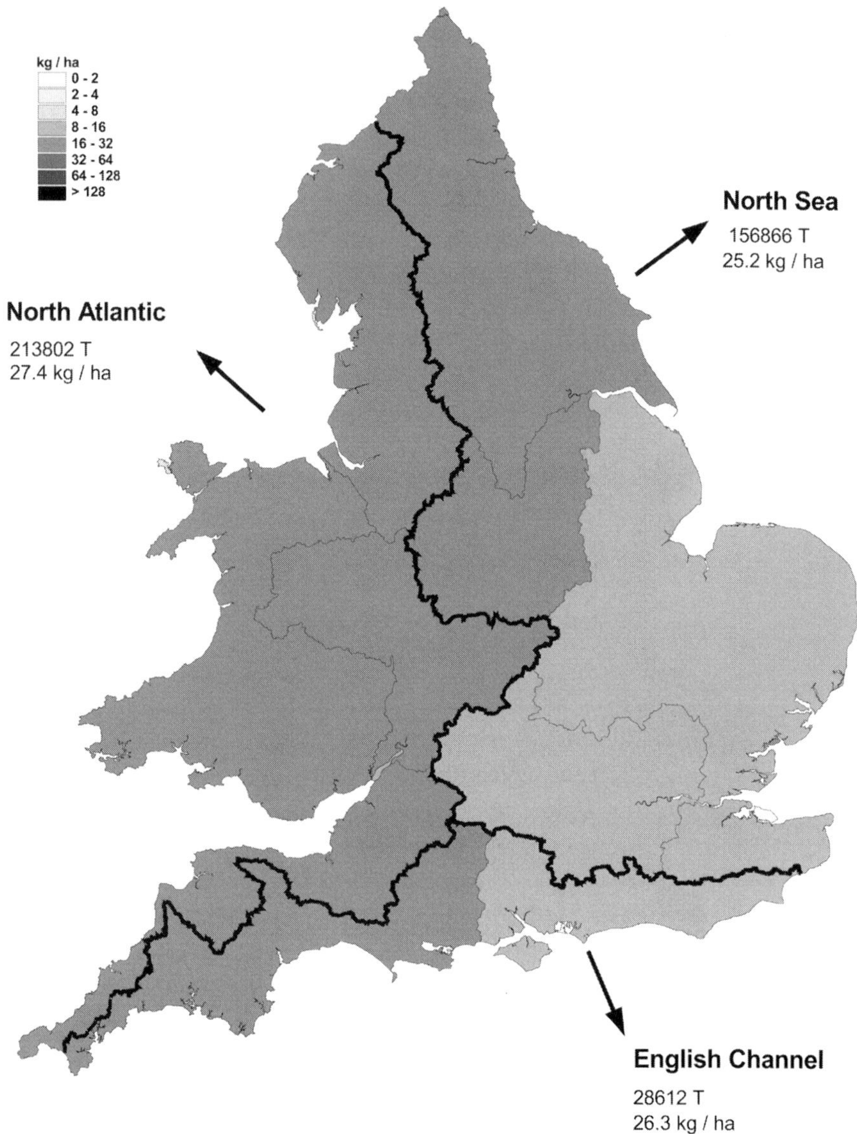

Figure 5. Export coefficient model prediction of TN export (1991) using 6 spatially distributed sets of export coefficients. Major catchment scale export rates aggregated from distributed parish scale TN export estimates. TN export to coastal waters calculated from aggregated major catchment scale TN export estimates.

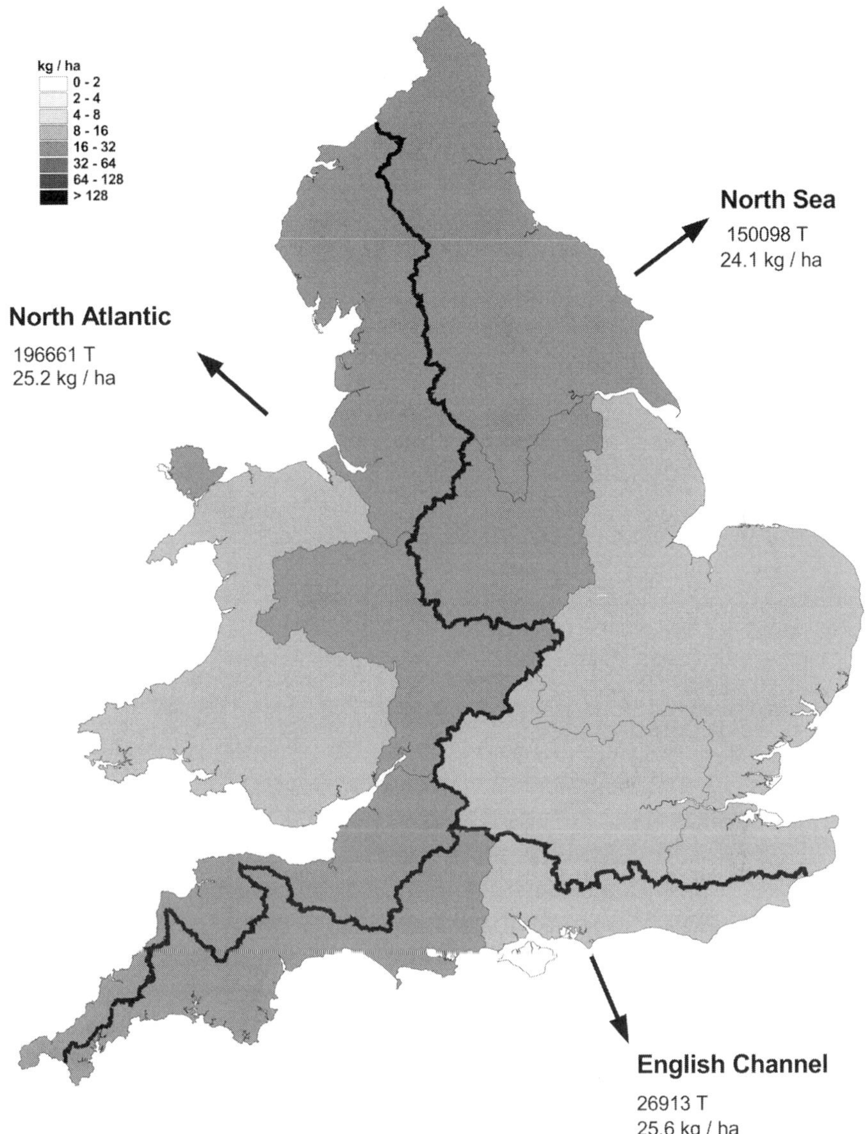

Figure 6. Export coefficient model prediction of TN export (1991) using 6 spatially distributed sets of export coefficients. Major catchment scale export rates aggregated from catchment scale TN export estimates calculated from catchment scale input data. TN export to coastal waters calculated from aggregated major catchment scale TN export estimates.

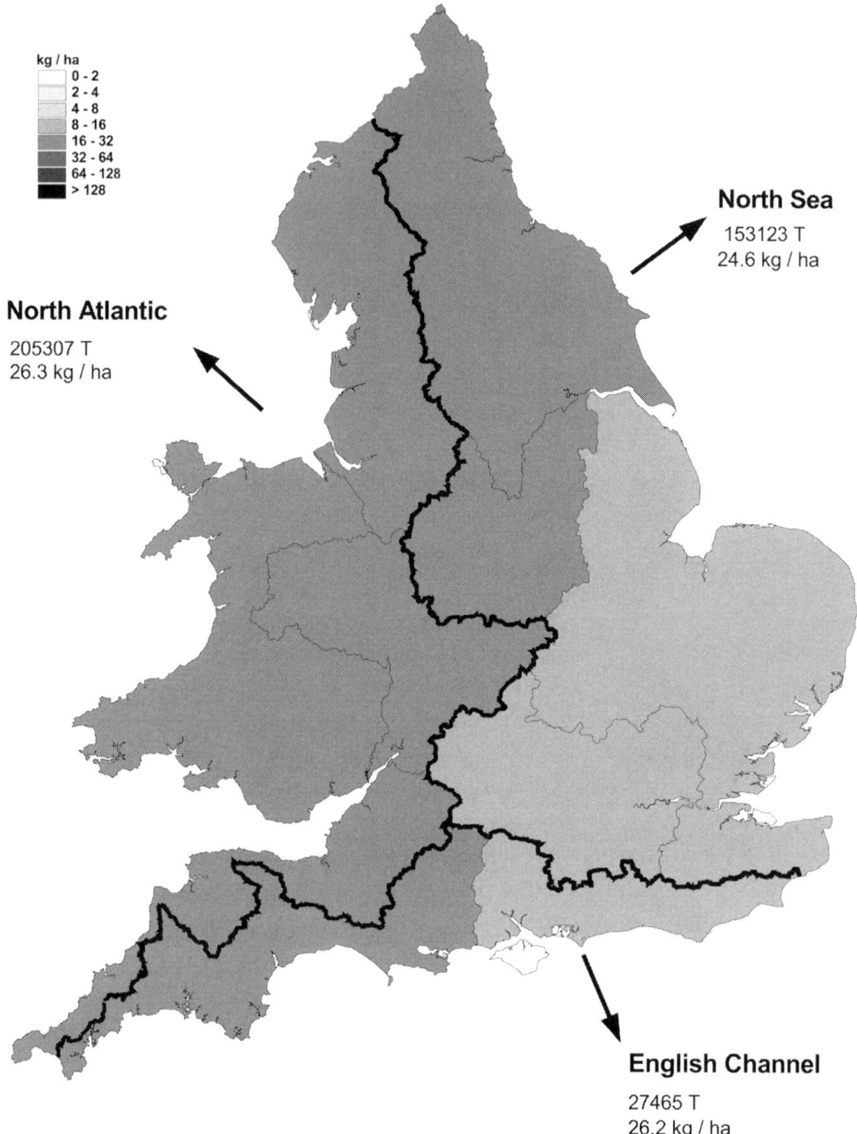

Figure 7. Export coefficient model prediction of TN export (1991) using 6 spatially distributed sets of export coefficients. Major catchment scale export rates aggregated from catchment scale TN export estimates, aggregated from distributed parish scale TN export estimates. TN export to coastal waters calculated from aggregated major catchment scale TN export estimates.

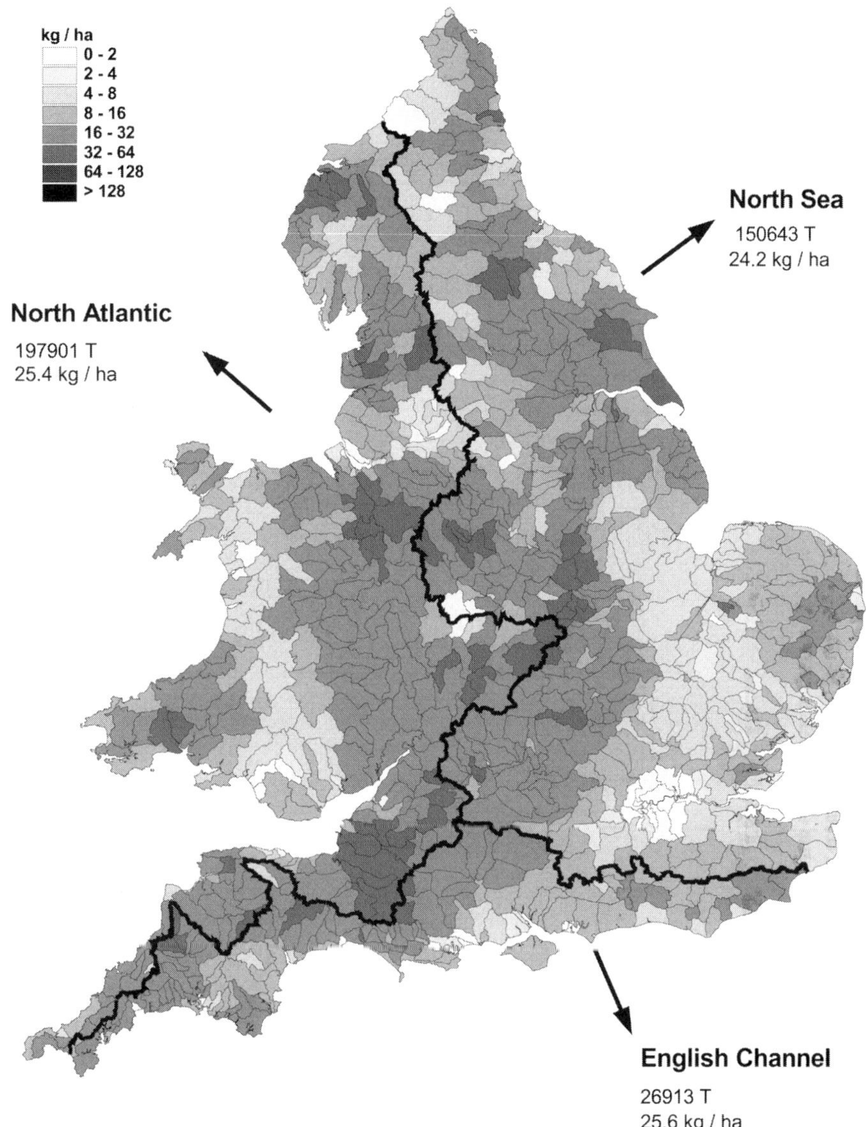

Figure 8. Export coefficient model prediction of TN export (1991) using 6 spatially distributed sets of export coefficients. Catchment scale export rates calculated from catchment scale input data aggregated from parish scale input data. TN export to coastal waters calculated from aggregated catchment scale TN export estimates.

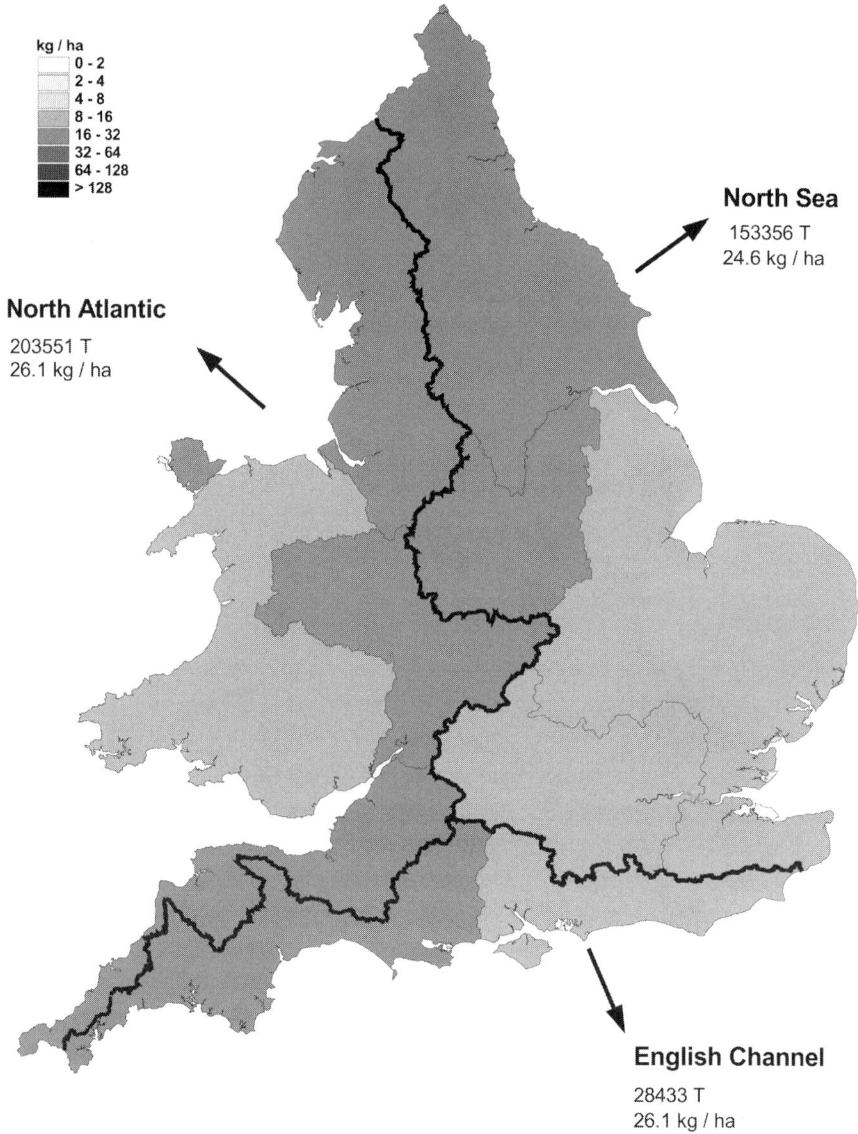

Figure 9. Export coefficient model prediction of TN export (1991) using 6 spatially distributed sets of export coefficients. Major catchment scale export rates calculated from major catchment scale input data aggregated from parish scale input data. TN export to coastal waters calculated from aggregated major catchment scale TN export estimates.

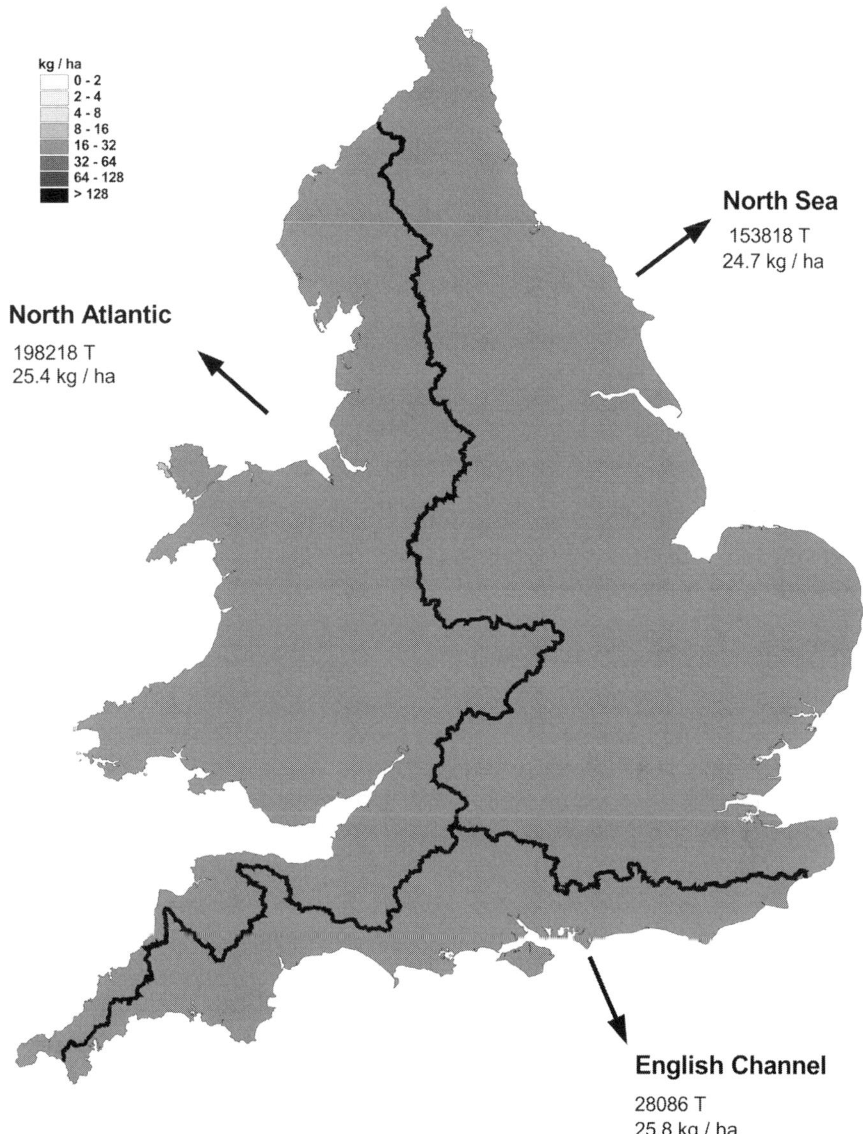

Figure 10. Export coefficient model prediction of TN export (1991) using 6 spatially distributed sets of export coefficients. TN export to coastal waters calculated from major drainage scale TN export estimates, based on aggregated parish scale input data.

error is generated in upscaling where modelling takes place at parish scale and parish scale TN export estimates are lumped directly into the catchment, major catchment or major drainage units, rather than the input data being lumped into coarser spatial units and modelling taking place at that scale. Greater errors are also generated where more steps are involved in aggregation and modelling. This is probably a function of the errors generated by modelling within a Geographical Information System (ARC-info GIS), since the maximum number of errors will be generated wherever two boundaries intersect, requiring the GIS to mathematically apportion either the input data or the modelled TN estimates between polygons.

Overall the changes in the estimates of mean annual TN export to coastal waters resulting from upscaling of the original model are relatively small. However, the loss of spatial resolution is more significant. Thus for the N flux estimates to coastal waters calculated using the original method, with TN export to coastal waters calculated from aggregated parish scale TN export estimates, the lower quartile Q1 = 8.14 kg N ha^{-1}, with the upper quartile Q3 = 64.0 kg N ha^{-1}, both for the North Atlantic, represent an interquartile range of 49.6 kg N ha^{-1} (Table 3, row 2, columns 2 and 7). For N flux estimates where parish scale TN export estimates were lumped into catchment units, those were lumped into major catchments and the results lumped into the 3 major drainage units, the lower quartile for North Sea drainages Q1 = 21.6 kg N ha^{-1}, with the upper quartile Q3 = 29.2 kg N ha-1, represent an interquartile range of 7.6 kg N ha^{-1} (Table 3, row 14, columns 2 and 7). Thus by spatial aggregation in the export coefficient model there was 7-fold decrease in spatial discrimination in the model. The impact of these procedures on the range of variation in N flux rates estimated by the model is summarised as a % change in N flux estimates for the interquartile range in Table 4. This suggests that where an estimate of the bulk N flux from England and Wales to its coastal waters is required, this can be estimated within a maximum error of 8% by lumping the parish scale input data into each of the 3 major drainage units, and running these summary data through the original export coefficient model. However, the loss of spatial resolution leads to an overestimate of the lower quartile of 212% and an underestimate of the upper quartile of 60.3% using this method (Table 4, row 23, columns 2 and 7). If an indication of the spatial origins and delivery zones is also required of the model, then the model needs to be run at the parish scale or at least with parish data lumped within individual watersheds and then run through the original model to retain credible spatial resolution. The impact of spatial aggregation on the range of variation in N flux estimates follows the same pattern as on the mean estimate of N flux estimate at each stage. Greater errors are generated where more steps are involved in modelling and aggregation, owing to the issue of boundary intersection and the problems associated with apportioning modelled data to output polygons. Errors are also higher where parish scale data were lumped into larger spatial units and then modelled than where the model was run at parish scale and the parish scale N flux estimates aggregated into larger spatial units. This probably reflects the problem associated with missing the coincidences between low nutrient retention capacity and high intensity agricultural production which would otherwise be captured by more accurate apportioning of input data to geoclimatic regions classes at the parish scale.

Two conclusions may be drawn from this analysis. First, it is intuitive that there will be a loss of spatial resolution when lumping export data from the parish scale to the major drainage scale, and the model cannot be expected to generated reliable estimates of N flux rates in small watersheds based on input data averaged at a larger scale. Second, and more importantly, is the fact that provided the landscape unitary approach is retained, then predicted regional flux rates do not change dramatically as inputs are lumped at progressively larger scales. This suggests that scaling up of such simple empirical watershed scale models to continental scale might be possible and provide a robust tool for generating regional to global scale N flux estimates in the future.

Results of conceptual lumping

Figures 11 to 14 illustrate row one of each of the tables (Lumped output (Parishes)) and show the effects of conceptual lumping on modelled estimates of N flux to coastal waters modelled at the parish scale, with parish scale TN export estimates then lumped to the 3 major drainage units draining to coastal waters. The original model parameterisation was modified by first reducing the number of land use units from 7 to 4 (amalgamated input data), and then to 2 (coarsely amalgamated input data), and then by removing the landscape unitary approach underpinning the model by running the model using only 1 set of export coefficients (originally those representing the mixed arable and dairying regions underlain by permeable bedrock as those reflecting the most typical agricultural practices across England and Wales). In this analysis the full range of livestock units (cattle, pigs, sheep, poultry) were maintained in each step, essentially because the only way to normalise the livestock populations would be on the basis of the per capita N production rate of each livestock type, and this would then preclude inclusion of issues relating to manure handling and stock management. Also, in U.K. conditions it would be conceptually impossible to exclude the livestock population from any valid representation of the sources of aquatic and atmospheric N flux. Conservatively estimated, livestock contribute approximately half of the total N flux

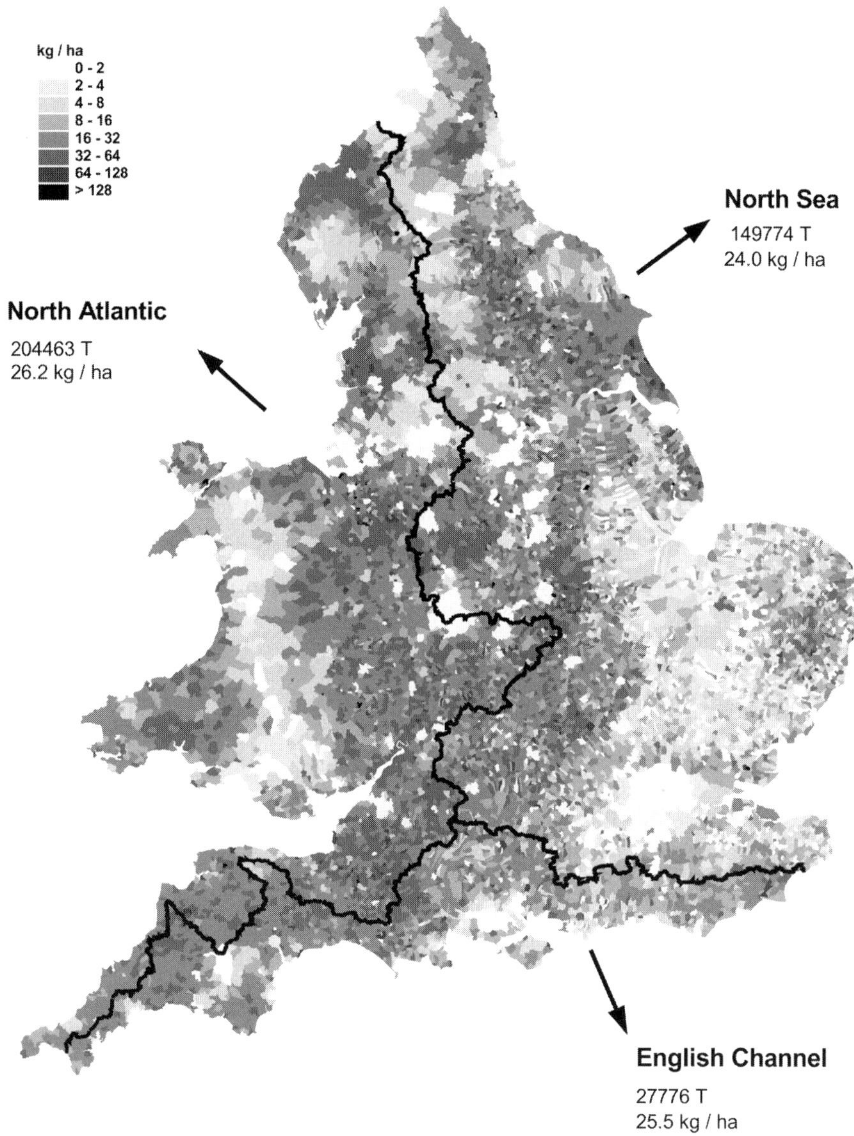

Figure 11. Export coefficient model prediction of TN export (1991) using 6 spatially distributed sets of export coefficients and amalgamated parish scale input data. TN export to coastal waters calculated from aggregated parish scale TN export estimates.

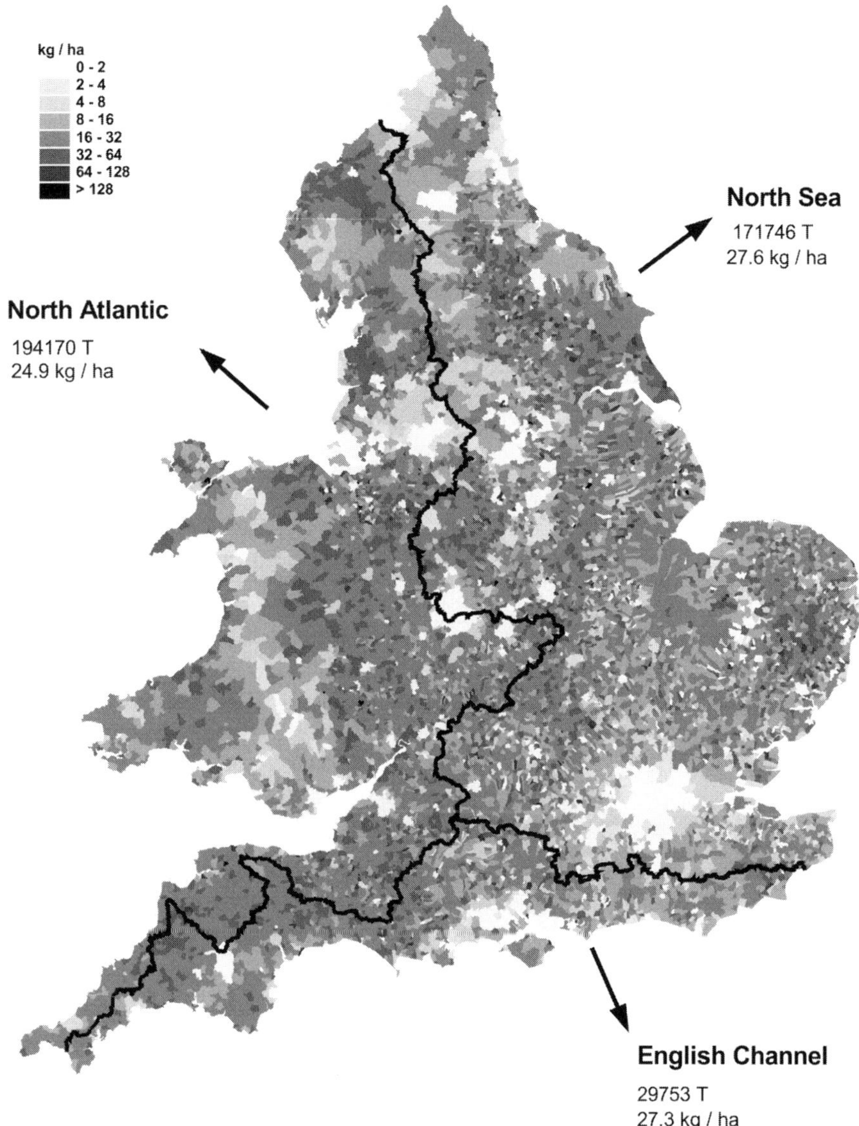

Figure 12. Export coefficient model prediction of TN export (1991) using 1 set of spatially aggregated export coefficients and amalgamated parish scale input data. TN export to coastal waters calculated from aggregated parish scale TN export estimates.

468

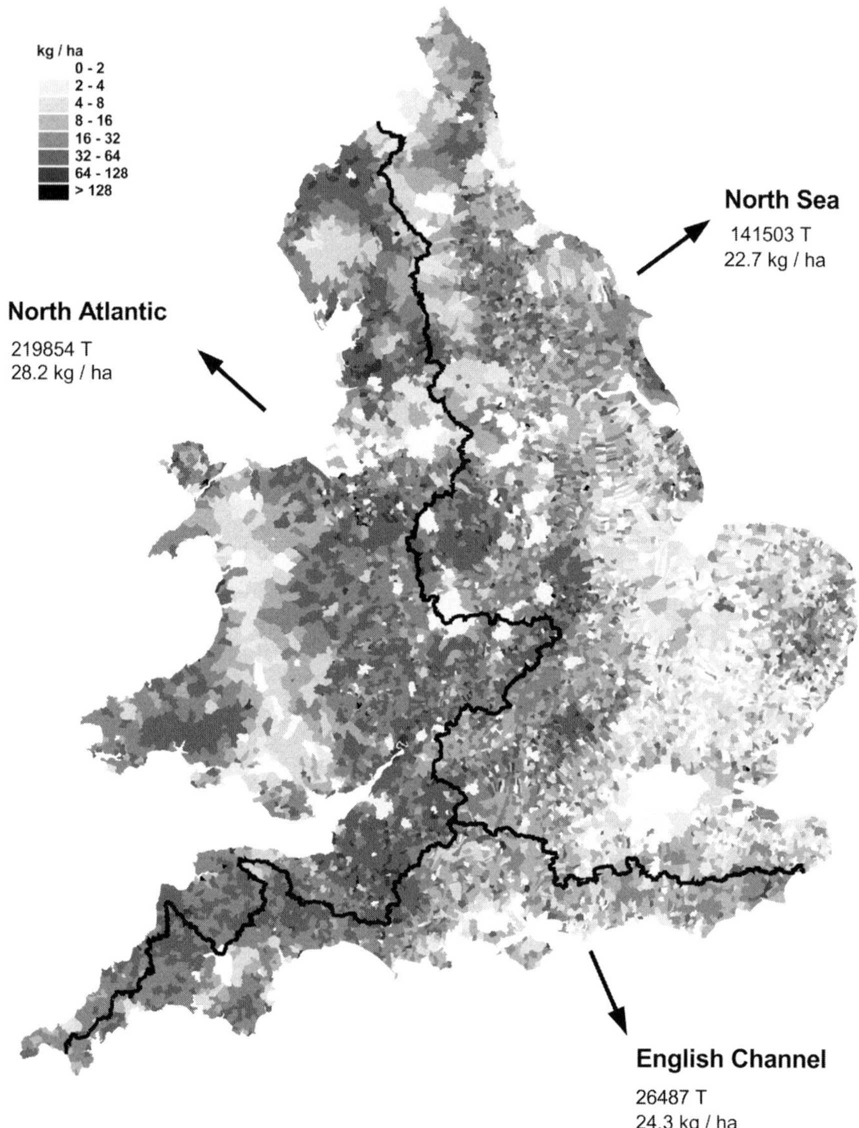

Figure 13. Export coefficient model prediction of TN export (1991) using 6 spatially distributed sets of export coefficients and coarsely amalgamated parish scale input data. TN export to coastal waters calculated from aggregated parish scale TN export estimates.

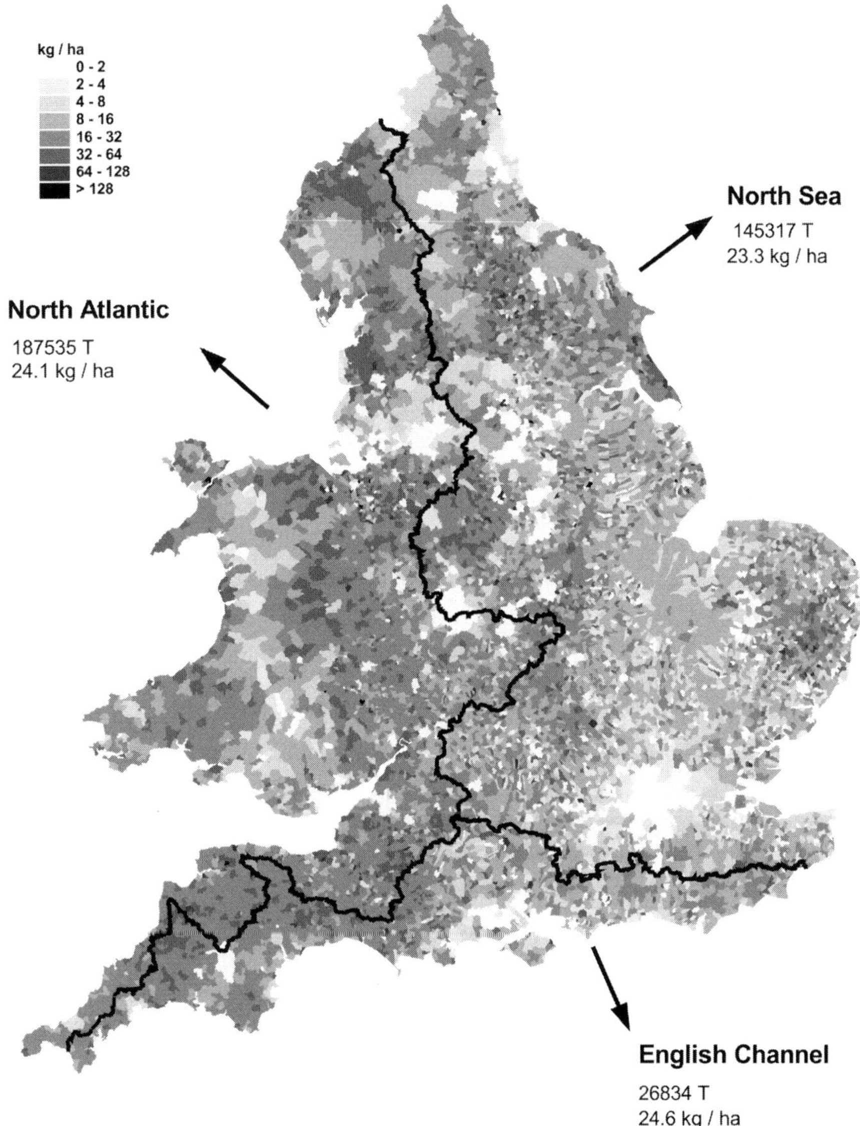

Figure 14. Export coefficient model prediction of TN export (1991) using 1 set of spatially aggregated export coefficients and coarsely amalgamated parish scale input data. TN export to coastal waters calculated from aggregated parish scale TN export estimates.

from England and Wales to coastal waters. Having accepted that they should not be excluded from the analysis, it was then impossible to generate any coarser units for livestock numbers beyond the species units already utilised in the original export coefficient model. The fact that livestock are accounted for on the same basis in each stage of conceptual lumping has an impact of the level of error generated from coarsening model structure with respect to the remaining land use units. Nevertheless, what is apparent from this analysis is that even with coarsely amalgamated land use units (agricultural and non-agricultural land) the model estimates of mean annual N flux to coastal waters have a maximum error of less than 10%, with the greatest errors generated for the North Sea drainage unit, and the lowest errors generated for N flux estimates to the North Atlantic. This reflects the fact that a much greater proportion of N flux to the North Atlantic is contributed by livestock and atmospheric N deposition and much less from fertiliser applications to agricultural land than in the North Sea drainage unit. The small error that results for the North Atlantic drainage unit reflects the predominance of livestock wastes and atmospheric N deposition in the N budget for this region and the fact that conceptual lumping was not possible for either of these source categories. Thus modifications to model parameterisation have a spatially variable impact, relative to the dominant nutrient sources in each drainage unit. The range of variation in N flux estimates is also little affected by coarsening the input data units for land use, with maximum errors of 3.56% in the lower quartile (Q1) and 7.50% in the upper quartile (Q3) resulting from using the original model, applied at parish scale, and run using coarsely amalgamated input data (see Table 4, row 2, columns 5 and 10).

A greater error is apparent where the landscape sensitivity of the model (the geoclimatic regions sub-model division) is removed, both in terms of the error associated with the mean annual estimate of N flux to coastal waters and in the range of variation in these estimates. This is most apparent in Figures 12 and 14 when compared with Figure 3 (original method). N flux estimates from parishes across the U.K. become homogenised, with marked increases in N flux estimates from upland areas such as in Wales and northern England, and in East Anglia, and lower N flux estimates for the south west. The order of errors associated with this coarsening of model structure are greater than for any of the categories of spatial aggregation in terms of the mean N flux estimates generated for the three major drainage units. They are also greater than the effects of amalgamation of input data units, particularly in terms of the range of variation in parish scale N flux estimates. The maximum error associated with removal of the landscape sensitivity of the model is an underestimate of 22% of the upper quartile Q3, and an overestimate of 12.2% of the lower quartile Q1. Thus the effect of removing landscape sensitivity from the export coefficient model is to reduce the range of variation in N flux estimates around the mean.

Error propagation in the export coefficient model and implications for regional to global scale N flux modelling

The overall effect of modifying the export coefficient model to make it more similar in terms of parameterisation and form to the existing regional to global scale models is illustrated where both spatial aggregation and conceptual lumping are combined. None of the existing regional to global scale models contains input data at less than major watershed scale, and only the Howarth et al. (1996) model contains any reference to livestock as a nutrient source category (implicitly accounted for in their model through estimates of food and feed imports by region). Thus, if livestock were to be excluded from the export coefficient model estimates of N flux to coastal waters then the model would underestimate by approximately 50%, before any spatial aggregation or conceptual lumping errors were introduced. The lack of adequate accounting for livestock as a N flux source in any of the existing regional to global scale N flux models generates significant uncertainty in model predictions, not least for subsistence economies where livestock and human wastes will be the dominant sources of N input to the land surface.

Our work suggests that the rates of riverine N flux estimated for England and Wales by simple, global scale regression type models may be significantly lower than is suggested by a validated catchment scale model which explicitly takes account of landscape sensitivity to N export. The reasons for this derive from two sources. First there will be errors in the input data used to drive these models. This reflects in part the fact that none of the existing regional to global scale models seem to provide an adequate accounting for the impact of livestock and livestock wastes on N cycling and flux, meaning that they will systematically underestimate N flux rates from land to ocean where livestock are a dominant part of the farm economy. Second there is the issue of the insensitivity of these simple regression models to spatial variations in the intrinsic nutrient retention capacity of smaller landscape units, as described by variations in the routing and efficiency of runoff from land to water. This may suggest that even if the global N flux from land to ocean predicted by these simple models is correct the N flux estimates at specific points of discharge to coastal waters will be incorrect for many oceans.

In terms of the effect of modelling N flux from land to water as a function of landscape sensitivity to N cycling processes, it is apparent from this analysis that the errors associated with bulk N flux estimates to coastal waters can be predicted within 10% of the original model estimates for the three major drainage units if the model is run at parish scale, with a maximum error of 21.6% associated with under-prediction of the upper quartile of the range for parish scale estimates. However, where it is run at major catchment or major drainage unit scale, the errors associated with the interquartile range are significant, with the lower quartile overestimated by over 200% and the upper quartile underestimated by over 60% when parish scale input data are lumped into the 3 major drainage units and then modelled. The message here seems to be that the incorporation of landscape sensitivity within national model structure is possible, defensible and vital to the production of sensible and robust estimates of N flux to estuaries and coastal waters. The way in which this has been incorporated in the export coefficient model, with definition of characteristic geoclimatic regions, allows account to be taken of regional scale variations in the sensitivity of the landscape to N cycling and flux. Existing regional to global scale models lacking this refinement have an additional element of uncertainty associated with their N flux predictions. Without this element within the model structure, the export coefficient model, at least, would provide an inaccurate and unreliable indication of the likely biogeochemical response to N loading on coastal waters.

Thus the conclusions of Becker and Braun (1999) regarding the necessity for subdivision of the land surface into smaller units displaying homogenous hydrological behaviour seem to hold true for the modelling of N flux to coastal waters at watershed to regional scale. It is also apparent from this analysis that the geoclimatic units defined within the export coefficient model need to be modelled separately using unit-specific export coefficient values, with the calculated unit N fluxes then aggregated to watershed or major drainage unit scale. This analysis also supports the arguments of Addiscott and Tuck (1996) and clearly demonstrates the point that averaging or lumping parish scale data into larger spatial units before running the export coefficient model does not give the same result as running the model and then averaging or lumping the resultant N flux estimates. The latter procedure generates fewer errors, both in terms of the mean and range of variation in N flux estimates generated by the model. Sivapalan and Kalma (1995) discussed the differences generated in model estimates as a result of lumping the entire mosaic of units across a landscape, as opposed to representing the land surface as a combination of units acting in parallel. This analysis has confirmed the principle that this generated different modelling outcomes and is thus a source of further error and uncertainty in model estimates. Thus as we scale up from parish scale to major watershed or landscape unit scale, in U.K. terms, the sequence of modelling steps does appear to be critical in determining the level of error and uncertainty associated with our ultimate regional scale N flux estimates. The principles that have emerged from this analysis have highlighted the areas where greatest error and uncertainty are introduced to regional scale N flux estimates. These principles can provide guidance on the steps necessary to minimise error propagation in the future development of regional to global scale N flux models.

If we accept, from this analysis, that regional to global scale modelling can be improved by adopting the principles emerging from watershed scale N flux modelling and from macroscale modelling in other related disciplines, then appropriate parameterisation of the models relative to the spatial scale at which the problem needs to be described will need to be re-visited. Further progress in regional to global scale N flux modelling will only be made if we build models not on a national or major watershed basis, but explicitly on a landscape unitary basis, taking account of the natural environmental controls of N flux rates as they vary between units. Another issue which then arises is how the landscape units can be determined for regional to global scale N flux modelling, or at what level of spatial resolution the landscape units should be defined. As discussed earlier, many of the national or major watershed data sets used by the existing regional to global scale N flux models have constrained these models to run at national or major watershed scale. However, many of the data required are available in national databases at a sub-national scale, or may be provided from remotely sensed images for certain model parameters. The current lack of readily available databases at landscape scale should not in itself limit the development of more accurate and more spatially explicit regional to global N flux models in the future.

Acknowledgements

This work was initiated as part of the International SCOPE N Project, which received support from both the Mellon Foundation, Cornell University and from the National Center for Ecological Analysis and Synthesis, Santa Barbara, California.

References

Addiscott TM (1998) Modelling concepts and their relation to the scale of the problem. Nutrient Cycling in Agroecosystems 50: 239–245

Addiscott TM & Tuck G (1996) Sensitivity analysis for regional scale solute modeling. In: Corwin DL & Wagenet RJ (Eds) Applications of GIS to the Modelling of Non-point Source Pollutants in the Vadose Zone (pp 153–162). Special Publication 48, Soil Science Society of America

Addiscott TM & Wagenet RJ (1985) A simple method for combining soil properties that show variability. Soil Sci. Soc. Am. J. 49: 1365–1369

Alexander RB, Johnes PJ, Boyer EA & Smith RA (2002) A comparison of models for estimating the riverine export of nitrogen from large watersheds. Biogeochemistry 57/58: 295–339

Arnell NW (1999) A simple water balance model for the simulation of streamflow over a large geographic domain. Journal of Hydrology 217: 314–335

Becker A & Braun P (1999) Disaggregation and spatial scaling in hydrological modelling. Journal of Hydrology 217: 239–252

Bergström S & Graham LP (1998) On the scale problem in hydrological modelling. Journal of Hydrology 211: 253–265

Blöschl G & Sivapalan M (1995) Scale issues in hydrological modelling: a review. Hydrological Processes 9: 251–290

Boulet G, Kalma JD, Braud I & Vaudin M (1999) An assessment of effective land surface parameterisation in regional-scale water balance studies. Journal of Hydrology 217: 225–238

Boyer EA, Goodale CL, Jaworski NA, Hetling L & Howarth RW (2002) Anthropogenic nitrogen sources and relationships to riverine nitrogen export in the northeastern U.S.A. Biogeochemistry 57/58: 137–169

Braud I, Dantas-Antonino AC & Vauclin M (1995) A stochastic approach to studying the influence of the spatial variability of soil hydraulic properties on surface fluxes, temperature and humidity. Journal of Hydrology 165: 283–310

Caraco NF & Cole JJ (1999) Human impact on nitrate export: an analysis using major world rivers. Ambio 28(2): 167–170

Entekhabi D & Eagleson PS (1989) Land surface hydrology parameterization for atmospheric general circulation models including subgrid spatial variability. Journal of Climate 2: 816–831

Famiglietti JS & Wood EF (1995) Effects of spatial variability and scale on areally averaged evapotranspiration. Water Resources Research 31(3): 699–712

Federer CA, Vörösmarty CJ & Fekete B (1996) Intercomparison of methods for calculating potential evaporation in regional and global water balance models. Water Resources Research 32(7): 2315–2321

Heathwaite AL & Johnes PJ (1996) Contribution of nitrogen species and phosphorus fractions to stream water quality in agricultural catchments. Hydrological Processes 10: 971–983

Howarth RW, Billen G, Swaney D, Townsend A, Jaworski N, Downing JA, Elmgren R, Caraco N & Lajtha K (1996) Regional nitrogen budgets and riverine N and P fluxes for the drainages to the North Atlantic Ocean: natural and human influences. Biogeochemistry 35: 75–139

Jaworski NA, Howarth RW & Hetling LJ (1997) Atmospheric deposition of nitrogen oxides onto the landscape contributes to coastal eutrophication in the Northeast United States. Environ. Sci. Technol. 31: 1995–2004

Johnes PJ (2000) Quantifying the non-point source contribution to nutrient loading on freshwaters in 32 U.K. catchments. Verh. Int. Verein. Limnol. 27: 1306–1309

Johnes PJ, Fraser A, Harrod T, Butterfield D & Withers PJ (2000) Predicting phosphorus loss from agriculture to water. Environmental R&D Newsletter 6, 9, Ministry of Agriculture, Fisheries and Food, London

Johnes PJ (1999) Understanding catchment history as a tool for integrated lake and catchment management. Hydrobiologia 395/396: 41–60

Johnes PJ & Hodgkinson RA (1998) Phosphorus loss from catchments: pathways and implications for management. Soil Use & Management 14: 175–185

Johnes PJ, Curtis C, Moss B, Whitehead P, Bennion HB & Patrick S (1998a) Trial Classification of Lake Water Quality in England and Wales: a proposed approach. R&D Technical Report E53, Environment Agency, Bristol

Johnes PJ, Bennion HB, Curtis C, Moss B, Whitehead P & Patrick S (1998b) Trial Classification of Lake Water Quality in England and Wales: a proposed approach. R&D Project Record E2-i721/5, Environment Agency, Bristol

Johnes PJ & Heathwaite AL (1997) Modelling the impact of land use change on water quality in agricultural catchments. Hydrological Processes 11: 269–286

Johnes PJ, Moss B & Phillips GL (1996) The determination of water quality by land use, livestock numbers and population data – testing of a model for use in conservation and water quality management. Freshwater Biology 36: 451–473

Johnes PJ (1996) Evaluation and management of the impact of land use change on the nitrogen and phosphorus load delivered to surface waters: the export coefficient modelling approach. Journal of Hydrology 183: 323–349

Johnes PJ & Burt TP (1991) Water quality trends and land use effects in the Windrush catchment: nitrogen speciation and sediment interactions, IAHS 203: 349–357

Larsson U, Elmgren R & Wulff F (1985) Eutrophication and the Baltic Sea: causes and consequences. Ambio 14: 9–14

Law CS, Rees AP & Owens NJP (1992) Nitrous oxide: estuarine sources and atmospheric flux. Estuarine Coastal Shelf Sci. 33: 301–314

Meybeck M (1982) Carbon, nitrogen and phosphorus transport by world rivers. Am. J. Sci. 282: 410–450

Nixon SW (1995) Coastal marine eutrophication: a definition, social causes and future problems. Ophelia 41: 199–219

Peierls BL, Caraco NF, Pace ML & Cole JJ (1991) Human influence on river nitrogen. Nature 350: 386–387

Seitzinger SP & Kroeze C (1998) Global Distribution of nitrous oxide production and N inputs in freshwater and coastal marine ecosystems. Global Biogeochemical Cycles 12(1): 93–113

Sivapalan M & Kalma JD (1995) Scale problems in hydrology: contributions of the Robertson Workshop. Hydrological Processes 9: 243–250

Smith RA, Schwarz GE & Alexander RB (1997) Regional interpretation of water quality monitoring data. Water Resources Research 33: 2781–2798

Stein A, Staritsky J, Bouma J, van Eijnsbergen AC & Bregt AK (1992) Simulation of moisture deficits and real interpolation by universal cokriging. Water Resources Research 27: 1963–1973

Troutman BM (1983) Runoff predictions, errors and bias in parameter estimation induced by spatial variability of precipitation. Water Resources Research 19(3): 791–810

Turner RE & Rabelais NN (1991) Changes in the Mississippi River water quality this century. Bioscience 41: 140–147

Uncles RJ, Fraser AI, Butterfield D, Johnes PJ & Harrod TR (in press) The prediction of nutrients into estuaries and their subsequent behaviour. Hydrobiologia: (in press)

Valiela I, Bowen JL & Kroeger KD (submitted) Assessment of models for estimation of land-derived nitrogen loads to shallow estuaries. Applied Geochemistry, submitted February 2001

van Breemen N, Boyer EA, Goodale CL, Jaworski NA, Paustian K, Seitzinger SP, Lajtha K, Mayer B, van Dam D, Howarth RW, Nadelhoffer KJ, Eve M & Billen G (2002) Where did all the nitrogen go? Fate of nitrogen inputs to large watersheds in the northeastern U.S.A. Biogeochemistry 57/58: 267–293

Vitousek PM, Aber JD, Howarth RW, Likens GE, Matson PA, Schindler DW, Schlesinger WH & Tilman DG (1997) Human alteration of the global nitrogen cycle: sources and consequences. Ecological Applications 7: 737–750

Vörösmarty CJ, Federer CA & Schloss AL (1998) Potential evaporation functions compared on U.S. watersheds: possible implications for global-scale water balance and terrestrial ecosystem modeling. Journal of Hydrology 207: 147–169

Whitehead PG, Wilson EJ & Butterfield D (1998) A semi-distributed Integrated Nitrogen model for multiple source assessment in Catchments (INCA): Part 1 – model structure and process equations. Sci. Tot. Environ. 210–211: 547–558

Biogeochemistry **57/58**: 477–516, 2002.
© 2002 *Kluwer Academic Publishers. Printed in the Netherlands.*

Policy implications of human-accelerated nitrogen cycling[†]

ARVIN R. MOSIER[1]*, MARINA AZZAROLI BLEKEN[2], PORNPIMOL CHAIWANAKUPT[3], ERLE C. ELLIS[4], JOHN R. FRENEY[5], RICHARD B. HOWARTH[6], PAMELA A. MATSON[7], KATSUYUKI MINAMI[8], ROZ NAYLOR[7], KIRSTIN N. WEEKS[6] & ZHAO-LIANG ZHU[9]
[1]*USDA/ARS, Fort Collins, CO 80522, U.S.A.;* [2]*Agricultural University of Norway, Aas, Norway;* [3]*Thailand Department of Agriculture, Bangkok, Thailand;* [4]*Center for Agroecology and Sustainable Food Systems, University of California, Santa Cruz, CA, U.S.A.;* [5]*CSIRO, Canberra, ACT, Australia;* [6]*Dartmouth College, Hanover, NH, U.S.A.;* [7]*Stanford University, Palo Alto, CA, U.S.A.;* [8]*NIAES, Tsukuba, Japan;* [9]*Chinese Academy of Science, Institute of Soil Science, Nanjing, China (*author for correspondence, e-mail: amosier@lamar.colostate.edu)*

Key words: fertilizer, food production, fossil fuel combustion, mitigation, N, NO_x, N_2O

Abstract. The human induced input of reactive N into the global biosphere has increased to approximately 150 Tg N each year and is expected to continue to increase for the foreseeable future. The need to feed ($\sim$125 Tg N) and to provide energy ($\sim$25 Tg N) for the growing world population drives this trend. This increase in reactive N comes at, in some instances, significant costs to society through increased emissions of NO_x, NH_3, N_2O and NO_3^- and deposition of NO_y and NH_x.

In the atmosphere, increases in tropospheric ozone and acid deposition (NO_y and NH_x) have led to acidification of aquatic and soil systems and to reductions in forest and crop system production. Changes in aquatic systems as a result of nitrate leaching have led to decreased drinking water quality, eutrophication, hypoxia and decreases in aquatic plant diversity, for example. On the other hand, increased deposition of biologically available N may have increased forest biomass production and may have contributed to increased storage of atmospheric CO_2 in plant and soils. Most importantly, synthetic production of fertilizer N has contributed greatly to the remarkable increase in food production that has taken place during the past 50 years.

The development of policy to control unwanted reactive N release is difficult because much of the reactive N release is related to food and energy production and reactive N species can be transported great distances in the atmosphere and in aquatic systems. There are many possibilities for limiting reactive N emissions from fuel combustion, and in fact, great strides have been made during the past decades. Reducing the introduction of new reactive N and in curtailing the movement of this N in food production is even more difficult. The particular problem comes from the fact that most of the N that is introduced into the global food production system is not converted into usable product, but rather reenters the biosphere as a

---

[†] Reprinted from *Biogeochemistry* Volume 52, pp. 281–320, 2001.

surplus. Global policy on N in agriculture is difficult because many countries need to increase food production to raise nutritional levels or to keep up with population growth, which may require increased use of N fertilizers. Although N cycling occurs at regional and global scales, policies are implemented and enforced at the national or provincial/state levels. Multinational efforts to control N loss to the environment are surely needed, but these efforts will require commitments from individual countries and the policy-makers within those countries.

Introduction

This paper provides a view of some of the complexities of release of newly fixed, reactive nitrogen and national and international environmental policy. To introduce the topic we briefly discuss some of the issues related to human-induced changes to the nitrogen cycle and how these changes relate to policy issues. We first look at some of the main concerns that result from introduction of reactive N into the biosphere through food production and fossil fuel consumption. We discuss some of the changes in N use distribution globally and the impact on reactive N production of changes in human diet. We also explore how the decoupling of cereal grain and livestock production systems contribute to increased reactive N production and changes in regional and global redistribution of reactive N. In the final section we discuss policy approaches related to N use in agriculture and in fossil fuel consumption.

Characteristics of change in the nitrogen cycle

Nitrogen regulates numerous essential ecological and biogeochemical processes, including species composition, diversity, population growth and dynamics, productivity, decomposition, atmospheric chemistry, and nutrient cycling of many terrestrial, freshwater, and marine ecosystems. While human activities have altered the N cycle in a number of ways, the most fundamental change is the dramatic increase in biologically available N, also termed 'reactive N'. Human activities have more than doubled the rate of transfer of N from the highly abundant but biologically unavailable form di-nitrogen (N_2) in the atmosphere to available forms such as ammonium (NH_4^+), nitrite (NO_2^-), and nitrate (NO_3^-) in the biosphere (Smil 1999; Vitousek et al. 1997; Galloway et al. 1995; Vitousek & Matson 1993).

Prior to extensive human alteration, the primary pathways for transfer from inert to available forms of N were biological N fixation by specialized bacteria (accounting for around 100 Tg y^{-1} in terrestrial ecosystems and 30–300 Tg y^{-1} in marine systems) and lightning fixation (accounting for up to 10 Tg y^{-1}) (Vitousek et al. 1997). A number of anthropogenic pathways have now more than doubled the amount of reactive N coming into the

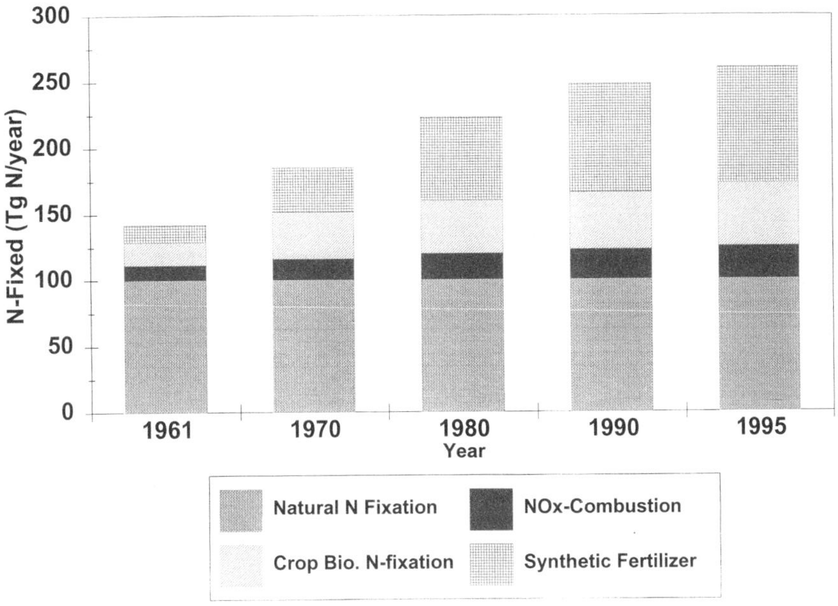

Figure 1. An estimate of global N fixation for 1961–1995 (FAO 1999; Galloway et al. 1995).

biosphere each year. These pathways include industrial fixation of N for use as fertilizers, cultivation of crops that fix N symbiotically, and mobilization and fixation during fossil fuel combustion, and have been the subject of a number of thorough reviews (e.g., Smil 1990, 1991, 1999; Vitousek & Matson 1993; Ayers et al. 1994; Galloway et al. 1995; Vitousek et al. 1997).

Current industrial fixation of N for fertilizer is approximately 85 Tg y^{-1}, a dramatic increase since 1961 (Figure 1). Until the late 1970s, most of the industrial N fertilizer was applied in developed countries, but use there has stabilized or declined while it has increased in developing countries (Figure 3). Animal manure and other organic residues are also used as fertilizer in crop production world-wide, and account for more N application than does industrial fertilizer (Bouwman 1997). Manure and organic residues represent recycling of already fixed N rather than new fixation of N; nevertheless, their management can have significant effects on the mobility of N and contribute to alterations in biospheric processes due to N use in human activities.

In addition to the use of fertilizers, agriculture has increased available N via the production of leguminous crops. Leguminous crops and forages, such as soybeans and alfalfa, support symbiotic N-fixing organisms, and their fixation rates typically far outstrip rates found in the natural systems that the crops replaced. Galloway et al. (1995) estimate that between 32 and 53 Tg N is fixed by crops annually.

The third major anthropogenic source of available N is via the burning of fossil fuels. Fossil fuel combustion inadvertently oxidizes atmospheric N_2 and also transfers some small amount of N from geologic reservoirs to available forms. Approximately 21 Tg N y^{-1} is converted to nitrogen oxides, which then react in the atmosphere and/or deposit to downwind ecosystems in gaseous, solution or particulate forms (Delmas et al. 1997).

Altogether, human activities cause the fixation of approximately 150 Tg N y^{-1} in terrestrial ecosystems, equivalent to biological N fixation by non-anthropogenic processes on land. In addition, land clearing, biomass burning, and drainage of wetlands all contribute substantially to the mobility of N in the biosphere, leading to reductions in long-term storage of N in soil organic matter and vegetation, and increasing fluxes to air and water.

These changes in available N, caused both directly and indirectly by human activities, affect the biosphere in many ways. They are associated with increased emission, transport, and deposition of a number of trace gases, including nitric oxide (NO) and ammonia (NH_3), both involved in air chemistry and downwind deposition, and nitrous oxide (N_2O), a greenhouse gas. Anthropogenic activities contribute significantly to all three of these gases. Deposition of nitrogen oxides, ammonia, organic N compounds, and other forms of N have consequences for downwind terrestrial and aquatic ecosystems, and may lead to increased production and decomposition, changes in species composition and biodiversity, and ultimately N saturation, with declining production and C storage and accelerated N losses (Aber et al. 1995, 1998; Galloway et al. 1995; Vitousek et al. 1997).

These anthropogenic changes also have led to major increases in N loading of aquatic systems over time. For example, nitrate concentrations in major rivers of the northeastern US have increased 3- to 10-fold since the early 20th century. Howarth et al. (1996) suggest that total riverine fluxes from most of the temperate zone land systems surrounding the North Atlantic have increased 2- to 20-fold since pre-industrial times. These elevated nitrogen concentrations and fluxes hold human health concerns due to pollution of drinking water, and also affect downstream ecosystems through acidification and eutrophication (Vitousek et al. 1997).

Considerations for policy development

Clearly, the dramatic changes in the global N cycle outlined above hold significant consequences for the way the Earth system functions. Increased nitrogen availability at the global scale has affected and will continue to affect terrestrial and aquatic systems alike. Some of those consequences are playing out as air (ozone formation due to NO_x emissions from soils) and water pollution problems (enrichment in nitrate from leaching and runoff from cropped field and domestic lawns) at local scales; others are reflected in regional changes in net primary production and eutrophication; still others are global in scale. As the source of multi-dimensional environmental problems, global change in the N cycle is drawing interest among managers and policy makers. There are, however, a number of characteristics and dynamics of this change that make the problems stemming from global change in N especially difficult to solve.

The first and most central characteristic to be considered in policy development is the fact that N use is closely tied to essential human endeavors such as the provision of food and energy. Fertilizer use has been an essential component of the Green Revolution, the set of technologies that dramatically increased food production in developing countries during the period between 1960 and 1980. Many areas of the world still do not use enough fertilizer to maximize crop yields, and most analysts suggest that fertilizer use will continue to grow as food production does, in order to keep pace with a still-rapidly increasing human population. Use of N for agriculture, whether in the form of synthetic or organic fertilizer, is not substitutable, and straightforward technological changes are unlikely to provide a replacement. Plants will always require a relatively large amount of N to carry out their photosynthetic processes, and one key to maintaining adequate food supplies will be supplying plants with adequate N (Matson et al. 1997). This characteristic sets fertilizer N apart from those environmental problems that can be solved by technological substitutions; for example, chlorofluorocarbon (CFC)-caused reduction in stratospheric ozone is being dealt with effectively by substituting non-ozone depleting chemicals in the myriad industrial processes that in the past relied on CFCs.

Global change in the N cycle is also linked to the use of fossil fuel energy, and thus is likely to increase dramatically over the next several decades, unless there is a concerted effort to control fossil fuel consumption. Galloway et al. (1994) suggest that the production of NO_x from fossil fuels will double over the next several decades, reaching approximately 46 Tg y^{-1} by 2020. In this case, technological change that either increases efficiency of fuel combustion or removes nitrogen oxides from the exhaust stream could reduce the total amount of N emitted, but complete solutions are closely linked to the development of non-polluting alternative energy sources.

The second characteristic of N that is critical to policy making is the fact that N is a highly mobile element – as one publication pointed out, it seems to 'hopscotch' around the globe (Galloway et al. 1995), moving through air and water, across political and geographical boundaries. As a result, sources and sinks (or cause and effect) are often widely separated. Thus, eutrophication in the Gulf of Mexico is linked to fertilizer use in the

482

Mississippi Valley (Downing et al. 1999), and N deposition and acid rain in Scandinavia are linked to fossil fuel burning and agriculture in nations to the south (Abrahamsen & Stuanes 1998). One consequence of this mobility is that policies to solve environmental problems associated with it must often be multi-national in scale.

Finally, a third characteristic of N that affects policy making and management approaches is the fact that changes in N are interactive with other global changes. To completely understand the effects of N additions to ecosystems, one must understand how those additions interact with elevated CO_2, with land use change, with biological invasions, and with other biogeochemical changes. For example, attributing forest dieback to N deposition alone has been quite difficult, because many forests are also being subjected to increased exposure to tropospheric ozone. Likewise, attributing increased forest growth to N deposition is complicated by the fact that climate change and elevated CO_2 are happening simultaneously. No policies to date are comprehensive enough to address the multiple and interacting changes that are occurring globally.

Nitrogen use in food production

Trends in fertilizer use and food production

Since 1950, N input into global crop production has greatly increased as have crop production and human population. In 1950 synthetic fertilizer N input comprised ~7% of total N input of ~56 Tg N. In 1996 synthetic N input was ~43% of the total N input (including biological N-fixation in crops) of 190 Tg for global crop production (FAO 1999; IPCC 1997). Animal waste used as fertilizer was an estimated 37 Tg in 1950 compared to ~65 Tg N in 1996. Globally, synthetic fertilizer N consumption is expected to grow relatively constantly at the rate of 1.6 Tg N y^{-1} between 2000 and 2020 (Bumb & Baanante 1996). In 1961 FAO began compiling world population, crop production and fertilizer use statistics (FAO 1999) and since that time global human population increased from ~3.1 billion to ~5.8 billion in 1996 (a 1.9 fold increase and an annual growth rate of ~2.5% y^{-1} using 1961 as the base time) (Figure 2). During this time world cereal grain production increased from ~880 Tg to ~2070 Tg, a 2.5 fold increase representing an annual gain of over 4.1% y^{-1}. Synthetic fertilizer N consumption increased from 11.6 Tg to almost 83 Tg in 1996. This 7.1-fold increase in fertilizer N use was the result of very rapid expansion in use between 1961 and 1980 (an annual increase of ~22%).

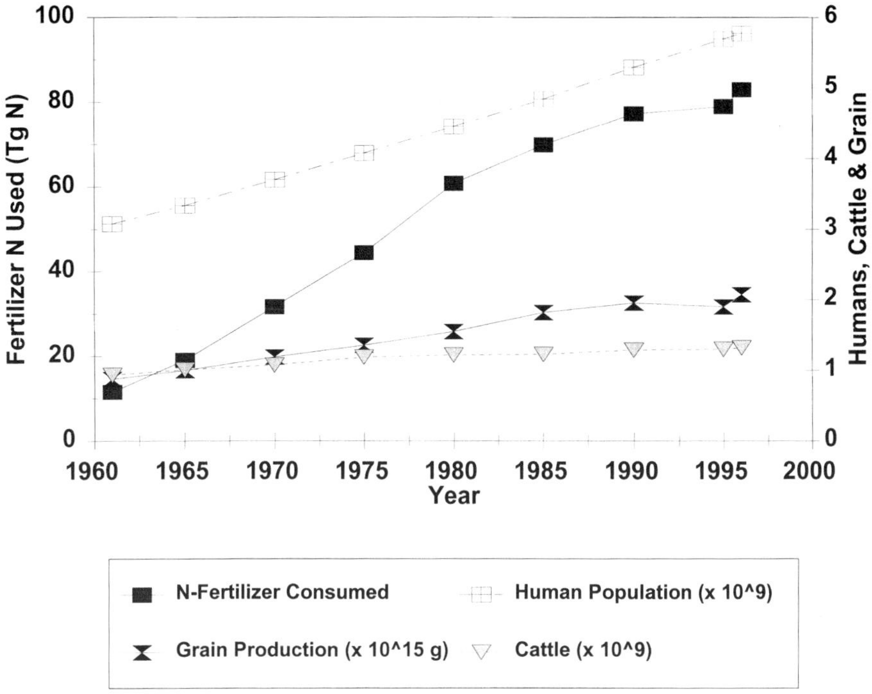

Figure 2. World human population, grain production, N-fertilizer consumption and cattle population (FAO 1999).

The continued increase in food production globally over the past 35 years has been remarkable. Particularly, since land area used to produce grain decreased from 0.2 to 0.12 ha per person between 1965 and 1994 while global grain production per person increased 16%. Agricultural production increase has come through the use of new crop varieties which need increased N-fertilization, pesticide use, irrigation and mechanization. About 40% of the large annual increase in crop production during this 35-y period is attributed to the increase in use of synthetic fertilizer N (Brown 1999).

Even though agricultural production has increased dramatically, fertilizer N use efficiency remains relatively low. Crops typically take up only 40–50% of the total organic and inorganic N added during each cropping season (Bleken & Bakken 1997; NRC 1993; Olsthoorn & Fong 1998). Since fertilizer N is not used efficiently in most parts of the world, N use in excess of crop requirements leads to losses to the environment through volatilization and leaching. Improved efficiency of fertilizer use can be attained, at least in part, through improved management, using current technology, to reduce N input into cropping systems without decreasing production (Peoples et al.

484

1995; Cole et al. 1996; De Jager et al. 1998; Hendriks et al. 1998; Matson et al. 1998; Mosier et al. 1998; Downing et al. 1999)

Shifts in global use of fertilizer N and crop production

The need for continued improvement in management of N in crop production is further enhanced by the changes in the global distribution of N-fertilizer use in the past few decades. Management techniques may or may not be applicable to the vast array of crop production systems in use globally, since much of the research on fertilizer N management techniques has been conducted in the developed part of the world. According to the FAO terminology of developed and developing countries (FAO 1999), fertilizer N use and grain production have declined in developed countries since 1985 (Figure 3). Fertilizer N consumption peaked in the developed part of the world in 1985 at 38.5 Tg and dropped to 28 Tg in 1994. The data are not shown here, but a close look at the use of N in developed countries shows that the decrease is due mainly to lower consumption in Eastern Europe and the former Soviet Union (FAO 1999). Since 1995 N use in North America and other parts of the developed world began to increase again. In the developing part of the world both grain production and N-fertilizer use have continued to increase at near linear rates over the past three decades. Cereal grain production increased from ~400 in 1961 to ~1200 Tg in 1996 while fertilizer N consumption increased from 2.2 to ~53 Tg (a 24 fold increase). On a per capita basis fertilizer N use in the developing world increased from ~1.1 kg N y^{-1} per person in 1961 to 11.6 kg N y^{-1} per person in 1996. During the same period human population in the developing world increased from ~2.1 billion to ~4.6 billion.

Human population in the developed world increased from 0.98 to 1.29 billion while grain production increased from 481 in 1961 to 913 Tg in 1990. In 1996, 1997 and 1998 developed country grain production was 867, 906 and 856 Tg, respectively (FAO 1999). On a per capita basis fertilizer N use in the developed part of the world was 9.6 and 23.3 kg N y^{-1} in 1961 and 1996, respectively.

Increases in N loss resulting from changes in human diet and food production systems

Increased N demand due to changes in human diet

Along with increasing fertilizer N use, continued high intake of animal protein in developed countries and changes in the diet of people in developing

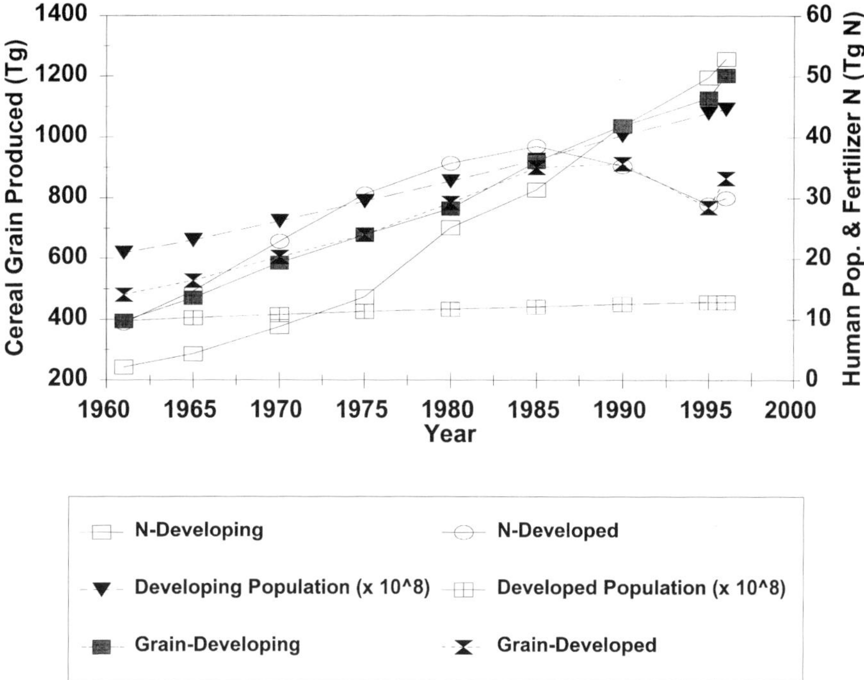

Figure 3. Distribution of N fertilizer use, grain production and human population in developed and developing parts of the world (FAO 1999).

countries will likely lead to greater N losses from global food production in the future. The first aspect of changes in food production concerns increasing meat consumption by people globally and resulting need for increased N input into food production (Figures 3, 4, 5 & 6).

In the developed part of the world total meat and cereal grain production were high in 1980 and have increased slightly since that time (Figures 3 & 4). In contrast, in the developing world, total meat production increased from ~40 Tg in 1980 to more than 110 Tg in 1998 (Figure 4). Grain production increased from ~800 to ~1200 Tg during this time (Figure 3). Improved economic conditions, particularly in East Asia, appear to have stimulated meat consumption at a rate greater than population increase (Figures 3 & 4). Even with the very rapid increase in meat production in the last decade, per capita protein consumption in the developing world remains much lower than in developed countries. Badiane and Delgado (1995) note that increased animal production may be necessary in countries that are poor and have a large rural population, as in Sub-Saharan Africa, to meet food needs and for the economic development of the country.

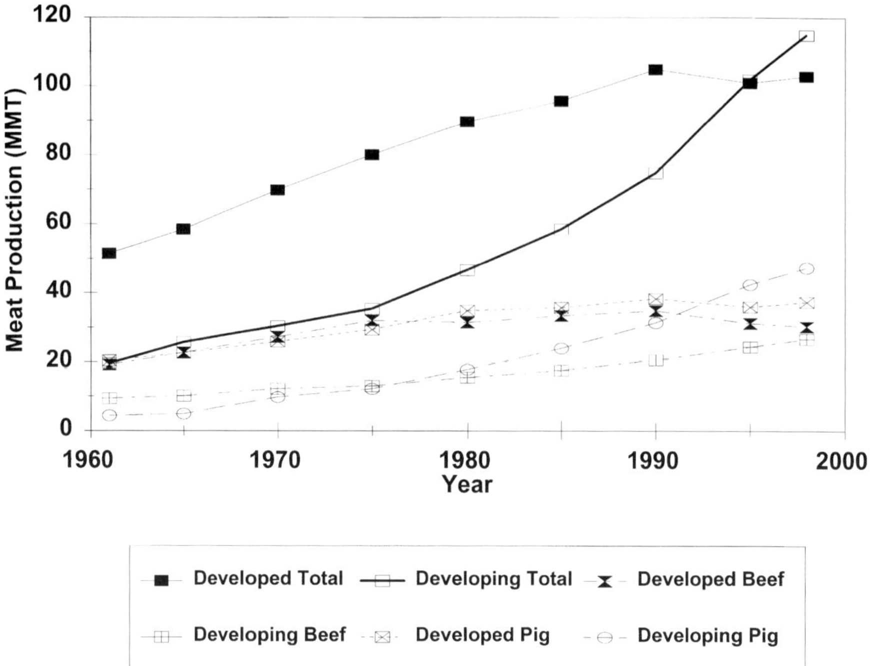

Figure 4. Changes in meat production since 1961 in developed and developing countries (FAO 1999).

The average protein supply per person in the developed countries is presently $\sim$100 g d^{-1}, while in the developing countries it is only $\sim$65 g d^{-1} (Figure 5). Protein is used because there is a direct proportionality between protein and nitrogen composition of food (ca 0.16 g N per 1 g protein). On average in 1995, developed countries consumed $\sim$55% of total protein from animal sources while developing countries derived $\sim$25% of total protein from animals (Figures 5 & 6). Protein consumption was highest in the USA and Western Europe, $\sim$70 and $\sim$60 g animal protein person^{-1} d, respectively. Protein consumption in the Former USSR declined markedly during the past decade, mainly due to the decrease in animal protein. In developing countries, the greatest change in animal protein consumption has occurred in China where the consumption of meat products has increased 3.2 fold since 1980. In Sub-Saharan Africa there has been no increase in either total or animal protein supply during the past 30+ years (Figures 5 & 6).

The reason for focusing on the consumption of animal protein is that more N is needed to produce a unit of animal protein than an equal amount of plant protein. This implies greater N loss from the soil than if crops were used directly as human food. Using information from Western Europe, Bleken and

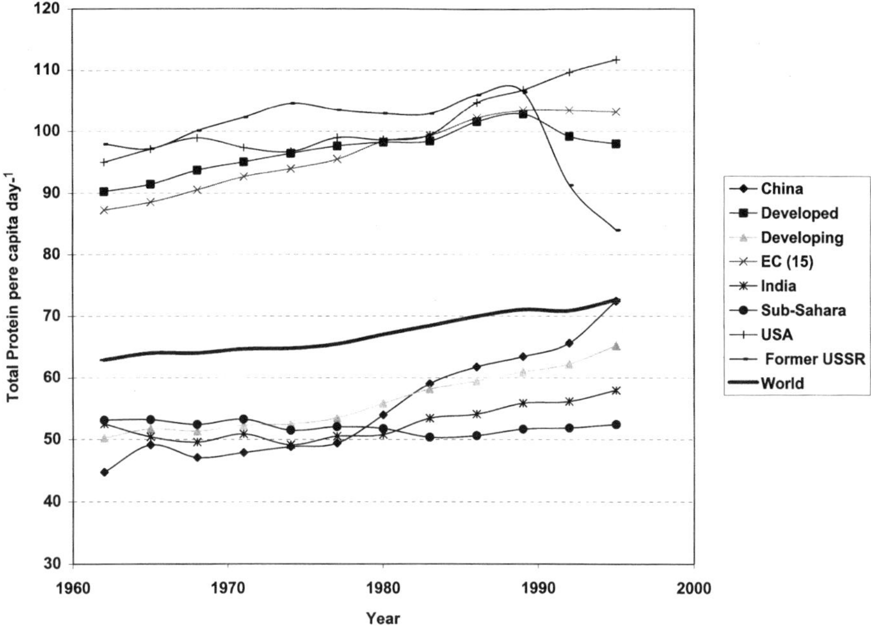

Figure 5. Human consumption of total protein in various parts of the world (FAO 1999).

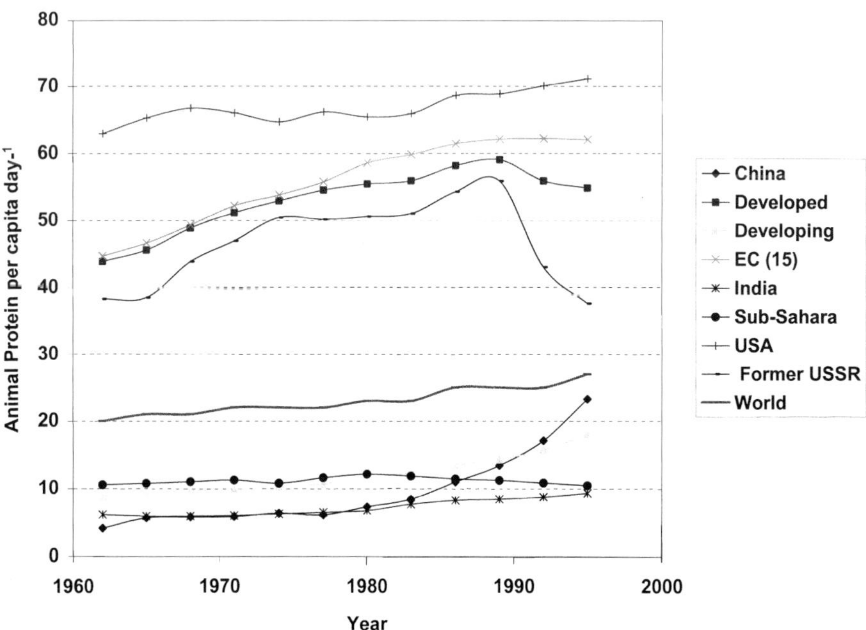

Figure 6. Human consumption of animal protein in various parts of the world (FAO 1999).

488

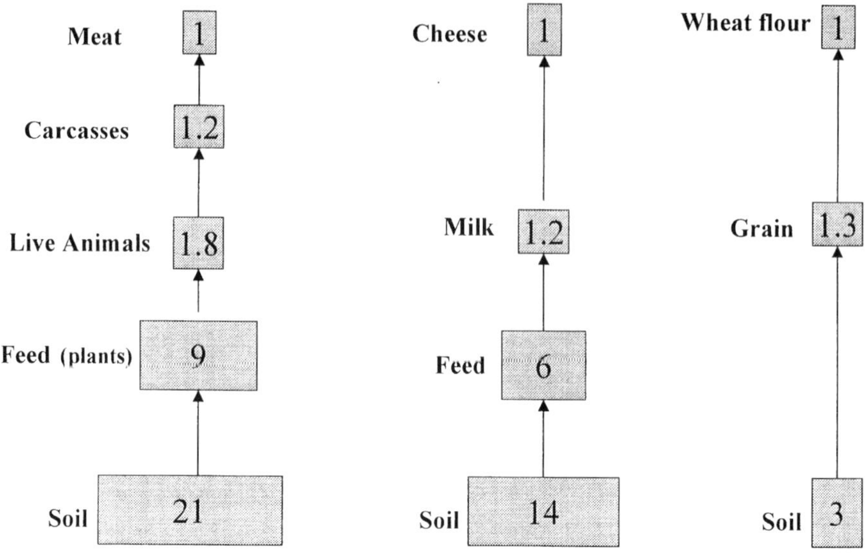

Figure 7. A diagram of flow of reactive N from application to soil through intermediary plant and animal products to produce a unit of edible protein N. The numbers indicate of amount of N. One g N corresponds to ~6.25 g protein, ~35 g meat, ~200 g milk or ~50 g wheat flour (Bleken & Bakken 1997).

Bakken (1997) found that about 3 g N must be supplied to the soil (sum of fertilizer, biological fixation and atmospheric deposition) to produce wheat flour containing 1 g N (~6.3 g protein). In order to obtain an equal amount of animal protein (weighted average of milk and several kinds of meat) the soil must receive 21 g N. Taking into account recycling of manure and recycling of animal products as feed, the net external N input to the agricultural system necessary to produce this amount of edible animal proteins is ~16 g N (Figure 7). This is more than five times the requirement for production of vegetable protein. This illustrates the fact that even though part of the N in animal feed can be recycled for fertilizer in crop production as animal manure, large losses occur from volatilization of NH_3 and from nitrate leaching and runoff and N_2O emissions following mineralization. Based upon the extra N required to produce animal protein (Figure 7), continued high animal protein consumption in developed countries and changes to animal protein-based diet in developing countries will likely increase N input and N losses into food production.

Moderating this increase by decreasing the average amount of animal protein consumed in developed countries is one mechanism of limiting part of the expected increased N requirement in food production. Bleken (1997) analyzed the relation between human diet and global N need for food production. Her analysis indicates that the total N input for diets with high animal protein intake is almost twice as high as the N needed for diets of moderate animal protein. Her analysis was based on a historical development of food consumption for different countries based on FAO (1999) data. One example of a country with a good food supply and moderate consumption of animal protein is Italy in 1963. At that time food supply was adequate to ensure sufficient nutrition to all groups of society (Bleken 1997). Total protein consumption was 85 g per capita d^{-1}, and consumption of animal protein was 32 g. This is considerably lower than the present average of the developed countries, and yet much higher than the average of developing countries. Another example is Japan, where animal protein consumption has traditionally been low, although it has increased from 25 g in 1963 to 54 g animal protein per capita d^{-1} in 1995. In the same period the total protein consumption has increased from 73 g to 96 per capita d^{-1}.

Increases in N loss due to the decoupling of livestock and crop production

The increase in consumption of animal products worldwide has been accompanied by an intensification of animal production in limited regions. This production localization is generating a second important contribution to increased N loss that results from the disassociation of crop and livestock production. In some European countries livestock feed is imported from other continents (Bouwman & Booij 1998). This spatial intensification of livestock production has caused a high concentration of animal manure on little land, with the consequence that manure has been regarded as waste rather than as a plant nutrient resource.

The relationship of livestock and crop production in the U.S. is an example where the dietary consumption of meat has been at a relatively high level for many decades. As a result of domestic consumption and exports, U.S. livestock production has historically been a large part of U.S. agricultural production. Meat production in the U.S. (beef, pork, poultry, lamb etc.) increased from about 90 kg/person in 1963 to about 125 in 1995 (FAO 1999). During this time there has been a shift in the location of U.S. livestock production, mainly to the Southern and Atlantic states from the Midwest and Great Plains. Much of this shift in resulting N deposition can be attributed to the increase in poultry and hog production in the South and Atlantic coastal sections of the country. There were increases in livestock production of 140% in the North Atlantic region, 270% in the South Atlantic region, 160% in the Midsouth and 140% in the Southern Plains from 1963 to 1995. At the same time livestock production decreased 40% in the Midwest and 30% in the Northern Plains where more than 70% of U.S. maize and wheat is produced (USDA 1997).

Table 1. Nitrogen balance sheet for global agriculture in 1994 (Van der Hoek 1998).

N input	Tg N	Products	Tg N
Fertilizer	74	Animal	12
N-fixation	50	Crops	40
Feeds	10	Surplus[1]	90
Unaccounted	10		
Total	140	Total	140

[1] Surplus is defined as the difference between input and output in animal or crop products.

Because of the centralization of livestock production into regions that produce relatively little animal feed, the area of crop production located in close proximity to the intensive animal production systems are not adequate to carry the animal waste input load. This disassociation between animal and crop production generates transportation costs for the manure that are greater than the N or other nutrient value of the waste (NRC 1993). Of the $\sim$11 Tg N in animal waste excreted in the U.S. in 1990, an estimated 34% was returned to cropped fields for use as fertilizer (NRC 1993). The remainder of the N is stored in lagoons or solid piles (Smil 1999) or distributed elsewhere partly through NH_3 volatilization (an estimated 25% of N excreted in an open cattle feedlot is directly volatilized as NH_3) (Hutchinson & Viets 1969) surface runoff, leaching and wind erosion. Most of the volatilized NH_3 is deposited near the feedlot but significant amounts can be converted to aerosols and transported 1000+ km (Ferm 1998). Much of the remaining 'unused' N eventually finds its way into ground and surface waters (Downing et al. 1999).

Van der Hoek (1998) estimated that 63% of the annual N input into food production was not converted into usable product. This 'surplus N', defined as the difference between input and output, is either lost to the environment or accumulates in the soil (Table 1).

Increased N loss resulting from changes in crop production techniques

Traditionally, crop production, animal production and human consumption were closely linked geographically in many parts of the world. Countries, such as China, have long histories of intensive human and animal food production that was conducted such that livestock and human wastes were returned to the field. This practice kept the N cycle more tightly linked within the food production system, thus minimizing N losses and maintaining soil fertility. In the past few decades the increased demand placed on food production from growing human populations coupled with increased consumption of animal protein has led to significant changes in crop production systems. N input into crop production from synthetic fertilizer has increased while N from animal waste has become relatively less important. This aspect is important for P and K and minor nutrient supply and maintaining soil organic matter (Zhu 1998a).

An important example is the current trend of livestock production in China where, based on data from each Province (Almanac of Agriculture in China 1982, 1997), the ratio of animal production (number of cattle + pigs + sheep) to that of cereal grain (rice + maize + wheat) produced increased from 0.82 to 1.58 between 1981 and 1996. Thus animal production was increasing more rapidly than cereal grain production. Even though livestock production is increasing, utilization of animal manures for fertilizers appears to be on the decline. To illustrate this point we use two examples from China. The first example is highlighted in Box 1 where changes in N management by Chinese villages in the Tai Lake region of China are documented.

Jiangsu Province example. The second example comes from Changshu in Jiangsu Province (Figure 8). Manure N input decreased from 50 to 20 kg N ha^{-1} between 1986 and 1997 while synthetic fertilizer N input increased from 113 to 274 kg N ha^{-1}. Total N input increased from 160 to 290 kg N ha^{-1}, an 80% increase while total grain yield ha^{-1} increased only 8%. Within Jiangsu Province between 1981 and 1996, the number of cattle, pigs and sheep sold increased by 83, 50 and 595%, respectively (Almanac for Agriculture of China 1982, 1997). With less manure N being used for crop production and more livestock being grown, where is the increased animal waste N going? A nutrient budget analysis of a large watershed in Southeastern China by Yan et al. (1998) suggests that much of the excess N is moving into aquatic systems. The nitrate content in the water of rivers and lakes in regions that have high N input and high crop production and has increased several fold during last 10 years (Zhu 1998a; Zhuang et al. 1995), due to the high losses of N from croplands (Zhu 1997, 1998b; Li et al. 1998).

Options for limiting N input into crop production

It is clear from many reports that when fertilizer N is applied in an amount needed by the crop for near optimum production, and at the time that the plants use the N, that N losses are relatively small. As an example, Broadbent and Carlton (1978) conducted a 3-year series of fertilizer N utilization trials in irrigated maize using ^{15}N-depleted ammonium sulfate fertilizer. Their study on a sandy loam soil in central California was supplemented with leaching,

Since 1949, fertilizer N inputs have increased more than five-fold in the densely populated alluvial flood plain of the Tai Lake Region. For centuries, farmers used less than 100 kg N ha^{-1} y^{-1} from legume green manure, canal sediments, manure and crop residues to sustain 4 mg ha^{-1} rice yields in their N-limited rice/wheat double cropping systems (Ellis & Wang 1997). Starting in the 1960s, synthetic N applications began to intensify, pushing paddy fertilizer inputs to nearly 500 kg N ha^{-1} y^{-1} by the late 1980s, doubling rice yields in support of doubled human populations. Over the same period, nitrate pollution and eutrophication became common (Ma 1997). By integrating household and landscape data collected on-site from 1993 to 1996 with historical data from reference materials and village elder interviews, the long-term impacts of these developments are being reconstructed in a typical floodplain village of the region. This village-scale analysis helps identify both the sources of large-scale environmental problems and pathways toward their solution. (Ellis et al. 2000a; Ellis et al. 2000b).

Synthetic N increased from 0% of paddy fertilizer N in 1930 to more than 80% in 1994, displacing traditional organic inputs and altering N storage and cycling. When communal agriculture ended in 1982, synthetic N replaced the traditional labor-intensive practice of harvesting canal sediments for fertilizer. Short of a political intervention, these unharvested sediments will completely fill most village canals within 25 years, increasing flood risk and impeding irrigation and transport. Another traditional fertilizer, nightsoil (human manure), was applied mostly to paddy land in the past, at rates rarely exceeding 40 kg N ha^{-1} y^{-1}. Now, to save labor, most nightsoil is applied to small upland plots near houses at rates greater than 200 kg N ha^{-1} y^{-1}, transforming an essential fertilizer into an excess nutrient source in the drier areas of the village most susceptible to nitrate leaching. Overall, changes in N management and land use have increased N storage in village soil and sediment by about 25% from 1930 to 1994, or by 1.4 mg N ha^{-1} on average. Accumulation in sediment accounts for about half of the N buildup, with the remainder due to a 20% increase in agricultural soil N concentration caused by the intensive use of N fertilizers.

Sustaining high yields is critical for village food security. Paddy land availability, already low in 1930 (0.11 ha person^{-1}) was halved to just 0.051 ha person^{-1} by 1994 and is still declining. Annual paddy N inputs to rice/wheat systems averaged 480 kg N ha^{-1} y^{-1} in a 1994 sample of 50 village households, varying between 210 and 850 kg N ha^{-1} y^{-1}. As inputs greater than 500 kg N ha^{-1} y^{-1} do not increase yields, N inputs by nearly half of village households exacerbate environmental pollution without any possible benefit to farmers. Paddy N losses could be reduced significantly if only the 48% of households with above average N loading reduced their inputs to the average.

Box 1. Changes in N management by villages in the Tai Lake Region of China.

denitrification and modeling efforts. Their results and synthesis of these results by Legg and Meisinger (1982) show that maximum N use efficiency was found at the same fertilizer rate needed to obtain maximum yield. When N was applied in excess of this amount, large amounts of nitrate accumulated in the soil profile that were subject to leaching. Nitrification/denitrification losses ($N_2 + N_2O + NO_x$) were estimated to be a relatively constant $\sim$22% of N applied.

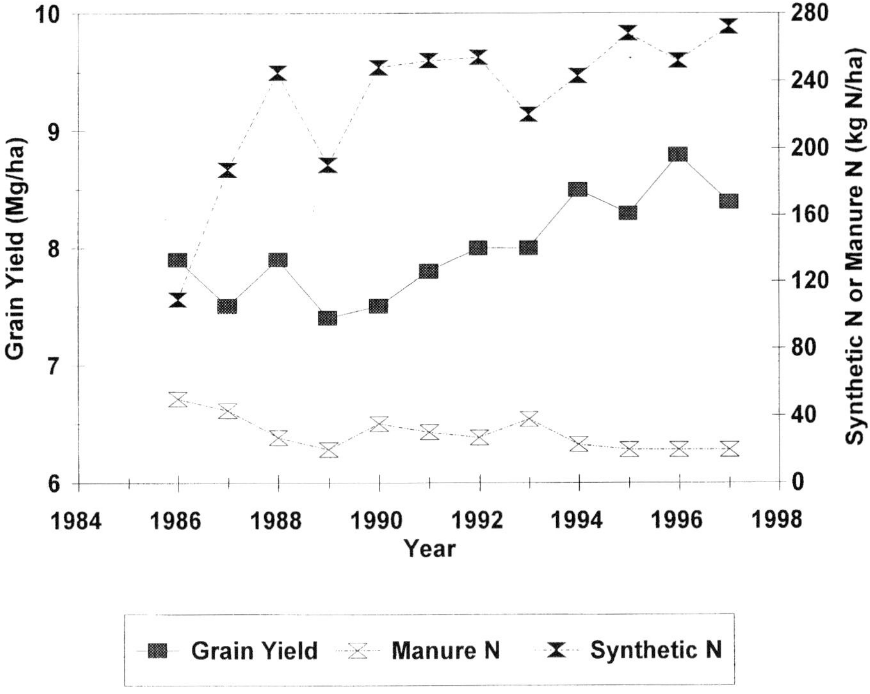

Figure 8. Animal manure N and synthetic N input into grain production in Changshu, Jiangsu Province, China (Zhu, Z.L., unpublished data).

These studies also showed that significant N losses through denitrification and leaching can be expected even at 'optimal' N application rates. The potential to conserve almost one third of the fertilizer N plus mineralized soil N still exists for most cropping systems. Management techniques to limit N input while maintaining crop production need to continue to be incorporated into crop production systems. Some management options that have proven useful in reducing N input into cropping systems without decreasing production follow (Peoples et al. 1995; Cole et al. 1996; De Jager et al. 1998; Hendriks et al. 1998):

- Match N supply with crop demand by: Using soil/plant testing to determine fertilizer needs; optimizing split application schemes; or matching N application to reduced production goals in regions of overproduction

- Tighten N flow cycles by: Integrating animal and crop production systems to utilize animal manure in crop production; maintaining plant residue N on the production site; or minimizing fallow periods to limit mineral N accumulation

- Use advanced fertilization techniques such as: Controlled release fertilizers; place fertilizers below the soil surface; foliar application of fertilizers; nitrification inhibitors; or match fertilizer type to seasonal precipitation
- Optimize tillage, irrigation and drainage
- Optimize animal protein in human diet

The first option aims at a more efficient use of N in agriculture and generally applies to intensively managed types of agriculture. These strategies include a better matching of N supply with crop demand and optimizing split application schemes. An example of a system where a change in fertilizer management decreased N demand, decreased gaseous N losses and improved profit to the farmer is provided by Matson et al. (1998) from their study in irrigated wheat production in west-central Mexico. This study shows that although it is unlikely that losses can be completely eliminated that management can improve efficiency. The amount of fertilizer N required to produce the same crop was decreased 28%, thereby, decreasing N losses (nitrate leaching, NO_x and N_2O emissions) and increasing farmer profit. In areas of high input agriculture it has been assumed that fertilizer use efficiency could be improved by 20% (Cole et al. 1996; Peoples et al. 1995).

A combination of these first mitigation options are being considered for improving N utilization in China (Zhu 1998a): (1) to maintain the high yield level with a reduced rate of N application through improvement in N application techniques. Three mechanisms to promote this strategy are: (a) to optimize the rate of N application (Zhu 1997, 1998b). For example, in set of field experiments for flooded rice conducted in Wujin county, south Jiangsu province, Cui et al. (1998) found that the optimal rate of N application for maximum economic benefit and minimum leaching loss was 220–260 kg N ha^{-1} for a grain yield of around 7.5 t ha^{-1}, (b) to optimize timing of N application, by applying little N in the early stage of growth and to top-dress the major portion of fertilizer N at vigorous growth stages. Many ^{15}N-microplot field experiments have found that the N loss in from a late application was much less than from an early application (Zhu 1997) and (c) to deep-place the fertilizer N into the soil, which can reduce ammonia volatilization substantially (Zhu 1997). (2) A second choice is to reduce the yield target in the high-yielding regions. This assumes that the yield loss can be compensated by the yield increase in other regions of low and moderate productivity where the rate of N application is less than 150–180 kg N ha^{-1} crop^{-1}. This strategy could only be adopted by policy-makers if the increase in yield in the low yielding regions can be realized through improvement in irrigation, land leveling and the increase in the rate of N application and balanced fertilization (P and K additions in addition to N).

The second type of fertilization management option is applicable to most areas of the world. As we have shown in examples in China and the USA, the tightening N flow cycles through reintegration of animal and crop production systems could be used to decrease the demand for synthetically fixed N. Utilizing plant residue in N recycling to both increase soil fertility but also to reduce erosion could be increased in many parts of the world (Lal et al. 1998). Tightening N cycles could reduce the need for synthetic N input by another 10 to 20% if fully utilized (Cole et al. 1996).

The third and fourth options for decreasing N losses from agriculture include new technological advances and management. The development of control release fertilizers which can release fertilizer components at a rate that matches crop demand while protecting the remainder of the fertilizer from release (Minami et al. 1994; Shoji & Kanno 1994). Managing N input so that the fertilizer type applied helps decrease N loss. For example, in a seasonally wet pasture applying an ammonium-based fertilizer during the wet period and a nitrate-based fertilizer during the dry period significantly decreased N losses (McTaggart et al. 1997). Full use of these techniques have the potential to improve N utilization globally by 10–20% (Cole et al. 1996).

Finally, a change in the human diet may be an effective way to reduce the total amount of N cycled through the agricultural system. Bleken (1997) analyzed the relation between N needed for food production and the daily consumption of animal proteins. Her analysis indicates that total N needed for diets with high animal protein intakes (comparable to many industrialized countries today) are almost twice as high as the N needed for the average diets in Italy 1963 or Turkey 1993. Based on her analysis, we assume that in high-N input regions per capita N need for food production may be reduced by 45%, which would reduce present-day N inputs by about 15% worldwide.

Fossil fuel combustion and NO_x

Trends in NO_x emissions

Fossil fuel combustion is a major source of NO_x inputs to the atmosphere, accounting for worldwide emissions of some 21 Tg N in 1985 (Benkovitz et al. 1996) (Table 2). The source of these emissions may be disaggregated into two components. Thermal NO_x is generated by the oxidation of diatomic nitrogen as a by-product of combustion. Fuel NO_x is formed when the nitrogen contained in the organic compounds that comprise fossil fuels is released to the atmosphere. Although thermal NO_x is dominant for fuels with low nitrogen content such as natural gas and petroleum distillates, fuel NO_x accounts for 50–90% of the emissions associated with heavier fuels such

Table 2. Worldwide NO$_x$ emissions from fossil fuel combustion in 1985 (Benkovitz et al. 1996).

	Tg N yr^{-1}
North America	6.24
Former USSR (stationary sources)	0.50
Europe, Middle East, North Africa	6.07
South Africa	0.22
Australia	0.21
25 Asian countries	4.20
Rest of world	3.56
Total	21.00

as residual oil and coal, which typically contain between 0.3% and 3.0% nitrogen by weight (Seinfeld 1986). Both pathways are strongly affected by combustion technologies and the details of equipment operation.

Estimating the anthropogenic emissions of NO$_x$ that are linked to fossil fuel consumption is an imprecise art. Emissions are often not subject to direct measurement but are instead inferred from data on fuel use. In the simplest case, the consumption of various energy carriers (e.g. hard coal, lignite, gasoline, residual fuel oil, natural gas, etc.) is multiplied by average emission coefficients that are derived from field observations and/or laboratory studies (Müller 1992). This approach is useful in generating order-of-magnitude emission estimates, but is unable to account for the role of specific technologies in mediating the relationship between fuel use and NO$_x$ emissions. This method, however, is often the best that can be done to estimate NO$_x$ emissions in developing countries, where disaggregated data on the disposition of fuel consumption by end use or process are often of low quality or are entirely lacking.

Relatively detailed data are available for the advanced industrial nations of North America, Europe, Oceania, and Japan, where pollution control has been a long-standing policy priority. In the United States, for example, emissions of NO$_x$ from stationary and mobile sources are regulated under the Clean Air Act, the main provisions of which date to 1970. This statute requires the monitoring of emissions from power plants and industrial facilities as well as detailed estimation of the emissions generated by automobiles and commercial vehicles. Given their role in the process of acid deposition, NO$_x$ emissions have come under close scrutiny in both North America and Europe, where emissions abatement is mandated under the Convention on Long-

Range Transboundary Air Pollution. According to Bergesen and Parmann (1997), 21 out of 25 nations that signed onto this agreement have met their commitment to stabilize NO_x emissions at 1987 (or, for the United States, 1978) levels. These targets have been achieved using a combination of economic incentives and technology-based standards.

At a global level, NO_x emissions have grown exponentially throughout the 20th century (Holland & Lamarque 1997). Although emissions dipped briefly during the Great Depression and the energy shocks of the 1970s, they grew from near zero in 1900 to over 20 Tg N in the 1990s. According to the Global Emissions Inventory Activity project (Benkovitz et al. 1996), worldwide NO_x emissions in 1985 were distributed between regions as shown in Table 2.

Unfortunately, time series data on global NO_x emissions are not readily available from official sources, although the Organization for Economic Cooperation and Development (1997) publishes periodic statistics on pollutant emissions for member countries, which include the industrial democracies of North America, Europe, Oceania, and Japan.

Options for decreasing NO_x emissions

A range of approaches to NO_x emissions control has been implemented. At the most general level, emissions may be reduced by curtailing energy use, by switching from fossil fuels to alternative energy sources, or by altering the technologies through which fossil fuels are harnessed to provide heat and power. It is well-recognized that energy use is an important contributor to human well-being, providing the services necessary to heat and cool buildings, operate vehicles and equipment, provide lighting, and drive industrial processes. Since the 1970s, however, major industrial economies have attained substantial economic growth without commensurate increases in energy use (Schipper & Meyers 1992). In part, reductions in the ratio of energy use per unit of economic activity were driven by a behavioral response to the energy price increases of the 1970s and early 1980s. More generally, however, the efficiency of energy use has been enhanced by a range of technologies related to buildings, transportation, and industrial production. Given current energy prices and the lack of interest in more stringent energy performance standards, progress towards enhanced energy efficiency has stalled in recent years. The Intergovernmental Panel on Climate Change (1996), however, estimates that worldwide energy use could be reduced by some 10–30% through the full implementation of cost-effective, energy-efficient technologies.

Opportunities to reduce NO_x emissions through reduced reliance on fossil fuels is a second possible control strategy. Of course, renewable energy sources such as biomass, animal power, and small-scale hydro have long histories in human societies and are still extensively used in developing countries. In the industrialized world, significant progress has been made in the use of solar and wind energy to provide significant flows of power at low environmental cost. These technologies, however, are characterized by high capital costs and reliability constraints that limit their general adoption (Holdren 1990). In a similar vein, nuclear power and large-scale hydroelectric dams – although free from the emissions of traditional air pollutants that are associated with fossil fuel combustion – face significant issues of cost, adaptability, and environmental impacts. These technologies are best suited to the generation of electric power, which constitutes a poor substitute for liquid fuels in transportation. Nuclear power raises questions of reactor safety, waste disposal, and weapons proliferation, while hydroelectric dams pose unique threats to natural environments and the people who populate them.

A broad range of technologies exists for reducing the amounts of NO_x generated by the combustion of specific fossil fuels. Because diatomic nitrogen is a chemically stable compound, its oxidation to NO_x involves high activation energies that are typically achieved only in regions of peak flame temperature. Since the relationship between flame temperature and emissions is strongly nonlinear, steps to reduce temperature are a major pathway to pollution control. In addition, NO_x formation is generally favored by the presence of excess oxygen in combustion gases. Carefully balancing the ratio of fuel to combustion air is thus also important in emissions control. Both combustion temperatures and fuel-air ratios are addressed through the use of flue gas recirculation and water injection in natural gas and oil boilers. These technologies achieve typical emissions reductions of 70–90% in practical applications (Seinfeld 1986). In coal-fired power plants the use of staged combustion methods is more common. This approach, in which fuel is burned in sequential stages in conditions of low oxygen availability and low combustion temperatures, reduces emissions by 30–50%. Emerging technologies involving advanced burner designs and the removal of NO_x from stack gases also have the potential to reduce energy-related NO_x emissions from stationary sources.

The technologies used to control NO_x emissions from vehicles are similar in principle but different in the details. In this context, pollution control technologies focus on achieving appropriate fuel-to-air ratios, exhaust gas recycling, and the removal of pollutants from exhaust gases using catalytic converters. Although technologies have improved substantially in the industrialized nations given the environmental regulations that emerged in the 1970s and 1980s, growth in transportation activity has offset these improvements so that NO_x emissions have been relatively stable. In addition, there is concern that catalytic converters actually increase emissions of nitrous oxide (N_2O), a greenhouse gas that generates a range of environmental harms (Dasch 1992). In developing countries that lack effective regulatory mechanisms, emissions of NO_x and other energy-related pollutants continue to pose serious threats to urban air quality and ecosystem health.

Policy approaches

Fertilizer N and livestock production

Given the characteristics of change in the N cycle discussed above, what policy approaches are available to mitigate or prevent environmental problems associated with fertilizer N use? In reviewing the policy options, three principles should be emphasized. First, although N cycling occurs at regional and global scales, policies are implemented and enforced at the national or provincial/state levels. Multinational efforts to control nitrogen loss to the environment are surely needed, but these efforts will require commitments from individual countries and the policy-makers within those countries.

Second, there is a wide range of direct and indirect policy instruments available to alter fertilizer N use at the national level, ranging from fertilizer subsidies, taxes, and regulations to exchange rate controls. Policies influencing the use and loss of fertilizer N are thus made by several institutions – often concurrently – including the Ministry (or Department) of Agriculture, the Central Bank, the Finance Ministry, and the Ministry of Environment and Natural Resources. These institutions may have opposing objectives, and communication between the groups is often poor or non-existent. Moreover, external advisory bodies, such as the World Bank, may play an important role in designing or endorsing policies related to fertilizer use.

Finally, officials implementing policies at the national level are usually different from the people engaged in international activities that deal with global N issues, such as the Framework Convention on Climate Change, the World Health Organization, and the World Trade Organization. Again, good communication between people within these international and national institutions is required to make global or regional initiatives on fertilizer N use and loss more effective.

While the following section is not a complete review, we will attempt to illustrate the policy dimensions of change in the nitrogen cycle by providing examples that have been employed at international, national and local/regional scales. At the national level, there are four sets of direct policy instruments that can be used to control N use and loss within the agriculture and livestock sectors: 1) subsidies and taxes on N fertilizer; 2) regulations on nitrate losses from agriculture and livestock activities; 3) investments in technology to increase N uptake by plants and reduce N loss (e.g., breeding efforts, nitrification inhibitors); and 4) expenditure on extension services and knowledge transfer to enhance the efficiency of N use through improved rates, timing, and placement of fertilizers. There are also several indirect policy measures that affect nitrogen use and loss within countries, which may be complementary to – or competitive with – direct policy instruments. Examples of such measures include commodity subsidies on crops or livestock products, land tenure policies, conservation and set-aside programs, credit subsidies, irrigation/water subsidies, macroeconomic and trade policies (and their impacts on exchange rates and interest rates), and marketing policies.

Taxes and subsidies on nitrogen fertilizer are the most direct and widespread means to influence N use in agriculture. With the introduction of Green Revolution seed technologies in the late-1960s and early-1970s, many developing countries subsidized fertilizers in order to encourage their use in conjunction with the new varieties. In other countries, fertilizers remained unsubsidized, or even implicitly taxed. During the past decade, price regimes for fertilizers have varied from free market pricing in Thailand to fully controlled pricing in Nigeria (Bumb & Baanante 1996). In countries where fertilizer supplies have not been constrained, fertilizer subsidies have promoted rapid growth in both fertilizer use and food production; examples include China, India, Indonesia, and Mexico.

In some of these cases, subsidies have become financial burdens on the economy and have led to overuse of fertilizer N (Conway 1997). In India, for example, fertilizer subsidies amounted to roughly 3% of the national budget in the early 1990s (Bumb & Baanante 1996). This subsidy rate is high even by OECD standards; total agricultural subsidies (inputs and outputs) in the OECD countries were $3.62 billion, or 1.3% of GDP, in 1998. The longer-term trend for subsidies, at least in the OECD countries, remains downward as governments increasingly introduce market-oriented farm policies. In 1986–1988, farm subsidies in the OECD countries averaged 2.1% of GDP (OECD 1999).

Taxes and subsidies for fertilizers are measured by the difference between international and domestic prices (accounting for transportation, insurance, and other marketing costs); if it is less expensive for a farmer to import fertilizer than to buy it domestically, the fertilizer is taxed, and vice versa for subsidies. Ratios of international to domestic fertilizer N prices have ranged from about 0.25 on the low side (tax) to about 1.5 on the high side (subsidy) in recent years. For example, in 1997, many African countries (e.g., Ethiopia, Kenya, and Zimbabwe) had ratios significantly below one, while several countries in Asia and South America (e.g., India and Argentina) had

Table 3. Ratio of fertilizer N to crop prices at the farm-gate.

Country	Ratio of N: wheat prices[1]	Ratio of N: corn prices[2]	Ratio of N: rice prices[3]
Ethiopia	2.40	7.79	–
Kenya	4.80	7.30	–
Zimbabwe	2.40	4.90	–
China	1.90	–	1.35
India	2.10	–	1.21
Pakistan	1.85	2.50	1.03
Japan	–	–	0.22
Argentina	5.80	2.40	–
Brazil	4.58	3.20	–
Mexico	3.30	1.30	–
World	2.00	2.50	1.16

[1] 1994, CIMMYT (1996); [2] 1997, CIMMYT (1999); [3] 1990, IRRI (1995).

ratios above one. The ratios are calculated from data collected by CIMMYT (1996), IRRI (1995), and the World Bank.

In many countries, 'high' (or 'low') fertilizer prices (as measured by historical prices or world prices) are compensated to some extent by high (or low) crop prices. Moreover, both sets of prices are often distorted by over- or under-valued exchange rates. An important policy variable for the objectives of enhanced food production and rural incomes is thus the ratio between N fertilizer prices and output prices, measured in the local currency at the farm gate. This ratio directly affects expected farm profits and can be used to gauge the potential farm level impact of increasing fertilizer prices. For a given crop and year, this ratio varies widely among countries. Table 3 shows this variation for a subset of countries, commodities, and years.

The extent to which price policies should be used to force fertilizer producers and farmers to 'internalize external costs' of fertilizer N is debatable, and the answer depends importantly on the economic and policy context of the country. Given the essential role of nitrogen fertilizers in crop production in many regions of the world, increasing the costs to producers could lead to reduced yields and/or higher food prices for consumers. This outcome would be damaging for chronic food-deficient countries, such as those in Sub-Saharan Africa, whose fertilizer use is still less than one-fourth the world average. In these regions, increasing (rather than decreasing) nutrient use is a major objective of agricultural policy. In countries that are over-applying fertilizer N, one tractable policy is to reduce subsidies on fertilizers. If subsidies do not exist, alternative policy measures, such as regulations on the type, timing, and amount of fertilizer N applied, or taxes on imported fertilizer can be used to reduce N use and lower N losses.

In most developing countries, regulations on N use or loss are either non-existent or not enforced. Even in North America and Europe, monitoring and enforcement of non-point source pollution of nitrates policies is difficult, and there are few mandates on other forms of N loss. Both the U.S. and the European Union have regulations in place to limit N leaching losses from agricultural and livestock systems. These policies are compatible with the World Health Organization's recommendation that nitrate levels in drinking water should not exceed 50 milligrams of nitrate per liter of water (Bumb & Baanante 1996; Conway 1997). The Nitrate Directive of the European Union, formed in 1991, has as its major objective the control of net N supplies (supply minus uptake) to the soil beginning in 1999. Similarly in the U.S. the Clean Water Act was enacted in 1972 to improve national water quality. To renew impetus, the Clean Water Action Plan was developed jointly by nine Federal Agencies in 1998 to supplement the Clean Water Act. The Action Plan targets watershed protection as the main priority and provides communities with new resources to control pollution runoff and enhance natural resource stewardship (http://www.cleanwater.gov/action/). The Action Plan contains 111 key actions and was initiated with a 5-year funding proposal to provide approximately $2.3 billion in new funds. In 1999 the U.S. Congress funded $171 million of the $568 million requested for the Action Plan. The U.S. EPA estimates that 52% of the community water wells and 57% of the domestic water wells in the country contain nitrate (http://www.epa.gov/grtlakes/seahome/groundwater/src/overview.htm). In the U.S., individual states can implement their own N policy as long as it is compatible with national programs. The policies adopted by the state of Nebraska highlighted in Box 2 are an example.

Although regulations on N loss are not always effective, extension services and other forms of knowledge transfer have played an important role in helping farmers in North America and Europe to reduce N losses from agriculture. In these countries, fields are regularly tested, and farmers receive precise recommendations on application rates and timing, dependent on the soil type, irrigation practice, and previous crop. By contrast, N fertilizer applications often are not timed well to the plants' needs and flooded conditions in many developing countries. Many rice farmers in Asia, for instance, apply fertilizer directly to the water 1–3 weeks after transplanting, which results in large N losses to the atmosphere (Peoples et al. 1995). A critical policy need for the developing world is investment in and implementation of more

The Nebraska Department of Environmental Quality has been charged by new state legislation to require permitting of all livestock feeding operations larger than 1000 animal units. These permits require whole-farm nutrient management plans with easements for application of the expected manure load on neighboring farms if the amount of land owned by the livestock farm operation is not adequate to absorb the nutrient load in the manure. ('Adequate' is measured by a simple input-output formula taking into account manure production and the nutrient content of the manure, crop nutrient requirements for N, and an 'availability index' for the manure which relates to the rate at which the N it contains becomes available to the crop.)

Despite positive actions to reduce N loss from livestock, a more important problem is the loss of phosphorus, because the N/P ratio in manure is much lower than in plants. Thus, applying the manure at rates to satisfy crop N requirements leads to massive P accumulation in soil and a significant potential for P runoff to surface waterways. Eventually both the EPA and state regulators will need to focus more tightly on the P challenge of animal manure applications to farm land.

Box 2. Water quality policies adopted by the state of Nebraska, USA.

extensive training services on nitrogen use efficiency. Investments in fertilizer N management would complement further investments in plant breeding aimed at improving N uptake and allocation in crops.

Indirect policy measures may have an equal, if not larger, impact on N use efficiency and loss from agricultural systems. For example, macroeconomic policy that raises interest rates or reduces credit availability for farmers may lead to lower N application rates or greater N use efficiency. In many developing countries, fertilizer accounts for a large share of cash expenditure for small and medium farmers, and resource-poor farmers generally depend on borrowed funds to purchase fertilizer and other agricultural inputs. In Eastern Europe, the division of large state farms into small-holder units has led to a situation in which many small farmers do not have money or credit to purchase fertilizers; as a result, total fertilizer consumption in East Europe declined from 10 Tg of nutrients in 1986/87 to 3.7 Tg in 1996/97 (IFDC 1999).

A different type of land policy has influenced fertilizer rates in Western Europe. Land set-asides through the Common Agricultural Policy have contributed to an annual reduction in fertilizer consumption of 2.1% during the past ten years, with the largest reductions occurring in France, Germany, Italy, and the U.K. (IFDC 1999). The E.U. had an estimated set-aside rate of 10% for the 1998/99 harvest, double that of recent years. Although this program has dramatically reduced the area planted, it is not yet clear whether it has resulted in increased fertilizer use per hectare or increased fertilizer use efficiency (IFDC 1999).

It is clear from this discussion that no easy generalizations can be made about global nitrogen policy approaches. Each country has a different balance of policy objectives and budget constraints. National policies directly targeted toward nitrogen use and loss may also be confounded by other types of policies, including macroeconomic policies affecting exchange rates and interest rates. Some countries may choose to target nitrogen gases as part of a multi-gas approach towards reducing greenhouse gases within the framework of the Kyoto Protocol (Reilly et al. 1999); industrialized countries with already high rates of fertilizer N use and increasing N use efficiency would most likely enter into such an agreement. The agricultural policy of the Netherlands is an example and is highlighted in Box 3. For food deficit countries, particularly those which still underutilize fertilizer N, entering into an international agreement to control global N loss is improbable. The most interesting set of countries to watch with respect to international agreements on N loss include China and India, which are among the largest consumers of fertilizer N in the world. These countries have relatively strong agricultural sectors, but also escalating population- and income-driven food demands.

Energy sector

Efforts to reduce energy-related N emissions have been approached mainly through air pollution control policies. These policies were pioneered in the 1960s in response to perceived problems of urban air quality. Measures to improve the efficiency of energy use have also played a role (Box 4). Although it had long been recognized that emissions of particulates, SO_2, and NO_x imposed negative aesthetic impacts, the effects of these pollutants on human health were first recognized in the years following World War II. In 1948, an air pollution event occurred in the steel town of Donora, Pennsylvania that caused 6,000 illnesses and 20 fatalities in a population of 14,000 people (Snyder 1994). And in 1952, a smog event in the city of London caused 4,000 deaths over the course of a few days (Wise 1968). Public health officials came to recognize the cause-and-effect relationship between health risks and pollutant concentrations. Growing awareness of such risks led to calls for environmental regulation.

The United States Clean Air Act

In the United States, air pollution control policy is carried out under the auspices of the Clean Air Act (CAA), the major provisions of which date to 1970. The CAA requires the Environmental Protection Agency (EPA) to establish ambient air quality standards for so-called 'criteria pollutants' – particulates, SO_2, NO_x, tropospheric ozone, carbon monoxide, and lead – that

Dutch agriculture is among the most intensive in the world and typically has had high livestock stocking density and a high level of crop production per hectare. During the past 15 years the Dutch government has embarked on a policy to address the manure and ammonia problem (Policy 1999). The agricultural sector has invested heavily in efforts to implement environmentally friendly technology and management. In the Netherlands, agriculture contributes $\sim$30% of total phosphate and 75% of total N to surface waters. More than half of total acidification is traced to ammonia emissions, $\sim$90% of which are derived from agriculture. The main burden of phosphates and N from agriculture are from the over-application of livestock manure, the over-application of inorganic fertilizers and ammonia emissions.

In general, intensive livestock farms with large phosphate and mineral N and NH_3 losses constrained to a relatively limited land area, produce the greatest environmental risk. The standard unit for livestock density in the Netherlands is one dairy cow. From 1998 to 2002 a minerals accounting is obligatory only for farms with a livestock density of 2.5 livestock units (LU)/ha. In 2002 the standard will be lowered to 2 LU/ha. To limit N leaching, runoff, and NH_3 emissions, spreading manure in autumn and winter is banned. Use of techniques that minimize NH_3 emissions when manure is spread or stored is required.

Manure management policy: The Dutch animal manure policy is aimed at farms posing the highest environmental risks. A farm minerals accounting approach is used where an input-output book-keeping system relates total applications of fertilizers to production. If the losses, which are reported yearly, exceed the standards then a fine is imposed on the excess. Beginning in 1998, farms with more than 2.5 LU stocking rates are required to report mineral losses.

Ammonia policy: The Policy (1999) report indicates that of the ammonia losses to the atmosphere from Dutch agriculture $\sim$36% is derived from livestock housing and manure storage, $\sim$50% from manure spreading and $\sim$14% from waste deposition in the field from grazing animals. Such mitigation measures as low-emission application and covering manure storage tanks are suggested for implementation within the livestock production sector, which is indicated to account for 55% of agricultural NH_3 emissions. Other options proposed include selling less profitable animals, increasing milk yield per cow and improving animal feeds. Pig production is attributed a 30% share of agricultural NH_3 emissions and is decreased by the development of low-emission pig rearing units. The poultry sector accounts for $\sim$15% of total emissions and could be reduced by low-emission housing systems NH_3 policy. The NH_3 policy emphasis is on emission reduction. Farmers having stocking densities below 2 LU will deal only with basic mitigation measures while farmers having stocking densities over 2 LU will be obligated to construct low-emission housing. NH_3 policy targets a 70% reduction between 2000 and 2005 compared to 1980. Lowering emissions from manure application are expected to reduce emissions from slurry boom 50%, spreading harrows $\sim$60%, shallow injection $\sim$85% and deep injection $\sim$95%. Since small fines on N will likely not entice farmers to build low-emission housing, special requirements are made on housing systems to reduce NH_3 emissions.

Box 3. Policy on manure and ammonia emissions from agriculture in The Netherlands.

Since anthropogenic emissions of NO_x and other air pollutants are dominated by fossil fuel combustion, efforts to improve energy efficiency, measured in terms of the services obtained per unit of fuel consumption, have been a major focus of environmental policy. In the United States, the Corporate Average Fuel Economy (CAFE) standards, introduced in response to the 1970s oil shortage, require passenger cars to achieve an efficiency target of 27.5 miles per gallon (8.5 l/100 km), or 20.7 miles per gallon (11.35 l/100 km) for 'light trucks' (pickup trucks, minivans, and sport utility vehicles). It is widely recognized that these standards have led to substantial improvements in vehicle technology and reduced energy use. In a similar vein, the U.S. Appliance Efficiency Standards, which target the energy efficiency of refrigerators, washers, dryers, and other household equipment, are expected to save some 24 exajoules of energy from their initial adoption in 1990 through the year 2015 (Geller 1997).

The merits of policies to promote energy efficiency are not without controversy. Critics charge that the CAFE standards have forced consumers to purchase small vehicles that compromise performance while posing increased safety risks. Crandall and Graham (1989), for example, suggest that the vehicle size reductions induced by the CAFE standards may have caused as many as 3,900 fatalities for the 1989 model year alone. Greene (1998), in contrast, points to opportunities to enhance the safety of small cars through careful engineering. Although safer for their occupants, the large vehicles that are currently so popular impose heightened risks on third parties; a factor not considered by Crandall and Graham.

By way of comparison, the appliance efficiency standards provide evidence that well-designed policies can yield simultaneous energy and economic savings. These regulations, which are based explicitly on cost-effectiveness criteria, provide consumers with an estimated $1.8 billion in net annual savings (Geller 1997). More generally, the Intergovernmental Panel on Climate Change (1996) concludes that the full adoption of least-cost energy-efficient technologies could reduce energy use per unit of economic activity by some 10–30% in the industrial economies of Europe and North America.

Although energy efficiency can clearly contribute to the achievement of environmental policy goals, the links between reduced energy use and NO_x emissions are not entirely clear-cut. In the United States, the Clean Air Act requires passenger vehicles to meet uniform technology standards that specify maximum levels of pollutant emissions per distance traveled. Since all passenger cars are held to the same standards, NO_x emissions are partially decoupled from energy efficiency. Until recently, pickup trucks, minivans, and sport utility vehicles faced less stringent requirements than automobiles with respect to both fuel economy and emissions standards. In December of 1999, the Clinton administration issued new regulations under which these 'light trucks' will be held to the same emissions standards as automobiles. The disparity in fuel economy requirements remains in place, however, providing incentives for the production and sale of large, energy-intensive vehicles.

Box 4. Energy efficiency and NO_x.

are known to have serious health impacts on affected populations. The statute requires EPA to set standards at the level required 'to protect public health' based on the best available experimental and epidemiological data.

Air quality standards for criteria pollutants set national targets that are implemented and enforced through a two-part policy approach. First, the CAA requires state governments, in consultation with EPA, to construct so-called 'state implementation plans' (SIPs) to achieve air quality standards within the state's boundaries. A SIP imposes specific emissions standards on major stationary sources of pollution such as industrial facilities and power plants. State administrators are granted considerable flexibility regarding the criteria used to distribute required emissions reductions between sources provided that air quality achieves federal standards. Second, the federal government itself defines and enforces national emissions standards for both new stationary sources of pollution ('New Source Performance Standards') and transportation equipment. Under these rules, new power plants and industrial facilities must meet stringent, technology-based standards that set upper bounds on allowable emissions of criteria pollutants. In a similar vein, passenger automobiles must be equipped with catalytic converters and related pollution abatement measures.

The ambient air quality standard for NO_x is currently set at 0.053 ppm (USEPA 1999). This standard is reviewed at 5-y intervals to account for advances in scientific knowledge. In addition, NO_x emissions are regulated with an eye towards reducing ambient concentrations of tropospheric ozone. NO_x is a well-known ozone precursor that, in conjunction with hydrocarbon emissions, gives rise to photochemical smog. The impacts of the CAA on NO_x emissions in the United States have been quite substantial. Although emissions were rising rapidly prior to 1970 with the growth of energy use, transportation activity, and industrial production, U.S. NO_x emissions in 1995 were limited to 21.8 Tg – an increase of only 11% between 1970 and 1997 despite substantial growth in population and economic activity (USEPA 1998b). Given the success of State Implementation Plans, New Source Performance Standards, and emissions standards for new vehicles, only 13 U.S. cities violated air quality standards on ten or more days in 1997 (USEPA 1998c).

Despite such success, environmentalists have called for more aggressive policies to address the ecological impacts of energy-related NO_x emissions. Given the CAA's traditional focus on improving urban air quality, questions of long-range pollution transport – in which NO_x is involved as a precursor of both tropospheric ozone and acid deposition – have been difficult to address under the prevailing regulatory framework. On the one hand, state regulators have traditionally had no reason to consider the impacts of their own state's emissions on air quality beyond their own jurisdictions. On the other hand, a mechanism for achieving coordinated emissions reductions across state boundaries was, until recently, not embodied in the statute.

This problem was partially addressed by the 1990 Clean Air Act amendments, which aimed in part to reduce the ecological impacts of acid deposition caused by the long-range transport of SO_2 and NO_x. The amendments call for a 50% reduction in sulfur emissions from electrical generating stations in the Midwest and Northeast, along with smaller reductions in NO_x emissions from these same sources. The differential treatment of sulfur and nitrogen was based in part on the perception that SO_2 is ecologically more damaging than NO_x. In part, this approach was based on the belief that emissions abatement costs would be lower for sulfur than for nitrogen since SO_2 emissions could be achieved through a mere shift from high- to low-sulfur fuels. In September of 1998, addition, EPA promulgated a new regulation known as the 'NO_x SIP Call' to address interstate transboundary flows of nitrogen oxides in order to reduce tropospheric ozone pollution. Under this rule, 22 states in the Northeastern U.S. must reduce their NO_x emissions by an average of 28% during the summer months, the period for which ambient ozone concentrations are likely to exceed federal standards (USEPA 1998a).

The convention on long-range transboundary air pollution

While the Clean Air Act of 1970 began to address the human health impacts of U.S. urban air pollution, evidence of a more complex problem was turning up in Europe, where studies were linking forest dieback and the acidification of Scandinavian lakes to continental emissions of SO_2 and NO_x. When scientists established that air pollutants could be transported hundreds or even thousands of kilometers across national frontiers, it became clear that effective response strategies would require international cooperation. In 1972, the participants of the United Nations Conference on the Human Environment in Stockholm declared that nations must ensure that activities carried out in their territories do not result in environmental damage in other states. Consequently, a 1979 ministerial meeting of the U.N. Economic Commission for Europe (UNECE) was convened in Geneva to address the region's environmental problems. The result was the Convention on Long-Range Transboundary Air Pollution (LRTAP), an agreement that was signed by all 35 (now 44) member countries.

The goal of LRTAP is to 'limit and, as far as possible, gradually reduce and prevent air pollution [using] the best available technology which is economically feasible.' Establishing the convention-protocol model that has since been used in international agreements on ozone depletion and climate change, the agreement led to a series of protocols mandating targets and timetables for pollution abatement as well as the long-term funding of the European Monitoring and Evaluation Program (EMEP) in 1984. EMEP is a main source of data both on air pollution levels and transboundary fluxes, gathering data at 100 monitoring stations in 25 countries (see http://projects.dnmi.no/~emep/index.html). The first protocol was signed in Helsinki in 1985, where signatories agreed to reduce SO_2 emissions by at least 30% relative to 1980 levels. Although neither the United States nor the

United Kingdom signed this agreement, overall sulfur emissions in UNECE countries had been reduced by over 50% through 1993 (UNECE 2000a).

A series of subsequent protocols has addressed the deleterious impacts of NO_x (1988), volatile organic compounds (1991), heavy metals (1998), and persistent organic pollutants (1998). A second sulfur protocol (1994) stiffened the requirements of the Helsinki accord. The most recent agreement – the Protocol to Abate Acidification, Eutrophication and Ground-level Ozone (1999) – deals with multiple effects and multiple pollutants, with a special emphasis on SO_2, NO, volatile organic compounds, and NH_3. Both the NO_x and the 'Multiple Effects' Protocols deal with human effects on the nitrogen cycle.

With 25 original signatories and 27 ratifications, the Sofia Protocol Concerning the Control of Emissions of Nitrogen Oxides or their Transboundary Fluxes entered into force in 1991. This agreement obliges its parties to freeze NO_x emissions at 1987 (or, for the United States, 1978) levels by 1994. The agreement prescribes both technical means for reducing NO_x emissions using specific technologies and policy measures that include the establishment of national emissions standards for all new and existing sources of NO_x. As of 1994, most signatory nations had met or exceeded their abatement targets, with a total emissions reduction of 9% (UNECE 2000a).

Article 6 of the NO_x Protocol laid the foundation for future emissions reductions based on the so-called 'Critical Loads Approach.' A critical load is "a quantitative estimate of the exposure to one or more pollutants below which significant harmful effects on specified sensitive elements of the environment do not occur according to present knowledge." The article aims to ensure that critical loads of nitrogen are not exceeded due to anthropogenic emissions, assigning the tasks of evaluation and monitoring to individual nations. This framework set the stage for the Multiple Effects Protocol that was signed in Gothenburg in November of 1999.

The Multiple Effects Protocol sets differentiated emissions ceilings for its 29 parties based on the severity of the effects to which each contributes (both within its borders and abroad) and the cost-effectiveness of emissions abatement (UNECE 2000b). This approach builds on research carried out by the EMEP in the 1980s and 1990s. Through 2010, UNECE (2000b) estimates that this agreement will yield emissions reductions of 63% for SO_2, 41% for NO_x, 40% for volatile organic compounds, and 17% for NH_3 relative to 1990 levels. For NO_x, the policy approach taken by the Multiple Effects Protocol largely resembles the earlier Sofia Protocol, setting overall emissions ceilings and limiting pollution from both new and existing sources. Unlike the Sofia agreement, however, the Multiple Effects Protocol allows countries to choose

510

their own pollution control strategies as long as they attain the stipulated emissions targets.

In addition to NO_x, the Multiple Effects Protocol establishes measures to control NH_3 emissions, directing parties to publish and disseminate guides to good agricultural practices to abate emissions within their borders. These guidelines must address nitrogen management (taking the full nitrogen cycle into account), livestock feeding strategies, manure spreading techniques, manure storage, animal housing, and the prospects for limiting ammonia emissions from the use of mineral fertilizers.

According to UNECE (2000b), the full implementation of the Multiple Effects Protocol would: (a) reduce the land area affected by excess acidification from 93 to 15 million hectares; (b) reduce eutrophication by 35%; (c) cut the number of lives lost from exposure to ozone and particulates by 47,500 per annum; and (d) lower the exposure of vegetation to high ozone levels by 44%. The protocol will enter into force 90 days after it is ratified by at least 16 of its 29 signatories. At present, the agreement has a large and growing number of parties, and there is cause for optimism regarding its implementation.

Overall, LRTAP and its protocols have been quite effective in reducing air pollution. In addition, these agreements have fostered a large body of scientific research and established a framework for international cooperation.

Summary

The human induced input of reactive N into the global biosphere has increased to approximately 150 Tg N each year (Figure 1) and is expected to continue to increase for the foreseeable future. The need to feed and to provide energy for the growing world population drives this trend. This increase in reactive N comes at, in some instances, significant monetary and nonmonetary costs to society through increased emissions of NO_x, NH_3, N_2O and NO_3^- and deposition of NO_y and NHx.

In the atmosphere, increases in tropospheric ozone and acid deposition (NO_y and NHx) have lead to acidification of aquatic and soil systems and in reductions in forest and crop system production. Changes in aquatic systems as a result of nitrate leaching have led to decreased drinking water quality, eutrophication, hypoxia and decreases in aquatic plant diversity, for example.

On the other hand, increased deposition of biologically available N may have increased forest biomass production and may have contributed to increased storage of atmospheric CO_2 in plant and soils. Most importantly, synthetic production of fertilizer N has contributed greatly to the remarkable increase in food production that has taken place during the past 50 years.

In 1994 fuel combustion fixed $\sim$21 Tg of N while food production fixed approximately 120 Tg of N. There are many possibilities for limiting reactive N emissions from fuel combustion, and in fact, great strides have been made during the past decades. The control and reduction of the introduction of new reactive N and in curtailing the movement of this N in food production is even more difficult. The particular problem comes from the fact that most of the N that is introduced into the global food production system is not converted into usable product, but rather reenters the biosphere as a surplus (Table 1). Policy issue problems are further exacerbated by the fact that most reactive N species are highly mobile through atmospheric transport or flow through aquatic systems. Reactive N produced in one region or country may cross territorial boundaries. Policies for control are typically found on a state or national basis. There are examples, such as the U.S. Clean Air Act of 1970 and later amendments, that have helped minimize the increase in NO_x emissions from combustion. The U.S. NO_x emissions have increased by $\sim$6% while human population has increased by $\sim$30% during the past 30 years.

The difficulties for policy development to control unwanted reactive N release are confounded by the fact that much of the reactive N release is involved with food and energy production. Global policy on N in agriculture is further difficult because many countries need to increase the input of fertilizer N to increase food production to raise nutritional levels or to keep up with population growth. There is a wide range of direct and indirect policy instruments available to alter fertilizer N use at the national level, ranging from fertilizer subsidies, taxes, and regulations to exchange rate controls. Policies influencing the use and loss of fertilizer N are thus made by several institutions – often concurrently – including the Ministry (or Department) of Agriculture, the Central Bank, the Finance Ministry, and the Ministry of Environment and Natural Resources. These institutions may have opposing objectives, and communication between the groups is often poor or non-existent. Moreover, external advisory bodies, such as the World Bank, may play an important role in designing or endorsing policies related to fertilizer use.

Although N cycling occurs at regional and global scales, policies are implemented and enforced at the national or provincial/state levels. Multi-national efforts to control nitrogen loss to the environment are surely needed, but these efforts will require commitments from individual countries and the policy-makers within those countries. Officials implementing policies at the national level are usually different than the people engaged in international activities that deal with global N issues, such as the Framework Convention on Climate Change, the World Health Organization, and the World Trade Organization. Again, good communication between people within these international and national institutions is required to make global or regional initiatives on fertilizer N use and loss and fossil fuel consumption more effective.

Acknowledgments

This work was initiated as part of the International SCOPE N Project, which received support from both the Mellon Foundation and from the National Center for Ecological Analysis and Synthesis, a Center funded by NSF (Grant #DEB-94-21535), the University of California at Santa Barbara and the State of California.

References

Aber JD, Magill A, McNulty SG, Boone RD, Nadlehoffer KJ, Downs M & Hallett R (1995) Forest biogeochemistry and primary production altered by nitrogen saturation. Water, Air and Soil Poll. 85: 1665–1670

Aber J, McDowell W, Nadelhoffer K, Magill A, Berntson G, Kamakea M, McNulty S, Currie W, Rustad L, & Fernandez I (1998) Nitrogen saturation in temperate forest ecosytems. Bioscience 48: 921–934

Abrahamsen G & Stuanes AO (1998) Retention and leaching of N in Norwegian coniferous forests. Nut. Cycling in Agroecosys. 52: 171–178

Almanac for Agriculture of China (1982, 1997) (in Chinese)

Ayers RU, Schlesinger WH & Socolow RH (1994) Human impacts on the carbon and nitrogen cycles. In: Socolow RH, Andrews C, Berkhout R & Thomas V (Eds) Industrial Ecology and Global Change (pp 121–155). Cambridge University Press, New York, New York, USA

Badiane O & Delgado CL (1995) A 2020 Vision for Food, Agriculture, and the Environment in Sub-Saharan Africa. International Food Policy Research Institute, Washington D.C. 56 pp

Benkovitz CM, Scholtz MT, Pacyna J Tarras n L, Dignon J, Voldner EC, Spiro PA, Logan JA & Graedel TE (1996) Global gridded inventories of anthropogenic emissions of sulfur and nitrogen. J. of Geophys. Res. 101: 29,239–29,253

Bergesen HE & Parmann G (1997) Green Globe Yearbook. Oxford University Press, New York.

Bleken MA (1997) Food consumption and nitrogen losses from agriculture. In: Lag J (Ed) Some Geomedical Consequences of Nitrogen Circulation Processes. Proceedings of an International Symposium, 12–13 June, 1997 (pp 19–31). The Norwegian Academy of Science and Letters, Oslo, Norway

Bleken MA & Bakken LR (1997) The nitrogen cost of food production: Norwegian Society. Ambio 26: 134–142

Bouwman AF (1997) Long-term scenarios of livestock-crop-land use interactions in developing countries. FAO Land and Water Bulletin 6. FAO Rome. 145 p

Bouwmann AF & Booij H (1998) Global use and trade of feedstuffs and consequences for the nitrogen cycle. Nutrient Cycling in Agroecosystems 52: 261–267

Broadbent FE & Carlton AB (1978) Field trials with isotopically labeled nitrogen fertilizer. In: Nielsen DR & MacDonald JG (Eds) Nitrogen in the Environment (pp 1–43). Academic Press, Inc. New York

Brown LR (1999) Feeding nine billion. In: Brown LR, Flavin C & French H (Eds) State of the World 1999. A World Watch Institute Report on Progress Toward a Sustainable Society (pp 115–132). W. Norton & Company, NY

Bumb BL & Baanante CA (1996) The role of fertilizer in sustaining food security and protecting the environment to 2020. Food, Agriculture, and the Environment Discussion Paper 17. International Food Policy Research Institute. Washington, D.C. USA. 54 pp

CIMMYT (1996) World Wheat Facts and Trends 1995/96. D. F., CIMMYT, Mexico

CIMMYT (1999) World Maize Facts and Trends 1997/98. D. F., CIMMYT, Mexico

Cole CV, Cerri C, Minami K, Mosier A, Rosenberg N & Sauerbeck D (1996) Agricultural options for mitigation of greenhouse gas emissions. In: Watson RT, Zinyowera MC & Moss RH Moss (Eds). Climate Change 1995. Impacts, Adaptations and Mitigation of Climate Change: Scientific Technical Analyses. Published for the Intergovernmental Panel on Climate Change, Chapter 23 (pp 745–771). Cambridge University Press, Cambridge, UK

Conway G (1997) The Doubly Green Revolution: Food for All in the 21st Century. Cornell University Press

Crandall, RW & Graham JD (1989) The effect of fuel economy standards on automobile safety. Journal of Law and Economics 32: 97–118

Cui YT, Cheng X, Han CR & Li RG (1998) Nitrogen utilization efficiency of rice and nitrogen leaching in Taihu Lake watershed of south Jiangsu. J. of China Agric. Univ. 3(5): 51–54 (in Chinese)

Dasch, JM (1992) Nitrous-oxide emissions from vehicles. Journal of the Air and Waste Management Association 42: 63–67

De Jager D, Blok K & van Brummelen M (1998) Cost-Effectiveness of Emission-Reducing Measures of Nitrous Oxide in The Netherlands. Ecosystem Report M704. Utrecht, The Netherlands

Delmas R, Serca D, & Jambert C (1997) Global inventory of NO_x sources. Nut. Cycling in Agroecosys. 48: 51–60

Downing JA, Baker JL, Diaz RJ, Prato T, Rabalais NN & Zimmerman RJ (1999) Gulf of Mexico Hypoxia: Land and Sea Interactions. Council for Agricultural Science and Technology, Task Force Report No 134. Ames, Iowa. 44 p

Ellis EC & Wang SM (1997) Sustainable traditional agriculture in the Tai Lake Region of China. Agriculture Ecosystems and Environment 61: 177–193

Ellis EC, Li RG, Yang LZ & Cheng X (2000a) Long-term change in village scale ecosystems in China using landscape and statistical methods. Ecological Applications 10: 1057–1073

Ellis EC, Li RG, Yang LZ & Cheng X (2000b) Changes in village-scale nitrogen storage in China's Tai Lake Region. Ecological Applications 10: 1074–1089

FAO, United Nations Food and Agricultural Organization (1999) FAOSTAT: Agricultural Data, are available on the world wide web: (http://www.apps.fao.org/cgi-bin/nph-db.pl?subset=agriculture)

Ferm M (1998) Atmospheric ammonia and ammonium transport in europe and critical loads – A review. Nutrient Cycling in Agroecosystems 51: 5–17

Fertilizers and Agriculture (January 1998) Is the Global Nitrogen Cycle Out of Balance? International Fertilizer Industry Association, Paris. 1 p

Galloway J N, Levy H II & Kasibhatla PS (1994) Year 2020: Consequences of population growth and development on the decomposition of oxidized nitrogen. Ambio 23: 120–12

514

Galloway JN, Schlesinger WH, Levy H II, Michaels A & Schnoor JL (1995) Nitrogen fixation: Anthropogenic enhancement-environmental response. Global Biogeochem. Cyc. 9: 235–252

Geller H (1997) National appliance efficiency standards in the USA: Cost-effective Federal regulations. Energy and Buildings 26(1): 101–109

Greene DL (1998) Why CAFE worked. Energy Policy 26: 595–613

Hendriks CA, de Jager D & Blok K (1998) Emission Reduction Potential and Costs for Methane and Nitrous Oxide in the EU-15. Interim Report. Ecofys report M714. Utrecht, The Netherlands

Holdren JP (1990) Energy in transition. Sci. Am. 263: 157–163

Holland EA & Lamarque JF (1997) Modeling bio-atmospheric coupling of the nitrogen cycle through NO_x emissions and NO_y deposition. Nutrient Cycling in Agroecosys. 48: 7–24

Howarth RW, Billen G, Swaney D, Toronsend A, Joworski N, Lajtha Downing JA, Elmgren R, Caraco N, Jordan T, Berendse F, Freney J, Kudeyarov V, Murdoch P & Zhao-liang Z (1996) Regional nitrogen budgets and riverine N and P fluxes for the drainages to the North Atlantic Ocean: Natural and human influences. Biogeochem. 35: 181–226

Hutchinson GL & Viets FG (1969) Nitrogen enrichment of surface water by absorption of ammonia volatilized from cattle feedlots. Sci. 166: 514–515

IFDC (1999) International Fertilizer Development Center. World Fertilizer Supply/Demand Situation. IFDC, Muscle Shoals, Alabama, February 1999

IPCC (1996) Intergovernmental Panel on Climate Change. Climate Change 1995: Economic and Social Dimensions of Climate Change. Cambridge University Press, New York

IPCC (1995) Intergovernmental panel on climate change. In: Houghton JT (Ed) Climate Change 1994. Radiative Forcing of Climate Change and an Evaluation of the IPCC IS92 Emission Scenarios. Published for the IPCC, Cambridge University Press, Cambridge UK. 337 pp

IPCC (1997) Intergovernmental panel on climate change. In: Houghton JT, Meira Filho LG, Lim B, Trennton K, Mamaty I, Bonduki Y, Griggs DJ & Callander BA (Eds) Revised 1996 IPCC Guidelines for National Greenhouse Gas Inventories, Vol. 1–3

IRRI (1995) International Rice Research Institute. World Rice Statistics, Los Banos, The Philippines

Lal R, Kimble JM, Follett RF & Cole CV (1998) The Potential of U.S./Cropland to Sequester Carbon and Mitigate the Greenhouse Effect. Ann Arbor Press, Chelsa, MI. 128 p.

Legg JO & Meisinger JJ (1982) Soil nitrogen. In: Stevenson FJ (Ed) Nitrogen in Agricultural Soils. Agronomy Monograph No. 22 (pp 503–566). ASA-CSSA-SSSA, Madison, WI

Leuck D, Haley S, Liapis P & McDonald B (1995) The EU Nitrate Directive and CAP Reform: Effects on Agricultural Production, Trade, and Residual Soil Nitrogen. Report 255. Economic Research Service. U.S. Department of Agriculture, Washington, DC

Li QK, Zhu ZL & Yu TR (1998) Fertilizers and manures in the sustainable agriculture of China. In: Jiangxi Science and Technology Publishers (pp 120–129). (in Chinese) Nanchang

Ma LS (1997) Nitrogen management and environmental and crop quality. In: Zhu ZL, Wen ZX & Freney JR (Eds) Nitrogen in Soils of China, Developments in Plant and Soil Sciences 74 (pp 303–321) Kluwer Academic Publishers, Dordrecht, The Netherlands

Matson PA, Naylor R & Ortiz-Monasterio I (1998) Integration of environmental, agronomic, and economic aspects of fertilizer management. Science 280: 112–115

Matson PA, Parton WJ, Power AG & Swift MJ (1997) Agricultural intensification and ecosystem properties. Science 277: 504–508

McTaggart IP, Clayton H, Parker J, Swan L & Smity KA (1997) Nitrous oxide emissions from grassland and spring barley, following N fertiliser application with and without nitrification inhibitors. Biology and Fertility of Soils 25: 261–268

Minami K (1994) Effect of nitrification inhibitors and slow-release fertilizer on emission of nitrous oxide from fertilized soils. In: Minami K, Mosier A & Sass R (Eds) CH_4 and N_2O-Global Emissions and Controls from Rice Fields and Other Agricultural and Industrial Sources (pp 187–196). NIAES, Yokendo, Tokyo

Mosier AR, Duxbury JM, Freney JR, Heinemeyer O & Minami K (1998) Assessing and mitigating N_2O emissions from agricultural soils. Climatic Change 40: 7–38

Müller JF (1992) Geographical distribution and seasonal variation of surface emissions and deposition velocities of atmospheric trace gases. J. Geophys. Res. 97: 3787–3804

NRC (National Research Council) (1993) Soil and Water Quality, An Agenda for Agriculture, Committee on Long-Range Soil and Water Conservation, Board on Agriculture, National Research Council, National Academy Press, Washington, DC. 516 p

OECD (1999) Orginazation of Economic Cooperation and Development. Data reported in Economist, "Farm Subsidies", June 5

Olsthoorn CSM & Fong NPK (1998) The anthropogenic nitrogen cycle in the Netherlands. Nut. Cycling in Agroecosys. 52: 269–276

Organization for Economic Cooperation and Development (1997) OECD Environmental Data Compendium. OECD, Paris

Peoples MB, Mosier AR & Freney JR (1995) Minimizing gaseous losses of nitrogen. In: Bacon PE (Ed) Nitrogen Fertilization in the Environment (pp 565–602). Marcel Dekker, Inc

Policy (1999) Policy Document on Manure and Ammonia, The Netherlands. HYPERLINK http://www.minlnv.nl/international/policy/green/notutmm.html

Reilly J, Prinn RG, Harnisch J, Fitzmaurice J, Jacoby HD, Kicklighter D, Stone PH, Sokolov AP & Wang C (1999) Multi-Gas Assessment of the Kyoto Protocol. MIT Joint Program on the Science and Policy of Global Change. Report No. 45. 14 p

Schipper L & Meyers S (1992) Energy Efficiency and Human Activity. Cambridge University Press, New York

Scinfcld JH (1986) Atmospheric Chemistry and Physics of Air Pollution. John Wiley and Sons, New York

Shoji S & Kanno H (1994) Use of polyolefin-coated fertiizers for increasing fertilizer efficiency and reducing nitrte leaching and nitrous oxide emissions. Fertilizer Research 39: 147–152.

Smil V (1990) Nitrogen and phosphorus. In: Turner BL II, Clark WC, Kates RW, Richards JF, Mathews JT & Meyer WB (Eds) The Earth as Transformed by Human action (pp 423–436). Cambridge University Press, Cambridge, England

Smil V (1991) Population growth and nitrogen: An exploration of a critical existential link. Population and Development Review 17: 569–601

Smil V (1999) Nitrogen in crop production: An account of global flows. Global Biogeochem. Cyc. 13: 647–662

Snyder LP (1994) The death-dealing smog over Donora, Pennsylvania: Industrial air pollution, public health policy, and the politics of expertise. Environmental History Review 18: 117

United Nations Economic Commission for Europe (UNECE) (2000a) Convention on Long-Range Transboundary Air Pollution. Environment and Human Settlements Division, Geneva. http://www.unece.org/env/lrtap

United Nations Economic Commission for Europe (UNECE) (2000b) Protocol to Abate Acidification, Eutrophication and Ground-Level Ozone. Environment and Human Settlements Division, Geneva. http://www.unece.org/env/lrtap/pdfbrochure/PolluEng_intro.pdf

USDA (United States Department of Agriculture) (1997) Agricultural Statistics 1997. United States Government Printing Office, Washington DC

UNEP/UNIDO (United Nations Environment Programme/United Nations Industrial Development Organization) (1996) Mineral Fertilizer Production and the Environment. Technical Report No. 26. In collaboration with the International Fertilizer Industry Assoc., Paris. 150 pp

U.S. Environmental Protection Agency (1998a) Fact Sheet: Final Rule for Reducing Regional Transport of Ground-Level Ozone. Office of Air and Radiation, Washington DC

U.S. Environmental Protection Agency (1998b) National Air Pollutant Emissions Update, 1900–1997. Office of Air Quality Planning and Standards, Research Triangle Park, North Carolina

U.S. Environmental Protection Agency (1998c) National Air Quality and Emissions Trend Report, 1997. Office of Air Quality Planning and Standards, Research Triangle Park, North Carolina

U.S. Environmental Protection Agency (1999) National Ambient Air Quality Standards. Office of Air and Radiation, Washington, DC

Van der Hoek KW (1998) Nitrogen efficiency in global animal production. Environmental Pollution 102: 127–132

Vitousek PM & Matson PA (1993) Agriculture, the global nitrogen cycle, and trace gas flux. In: Oremland RS (Ed) The Biogeochemistry of Global Change: Radiative Trace Gases (pp 193–208). Chapman & Hall, New Youk, New York, USA

Vitousek PM, Aber J, Howarth RW, Likens GE, Matson PA, Schindler DW, Schlesinger WH & Tilman DG (1997) Human alteration of the global nitrogen cycle: Causes and Consequences. Issues in Ecology 1: 1–15

Wise W (1968) The World's Worst Air Pollution Disaster. Rand McNally, Chicago

Yan WJ, Yin CQ & Yu SM (1998) Nutrient budgets and biogeochemistry in an experimental agricultural watershed in Southeastern China. Biogeochem. (in press)

Zhu ZL (1997) The fate and management of chemical fertilizer nitrogen in agro-ecosystems. In: Zhu ZL, Wen QX & Freney JR (Eds) Nitrogen in Soils of China, Chap. 11. Kluwer Academic Publishers, Dordrecht/Boston/London

Zhu ZL (1998a) Fertilizers in relation to agriculture and environment. Exploration of Nature 17(4) 25–28 (in Chinese)

Zhu ZL (1998b) The present situation and problems of nitrogen fertilizer use and the strategies. In: Li QK, Zhu ZL & Yu TR (Eds) Fertilizers and Manures in the Sustainable Agriculture of China (pp 38–51). Jiangxi Science and Technology Publishers (in Chinese). Nanchang

Zhuang WL, Tian Z, Zhang N & Li K (1995) Investigation of nitrate pollution in ground water due to nitrogen fertilization in agriculture in north China. Plant Nutrition and Fertilizer Sciences 1: 80–87 (in Chinese)

Biogeochemistry **57/58**: 517–519, 2002.
© 2002 *Kluwer Academic Publishers. Printed in the Netherlands.*

Note added in proof *

The field of marine N_2 fixation has moved forward at a rapid pace since we prepared the final draft of this review article nearly two years ago. It is not possible to summarize all of the data, however, we believe that mention of a few papers will provide the reader with access to new lines of research. In addition, several recent review articles that deal, in part, with marine N_2 fixation have also appeared (Capone 2001, Codispoti et al. 2001; Cullen et al. 2002).

The most important recent publication, in our view, is the discovery of unicellular N_2-fixing cyanobacteria in the subtropical North Pacific Ocean (Zehr et al. 2001). The *in situ* expression of nitrogenase in this diverse, numerically abundant population of phototrophs suggests that environmental rates of N_2 fixation, based on extrapolation of *Trichodesmium* distributions and abundances – which is not a very precise method to begin with (Chang 2000) – may have been systematically underestimated in this environment. Likewise, the discovery of small (<10 μm) coccoid (non-heterocystous) cyanobacteria in the Baltic Sea also requires a recalculation of the impact of N_2 fixation as a source of new nitrogen in that habitat (Wasmund et al. 2001). More discoveries of novel N_2 fixing microorganisms are likely to follow in future years.

Recent studies of the oligotrophic regions of the western North Atlantic Ocean have focused on the dynamics of the dissolved P pools and the N:P stoichiometry of the ambient inorganic and organic substrate pools. Wu et al. (2000) and Cavender-Bares et al. (2001) presented evidence that the bioavailable pool of soluble reactive phosphorus in the western North Atlantic Ocean near Bermuda was <1 nM compared to values that were 1–2 orders of magnitude higher in the oligotrophic North Pacific Ocean near Hawaii. Wu et al. (2000) hypothesized that these regional differences were a result of a larger atmospheric flux of iron in the Atlantic Ocean. The iron-sufficient Atlantic would sustain enhanced N_2 fixation, resulting in the efficient and complete removal of bioavailable phosphorus (Wu et al. 2000). This model

---

* This note in proof refers to the article "Dinitrogen fixation in the world's oceans" by Karl et al., pp. 47–98 of this issue.

518

supports the suggestion that dust-derived iron may control N_2 fixation in the North Pacific Ocean, as discussed in the body of our review.

Consistent with the suggestion that the western North Atlantic Ocean may be iron-sufficient, Sañudo-Wilhelmy et al. (2001) have recently reported that the activities of field-collected *Trichodesmium* were independent of dissolved iron concentrations or the iron cell quotas of the colonies. They further concluded that the structural iron requirements for the growth of N_2-fixing microorganisms may be lower than previously calculated; a revised iron efficiency metabolic model has now been prepared (Kustka et al. 2002). Sañudo-Wilhelmy et al. (2001) also concluded that the phosphorus cell quota may control N_2-fixation rates in *Trichodesmium*, and at least one recent deterministic model for the control of marine N_2 fixation, has explicitly included both dissolved inorganic phosphorus concentrations and the nitrogen-to-phosphorus ratios in diazotrophs as key environmental controls (Fennel et al. 2002). Ongoing research at two open ocean time-series stations, BATS and HOT, will continue to focus on the rates, mechanisms, controls and ecological consequences of N_2 fixation (Orcutt et al. 2001; Church et al. 2002; Dore et al. 2002) during the next decade.

References

Capone DG (2001) Marine nitrogen fixation: what's the fuss? Curr. Opinion Microbiol. 4: 341–348

Cavender-Bares KK, Karl DM & Chisholm SW (2001) Nutrient gradients in the western North Atlantic Ocean: Relationship to microbial community structure and comparison to patterns in the Pacific Ocean. Deep-Sea Res. I 48: 2373–2395

Chang J (2000) Precision of different methods used for estimating the abundance of the nitrogen-fixing marine cyanobacterium, *Trichodesmium* Ehrenberg. J. Exp. Mar. Biol. Ecol. 245: 215–224

Church MJ, Ducklow HW & Karl DM (2002) Multiyear increases in dissolved organic matter inventories at Station ALOHA in the North Pacific Subtropical Gyre. Limnol. Oceanogr. 47: 1–10

Codispoti LA, Brandes JA, Christensen JP, Devol AH, Naqvi SWA, Paerl HW & Yoshinari T (2001) The oceanic fixed nitrogen and nitrous oxide budgets: Moving targets as we enter the anthropocene? Sci. Mar. 65 (Suppl. 2): 85–105

Cullen JJ, Franks PJS, Karl DM & Longhurst A (2002) Physical influences on marine ecosystem dynamics. In: Robinson AR, McCarthy JJ & Rothschild BJ (Eds) The Sea, Volume 12 (pp 297–336). John Wiley & Sons, Inc., New York

Dore JE, Brum JR, Tupas LM & Karl DM (2002) Seasonal and interannual variability in sources of nitrogen supporting export in the oligotrophic subtropical North Pacific Ocean. Limnol. Oceanogr., in press

Fennel K, Spitz YH, Letelier RM, Abbott MR & Karl DM (2002) A deterministic model for N_2 fixation at stn. ALOHA in the subtropical North Pacific Ocean. Deep-Sea Res. II 49: 149–174

Kustka A, Sañudo-Wilhelmy S, Carpenter EJ, Capone DG & Raven JA (2002) A revised iron use efficiency model of nitrogen fixation, with special reference to the marine N_2 fixing cyanobacterium, *Trichodesmium* spp. (Cyanophyta). J. Phycol., in press

Orcutt KM, Lipschultz F, Gundersen K, Arimoto R, Michaels AF, Knap AH & Gallon JR (2001) A seasonal study of the significance of N_2 fixation by *Trichodesmium* spp. at the Bermuda Atlantic Time-series Study (BATS) site. Deep-Sea Res. II 48: 1583–1608

Sañudo-Wilhelmy SA, Kustka AB, Gobler CJ, Hutchins DA, Yang M, Lwiza K, Burns J, Capone DG, Raven JA & Carpenter EJ (2001) Phosphorus limitation of nitrogen fixation by *Trichodesmium* in the central Atlantic Ocean. Nature 411: 66–69

Wasmund N, Voss M & Lochte K (2001) Evidence of nitrogen fixation by non-heterocystous cyanobacteria in the Baltic Sea and re-calculation of a budget of nitrogen fixation. Mar. Ecol. Prog. Ser. 214: 1–14

Wu J, Sunda W, Boyle EA & Karl DM (2000) Phosphate depletion in the western North Atlantic Ocean. Science 289: 759–762

Zehr JP, Waterbury JB, Turner PJ, Montoya JP, Omoregie E, Steward GF, Hansen A & Karl DM (2001) Unicellular cyanobacteria fix N_2 in the subtropical North Pacific Ocean. Nature 412: 635–638